ROUTLEDGE HANDBOOK OF THE UN SUSTAINABLE DEVELOPMENT GOALS RESEARCH AND POLICY

This handbook brings together a collection of seminal research on the Sustainable Development Goals (SDGs) and investigates the effectiveness of the 17 goals in achieving transformative change toward sustainable development.

As a collection of inter- and transdisciplinary research from around the world, this volume explores the applications, implications, and best practices of the goals at thematic, regional, and national levels, providing specific examples from a diverse range of places, such as Australia, Brazil, China, DRC, India, Italy, the Sahel region of Africa, and the USA, among others. The book serves as a mid-term evaluation of the SDGs, bringing to the fore comprehensive experiences and evidence related to the SDGs, whilst highlighting the interlinkages between the different goals. The handbook is divided into two parts: Part I brings together groundbreaking research to define, identify, and present conceptual frameworks for a sustainable future, whilst Part II focuses on the policies, practices, and implementation of the SDGs. The chapters identify key aspects missing from the 2030 Agenda, such as global power imbalances, cultural diversity, Indigenous rights, and unsustainable levels of consumption; they also critically evaluate the overall delivery and effectiveness of the SDGs whilst outlining potential future directions for the post-SDG 2030 Agenda.

This handbook is aimed at a diverse and global audience of academics and students of economics, business studies, political science, and development studies. It will also serve as a valuable reference for leaders in the industry, the public sector, civil society, and international policymakers keen to gain a better understanding of the SDGs.

Ranjula Bali Swain is Research Director at the Center for Sustainability Research (CSR), SIR, Stockholm School of Economics, and Professor of Economics at Södertörn University, Stockholm. She is affiliated with the Center for European Research in Microfinance (CERMi), at the Solvay Brussels School of Economics and Management. Bali Swain has worked at the International Labour Organisation, Geneva, and at Uppsala University and has been Visiting Professor at the University of California, Berkeley; Stanford University; Tufts University; University of Pretoria, South Africa; and UN Sustainable Development Solutions Network, New York, USA, to name a few. She has researched and published widely in the fields of sustainable development, environmental economics, energy, circular economy, sustainable finance, gender, and development economics. She has written and edited nine books, of which her edited volume *Interlinkages between the Sustainable Development Goals* was published in 2023.

Peter Dobers is Professor in Business Studies at Södertörn University and previous Dean of the Faculty Board between 2016 and 2022; Principal Editor of the journal *Business Strategy and the Environment* (BSE) between 2010 and 2024, currently with an impact factor of 12.5; President of the International Sustainable Development Research Society (ISDRS) between 2021 and 2023; and book series editor of Routledge/ISDRS Series on Sustainable Development Research between 2019 and 2024. He has been a board member of the Knowledge Foundation since 2022. He has published extensively in fields such as corporate social responsibility, urban development, and sustainable development. He has written and edited five books and eight special issues in *BSE*, *Corporate Social Responsibility and Environmental Management*, *Sustainable Development*, and *Scandinavian Journal of Management*.

ROUTLEDGE HANDBOOK OF THE UN SUSTAINABLE DEVELOPMENT GOALS RESEARCH AND POLICY

Edited by Ranjula Bali Swain and Peter Dobers

Routledge
Taylor & Francis Group
LONDON AND NEW YORK

earthscan
from Routledge

Designed cover image: Getty Images

First published 2025
by Routledge
4 Park Square, Milton Park, Abingdon, Oxon OX14 4RN

and by Routledge
605 Third Avenue, New York, NY 10158

Routledge is an imprint of the Taylor & Francis Group, an informa business

British Library Cataloguing-in-Publication Data
A catalogue record for this book is available from the British Library

Library of Congress Cataloging-in-Publication Data
Names: Bali Swain, Ranjula, editor. | Dobers, Peter, editor.
Title: Routledge handbook of the UN Sustainable Development Goals research and policy / edited by Ranjula Bali Swain and Peter Dobers.
Other titles: Routledge handbook of the United Nations Sustainable Development Goals research and policy
Description: Abingdon, Oxon ; New York, NY : Routledge, 2025. | Includes bibliographical references and index.
Identifiers: LCCN 2024042223 (print) | LCCN 2024042224 (ebook) | ISBN 9781032257037 (hardback) | ISBN 9781032258874 (paperback) | ISBN 9781003285472 (ebook)
Subjects: LCSH: Sustainable Development Goals (Project) | Sustainable Development Goals (Project)—Case studies. | Sustainable development. | Sustainable development—Case studies.
Classification: LCC HC79.E5 R67494 2025 (print) | LCC HC79.E5 (ebook) | DDC 338.9/27—dc23/eng/20250219
LC record available at https://lccn.loc.gov/2024042223
LC ebook record available at https://lccn.loc.gov/2024042224

ISBN: 978-1-032-25703-7 (hbk)
ISBN: 978-1-032-25887-4 (pbk)
ISBN: 978-1-003-28547-2 (ebk)

DOI: 10.4324/9781003285472

Typeset in Times New Roman
by Apex CoVantage, LLC

CONTENTS

CONTRIBUTORS

Virgínia Simão Abuhid graduated in biological sciences from the Pontifical Catholic University of Minas Gerais (1983), has a specialization in vertebrate zoology from the Catholic University of Minas Gerais (1986), and has a master's degree in geology, concentration area of paleontology, from the Federal University of Rio de Janeiro (1991). She is currently an assistant professor III at the Pontifical Catholic University of Minas Gerais, in the Department of Biological Sciences, teaching classes in the biological sciences course and working with teacher training. Extensionist, she coordinates, participates, and monitors projects at the Institute of Biological Sciences at PUC Minas, where she is the extension coordinator. She has worked in administrative and academic management in the Department of Biological Sciences at PUC Minas, where she has previously held course leadership and coordination, and is currently a member of the Course Board. She participated in the creation of and held different roles at the PUC Minas Museum of Natural Sciences. She is part of the team at the Integrated Regional Development Agency (ADERI) of the Archdiocese of Belo Horizonte and ANIMA PUC Minas, working with cultural, educational, and socio-environmental projects.

Ananya Acharya is an academic associate within the economics area at IIM Ahmedabad, where her research interests converge at the juncture of law and economics, gender, and labor studies. She completed her postgraduation in economics with a specialization in development studies from Symbiosis International University, Pune, and her undergraduate studies in economics at St. Xavier's College, Mumbai.

Nasrin Akter is currently working as an assistant professor in the Department of Land Management and Law, Jagannath University (JnU), Bangladesh. Prior to her appointment at the JnU, she was a lecturer in the Department of Law at the Noakhali Science and Technology University (NSTU), Bangladesh. She completed her bachelor's and master of laws degrees from the University of Dhaka. She aced the LLB and LLM examinations, garnering her slew of prestigious scholarships and honors, including the Prime Minister's Gold Medal in 2017. Her research works have been published in both national and international journals, with two publishers having recently published her research works as book chapters. In addition, she has already presented research papers in several international conferences.

Sabina Alkire holds a DPhil in economics, an Msc in economics for development, and an MPhil in Christian political ethics from the University of Oxford. Alkire directs the Oxford Poverty and Human Development Initiative (OPHI) at ODID. Her research interests include multidimensional poverty measurement and analysis, welfare economics, the capability approach, the measurement of freedoms, and human development. Together with Professor James Foster, she developed the Alkire–Foster (AF) method for measuring multidimensional poverty, a flexible technique that can incorporate different dimensions, or aspects, of poverty to create measures tailored to each context. With colleagues at OPHI, this has been applied and implemented empirically to produce a multidimensional poverty index (MPI). The MPI offers a tool to identify who is poor by considering the range of deprivations they suffer. It is used to report a headline figure of poverty (the MPI), which can be unpacked to provide a detailed information platform for policy design showing how people are poor nationally, and how they are poor by areas, by groups, and by each indicator.

Filipe Almeida is currently the president of Portugal Social Innovation, a pioneering government initiative that mobilizes European funds to promote social innovation and social entrepreneurship as well as develop the social investment market in Portugal. He is also a professor of the University of Coimbra and a researcher at the Center of Social Studies (CES/FEUC) and Center of Cooperative and Social Economy Studies (CECES/FEUC). He holds a PhD in business administration from EBAPE/FGV (Brazilian School of Public Administration and Business, Getúlio Vargas Foundation, Rio de Janeiro, Brazil) and a PhD in business management from the University of Coimbra. He is the author of several books and articles on business ethics, corporate social responsibility, and academic fraud.

M. Niaz Asadullah is a professor of development economics at Monash University Malaysia. Other active affiliations include as the head of the Southeast Asia cluster of the Global Labor Organization and visiting academics at the University of Reading, UK, and North South University, Bangladesh. Niaz is also the co-editor of several academic journals, such as the *Journal of Human Development and Capabilities*, *The Singapore Economic Review*, the *International Journal of Educational Development*, and *PLOS One*. Previously, he held academic/visiting positions at Harvard, Malaya, Dhaka, and BRAC Universities. Publication record includes 70+ peer-reviewed articles on MDGs, SDGs, gender, and education-related topics and numerous op-eds for media outlets, such as the Project Syndicate, the World Economic Forum, the Conversation, and the Diplomat. He has appeared on interviews with Al Jazeera, BBC World Service, and the Economist and consulted for UNESCO and the World Bank.

Reem Ashraf received a PhD in economics of education from Jawaharlal Nehru University, Delhi, India, in 2018. With a focus on advancing Sustainable Development Goals 4 and 5, her expertise spans across South Asia and Central Asia, focusing on educational inequality, gender, social norms, and public policy. Prior to her current role as an independent researcher, she worked at the National Council of Applied Economic Research, New Delhi. With a solid background in public policy consultancy, Reem has led and contributed to studies sponsored by various donor agencies, including selected governments, the World Bank, the Asian Development Bank, and UNICEF.

Mirey Atallah is the head of the Nature for Climate branch of the United Nations Environment Programme in Nairobi. She holds over 20 years of experience working on environment, climate change, REDD+, and biodiversity-related issues across the globe. Furthermore, she was the chief technical advisor for the national REDD+ fund of the Democratic Republic of the Congo until 2021.

Manasi B. is currently serving as an assistant professor at RV University, Bangalore, India. She holds a PhD from IFMR Graduate School of Business, Krea University, India. Her research interest lies in the broad area of development economics. She is currently working on energy poverty and related energy policies in the context of India. Her dissertation addresses the existence, and policy impact, of energy poverty among poor households. She is also partial toward industrial economics, where her research focuses on the growth of Indian pharmaceutical firms.

Karan Babbar is currently serving as an assistant professor at Centre for Development Studies, Jindal Global Business School (JGBS), OP Jindal Global University. He holds a PhD from the Indian Institute of Management, Ahmedabad. His research areas encompass a wide range of topics, including gender, health policy, health economics, health psychology, and development economics. With a specific focus on menstrual health and hygiene, domestic violence and women empowerment, LGBTQIA issues, and health and sanitation, he actively contributes to the academic discourse in these areas.

Bibhuti Ranjan Bhattacharjya is an assistant professor in the Department of Design, Indian Institute of Technology Roorkee, India. His primary research interests are product design, engineering for sustainable global development, and frugal innovation and frugality. He holds a PhD in agro- and rural technology and an MTech degree in mechanical engineering from the Indian Institute of Technology Guwahati, India. He has been associated with the government of India's national missions, like the Rural Technology Action Group and Unnat Bharat Abhiyan, since 2011.

Angela Bruckner is a project manager, administrator, and chartered accountant with a passion for science and a mission to preserve our natural environment for future generations. She holds a Master of Environment (Climate Change) degree, with a particular interest in pursuing further research in climate mitigation and adaptation in the context of international relations and security. Angela is the general manager and an associate of Blue Praxis and works with Australia's Superpower Institute on energy transition policy and practice.

Graeme Coles is the director of Full Flight Birds of Prey (www.fullflight.com.au/) at Miners Rest, Victoria, Australia. He has been working with raptors all his life. Full Flight has many captive-bred and rehabilitated birds used for conservation and education, as well as captive-bred, reared, and trained birds for research and wildlife management activity.

Svenja Damberg an associate professor in Sustainability at the Aalborg University, Copenhagen, Denmark. Her research field is sustainable innovation management, where she addresses topics such as sustainable consumption behavior, sustainable business models, and innovation management for a circular economy in organizations.

Pritha Dev is an associate professor in the economics area at the Indian Institute of Management Ahmedabad. Her research and teaching interests are at the intersection of gender, game theory, economics of networks, and experimental economics. Within gender, her work focuses particularly on issues related to health, employment, and empowerment. She was one of the founding chairpersons of the Gender Center at IIMA.

Haomiao Du is an assistant professor at the Department of International and European Law, Utrecht University. Haomiao's work mainly surrounds international and comparative climate law, covering both mitigation and adaptation. She also has rich experience on inter- and transdisciplinary climate change research.

Thomas Fenzl is a professor for quantitative methods and data analytics at the School of Management at the Carinthia University of Applied Sciences (CUAS). His research interests lie in the fields of empirical research methods and behavioural economics and include (1) the development, refinement, and application of techniques of qualitative content analysis and mixed methodology research designs; (2) decision-making and choice architecture in various domains of human behaviour; as well as (3) collective dynamics in terms of the interplay between motives/behaviour on the individual level and outcomes/behaviours on the aggregate level.

Morgane Fritz is full professor in sustainability management in supply chains at Excelia Business School, La Rochelle, France. Her research and teaching interests include supply chain sustainability, sustainable procurement, ethical supply chain management, business ethics, sustainable digitalization, Sustainable Development Goals, and stakeholder management.

Anton Ming-Zhi Gao is a professor at the National Tsing Hua University's Institute of Law for Science and Technology. He has devoted himself in energy and climate legal research and projects in Taiwan, including publishing books and articles, hosting local and international energy law conferences, and participating in international events. Over the past years, he has been very active in international publications involving investigations on global legal issues, as well as European, Asian, and Taiwanese energy laws. In recent years, he has remained one of the most productive scholars in the field of energy law in Taiwan. He is becoming the most productive law school professor in SCI journals in Taiwan.

Supriya Garikipati is a professor in sustainable development and serves as the director of the Centre for Sustainable Development, University College Dublin. She completed her doctorate in development economics from the University of Cambridge (UK), where she was the recipient of the Nehru–Cambridge and British Chevening scholarships. Prof. Garikipati's expertise lies in the political economy of growth and development, with a particular focus on gender and public policy. Her research encompasses development finance, social exclusion, female labor force participation, reproductive health, intra-household bargaining, and gender differences and discrimination.

Muhamad Sayuti Hassan is a senior lecturer at the Centre for International Law and Siyar (CILAS) at the Faculty of Law, Universiti Kebangsaan Malaysia, Bangi, Malaysia. He obtained his doctor of philosophy (PhD) degree from the Faculty of Law, UKM. His research interest is in international law and Indigenous issues, criminal law, and legislative studies.

Kazi Mukitul Islam is a doctoral candidate in the Department of Sociology at Boston University, USA. Previously, he served as a political advisor to the German Ambassador in Bangladesh, where he worked on issues related to the promotion of human rights among disadvantaged populations. His current research focuses on education policy and gender biases in textbooks in South and Southeast Asia. Publication record includes peer-reviewed papers in *PLOS One*, *The Review of Faith and International Affairs*, and the *Journal of Biosocial Science*.

Nishith Jardosh is a PhD scholar at the Indian Institute of Technology (IIT), Bombay, researching in the areas of solid waste, plastic waste, and policies to manage them. He has a background of over two decades in business management and runs two entities, Recube Circular Solutions and Stallion Barware, both focused on sustainable packaging and reducing plastic waste in the food industry.

Marijane Luistro Jonsson is a post-doctoral researcher at the Department of Neurobiology, Care Sciences, and Society at Karolinska Institutet (KI). Drawing on her social science background, she was a postdoctoral researcher and lecturer at the Stockholm School of Economics (SSE), focusing on the behavioral dimensions of sustainability. She is affiliated with the Center for Sustainability Research (CSR) at SSE and the Center for Social Sustainability at Karolinska Institute. Dr. Jonsson holds a PhD from SSE and has a rich multidisciplinary academic foundation, and her master's degrees in development studies from the University of Auckland (New Zealand), in business administration from Ateneo de Manila University (Philippines), and in environmental management from Stockholm University (Sweden).

Stinne Skriver Jørgensen is chief advisor and team lead for the Data and Analysis team at the Danish Institute for Human Rights. She has worked extensively on quantitative analysis on varies issues regarding human rights and equal treatment, both in a Danish context and internationally, including research project on human rights and economics, salary structures and gender wage gap, poverty among immigrants, and data gaps concerning Indigenous peoples. She has an economics degree from Aarhus University and has previously been employed in the Statistical Division of the Central Bank of Denmark.

Uma Kambhampati is the head of school, School of Politics, Economics, and International Relations, and a professor, Department of Economics, University of Reading, UK. She has a doctoral degree in economics from the University of Cambridge, UK. Her research ranges across issues relating to childhood inequalities, particularly in relation to child work and schooling participation, inequalities between girls and boys, as well as inequalities across social and religious groups. She has also worked on women's empowerment and well-being, including women's labour market participation, life satisfaction, as well as nutritional adequacy. A third strand of her research relates to the issue of well-being and life satisfaction in the UK, including analysis of the extent to which conditions during childhood and youth influence the life satisfaction of adults.

Matthias Karmasin (Univ. Prof. DDr., MAE) is the director of the Institute for Comparative Media and Communication Studies of the Austrian Academy of Sciences and the University of Klagenfurt, where he is a full professor and the chair for communications sciences. He is a full member of the Austrian Academy of Sciences and an ordinary member of the European Academy of Sciences. Since 2023, Karmasin is also the dean of the Faculty of Social Sciences at the University of Klagenfurt. His research interests include media development, organizational communication, political communication, communication theory, media economy, media ethics, and media practice.

Vinish Kathuria is a professor and director at the Institute of Development Studies (IDS), Jaipur, India. He is on lien from the Indian Institute of Technology (IIT), Bombay. He is working in the area of sustainable development, renewable energy, and informal economy, with a focus on solid waste management.

Astha Kushwaha is a research assistant at the Gender Center, Indian Institute of Management Ahmedabad. Her research interests lie in development economics, with a focus on gender inequality, crimes against women, and empowerment. She completed her postgraduate in economics with a specialization in quantitative finance from Madras School of Economics, Chennai, and her undergraduation in computer science.

Matheus Gobbato Leichtweis is a lawyer and PhD candidate in international law at the Federal University of Rio Grande do Sul (UFRGS), Porto Alegre, Brazil. He holds a master's degree in legal studies at the Federal University of Rio Grande do Sul, and an LLM in environmental law at the University of Dundee, UK. His research interests are on international legal history and theory, law and development, and critical approaches to international law.

Stella Lemke is a senior associate scientist at the University of Lübeck, Department of Social Medicine and Epidemiology (Germany). She also works as a teaching fellow at the University of Klagenfurt and obtained her master's degree in psychology. Stella is a founding member of the Association for Supporting Qualitative Research ASQ.

Clara My Lernborg obtained her PhD from Stockholm School of Economics in 2019, with her thesis focus on corporate social responsibility in the Swedish fashion and textile industry. She has since worked with public and private entities on climate change, REDD+, and biodiversity-related projects in sub-Saharan Africa. Her research interests include corporate social responsibility, private governance, meta-organisations, sustainable mining, and circular fashion.

Giuseppe Mangano is a research fellow for ABITAlab, Interuniversity Center, at the Department of Architecture and Design (dAeD) of the Mediterranean University of Reggio Calabria. He has developed a qualified profile on the management of competitive issues and projects in the field of community policies and co-design on the paths of environmental design, with a focus on the themes of innovation and sustainable local development in urban and inland areas of territories. He is the operational coordinator and Responsible of Design team of ABITAlab, and former president of the Ass. Pensando Meridiano.

Jaqueline Silva Melo is a PhD student in psychology and master in administration from PUC Minas. She has a postgraduate degree in production engineering, smart and sustainable cities, and project management. She has a degree in physiotherapy and administration from PUC Minas. She is currently a senior analyst at the Dean of Extension at the Pontifical Catholic University of Minas Gerais, a professor of the specialization course in project management at ICE (Institute of Continuing Education) at PUC Minas and PUC Minas Virtual, and a professor and the coordinator of the specialization courses in humanized management and the MBA in management of third sector organizations at IEC/PUC Minas. She has CAPM certification from PMI (Project Management Institute).

Kenadi E. Mitchell is a 2024 JD candidate at Howard University School of Law, specializing in environmental law and environmental justice. She has engaged deeply with these fields through her participation in Howard's Environmental Research and Advocacy Clinic and a clerkship at the Environmental Law Institute. Kenadi attended COP28 in Dubai as an inaugural fellow of the Climate Leader Academy, sponsored by the US National Science Foundation, where she spoke on a UNFCCC panel about higher education's role in climate education. Her chapter, "A Comparative

Analysis of Disparities in Sustainability, Decision-Making, and Environmental Justice," leverages her academic achievements, including receiving the CALI Award in Title VI and the Environment, indicating the highest grade in her class. This work examines how legal frameworks can support communities facing environmental injustices and contribute to achieving the UN Sustainable Development Goals.

Bijon Kumer Mitra is the deputy director of the Integrated Sustainability Centre at the Institute for Global Environmental Strategies in Japan. He has more than 20 years of experience in the field of sustainable development, focusing on the application of an integrated approach. His research interests include water–energy–food–climate nexus, rural–urban linkage for collective resilience, climate change adaptation in water, and water environment management. He uses quantitative assessment frameworks to assess natural resource allocation trade-offs, aiming to provide guidance for optimal decision-making.

Jyoti Prasad Mukhopadhyay holds a master's degree in economics from the Delhi School of Economics (University of Delhi) and a PhD in economics from the Institute for Financial Management and Research (IFMR) affiliated with the University of Madras. Before pursuing his PhD, he worked with the Centre for Microfinance (CMF) at IFMR as a research associate, and also with the World Bank as a short-term consultant (STC) on an impact evaluation study in Chhattisgarh. He has participated in and presented research papers at various academic conferences in India and abroad. He is an associate professor of economics at the IFMR Graduate School of Business. Professor Mukhopadhyay has published research papers in reputed peer-reviewed journals, including *Energy Economics*, *Journal of Development Studies*, *Development Policy Review*, *Economic and Political Weekly*, among others.

Hiranya K. Nath is a professor of economics at Sam Houston State University in Huntsville, Texas (USA). Concurrently, he serves as an adjunct faculty at the Indian Institute of Information Technology Guwahati (IIITG), India, and holds the position of section editor for economics and econometrics at *PLOS One*. Professor Nath has held visiting and adjunct positions at the Indian Institute of Technology Guwahati (IITG), India; the Queensland University of Technology in Brisbane, Australia; and Tezpur University, India. His research focus encompasses macro- and monetary economics, information economy, and various developmental and transitional economic issues. His scholarly contributions have been featured in peer-reviewed journals, including *California Management Review*, the *Journal of Economic Behavior and Organization*, the *Journal of Money, Credit and Banking*, and the *Review of Development Economics*, among others. Professor Nath earned his PhD in economics from Southern Methodist University in Dallas, USA, and pursued his undergraduate- and masters-level studies at the University of Delhi. Additionally, he obtained an MPhil degree in economics from Jawaharlal Nehru University in New Delhi. Professor Nath's academic pursuits have taken him around the globe, where he has engaged with universities, presented his research findings, and imparted knowledge to students.

Rohaida Nordin is an associate professor at the Faculty of Law, Universiti Kebangsaan Malaysia (UKM), since February 2014. Her research interest is in international human rights law and international law and Indigenous peoples.

Cristina Poncibò is a full professor of comparative private law at the Law Department of the University of Turin, Italy, and a faculty member at the Georgetown Law Center for Transnational

Legal Studies (CTLS) in London. Cristina regularly acts as an expert for European institutions and international organizations and she is the scientific coordinator of the Master in International Trade Law, co-organized with ITC-ILO, in cooperation with Uncitral and Unidroit. Cristina is a graduate of the Universities of Turin (MA) and Florence (PhD) and was an associate in an international UK-based law firm. In her career, she has been a Marie Skłodowska-Curie fellow from the European Commission for EU Networks (Université Panthéon–Assas) and a Max Weber fellow at the European University Institute.

Nurul Hidayat Ab Rahman is a lecturer at the Faculty of Law, Universiti Kebangsaan Malaysia, with a background study in law, focusing mainly on the correlation between children's rights and the concept of sustainability. Her current research interests are sustainable development, child, and family law.

Saionara Reis is team leader on Data for Rights and Development at the Danish Institute for Human Rights. Previously, Saionara was the head of office and representative to the United Nations for Dianova International in New York, where she also worked with UNDP on monitoring and reporting of SDG 16. She has experience with NGOs in Ethiopia and Brazil, as well as with governmental institutions, including the former Brazilian Ministry of Women's Affairs. Previously, Saionara was also a legal advisor to the former vice president of the Inter-American Court of Human Rights.

Carolina Costa Resende is a post-doctorate student in administration at FEA-RP/USP (current) with the theme people management in the context of the circular economy. She has a PhD in psychology from the Catholic University of Minas Gerais (2014), with a sandwich doctorate from the Université Paris Descartes (Paris V), funded by CAPES (2012), and a master's degree in psychology (2007) and degree in psychology from PUC Minas (2003), with specialization in institutional analysis and schizoanalysis from the Félix Gattari Institute (2004). She has been a professor (adjunct professor IV) at PUC Minas since February 2009, coordinating university extension programs and projects, scientific initiation, and TCC guidance. She has experience in the areas of social psychology, organizational behavior, organizational and work psychology, and the inclusion of people with disabilities in the job market.

Maria João Ruela is currently the social affairs advisor for the president of the Portuguese Republic. She is a former journalist and anchorwoman in the first private TV channel in Portugal. She has a degree in communication sciences and a PhD in social sustainability and development. Her areas of interest are social innovation, sustainable development, homelessness, and the Portuguese diaspora. She is the author of two books on social themes.

Ulla Saari, DSc (Tech), is a senior university researcher at the Unit of Industrial Engineering and Management, Tampere University. Her research topics include sustainability-oriented technology and innovation management, sustainable production and consumption, sustainable business models, and transitioning to a sustainable circular economy.

Partha Sen obtained his PhD at the London School of Economics. He was a professor at the Delhi School of Economics from 1996 to 2011 (and reader in economics from 1990 to 1996). He has also held full-time appointments at the University of Illinois Urbana-Champaign, the London School of Economics, and the University of Bristol, and visiting appointments, among others, at the University of Michigan, HKUST, NUS, NTU, and City University of Hong Kong. His areas of interest are macroeconomics, international economics, and environmental economics.

Mohammad H. Sepahvand is a development economist from Lund University and the European Center for Advanced Research in Economics and Statistics in Brussels, with long-running experience of data collection and research on attitude formation and decision-making in Burkina Faso. He has been responsible for the Swedish government's bilateral technical assistance project in statistics in the country between 2013 and 2017. He holds a PhD in economics from Uppsala University.

João Simão has a degree in management, a master's degree in statistics and information management, and a PhD in management. He is an associate professor at Universidade Aberta and a researcher at the Center for Global Studies (CEG), Universidade Aberta, Portugal, and Centre for Public Administration and Public Policies (CAPP), Universidade de Lisboa, Portugal. His areas of interest are sustainable development and corporate social responsibility.

Örjan Sjöberg is a professor of economic geography, Department of Marketing and Strategy, Stockholm School of Economics. He is also affiliated with the SIR Center for Sustainability Research. A geographer with an interest in issues at the intersection of migration, labor markets, and urban change, he typically approaches these and other concerns with an eye on political economy aspects as may apply.

Raiana Schirmer Soares is an energy engineer and PhD candidate in the Energy and Environment Institute at the University of São Paulo. Her research explores energy poverty in Brazil, focusing on developing alternatives for its measurements and understanding the nuances of its impacts. Parallel to this, she works as a socio-environmental consultant on energy planning and just transitions.

Vibhas Sukhwani works as a policy researcher at the Institute for Global Environmental Strategies (IGES) in Japan. A civil engineer/urban planner from background, he holds a PhD in media and governance. With an interdisciplinary research expertise, he applies various qualitative and quantitative research tools to conduct research in the areas of urban, rural, and regional planning; water–energy–food nexus; and circulating and ecological sphere. He has authored more than 35 research publications, including several peer-reviewed journal articles, book chapters of international repute, and reports.

Zelina Sultana is presently doing her PhD under the School of Economics, Finance, and Law, Faculty of Business and Law, at Anglia Ruskin University, Cambridge, UK, and an associate professor (study leave), Department of Law, Jagannath University, Dhaka, Bangladesh. Since 2007, she has been in the teaching profession in many renowned private and public universities of Bangladesh and taught in undergraduate and postgraduate levels. She has successfully completed many national and international trainings on legal social science research method, on environmental law teaching and research, as well as on SDGs. Her research articles have been published in international and national renowned journals relating to environmental protection, human rights, gender, and labour issue. She has also participated in many national and international workshops, seminars, and conferences and presented papers on contemporary issues.

Armindo dos Santos de Sousa Teodósio holds a PhD in management from the Getúlio Vargas Foundation in Brazil. He is currently an associate professor in the postgraduate program in management at the Pontifical Catholic University of Minas Gerais. In addition to his teaching responsibilities, he also heads the Research Center on Ethics and Social Management and the extension program Knowledge and InovAction for Sustainability. Teodósio is part of the Economics and Management Sciences Institute at the ICEG Business School. His research is primarily dedicated to critical studies about sustainability, environmental justice, environmental education, public policies, and environmental movements.

Sebastian Thomas (PhD) is a sustainability scientist, educator, and entrepreneur with expertise in climate change policy and economics, environmental science, sustainable business strategy, carbon and energy management, and the blue economy. He is an adjunct professor at Curtin University and the director of Blue Praxis, a scientific consulting company that provides strategic advice and professional services to governments and the private sector. Sebastian also works with a major Australian corporation in the asset management sector, supporting the decarbonization of financial products and investment portfolios. He is a scuba diver, hiker, and parent, and is deeply committed to contributing to positive social change and a healthy natural world.

Joana Soler Torramilans is a data advisor in the team Data for Rights and Development at the Danish Institute for Human Rights. Previously, she has been a research assistant at the Swiss Center of Expertise in Life Course Research at the University of Lausanne, in Switzerland. Before that, Joana collaborated with the Diversitat i Orientació (Diversity and Guidance) research group at the Autonomous University of Barcelona, in Spain. She has a degree in social data science from Copenhagen University.

Denise Voci, PhD, is a senior scientist at the Department of Media and Communications of the University of Klagenfurt. She received her PhD in the field of media and communication studies, with a focus on media economics and cross-border media management. Her research interests include media economics and media management, currently with a specific focus on the transformation process caused by the COVID-19 pandemic in these areas, as well as motivation in public service media, media sustainability, and environmental and sustainability communication.

Robert Wallis is the professor of research development at Federation University in Victoria, Australia. He is an ecologist who has published widely in the field of mammalian ecology, diets of introduced species, wildlife management, and environmental education. He is the chair of Federation University's Animal Ethics Committee.

Franzisca Weder (Univ. Prof., Dr. habil) is a professor for international organizational and sustainability communication at the University of Business and Economics, Vienna (Austria), and a research fellow at the University of Queensland, Brisbane (Australia). Franzisca is researching, writing, and teaching in the areas of organizational communication and public relations, with a specific focus on sustainability communication and corporate social responsibility. She worked as a guest professor at the University of Alabama (USA), University of Eichstätt-Ingolstadt (GER), University of Waikato (NZ), RMIT (Melbourne, AUS), and University of Ilmenau (GER). Franzisca Weder is the chair of the International Environmental Communication Association (IECA).

Helene Møller Winterskov is a project officer in the team Data for Rights and Development at the Danish Institute for Human Rights. Helene is an expert in data processing and analysis and has extensively been involved in development, training, and application of text mining methodologies. Her focus has been on linking the SDGs and human rights through machine learning.

Izawati Wook is an associate professor at the Faculty of Syariah and Law, Universiti Sains Islam Malaysia (USIM). She obtained a PhD from Victoria University, Melbourne. She was also admitted as an advocate and solicitor of the High Court in Malaya. Her research interests are Indigenous peoples and law.

Xin Zhou is the research director of the Integrated Sustainability Centre, Institute for Global Environmental Strategies (IGES), Japan. She has over 25 years of experience in environmental and sustainability policy research. She has devoted to support SDG-integrated policymaking in a couple of countries in Asia and led the development of the SDG Interlinkages Analysis & Visualisation Tool. Her expertise covers energy system scenario analysis, input–output analysis, macroeconomic modelling, life cycle analysis, indicators, and natural language processing for text mining and content analysis.

PREFACE

The UN Sustainable Development Goals (SDGs) have defined the international development agenda (Agenda 2030) for 2015–2030. Around mid-term, through their implementation, in a post-pandemic world with ongoing conflicts (Russian–Ukraine; Israel–Hamas–Palestine, etc.), the *Routledge Handbook of the UN Sustainable Development Goals Research and Policy* critically evaluates the progress and the future of the SDGs. The handbook is unique in that it does not address the SDGs in silos. Instead, each chapter focuses on a unique goal and deliberates on its interlinkages with other SDGs, taking account of both the positive (synergies) and the negative interlinkages (trade-offs).

The idea for the handbook came about in late 2020, when Routledge reached out to the Board of ISDRS, the International Sustainable Development Research Society, to create an international handbook on the SDGs. The then ISDRS president, Peter Dobers, and Ranjula Bali Swain teamed up in 2021 to create a strong and conference-based process to attract contributions of high quality from around the world. At the 28th Annual ISDRS Conference in 2022 in Stockholm, the first on-campus conference after three years of COVID-19, hybrid sessions were held for the handbook track. While battling the aftermath of the pandemic, scholars from around the globe participated in the two-day international workshop for the handbook during October 2023.

This book brings together a collection of seminal research on the SDGs. Particular attention has been given to the selection of the chapters, to provide a balanced inter- and transdisciplinary perspective and geographical representation of the regions and scholars around the globe. The book is aimed at a diverse audience, and we hope it will serve as a valuable reference for those seeking to better understand the SDGs progress with an interlinkages perspective. We hope that this edited volume will also provide useful insights and examples from a diverse regional perspective, with possible applications and implications to support the SDGs' implementation. The handbook also attempts to connect the SDGs to other global initiatives and concepts like climate change action, the Paris Agreement, planetary boundaries and limits to growth, etc.

We would like to gratefully acknowledge and thank the Routledge/Taylor & Francis Group, particularly Grace Harrison, Rosie Anderson, and Matthew Shobbrook. We are particularly grateful to the contributing authors, who have generously shared their expertise and knowledge, time, patience, and efforts at the ISDRS 2022 Conference and the 2023 International Workshop, making this volume possible. Their insightful contributions have helped produce a valuable resource for policymakers, researchers, college students, practitioners, stakeholders in the private and public sectors, and civil society. Without many un-named funding sources for individual scholars, this would also not have been possible. Thanks also goes to the ISDRS Board for putting trust in us to run a specific SDG handbook session at ISDRS 2022 in Stockholm, Sweden. Bali Swain would like to thank the research grants received from the Swedish Research Council for Sustainability (Formas) and the Swedish Energy Agency. Dobers would like to thank the research funding of his chair at Södertörn University, provided by the Foundation for Baltic and East European Studies.

In conclusion, we would like to emphasize the importance of understanding the interlinkages among the SDGs for the achievement of sustainable development. The United Nations–based SDGs represent a science-based and global consensus on the most pressing challenges facing our world and provide a road map for addressing them. Like any large-scale initiative, they are subject to ongoing debate and refinement. Thus, interpretation and implementation of the SDGs can vary.

We hope this edited volume will contribute to the broader discussion on the SDGs' implementation and their impact. Our aspiration is that it will inspire further research, discussions, and collaborations to support the achievement of the SDGs, and a more sustainable future for us all, beyond Agenda 2030 and the current system of the SDGs.

Ranjula Bali Swain
Peter Dobers

1

THE SUSTAINABLE DEVELOPMENT GOALS PROMISE AND BEYOND

Ranjula Bali Swain and Peter Dobers

1.1 Introduction

Referred to as comprehensive, universal, anthropocentric, and utopian, the Sustainable Development Goals (SDGs) have been unanimously embraced by United Nations member states in 2015, signifying a worldwide pledge toward specific objectives to accomplish by 2030. The SDGs aim to eradicate poverty and injustice, establish socio-economic inclusion, and protect the planet and its biospheres. The SDGs function as a forum for a global dialogue on efficacious resolutions to core issues pertaining to poverty and ecological deterioration.

The absence of a lucid implementation blueprint leaves the decision-making for suitable actions primarily to individual governments, many of which confront resource constraints or lack a national governance or a majority of people strongly supporting important steps toward achieving the goals. Not necessarily political by themselves, the SDGs have become tools in political discourse and become increasingly discredited by some political forces. While the SDGs have been presented as a panacea to current challenges of civilization, studies suggest there exists a potential inconsistency in the SDGs (Niklasson, 2019), particularly between the socio-economic development and the environmental sustainability goals.

1.2 Critiques of the Sustainable Development Goals

The SDGs represent a science-based and global consensus on the most pressing challenges facing our world and provide a road map for addressing them. Like any large-scale initiative, they are subject to ongoing debate and refinement. SDGs have been criticized for being inconsistent, difficult to quantify and implement. Critiques also raise questions on the measurability and monitoring of the broadly framed SDGs. The goals are non-binding, with each country being expected to create their own national or regional plans. Moreover, the source(s) and the extent of the financial resources and investments for the SDGs are ambiguous.

Despite their label as Sustainable Development Goals (SDGs), they are often perceived more as aspirations than concrete objectives (Koehler, 2023). The absence of a robust accountability framework means that responsibility for their achievement remains vague when nations or the international community fail to meet targets when meeting at so-called Conferences of the Parties, where representatives from many nation-states agree on steps to move forward.

DOI: 10.4324/9781003285472-1

The economic fallout from the COVID-19 pandemic has widened the already-substantial gap in funding for the SDGs, transforming it into a gaping funding challenge (Park, 2022), albeit there are ideas and principles for the transparency and interdependence of global funding (Sachs and Schmidt-Traub, 2017) or the role of fintech in funding sustainable development (Michael and Latkovska, 2021).

Critics contend that the SDGs overlook deep-rooted inequalities inherent in the global system (Driel et al., 2023; Lorenzo, 2024). The prevailing world order disproportionately benefits a wealthy minority. Addressing sustainable development necessitates a profound overhaul of systems, such as trade regulations, governance beyond nation-states, and abstaining of power by the influential, which for instance will further be put on trial now that we know of the Donald Trump presidency in the USA 2025–2029.

Despite their noble intentions, the UN-adopted SDGs face criticism for inconsistency, complexity (Fu et al., 2019), and the difficulty in quantifying, implementing, and monitoring progress (Costales and Zeyen, 2022; Greenland et al., 2023).

1.3 Missing Parts of the Sustainable Development Goals

And there are areas that the current system of Sustainable |Development Goals do not address enough, or not at all. The SDGs do not directly address issues related to global power imbalances, such as the influence of multinational corporations or the super-rich such as Elon Musk on local economies and the impact of global trade agreements on developing countries. We need to acknowledge that partnerships for SDGs also underlie power imbalances, harbor conflicting interests, and at times have limited democratic representation (Horan, 2021) when, for instance, crowdfunding helps leverage sustainability-promoting activities (Gellers, 2016).

While the SDGs mention the importance of inclusive societies, they do not explicitly address the rights and contributions of Indigenous peoples and the need to preserve cultural diversity and traditional knowledge, like, for instance, the significance of Myanmar Indigenous rights (Liljeblad, 2022) and Sámi rights (Cambou and Ravna, 2024; Larsen and Raitio, 2023; Dahlström et al., 2021).

The SDGs call for peace and justice, but they do not explicitly address the role of the military–industrial complex and arms trade in fueling conflict and instability. There are many war-ridden areas in the world that could be mentioned here: As mentioned before, we see the Russian war in Ukraine well into its third year, with its wide-ranging carbon emissions (Sasmoko et al., 2023); the wider Middle East war in general and the Israeli–Hamas–Palestine conflict in particular; the long-ongoing Armenia–Azerbaijan conflict; or the Myanmar situation.

As we have seen in previous studies (Dobers and Halme, 2009), Africa is an especially war-ridden continent (Akinola, 2023) that challenges governance ambitions in line with Agenda 2030 based on climate crises–based migration, linking also to war, conflicts, and security issues. Military conflicts, and functioning arm trade treaties, are thus of high importance (Nave, 2019) yet missing in the current system of the SDGs, albeit somewhat addressed by SDG 16 (Milton, 2021).

Also, the SDGs aim to ensure sustainable consumption and production patterns. The relation between sustainable consumption and growth can be questioned, and attempts are made to show some complementarities (Salimath and Chandna, 2021; Kreinin and Aigner, 2022). However, they do not directly address the issue of overconsumption in developed countries and its impact on global sustainability.

Further, the SDGs focus on preserving biodiversity and ecosystems but do not explicitly mention animal rights (Schapper and Bliss, 2023), wild species (Fromentin et al., 2023), and the welfare of the one health that intersects animal, human, and plant health (Verniers and Brels, 2021). These aspects have also been suggested as the SDG 18, for instance (Visseren-Hamakers, 2020).

When the SDGs and Agenda 2030 were formulated, the world was heavily invested in digital technologies already, when considering how economies and the lives of many had been impacted by digital technologies, social media apps, and digital services. But the SDG system neglected to include digitalization (Mondejar, 2021), and artificial intelligence was not really on the radar for many but currently is very important to relate to sustainable development as well (Sætra, 2021; Visvizi, 2022; Raper et al., 2022; Nasir et al., 2023; Lau et al., 2023).

Another good overview of how targets of the 17 SDGs can be improved in light of biophysical and planetary limits can be found in the book *Ecological Limits of Development: Living with the Sustainable Development Goals* (Kish and Quilley, 2022).

1.4 From Sustainable Development Goals to Inner Development Goals

When hosting the annual conference 2022 in Stockholm for ISDRS, the International Sustainable Development Research Society (ISDRS), during the "Stockholm+50" month of June 2022, we had as one of the guest speakers in one of our panels Jakob Trollbäck, the founder and CEO of the New Division. While at the New York offices of this company, Mr. Trollbäck linked up with film director Richard Curtis to discuss how to make the Sustainable Development Goals famous. Curtis and Trollbäck recognized the importance of creating a communication system that could resonate universally and inspire action.

Insights from UN ambassadors and non-governmental organizations (NGOs) played a crucial role in shaping this endeavor. They emphasized the need for icons that could be easily understood across cultures and languages. The resulting designs and color palette were carefully crafted to

Figure 1.1 The 17 United Nations Sustainable Development Goals (SDGs).

ensure cross-cultural relevance and symbolism. One of the key visual elements is the circular logotype for the master brand, which symbolizes unity and utilizes the full color palette from the icon system. Its symbols and colors have become widely known as the 17 SDGs.

The outcome of this collaborative effort was a comprehensive communication system that received approval from all 193 member nations. Today, it stands as a universally embraced tool and a symbol of hope and a catalyst for change, motivating individuals worldwide to pursue the global goals by 2030. In addition to the 17 main goals, there are 169 targets connected to these goals. These targets define specific measures necessary to achieve the broader objectives. To make the agenda even more accessible, the visual language was extended to include these targets. Each target is summarized in a short understandable headline, and powerful visual symbols were created to represent them. By the use of bright colors and bold typography, families of symbols were associated with the main goals, expressing determination and optimism. (See more at the United Nations web page for the SDGs: https://sdgs.un.org/goals.) So when asked if he could speak to the ISDRS 2022 annual conference, Mr. Trollbäck was happy to do so but insisted on speaking about a more recent undertaking of his, namely, the inner development goals (IDGs). As shown on the ISDRS 2022 playlist on YouTube, Trollbäck speaks out about these new goals[1] about 59:30 minutes into the video.

He talks vividly about empowering people to pursue the SDGs. About 1,000 persons replied to a survey for what capacities and skills we need to fulfil the SDGs. And then Mr. Trollbäck briefly presented 23 skills in five categories. The background, method, and IDG framework have been documented in a report (Jordan, 2021). Our understanding of the IDGs is that they foster an understanding for what skill sets may be needed for working toward sustainable development and achieving many of the promises set forth by the SDG framework. The IDGs are about an inner growth as individuals, nurturing also the existential aspects of life on the planet and enabling us to grow as individuals beyond a material and economically based growth, empowering business models and people toward dematerialization (Dobers and Wolff, 1999; Sun, 2000). There is something attractive to the collaborative and explorative concept of skill sets for fostering the achievement of the SDGs, where the IDGs are just one system of capacities and skills of individuals (Ankrah et al., 2023; Shtaltovna et al., 2024) among similar ideas, such as ethics, responsibility, and sustainability competences, recently being explored in academia (Laasch et al., 2023).

1.5 About the Routledge SDG Handbook

Two-thirds through Agenda 2030, this edited volume is an assessment of Agenda 2030 and

- investigates if these 17 SDGs have been effective in achieving the transformative change toward sustainable development;
- investigates thematic, regional, and country levels and provides examples of possible applications, implications, and best practices;
- identifies some missing aspects of Agenda 2030 and critically evaluates SDGs;
- employs inter- and transdisciplinary research from across the various continents to present experience and evidence on SDGs; and
- suggests building blocks for the post-SDG 2030 Agenda.

The volume is aimed at a diverse audience, including policymakers, university researchers, college students, practitioners, stakeholders in the private and public sectors, and civil society.

Part 1 of the handbook brings together seminal research to define, identify, and present conceptual frameworks for a sustainable future. The handbook is not SDG- and target-specific; instead, individual chapters maintain the focus on one or two main SDGs, while investigating their mid-term progress from the interlinkages perspective and the connections with other SDGs. Part 2 of the handbook focuses on the policies, practices, and implementation part of the SDGs.

The sustainable goals inherently and emphatically recognize development to be multidimensional, encompassing economic, social, and environmental aspects of life. SDG 1 particularly emphasizes ending poverty in all its forms everywhere. In Chapter 2, Alkire investigates the progress in multidimensional poverty, past the mid-point of the SDGs in a post-pandemic world driven by crises and uncertainty. The chapter systematically sets out the ways that measure multidimensional poverty indices (MPIs) and their engagement with Agenda 2030 and summarizes their insights and policy tools.

Leichtweiss and Soares delve into the crucial role of the 5Ps ("people," "planet," "prosperity," "peace," and "partnership") within the framework of Agenda 2030. Chapter 3 further explores their relationship with the SDGs. As an illustrative example, the chapter examines SDG 7, which focuses on affordable and clean energy. While traditionally associated with *prosperity*, energy is also central to the dimensions of *people*, *planet*, *peace*, and *partnership*. By recognizing this interconnectedness, we can better address the complex challenges posed by the SDGs. It is a critical and interdisciplinary endeavor that problematizes mainstream interpretations of this relationship, seeking to understand how the 5Ps can be better used to guide action toward the implementation of the goals. Further, the chapter shows ways to overcome fragmentation, strengthen integrated links between the SDGs, and better detect synergies and trade-offs between them, but also the external connections within the legal-political environment that shapes their implementation.

The main development patterns and challenges that lie ahead in reaching the targets set in the Sustainable Development Goals (SDGs), with a particular focus on SDG 1 (No Poverty), 2 (Zero Hunger), and 5 (Gender Equality), are explored in Chapter 4 by Sepahvand. The chapter examines how the SDGs are impacted by the economic and societal factors caused by the crises and shocks, such as political instability, violent conflict, and the COVID-19 pandemic, in the Sahel region of Burkina Faso in Africa. In a world committed to end poverty and hunger and "leave no one behind," what are the challenges and opportunities to implement it? Chapter 4 aims to address these issues and the enormous challenges from climate change, political insecurity, and violent conflict while accounting for the progress in the efforts to reduce hunger and food insecurity in Africa's Sahel region.

Sustainability is increasingly established as a master frame for transformation in public communication processes and influences social practices, attitudes, and behavior on an individual level by adding value and morality to it (i.e., differentiating "good" from "bad" consumption behavior). The question of how far sustainability is a normative principle, and the Sustainable Development Goals the related normative framework that guides and potentially moralizes food-related decision-making, is explored and answered with empirical data from an exploratory qualitative study. In Chapter 5, Weder, Lemke, and Fenzl conduct three series of guideline-based interviews with national and international respondents from Austria (25) and national respondents from Australia (5) and Indonesia (5) to gather information on individual perceptions of food choices and eating behaviors (Weder et al., 2025). Interview data was analyzed using inductive category formation. Based on the results of the study, they hypothesize that sustainability is not a defining moment in decision-making, however has the character of a moral imperative related to a specific eating behavior. Despite methodological limitations, outlined at the end of the chapter, the research insights challenge existing literature on food consumption and environmental communication which treats sustainability as a "new" or "normalized," and therefore established, norm and ask for more empirical research on defining non-mediated communicative moments leading to behavior changes.

Energy poverty is a vital global concern affecting the lives of millions across the world. Deprivation of access to clean, affordable, and reliable energy sources is acute in developing countries like India. The consequences of being energy-poor are reflected in the quality of life via different means, including educational outcomes. In this study, they examine the impact of energy poverty on educational outcomes among Indian households. In Chapter 6, Manasi B. and Mukhopadhyay measure energy poverty through electricity access, clean cooking fuel access, and a multidimensional energy poverty index. They find a significant negative impact of energy poverty on various educational outcomes through different channels.

Chapter 7 assesses the interactions among SDGs for the case of women's economic empowerment. For illustration purposes, Niaz, Islam, Kambhampati, and Ashraf single out two forms of hidden gender inequalities within education (i.e., gender bias in learning materials and educational leadership) that may potentially undermine interlinkages between SDGs 4 and 5. They present descriptive secondary evidence on the broken and intact interlinkages between female schooling and women's employment, followed by a discussion of the policy lessons from South Asia and Southeast Asia—two regions with contrasting experience in harnessing synergistic developments and positive interlinkages for women's development. They conclude that if gender inequality within the production of education is overlooked, schooling girls will not be enough for economic empowerment of women and may even reinforce the broken links between SDGs 4 and 5.

Dev, Acharya, and Kushwala examine the pursuit of achieving gender equality (SDG 5), in the context of India, in Chapter 8. Considering the measures implemented by both the central and the state governments, they use data from two rounds of the National Family Health Survey (NFHS). They assess the policy impact on achieving the SDG 5 target on women's empowerment in India. Additionally, they explore the theoretical connections between SDG 5 and other Sustainable Development Goals (SDGs). Through empirical research, their study establishes interconnections between SDG 5 indicators across time. They consider how district-level averages of empowerment measures related to decision-making and freedom of movement from NFHS-4 influence individual-level outcomes regarding employment, education, and family planning in the subsequent NFHS-5 round. Their chapter underscores the significance of tailoring data surveys and refining target visions to align with worldwide diverse cultural contexts.

The intricate interplay between water, energy, and food resources and their management plays a pivotal role in advancing the Sustainable Development Goals (SDGs). Zhou, Mitra, and Sukhwani explore the critical interlinkages among these essential components and their profound impact on achieving the SDGs in Chapter 9. Water resources, as a fundamental pillar, not only sustain life but also underpin agriculture, industry, and energy production. Energy, in turn, is indispensable for water management and food production. Simultaneously, agriculture, as a primary source of food, demands water and energy inputs. These intricate connections form the nexus that is vital for addressing the SDGs, especially those related to poverty reduction (SDG 1), zero hunger (SDG 2), clean water and sanitation (SDG 6), affordable and clean energy (SDG 7), and sustainable cities and communities (SDG 11). To illustrate the profound implications of this nexus for SDG advancement, they present a case study of the Ganges River Basin. The Ganges, spanning multiple countries in South Asia, serves as a representative of the complex interdependencies among water availability, energy security, and food production. By applying climate change scenarios and integrated assessment methodologies, they scrutinize the future challenges and opportunities within this basin. Their findings highlight the necessity of holistic and synergistic approaches to resource management. These insights underscore the importance of embracing the water–energy–food nexus as a guiding framework for policy development and implementation to catalyze progress toward the SDGs. By addressing this nexus with a comprehensive perspective, the transformative pathways for sustainable development may be unlocked.

Protection of water resources is critical to accomplish SDG 6, notably Target 6, linked to clean water and, indirectly, the other Earth goals. The extant plans, policies, and institutional protection mechanisms for water resource conservation and management in Bangladesh tend to be weak and to exhaust ecosystems and human health and accelerate poverty rate. In Chapter 10, Sultana and Akter attempt to analyze Bangladesh's existing legislation and policies for the conservation and management of water resources to balance the environment and attain the SDGs, as well as their regulation, by its environmental and development-related policies and plans. The chapter focuses on the limitations of existing instruments to find out the robust solution for water resources conservation. It concludes that excessive pollution, demand for fresh water, lack of synergy between multiple institutions, and lack of coordination with other SDGs remain a persistent concern for Bangladesh. This is detrimental to maintaining environmental equilibrium, the SDGs, and other relevant national targets.

Bali Swain and Sjöberg, in Chapter 11, examine if markets can promote sustainable development. They discuss the challenges posited by climate change and sustainable development, and the limitations of the market system. The chapter further discusses if the markets should be discarded or if their shortcomings can be overcome. Or can markets become sustainable in the sense of fostering sustainability outcomes?

Damberg, Fritz, and Saari explore how the principal values and drivers of born-sustainable companies are built into the core of their business models and are crucial to achieve some of the critical SDGs. Chapter 12 analyzes how born-sustainable and sustainability-driven companies present themselves, including how they have built their businesses, by focusing on their core sustainability values. These components offer guidance for companies that need to develop their business models to become more sustainable in alignment with the SDGs. They apply a case study methodology to assess companies from the clothes and apparel to cosmetics and food industries and use secondary data. They present a conceptual model of how these case companies build their business models on a sustainable foundation that promotes key SDGs. This helps build more sustainable business and practices, which will help ensure overall well-being in society and have a positive impact on the natural environment.

Governments worldwide are grappling with the urgent challenges posed by climate change and the looming specter of ecological collapse, threats that imperil social stability and economic continuity. Simultaneously, the private sector is navigating its role in addressing these global crises while adhering to its core principles of investment and return on investment. According to Thomas and Bruckner in Chapter 13, this delicate balance requires innovative approaches that align financial interests with environmental and social imperatives. Global movements are gaining momentum, advocating for the integration of climate and nature considerations into financial decision-making. These movements recognize that financial institutions and organizations wield significant influence over sustainability outcomes. As a result, the way businesses invest and conduct their operations is undergoing a profound shift.

Innovations are a critical to long-term growth. Frugal innovation strategically reduces resource utilization, including raw materials, energy, and financial resources, enhancing technology's affordability and accessibility while lessening ecological impact. This connection positions frugal innovation at the intersection of economic, social, and ecological sustainability. Bhattacharjya and Nath comprehensively review the existing literature in Chapter 14, revealing strong evidence affirming a positive relationship between frugal innovation and sustainability. Limited cases indicate negative impacts. Frugal innovation significantly contributes to multiple Sustainable Development Goals (SDGs), particularly SDGs 8 (Decent Work and Economic Growth), 3 (Good Health and Well-Being), 10 (Reduced Inequalities), 12 (Responsible Consumption and Production), and 9 (Industry, Innovation, and Infrastructure), as evidenced by the cases studied in this chapter.

A policy to fight climate change involves making fossil fuels more expensive. A justification for implementing the proposed border carbon adjustment (BCA) tariff is that stringent environmental regulation (generally) makes footloose polluting industries move to countries with lower regulation. Even if the relocation does not occur, strict climate change policies cause carbon-intensive sectors to contract in countries where carbon is priced higher. Existing firms' output changes, causing "carbon leakages". Such a tax would correct this. Sen, in Chapter 15, argues that there are problems with a BCA of measurement, compatibility with trade laws, and equity issues (underlying the common but differentiated responsibilities). The attainment of SDGs 13 (climate change) and 17 (global partnership for sustainable development) especially would receive a setback.

Sustainable Development Goal 15 involves the development of terrestrial and freshwater ecosystems in a way that does not deleteriously affect resource use for future generations. Through a series of targets, it aims for no loss of biodiversity, sustainable forest management, increased coverage of protected areas, halting desertification, and reducing land degradation. Progress to achieving these targets by 2030 has been slow and patchy across the world. The business sector is becoming aware that its future prosperity and sustainability depend on healthy ecosystems and the services these provide. Many progressive and innovative corporations have committed to becoming nature-positive and ensuring their operations leave a better environmental footprint. However, there are others that have been accused of greenwashing, in which their claims of sustainable practices are without foundation. Wallis and Cole present a case study in Australia in which damage to agricultural production has been averted using a humane and effective use of raptors to disperse large flocks of native birds that threatened to consume a large almond crop. The technique has been used to solve problems caused by birds not only in agricultural production but also in infrastructure damage and disruption to sporting and cultural activities. In Chapter 16, they find that the technique works best when the period of damage activity is discrete, the area small, and the financial risk of damage by the birds high. However, the method is labor-intensive and does not work in certain other circumstances. The chapter also discusses the potential improvement in the effectiveness of the technique if lethal take of some of the pest birds is encouraged. The case study has relevance to other SDGs besides 15, such as SDGs 2 and 12.

Over the past 20 years, there has been a significant surge in global plastic consumption, leading to a corresponding rise in pollution. Addressing this plastic pollution has become a top priority on both global and local scales. Among the international initiatives aimed at tackling environmental challenges, the Sustainable Development Goals (SDGs) stand out, albeit with a relatively minor emphasis on plastic waste. Chapter 17, by Kathuria and Jardosh, delves into the origins of the plastic pollution crisis and evaluates the potential effectiveness of the SDGs in addressing it. Ultimately, while the SDGs have influenced the discourse and regulatory frameworks, their tangible impact on plastic waste management remains modest, reflecting broader patterns observed across various SDG-related issues.

The focus on social innovation and how entrepreneurs and government officials are using it indicates an increased ability of civil society to creatively intervene in response to long-standing social issues, like unemployment, and new challenges caused by digitalization, climate change, and environmental sustainability. This attention highlights the potential for positive change through social innovation. From this perspective, social innovation can function as a tool for sustainable development. It is a proposal for development that stems from cross-sector partnerships and entrepreneurial initiatives designed to create social and environmental impact. Moreover, it can contribute to the achievement of the Sustainable Development Goals outlined in the 2030 Agenda published by the United Nations in 2015. This fact has served as a narrative element to shift the development paradigm within the European Union. Specifically, it is a key element of the Europe 2020 strategy aimed at mitigating the economic crisis, climate change, and other global issues, collectively referred to as "grand challenges." Ruela, Simão, and Almeida, in Chapter 18, aim to theoretically support the points of contact between social innovation and sustainable development through a literature review. It discusses the significance of social innovation's contribution to implementing the United Nations 2030 Agenda and achieving the Sustainable Development Goals at the local level.

Chapter 19 focuses on the intricacies of the goals SDG 16 (Peace, Justice, and Strong Institutions) and SDG 17 (Partnerships for the Goals), particularly in an age of disinformation. Often depicted to be enabler goals which serve as tools to achieve other goals, they are often taken for granted to uphold and synergize with other goals. What are neglected are the complicities and trade-offs in the decision-making process involved in both goals, especially when there are threats involved, such as the spread of disinformation. By taking the case of how SDG agents decide in handling information, Jonsson illustrates the individual intricacies and dilemmas involved in actualizing strong institutions and partnerships. Various simulation scenarios are presented to show how the interplay of cooperation, organized fake news machineries, and sanctions affects the knowledge commons, exposing how SDG attainment can become more elusive in the presence of disinformation.

Chapter 20 critically analyzes the law in theory and practice on free, prior, and informed consent (FPIC) as part of the right to self-determination of Indigenous peoples in ASEAN by looking at international law standards. Hassan, Nordin, Wook, and Ab Rahman evaluate the issues and challenges in applying the principle of FPIC in Malaysia, the Philippines, and Indonesia according to domestic law compared to international law standards. The study found that the FPIC in ASEAN member states remains a regime of unfulfilled promises. This chapter suggests several legal reforms to improve the recognition of FPIC as part of the right to self-determination of the Indigenous peoples. It will substantially complement the realization of the SDG's central pledge to "leave no one behind" (LNOB) and Goal 16 in promoting a just, peaceful, and inclusive society. It is vital to advance the right of Indigenous peoples, thus assisting ASEAN member states in achieving Targets 16.7 and 16.10 of SDG 16.

The remaining chapters in the handbook's Part 2 are oriented toward policy and practice.

COVID-19 has caused numerous casualties worldwide. Nonetheless, it has also led to the recovery of the global environment due to an economic slowdown. However, this is not the case in Taiwan. With almost no lockdown in early 2020, Taiwan's economic growth has improved. Consequently, concerns about revising the Climate Change Act of 2015 to improve carbon emissions reduction from mid-April 2020 emerged. The main objective was to implement a carbon fee scheme. Such measure is related to the EU's proposed carbon border adjustment mechanism (CBAM) under the EU Green Deal since March 2020. After two legislative proposals provided by the Environmental Protection Administration (EPA) in 2020 and 2021, the cabinet adopted its version in April 2022, which was then submitted to Parliament and adopted by Parliament in January 2023. Parallel to this development, the government adopted "Taiwan's Pathway to Net-Zero Emissions in 2050" on March 30, 2022, and announced the detailed action plans. These measures are intended to address net-zero, EU CBAM, and SDG 13. In Chapter 21, Gao aims to provide an overview of this legislative development and the key issues in advancing carbon emission reduction and meeting CBAM's need.

In Chapter 22, Mitchell aims at investigating the relationship between the Sustainable Development Goals (SDGs) and contract law in a comparative and interdisciplinary perspective. In fact, scholars are still at the beginning of researching how closely concepts, methodologies, and instruments of private law, including but not limited to contract law, can contribute to the greatest challenges of our time. Precisely, this exploration grounds on three main issues. Mitchell aims at discussing whether contract law may contribute to enhancing and governing the SDGs. In this respect, it considers three case studies, consisting of (1) the standardization and certification processes of SDG indicators, (2) the agreement concerning the enforcement of some of the SDGs included in the Bangladesh Accord, and (3) the practice of incorporating sustainability contractual clauses (SCCs) into international supply chain agreements. Second, it analyzes the nature and scope of the interplay of contract law and sustainable development (SD) and the enforcement problem of contractual clauses dealing with the goals of SD. Third, the chapter considers the potential need and possibilities for contract law to respond to the public challenges of the SDGs and to come up with new orientations.

The UN has encouraged national governments to build their own National Sustainable Development Strategies to align with Agenda 2030 and sustainability policies. To make this possible, an integrated approach that includes methodologies and operational tools of co-design, with a focus on environmental, social, and economic sustainability and involving communities and public investments, is necessary. At the metropolitan level, RC Metro Citizens in Transition has been undertaken to build the Metropolitan Agenda and align with the National Strategy for Sustainable Development. Through knowledge transfer and capacity building, the project has involved a wide range of actors, resulting in progress toward multiple SDGs. The medium-term results and educated community aim to project themselves on integrated trajectories of the Metropolitan Strategic Plan and launch innovative sustainable projects. The ultimate goal is to align with the provisions of the Recovery and Resilience Plan and to positively impact the indicators related to knowledge and skills on the issues of ecological and digital transition of young people under 35. Poncibò discusses this in Chapter 23.

For Agenda 2030 and sustainability policies, the UN has encouraged national governments to build their own National Sustainable Development Strategy capable of "territorializing" the SDGs at local level. If, on one hand, global initiatives, "in the context of policies for Climate Change and Sustainable Development, change the paradigms of the transition to economic and social, as well as environmental scenarios of the Fourth Industrial Revolution," on the other it seems necessary to adopt an integrated approach to generate programs and projects for the "dissemination of urban and territorial issues, experimenting best practices . . . , pursuing processes of involvement of communities and engaging public investments" (Nave, 2019). It means applying in a "trans-disciplinary" way methodologies and operational tools of co-design in multi-level and multi-actor project experiences (Schütz et al., 2019) in order to co-design actions in which the aspects related to environmental, social, and economic sustainability are closely related to the ability to transfer them to the communities, preparing the necessary field of knowledge and skills on contemporary scenarios for sustainable development and digital and ecological transition according to "fair and equitable" mechanisms (Pianta and Lucchese, 2020). According to Mangano in Chapter 24, this trajectory is currently being experimented at local level with the project "RC Metro Citizens in Transition," aimed at the construction of the Metropolitan Agenda in the implementation of the National Strategy for Sustainable Development (NSDS) (Mangano). Through knowledge transfer and capacity-building actions, like "Knowledge Metrocity" (seminars and co-design workshops on sustainability, innovation, and enabling technologies), which has involved an open "civic ecosystem" (professionals, university and high school students, institutions, associations, citizens, etc.), it has been possible to act on the targets of SDGs 4 (Quality Education), 7 (Clean and Accessible Energy), 9 (Business, Innovation, and Infrastructure), 11 (Sustainable Communities and Cities?" 12 (Responsible Production and Consumption), and 13 (Climate Action) and activate the monitoring processes of the projects in accordance with the set of the 43 indicators shared by the National Sustainable Development Table for the "Planet" and "Prosperity" areas of the NSDS. The results obtained in the medium term, and the creation of a community educated on the issues of transition, have the ambition to project themselves on identified integrated trajectories of the Metropolitan Strategic Plan, with particular reference to the possibility of launching innovative experimentations with high regenerative rate in territorial areas, such as the inner areas of the metropolitan city of Reggio Calabria, in the implementation of the provisions of the Recovery and Resilience Plan (RRP).

Voci and Karmasin, in Chapter 25, aim to address the crucial absence of the media and communications field in the SDG implementation strategy, emphasizing the substantial advantages and potentials linked to its inclusion for the SDGs' achievement. The only mention of media and communication can currently be found in Target 16.10: "Ensure public access to information and protect fundamental freedoms." An unfortunate formulation, we argue, as it fails to recognize the fundamental role of responsible media and communications in the creation of the public sphere and—moreover—the maintenance of democracies. Thus, they aim to demonstrate that "public access to reliable information" is not just a prerequisite but also an enabler to "protect fundamental freedoms," by arguing that media and communications' contribution to sustainable development extends beyond information transmission. Indeed, public access to reliable information requires responsible sustainability communication, including responsible information production and consumption. This, in turn, requires support from media education, media literacy, and the existence of a strong, independent Fourth Estate. Media literacy is also a crucial tool to combat the growing issue of politicization of the sustainability discourse. As the discussion around sustainability and its development transitions from the scientific to the political realms, the primary communicators shaping this narrative have shifted from scientists to politicians. Rather than leading the "SDG agenda," politicians often prioritize their political agendas, aiming to secure votes during elections. In this context, media literacy assumes a key role by ensuring that individual behavioral decisions within a media-literate society are founded on a balanced and thoughtful examination of information and scientific facts, rather than being influenced by political opinions and preferences. Chapter 25 concludes that it is essential to underscore that responsible sustainability communication and media literacy are not realizable without the presence of a strong and independent Fourth Estate, as they are interlinked necessities. Placing these elements at the forefront of sustainable development strategies is crucial to fostering a media landscape that facilitates the dissemination of accurate information, encourages responsible citizen engagement, and holds stakeholders accountable. This, in turn, safeguards fundamental freedoms and significantly contributes to the achievement of the SDGs and the building of a more sustainable future.

The achievement of net-zero emissions is in line with the Sustainable Development Goals (SDGs), mainly SDG 7 (Affordable and Clean Energy) and SDG 13 (Climate Action). In the debate on various options to achieve carbon neutrality before 2060 in China, scholars and practitioners have been mainly focusing on the decarbonization of the coal power sector and the transition to renewable energy, while the development and employment of carbon dioxide removal (CDR) technologies to remove CO_2 from the atmosphere have not yet received sufficient attention. The corresponding legal research remains a gap. In Chapter 26, Du aims to explore the regulation of three categories of CDR techniques in China and the role of CDR in the mitigation policy portfolio for carbon neutrality. The chapter first describes the status quo of the policy setting of the commitment to carbon neutrality and the features of the main CDR techniques. Then, via examining the current legal and policy frameworks for CDR in China, three regulatory gaps are identified: the underdevelopment of explicit regulation on CDR, the inconsistency with international law in regulating ocean-based CDR, and a lack of financial and market mechanisms. Third, this chapter provides suggestions on the future development of CDR in contributing to the achievement of carbon neutrality in China. The chapter concludes that the supplementary role of CDR in achieving carbon neutrality needs to be recognized by future climate policy. The implementation of different CDR techniques should be managed appropriately to avoid the unbalancing between carbon neutrality and the other Sustainable Development Goals.

Universities face major challenges in producing knowledge capable of promoting sustainability in the contemporary world. The relationship between university, civil society, market, and government actors is marked by the challenge of recognizing valid knowledge for the change toward sustainability. One of the strategies to overcome these challenges involves the development of socio-environmental programs and projects capable of linking researchers and students with communities. In the Brazilian reality, a series of devices standardize the activities related to the interaction between university and society, seeking the construction of collaborative projects, which allows the materializing of an agenda of social inclusion, citizenship, and humanist education. This agenda of activities brings the need for important changes in the way teaching, research, community relations, and internationalization activities are developed by universities. In Chapter 27 Melo, dos Santos de Sousa Teodósia, Costa Resende, and Simão Abuhid analyze the experience of a Brazilian university through a qualitative case study. The results point to the possibilities, risks, and pitfalls of modernizing the role of the university as one of the relevant actors for the promotion of sustainability in contemporary societies.

Aligned with the objectives of Sustainable Development Goals (SDGs) 3 (Good Health and Well-Being) and 5 (Gender Equality), there is an expanding global emphasis on championing optimal menstrual health. Recognized not only as a public health priority but also as a fundamental human rights matter, achieving quality menstrual health is essential to realizing the broader aims set by the SDGs. Despite this, many low- and middle-income countries continue to primarily subsidize disposable sanitary pads. While intended to support women, this approach inadvertently impedes SDG progress by contributing to environmental waste and excluding numerous marginalized women who cannot afford these pads, thereby exacerbating gender inequities. In light of the pivotal role of sustainability in the SDGs, particularly concerning SDGs 3 and 5, it is paramount to identify solutions that address menstrual health needs while also curtailing environmental impact. In Chapter 28, Babbar and Garikipati discuss how one such promising solution are reusable menstrual products, notably menstrual cups. However, their adoption in several low- and middle-income nations, including India, faces obstacles rooted in deep-seated taboos and misconceptions about menstruation and reproductive health, especially since these products necessitate vaginal insertion. The chapter utilizes comprehensive data from the National Family Health Survey, India's counterpart to the Demographic Health Survey, spanning 2019–2021. Through this analysis, the authors aim to shed light on menstrual cup adoption patterns and the socio-demographic dynamics influencing their use. They conclude by outlining policy implications and suggesting routes to bolster menstrual cup adoption in low- and middle-income contexts, advancing the principles of SDGs 3 and 5.

In Chapter 29, Lernborg and Atallah explore the critical nexus between biodiversity and its financing in achieving the Sustainable Development Goals. Focused on the Democratic Republic of the Congo, a biodiversity-rich region confronting pressing challenges like deforestation and habitat loss, the chapter scrutinizes its potential contribution to SDG 15 amidst new global commitments. It offers an overview of biodiversity financing, underscoring the importance of payment for ecosystem services (PES) as a tool for incentivizing sustainable land management practices and biodiversity conservation in developing regions. Drawing on global and local case studies, the chapter highlights PES's potential to drive economic development while preserving ecosystems yet acknowledges its limitations. Through a review of PES success factors and an examination of national initiatives, it elucidates both the opportunities and hurdles associated with implementing such programs in the Global South.

The 2030 Agenda is grounded in international human rights norms, and the Sustainable Development Goals (SDGs) seek to realize the human rights of all. The interlinkage between the SDGs and human rights are substantial, but how do they intersect in practice, and how much attention are states paying to one another when it comes to monitoring and reporting to respective international bodies? Reis, Skriver Jørgensen, Møller Winterskov, and Soler Torramilans showcase examples and tools which facilitate the operationalization of an interconnected approach between SDGs and human rights in Chapter 30. In addition, the research will investigate the prevalence of human rights issues in the voluntary national review (VNR) and the prevalence of the SDGs in states' report to the Universal Periodic Review (UPR), which is the universal human rights monitoring mechanism. This analysis intends to uncover examples of good practices and shortcomings in how states are reporting on progress to the SDGs and human rights mechanisms, and if there are lessons to improve coherence and efficiency in their reporting practices to international bodies. The lessons are drawn from employing artificial intelligence to support the text mining of the VNRs and UPRs from 2016 until 2023.

This handbook investigates the effectiveness of the 17 SDGs in achieving transformative change toward sustainable development from the interlinkages perspective. It explores the application, implications, and best practices of these goals at thematic, regional, and country levels, providing specific examples where possible. Additionally, it identifies missing aspects of Agenda 2030, critically evaluates the SDGs, and suggests future directions for the post–SDG 2030 Agenda. By employing inter- and transdisciplinary research from various continents, the handbook is a midterm evaluation of the SDGs and provides comprehensive experiences and evidence related to the SDGs and their interlinkages.

Note

1 www.youtube.com/watch?v=IObpvWs5zeo&list=PLu-70cD5hTjNf2X3ppxTBx6444mqo86pA&index=5.

References

Akinola, A. O. (2023). *Contemporary Issues on Governance, Conflict and Security in Africa*. Cham: Palgrave Macmillan. https://doi.org/10.1007/978-3-031-29635-2

Ankrah, D., Bristow, J., Hires, D., & Henriksson, J. A. (2023). Inner Development Goals: From Inner Growth to Outer Change. *Field Action Science Reports. The Journal of Field Actions*, *25*, 82–87. https://journals.openedition.org/factsreports/7326

Cambou, D., & Ravna, Ø. (2024). *The Significance of Sámi Rights. Law, Justice, and Sustainability for the Indigenous Sámi in the Nordic Countries*. New York: Routledge. https://doi.org/10.4324/9781003220640

Costales, E., & Zeyen, A. (2022). The Sustainable Development Goals, Complexity, and Errors of the Third Kind. In E. Costales & A. Zeyen (Eds.), *Social Entrepreneurship and Grand Challenges. Navigating Layers of Disruption from COVID-19 and Beyond* (pp. 99–116). Cham: Palgrave Macmillan. https://doi.org/10.1007/978-3-031-07450-9_6

Dahlström, Å. N., Dahlin, J., & Tunón, H. (2021). Pathfinders for the Future? Indigenous Rights and Traditional Knowledge in Sweden. *Sustainability, 13* (11195), 1–23. https://doi.org/10.3390/su132011195

Dobers, P., & Halme, M. (2009). Corporate Social Responsibility and Developing Countries. *Corporate Social Responsibility and Environmental Management, 16* (5), 237–249. https://doi.org/10.1002/csr.212

Dobers, P., & Wolff, R. (1999). Eco-Efficiency and Dematerialisation. Scenarios for New Industrial Logics in Recycling Industries, Automobile and Household Appliances. *Business Strategy and the Environment, 8* (1), 31–45. https://doi.org/10.1002/(SICI)1099-0836(199901/02)8:1<31::AID-BSE178>3.0.CO;2-2

Driel, M. V., Biermann, F., Vijge, M. J., & Kim, R. E. (2023). How the World Bank Engages with the Sustainable Development Goal on Reducing Inequalities: A Case of Organizational Jiu-Jitsu. *Global Studies Quarterly, 3* (3), 1–13. https://doi.org/10.1093/isagsq/ksad035

Fromentin, J.-M., Emery, M. R., Donaldson, J., Balachander, G., Barron, E. S., Chaudhary, R. P., Danner, M.-C., Gasalla, M. A., Hallosserie, A., Halmy, M., Hicks, C., Kieling, D., Park, M. S., Parlee, B., Rice, J., Ticktin, T., & Tittensor, D. (2023). Status, Challenges and Pathways to the Sustainable Use of Wild Species. *Global Environmental Change, 81* (102692). https://doi.org/10.1016/j.gloenvcha.2023.102692

Fu, B., Wang, S., Zhang, J., Hou, Z., & Li, J. (2019). Unravelling the Complexity in Achieving the 17 Sustainable-Development Goals. *National Science Review, 6*, 386–388. https://doi.org/10.1093/nsr/nwz038

Gellers, J. C. (2016). Crowdsourcing Global Governance: Sustainable Development Goals, Civil Society, and the Pursuit of Democratic Legitimacy. *International Environmental Agreements, 16*, 415–432. https://doi.org/10.1007/s10784-016-9322-0

Greenland, S. J., Saleem, M., Misra, R., Nguyen, N., & Mason, J. (2023). Reducing SDG Complexity and Informing Environmental Management Education via an Empirical Six-Dimensional Model of Sustainable Development. *Journal of Environmental Management, 344* (118328), 1–11. https://doi.org/10.1016/j.jenvman.2023.118328

Horan, D. (2021). Towards a Portfolio Approach: Partnerships for Sustainable Transformations. *Global Policy, 13*, 160–170. https://doi.org/10.1111/1758-5899.13020

Jordan, T. (2021). *Inner Development Goals: Background, Method and the IDG Framework*. Stockholm: Growth that Matters AB.

Kish, K., & Quilley, S. (2022). *Ecological Limits of Development. Living with the Sustainable Development Goals*. New York: Routledge. https://doi.org/10.4324/9781003087526

Koehler, J. K. L. (2023). Not All Risks Are Equal: A Risk Governance Framework for Assessing the Water SDG. *International Environmental Agreements, 23*, 179–189. https://doi.org/10.1007/s10784-023-09617-7

Kreinin, H., & Aigner, E. (2022). From "Decent Work and Economic Growth" to "Sustainable Work and Economic Degrowth": A New Framework for SDG 8. *Empirica, 49*, 281–311. https://doi.org/10.1007/s10663-021-09526-5

Laasch, O., Moosmayer, D. C., & Antonacopoulou, E. P. (2023). The Interdisciplinary Responsible Management Competence Framework: An Integrative Review of Ethics, Responsibility, and Sustainability Competences. *Journal of Business Ethics, 187*, 733–757. https://doi.org/10.1007/s10551-022-05261-4

Larsen, R. K., & Raitio, K. (2023). Protected Areas and Indigenous Rights in Sápmi: An Agonistic Reading of Conflict and Collaboration in Land Use Planning. *Journal of Environmental Policy & Planning, 25* (3), 342–354. https://doi.org/10.1080/1523908X.2022.2137483

Lau, P. L., Nandy, M., & Chakraborty, S. (2023). Accelerating UN Sustainable Development Goals with AI-Driven Technologies: A Systematic Literature Review of Women's Healthcare. *Healthcare, 11* (401), 1–16. https://doi.org/10.3390/healthcare11030401

Liljeblad, J. (2022). *Indigenous Identity, Human Rights, and the Environment in Myanmar Local Engagement with Global Rights Discourses*. London: Routledge. https://doi.org/10.4324/9781003133728

Lorenzo, J. A. P. (2024). Differential Treatment and Inequalities under the Sustainable Development Goals: Beyond Preferential Market Access. *Law and Development Review, 17* (1), 117–154. https://doi.org/10.1515/ldr-2023-0031

Michael, B., & Latkovska, T. (2021). The FinTech Dividend: How Much Money Is FinTech Likely to Mobilize for Sustainable Development? *Global Policy, 12* (5, Nov), 677–688. https://doi.org/10.1111/1758-5899.12994

Milton, S. (2021). Higher Education and Sustainable Development Goal 16 in Fragile and Conflict-Affected Contexts. *Higher Education*, *81*, 89–108. https://doi.org/10.1007/s10734-020-00617-z

Mondejar, M. E., Avtar, R., Diaz, H. L. B., Dubey, R. K., Esteban, J., Gómez-Morales, A., Hallamh, B., Mbungu, N. T., Okolo, C. C., Prasad, K. A., She, Q., & Garcia-Segura, S. (2021). Digitalization to Achieve Sustainable Development Goals: Steps Towards a Smart Green Planet. *Science of the Total Environment*, *794*, 1–20. https://doi.org/10.1016/j.scitotenv.2021.148539

Nasir, O., Javed, R. T., Gupta, S., Vinuesa, R., & Qadir, J. (2023). Artificial Intelligence and Sustainable Development Goals Nexus via Four Vantage Points. *Technology in Society*, *72*, 1–7. https://doi.org/10.1016/j.techsoc.2022.102171

Nave, E. (2019). The Importance of the Arms Trade Treaty for the Implementation of the Sustainable Development Goals. *Journal of Conflict & Security Law*, *24* (2), 297–324. https://doi.org/10.1093/jcsl/krz010

Niklasson, L. (2019). *Improving the Sustainable Development Goals. Strategies and the Governance Challenge*. New York: Routledge.

Park, J. (2022). How Can We Pay for It All? Understanding the Global Challenge of Financing Climate Change and Sustainable Development Solutions. *Journal of Environmental Studies and Sciences*, *12*, 91–99. https://doi.org/10.1007/s13412-021-00715-z

Pianta, M., & Lucchese, M. (2020). Rethinking the European Green Deal: An Industrial Policy for a Just Transition in Europe. *Review of Radical Political Economics*, *52* (4). https://doi.org.till.biblextern.sh.se/10.1177/0486613420938207

Raper, R., Boeddinghaus, J., Coeckelbergh, M., Gross, W., Campigotto, P., & Lincoln, C. N. (2022). Sustainability Budgets: A Practical Management and Governance Method for Achieving Goal 13 of the Sustainable Development Goals for AI Development. *Sustainability*, *14* (4019). https://doi.org/10.3390/su14074019

Sachs, J. D., & Schmidt-Traub, G. (2017). Global Fund Lessons for Sustainable Development Goals. Transparency and Independence Underpin Successes. *Science*, *356* (6333), 32–33. https://doi.org/10.1126/science.aai9380

Sætra, H. S. (2021). AI in Context and the Sustainable Development Goals: Factoring in the Unsustainability of the Sociotechnical System. *Sustainability*, *13* (1738), 1–19. https://doi.org/10.3390/su13041738

Salimath, M. S., & Chandna, V. (2021). Sustainable Consumption and Growth: Examining Complementary Perspectives. *Management Decision*, *59* (6), 1228–1248. https://doi.org/10.1108/MD-12-2016-0934

Sasmoko, Imran, M., Khan, S., Khan, H. U. R., Jambari, H., Musah, M. B., & Zaman, K. (2023). War Psychology: The Global Carbon Emissions Impact of the Ukraine-Russia Conflict. *Frontiers in Environmental Science*, *11*. https://doi.org/10.3389/fenvs.2023.1065301

Schapper, A., & Bliss, C. (2023). Transforming Our World? Strengthening Animal Rights and Animal Welfare at the United Nations. *International Relations*, *37* (3), 514–537. https://doi.org/10.1177/00471178231193299

Schütz, F., Heidingsfelder, M. L., Schraudner, M. (2019). Co-Shaping the Future in Quadruple Helix Innovation Systems: Uncovering Public Preferences Toward Participatory Research and Innovation. *She Ji: The Journal of Design, Economics and Innovation*, *5* (2). https://doi.org/10.1016/j.sheji.2019.04.002

Shtaltovna, Y., Carreon, V. R., Lindencrona, F., & Donald, W. E. (2024). Cognitive Skills Within the Inner Development Goals (IDG) Framework: Empowering Sustainable Careers and Sustainable Development. *GILE Journal of Skills Development*, *4* (1), 74–94. https://doi.org/10.52398/gjsd.2024.v4.i1.pp74-94

Sun, J. W. (2000). Dematerialization and Sustainable Development. *Sustainable Development*, *8*, 142–145. https://doi.org/10.1002/1099-1719(200008)8:3<142::AID-SD139>3.0.CO;2-H

Verniers, E., & Brels, S. (2021). UNCAHP, One Health, and the Sustainable Development Goals. *Journal of International Wildlife Law & Policy*, *24* (1), 38–56. https://doi.org/10.1080/13880292.2021.1923731

Visseren-Hamakers, I. J. (2020). The 18th Sustainable Development Goal. *Earth System Governance*, *3* (100047). https://doi.org/10.1016/j.esg.2020.100047

Visvizi, A. (2022). Artificial Intelligence (AI) and Sustainable Development Goals (SDGs): Exploring the Impact of AI on Politics and Society. *Sustainability*, *14* (1730), 1–3. https://doi.org/10.3390/su14031730

PART I

Research

2
MULTIDIMENSIONAL POVERTY INDICES AND SDGS

Sabina Alkire

2.1 Introduction

The Sustainable Development Goals inherently and emphatically recognize development to be multidimensional, encompassing the economic, social, and environmental aspects of life. And the opening sentences of the foundational document *Transforming Our World* (UN 2015) refer to poverty as the "greatest global challenge." In 2025, we probe this theme of multidimensional poverty, past the mid-point of the SDGs, in a post-pandemic world riven by crises and uncertainty. Poverty, in all its forms and dimensions, remains in the balance, and the need for incisive, cost-effective, yet authentic tools to redress human disadvantage remains.

This chapter systematically sets out the ways that measures of multidimensional poverty (MPIs) engage the SDG agenda and summarizes their insights and uses as policy tools. It is structured as follows.

Section 1 introduces the measure of multidimensional poverty, how it resonates with Agenda 2030. It also signals how Agenda 2030 has catalyzed research and new methodologies, for example, related to children, gender, and intra-household analyses, interlinkages, and the environment.

Section 2 introduces two prominent measures of multidimensional poverty. National MPIs are reported as SDG 1.2.2 and use dimensions and indicators that are contextually chosen (so cannot be compared across countries). The global MPI uses the same structure (three dimensions and ten associated indicators) across 110 countries and permits comparisons.

Section 3 provides insights from the global MPI for the SDG agenda, including its prominence in international SDG-related bodies and its insights for SDG 1 monitoring. The global MPI is jointly produced with UNDP and OPHI; provides detailed information platforms for 6.1 billion people living in 110 countries, of which 1.1 billion are poor; and is disaggregated by 1,281 subnational regions, age cohorts, gender of the household head, and rural–urban areas. The global MPI covers poverty trends for over 5.5 billion people and 81 countries, with dimensional and disaggregated details as well.

Section 4 provides a briefing on national MPIs, their extent, their reporting as SDG 1.2.2 and in VNRs, and very importantly, their policy uses. Multidimensional poverty measures from Eurostat are available for 543 million people in 34 European countries. National MPIs are available for 2.93 billion people in 44 developing countries. Almost all national MPIs are extensively disaggregated.

DOI: 10.4324/9781003285472-3

2.2 Multidimensional Poverty Index: Definition and Early Uses

2.2.1 What Is Poverty?

Poverty as a condition has also traditionally been defined by economists in monetary terms (Atkinson 2019). But participatory studies and bottom-up approaches to poverty articulate poverty as different dimensions (Narayan et al. 1999) that are connected to each other. As Nobel laureate in economics Amartya Sen observed, "[t]here can also be some 'coupling' of disadvantages between different sources of deprivation, and this can be a critically important consideration in understanding poverty and in making public policy to tackle it" (2009: 256). These insights matter—to policy and to partnership with the protagonists. As Pope Francis put it, " '[t]he poor' and 'the excluded' are real people. Instead of viewing them from a merely technical or functional standpoint, it is time to let them become protagonists in their own lives and in the fabric of society as a whole. Let us not think *for* them, but *with* them." The insights of the protagonists and other actors are used to identify a core set of poverty indicators, and their importance.

2.2.2 Measuring Multidimensional Poverty: Methodology

Multidimensional poverty indices in the counting tradition begin very simply and transparently by counting the deprivations that poor people experience and policy actors have prioritized (Alkire et al. 2015: Ch 6). These "counting approaches" to poverty measurement have been used in Europe and Latin American since the 1970s and are widely used in civil society and non-governmental organizations (Alkire et al. 2015: Ch 4). The counting approach recognizes that, in the lives of the poorest, different disadvantages "cluster" (Wolff and DeShalit 2007). Some deprivations might be more important than others, and weights will reflect this, although any measure used for policy should be robust to a range of weights (Alkire et al. 2022).

So how does measurement proceed? This chapter focuses on a widely used measure, namely, the adjusted headcount ratio methodology (Alkire and Foster 2011; Alkire Foster Seth Santos Roche and Ballon 2015). A set of indicators that reflect poverty is selected, and each person is coded as deprived or not deprived in each indicator. The indicators in which a person is deprived are summed into that person's weighted deprivation score. A higher deprivation score indicates a more difficult condition—it is worse to be deprived in 80% of possible indicators than 40%.

Next, a poverty cutoff is set, and each person whose deprivation score is at least that value is identified as multidimensionally poor, and the rest are non-poor. Immediately after identification, it is possible to know the percentage of people who are poor—a headcount ratio or *incidence* (H). But going beyond this, the next step is to compute the average deprivation score among all poor people—that is called the *intensity* of poverty (A). Multiplying both of these—the incidence times intensity—gives the MPI value (MPI = H × A). As MPI can be disaggregated (data permitting), we can see if MPI levels are highest among Indigenous people or among children or in remote regions.

The next step for policy design is to use the MPI to guide actions. For this they need to know how many people have which deprivations, and how deprivations are interlinked. This information is provided by returning to each poor person's deprivation profile indicator by indicator. Maybe in rural areas, poverty usually includes ramshackle schooling, lack of schooling, indoor air pollution, and malnutrition, whereas in urban areas, it is more common to have overcrowded housing, poor health, and the threat of violence. This information is tangible, actionable. In terms of measurement, two statistics provide this information. The first counts number (or share of the population) of people who are poor and are deprived in each indicator (*censored headcount ratio*). Note that the MPI value is equal to taking all weighted deprivations of all poor people and adding them up. So the statistic figure shows the indicator composition of MPI (*the absolute or percentage contributions of each indicator to poverty*)—it considers weights as well as the censored headcount deprivations. Why are both of these statistics vital for policy? Because if any deprivation of any poor person ends, MPI goes down. Every time. And if a higher-weighted deprivation is solved, MPI goes down more steeply.

These statistics—the level of MPI, H, and A, and the indicator composition of MPI—which are available at the national level and are nearly always disaggregated by subgroups, form an *information platform*. So while the MPI itself is just one number, it is always reported alongside these other statistics in an information platform—so a range of actors from the most local to the national levels have what they need to plan concrete actions.

2.2.3 Early Applications of the MPI

Using the MPI methodology, in 2009, in preparation for the 2010 *Human Development Report*, which was the 20th anniversary volume of this august and policy-salient report, UNDP's Human Development Report Office (HDRO) and OPHI jointly designed the original global multidimensional poverty index. The global MPI measures acute multidimensional poverty in over 100 developing countries. It was first launched in 2010 (Alkire & Santos 2014; UNDP 2010) and subsequently updated nearly annually and, as will be covered in Section 4, was adjusted in 2018 to align with the SDG indicator framework.

The MPI methodology, which is called adjusted headcount ratio, or dual cutoff (Alkire and Foster 2011; AFSSRB 2015), extends the FGT (Foster–Greer–Thorbecke) methodology in monetary poverty to multidimensional space. The same methodology can be used with different dimensions and indicators, weights and cutoffs, depending on the context. This methodology motivated Mexico's multidimensional poverty measure, launched in 2009, and was implemented in Bhutan and Colombia's official national MPIs, launched in 2010 and 2011. By 2013, so many countries had launched or were exploring national MPIs that a South–South network, the Multidimensional Poverty Peer Network, was launched and grew to 45 countries by 2015, the beginning of the SDG era. These national initiatives were advancing in parallel to discussions about the post-2015 agenda, and interchanges between them were evident.

2.2.4 Multidimensional Poverty Indices in the SDGs: A Brief History

Building on and reflecting the renewed national interest in multidimensional poverty, Agenda 2030 incorporated multidimensional poverty indicators, and in the early years of the SDGs, pivotal international efforts served to consolidate and clarify their value added. This section charts some of these steps.

2.2.5 Transforming Our World ***and Early Agenda 2030 Documents***

The first goal of the Sustainable Development Goals (SDG 1) is to end poverty in all its forms and dimensions. The second sentence of the pivotal SDG document, *Transforming Our World: The 2030 Agenda for Sustainable Development,* reads: "We recognise that eradicating poverty in all its forms and dimensions, including extreme poverty, is the greatest global challenge and an indispensable requirement for sustainable development" (UN 2015). The explicit mention of poverty as having multiple forms and dimensions recurs throughout the document, marking a clear departure from the Millennium Development Goals' focus on monetary poverty alone.

The first goal of 17—to "[e]nd poverty in all its forms everywhere"—echoes the priority of poverty as the "greatest global challenge." Within SDG 1, Target 1.1 is to end extreme monetary poverty. The 2nd out of 169 targets, Target 1.2, addresses multidimensional poverty. It aims to "reduce at least by half the proportion of men, women and children of all ages living in poverty in all its dimensions according to national definitions." The other SDG 1 targets refer to other non-monetary aspects of poverty.

The inclusion of multidimensional poverty had support from different angles. Well before Agenda 2030 was negotiated, for example, Africa's Agenda 2063 already had adopted the global MPI to track progress in Africa, rather than the $1.25 monetary metric. Moving within the UN system, in December 2014, UN Secretary General Ban Ki Moon wrote, "Poverty measures should reflect the multi-dimensional nature of poverty."[1] A United Nations General Assembly resolution in December 2014 also underlined "the need to better reflect the multidimensional nature of development and poverty" and invited member states and others "to consider developing complementary measurements . . . that better reflect that multidimensionality" (A/RES/69/238 paragraph 5). Turning to financing, the Addis Ababa Accord of 2015[2] called on the United Nations and international financial institutions to "recognize the multi-dimensional nature of poverty" (UN 2015). So the call for more adequate measures had been growing also within the actors of the post-2015 world—in parallel with the national and global experiences.

The Republic of Colombia played a particularly pivotal role in joining national MPI work with the SDGs. Not only was Colombia a pioneer in launching a national MPI; it was, and is, also perhaps the country most associated with policy uses of the MPI. Colombia was also actively engaged in Rio+20. The 2023 book *The Battle Against Poverty* by Nobel Peace Prize laureate and former president of Colombia Juan Manuel Santos, devotes one chapter to how Colombia—as a key actor in shaping the SDGs after Rio—advocated this emphasis in the SDGs.

2.2.6 The World Bank's 2017 Commission Report: Monitoring Global Poverty

In 2015, the World Bank's chief economist, Kaushik Basu, set up a commission on monitoring global poverty under the leadership of the late and deeply respected Sir Tony Atkinson. The Atkinson Commission Report *Monitoring Global Poverty*, launched in October 2016, recommended that a multidimensional poverty indicator using the AF adjusted headcount ratio methodology be used to measure global poverty alongside the $1.90/day extreme income poverty indicator and other monetary and complementary measures. The report suggested that a global non-monetary multidimensional measure should cover indicators relating to health, nutrition, education, living standards, work, and violence.

Of the report, Sir Tony Atkinson (Fellow, Nuffield College, Oxford; Centennial Professor, London School of Economics; and Chair of the commission) said:

> Let me start from the Sustainable Development Goals because they are an important part of the context. And I would like to stress . . . the fact that they refer . . . to all dimensions of poverty. It is a multidimensional approach to the problems that we face.

Atkinson observes that "that we perhaps have become too narrowly focused in the way economists . . . approach the definition of poverty." Hence, to incorporate the measurement of non-monetary aspects of poverty—and urge politicians to use non-monetary indicators for performance monitoring—Atkinson brings in Amartya Sen's capability approach.

> The Capability Approach developed by Amartya Sen, who was a member of the Commission, and Martha Nussbaum and widely influential in the work of OPHI here, says we should . . .

> probe beneath the levels of consumption to see what consumption is designed to bring about and relate it to the capacity to function in a particular society.[3]

But Atkinson recognized that a dashboard with six dimensions (health, nutrition, education, living standards, work, and violence) would overlook their interlinkages. Do those living in ramshackle houses also suffer from undernutrition? Are their children out of school? The report thus recommended that the World Bank's non-monetary poverty portfolio should "include a multidimensioned poverty indicator based on the counting approach" (he specified that this should not include an indicator of monetary poverty) and further envisaged "the counting approach as being implemented in terms of the adjusted head count ratio, and its constituents of the head count and average breadth of deprivation" (2017: 170–171).

The report also presented, cited, and analyzed existing national multidimensional poverty indices—both in Europe and, following the adjusted headcount ratio method, in developing countries. It proposed that country poverty reports henceforth consider the level and trends of national and global monetary and multidimensional poverty measures in order to obtain a more precise overview.

Thus, alongside the SDG focus on poverty in all its dimensions, the World Bank Commission on Monitoring Global Poverty explicitly endorsed the need for global and national MPIs using the adjusted headcount ratio methodology (but called for more dimensions, for example, insecurity).

2.3 The Global MPI in the SDGs

The global MPI is a measure of acute poverty, produced by the Oxford Poverty and Human Development Initiative (OPHI) and the United Nations Development Programme (UNDP) since 2010. Its indicators were revised in 2018 to better align with the SDGs (Alkire and Jahan 2018; Alkire and Kanagaratnam 2021). It complements the \$1.90/day poverty measure, and related monetary metrics, by profiling acute deprivations in ten non-monetary indicators related to health, education, and living standards. Going beyond the extreme monetary poverty measures, the global MPI offers extensively disaggregated data—for example, by subnational regions, age cohort, and rural and urban areas. Like the \$1.90/day monetary poverty measure, the global MPI provides international comparisons, identifying patterns of poverty and progress made toward the SDGs. Its component indicators provide information on advancements or shortcomings related to SDGs 2, 3, 4, 5, 6, 7, and 11. The global MPI directly reflects the SDG emphasis on understanding interlinked deprivations and responding with integrated policies

The global MPI is not an SDG indicator, but it does summarize many SDG indicators, and themes like interlinkages and integrated policies; and it alone can compare trends in multidimensional poverty over time for developing countries. Furthermore, several prominent documents use it for scene setting and monitoring, two of which are briefly presented in what follow.

2.3.1 The Third United Nations Decade for the Eradication of Poverty (2018–2027)

In 2018, the Third United Nations Decade commenced, calling for "accelerating global actions for a world without poverty," in line with the 2030 Agenda for Sustainable Development, and put forward "an inter-agency, system-wide plan of action for poverty eradication" for ensuring effectiveness and coordination of UN efforts to achieve progress in eradicating poverty in all its forms and dimensions.[4] The Third UN Decade for the Eradication of Poverty uses global monetary and multidimensional poverty indices to track trends.

In particular, the Third UN Decade uses the global multidimensional poverty index (MPI) in three ways. First, in line with the Sustainable Development Goals (SDGs), the Third UN Decade agenda's pledge to end poverty in *all its forms* recognizes poverty as multidimensional: "Ending poverty is not just about people's income but also their access to basic needs, such as quality education and health care, clean water and sanitation, decent housing and security."[5] The global MPI includes indicators of education, nutrition, child mortality, water, sanitation, housing materials, hence supports the Third UN Decade's aim. Specifically relevant is the Decade's thematic policy action area focused on human capability development.[6]

The Third UN Decade agenda also reiterates the SDG pledge *to leave no one behind* and identifies "reducing inequality" as one of its policy areas.[7] Citing the number of people and of children who are poor according to the global MPI, the document observes that "the poorest of the poor are subject to overlapping dimensions of poverty and inequalities." Because the global MPI levels and trends are extensively disaggregated, inequality can be visible between rural and urban areas, subnational regions, age cohorts, as well as by disability status, ethnicity, gender of the household head, and so on. Disaggregated data on MPI trends shows whether reduction among the poorest groups was faster than average—hence whether they are, or are not, being left behind.[8]

The Third UN Decade also calls for an inter-agency approach and *integrated, multi-sectoral policies*: "The overarching objective regarding non-income dimensions of poverty is to promote integrated approaches and policy frameworks to address multidimensional poverty. Several United Nations development system organizations have specialized mandates" (point 47). The MPI information platform is very useful as it visibilizes interlinked deprivations.

2.3.2 Global Sustainable Development Report 2019

The 2019 Global Sustainable Development Report (GSDR), *The Future Is Now*, commissioned by the Secretary General of the United Nations, highlighted six key "entry points" and 20 corresponding calls to action to leverage interlinkages and accelerate progress across all 17 Sustainable Development Goals. The authors used the global MPI to observe that, in 2018, 1.3 billion persons were poor, and recognize that the MPI presented a "more sobering picture" than the extreme monetary poverty measure, which it cited in parallel.

Again using the global MPI as an example, the report stresses the need to consider "overlapping and concentrated deprivations":

> Income poverty, poor health, low levels of education, lack of access to water and sanitation, and other deprivations tend to overlap. . . . This can be illustrated using the [global] multidimensional poverty index (MPI) which captures the severe deprivations that each person faces with respect to education, health and living standards. . . . Women, older people, ethnic and racial minorities, certain religious groups, indigenous peoples, persons with disabilities, children and other groups are below the population average in many contexts across measures of well-being. Of the [1.3 billion] multidimensionally poor, 1.1 billion live in rural areas and almost half are children. . . . Not only are those deprivations overlapping, but they are interdependent and mutually reinforcing. For example, people who lack access to safely managed drinking water, sanitation services, and hygiene services are likely, as a consequence, to suffer health deprivations, especially diarrhoeal diseases, which can cause severe illness and death, particularly in infants and children.
>
> (p. 39)

In advocating an "integrated path to transformation," the report recommends that alongside the global MPI, countries should build national MPIs: "Countries should measure and address poverty in a multidimensional way, with special attention to those dimensions that are the most pertinent in their context and according to their own definitions" (p. 48; see also Box 2.12).

Thus, the Third UN Decade and the Global Sustainable Development Report used the global MPI to track poverty across the developing world in parallel to the extreme monetary poverty measure and emphasized the need to look at overlapping and concentrated deprivations.

2.3.3 Global MPI 2023 Trends

Trends in the global MPI are reported for 81 countries, covering over 5 billion people, in Table 2.1, the joint UNDP–OPHI 2023 global MPI report (2023). Data on the level and trends of poverty, and of its component indicators, are disaggregated by subnational region, age group, and rural urban area in the online OPHI tables.[9] Unfortunately in 2023, still, most countries' datasets were pre-pandemic so did not make visible the impact of the pandemic and subsequent challenges on poverty.

It should be noted that OPHI estimate strictly harmonized trends over time for over 200 datasets dated 2010 to 2021/2022. The years and periods between datasets vary by country, and we do not yet extrapolate or nowcast. Because of this, we publish national and subnational trends using rigorously harmonized datasets but do not release global trends.

Before assessing trends, it may be useful to observe a number of countries that already have very low levels of poverty. In 37 of the 110 countries, less than 3% of the population was living in multidimensional poverty in the most recent dataset. Countries with such low levels of poverty can improve yet are already very near to the goal of eradicating acute multidimensional poverty.

Over time, there has been a visible net reduction in the number of poor persons. But SDG 1.2 is rightly evaluated at the country level, and country trends show very mixed results. In 2023, 72 of the 81 countries with data on trends significantly reduced the global MPI in absolute terms during at least one period; 24 countries reduced poverty significantly in two periods.[10] Nine countries had no significant reduction or increase in poverty in any period,[11] and in two countries, Guinea and the Central African Republic, poverty increased.

The three countries having the fastest absolute reduction of the incidence of MPI were Sierra Leone (2013–2017), Timor L'este (2009/2010–2016), and Togo (2013/2014–2017). And those in which the greatest number of people left multidimensional poverty are India (2005/2006–2019/2021), China (2010–2014), and Bangladesh (2014–2019). These show what was possible.

Fastest	Sierra Leone (2013–2017)	74%–58% in 4 years
	Timor Leste (2009/2010–2016)	70%–47% in 6.5 years
	Togo (2013/2014–2017)	55%–43% in 3.5 years
Largest Number of Poor	India (2005/2006–2019/2021)	55%–16% in 15 years **415 M out of poverty**
	China (2010–2014)	9.5%–4.2% in 4 years **69 M out of poverty**
	Bangladesh (2014–2019)	38%–24% in 5 years **19 M out of poverty**

Table 2.1 25 countries that halved their global MPI value in under 15 years

Arab States (1)	*East Asia and the Pacific (6)*	*Europe and Central Asia (5)*	*Latin America and the Caribbean (7)*	*South Asia (2)*	*Sub-Saharan Africa (4)*
Morocco	Cambodia China Indonesia Mongolia Thailand Vietnam	Albania Kyrgyzstan North Macedonia Serbia Turkmenistan	Plurinational State of Bolivia Dominican Republic Guyana Honduras Nicaragua Peru Suriname	India Nepal	Gabon Congo Lesotho Sao Tome and Principe

Source: UNDP and OPHI (2023).

Fully, 25 of the 81 countries halved their global MPI value and did so within 4–12 years. Hence, the SDG Target 1.2 to halve multidimensional poverty within 15 years is possible. Successful countries are located in every included world region.

Countries of different sizes halved poverty, ranging from small island states, such as Sao Tome and Principe, to populous countries, like India, China, and Indonesia. Cambodia halved its MPI in a 7.5-year period which spanned the COVID-19 pandemic (2014–2021/2022). Countries with different starting levels of poverty also halved poverty. For 17 of the 25 countries that halved their MPI, less than one quarter of their population was poor in the initial period, while in India and the Congo, the initial incidence was over half (55% and 54%, respectively).

But while 25 countries halved MPI, 56 countries did not—either because data are not available for sufficient time periods to show such progress or because their pace of reduction was too slow.[12]

In terms of leaving no one behind, recall that children are poorer than adults in nearly every country. Yet in half of the countries (42 of the 81 countries), it was either children had no significant poverty reduction or multidimensional poverty among children fell more slowly than among adults in at least one period. So children are being left behind quite visibly—a troubling trend.

Also, it might be assumed that if the incidence of poverty falls, the fiscal resources required to address poverty will also fall, because of lower numbers of poor people. But that is not always the case: In 15 countries that reduced MPI significantly, the number of poor people increased—population growth outpaced poverty reduction. For example, in Ethiopia and Nigeria, the number of poor people increased by 5 million people each (during 2011–2016 for Ethiopia and 2013–2016/2017 for Nigeria) though each reduced MPI, while in the Democratic Republic of the Congo, the number of poor people increased by 8 million in 6.5 years.

Furthermore, sadly, none of the poorest 37 countries[13] halved their MPI, even though some, like Sierra Leone and Togo, had periods with stellar progress. Thirty-three of these 37 countries are in sub-Saharan Africa. In these contexts, the global MPI sounds a ringing call to action: Without concerted investment in the poorest countries, they will be left behind in terms of poverty in its many forms.

2.4 National MPIs in the SDGs

The SDG indicators were approved in July 2017—a list of 232 distinctive indicators proposed by the UN Statistics Commission, who developed these with an inter-agency expert group on SDG indicators co-chaired by two national statisticians. The process of indicator development clearly echoed what has become a pivotal phrase within the SDGs of the need to

1. leave no one behind;
2. have measures that can be disaggregated;
3. break siloes;
4. inform integrated, multi-sectoral policies; and
5. recognize the priority of eradicating poverty in all its forms and dimensions.

2.4.1 Multidimensional Poverty in Voluntary National Reviews

Every year, a set of countries are invited to report their progress toward the SDGs in voluntary national reviews. To date, 52 countries have updated the HLPF on multidimensional poverty in their countries, either documenting progress or signaling plans for its measurement or reduction.

2.4.2 National MPIs Reported as SDG Indicator 1.2.2

National MPIs are reported against a third of those 232 indicators. SDG Indicator 1.2.2 is quite distinctive. Recall that the first SDG indicator is $1.25/day poverty, the second are national monetary poverty measures, and the third, SDG 1.2.2, is "the proportion of men, women and children of all ages living in poverty in all its dimensions according to national definitions."

Of all 232 SDG indicators, there is exactly one SDG indicator for which countries are the custodian agency, and that is SDG Indicator 1.2.2. For a number of years, SDG 1.2.2 was never reported, which created a complex situation (the president of one country sent national MPI data to the UN statistics division directly, because there was no way to post it into the global MPI database). The countries used the MPPN to request a channel for reporting be opened, as countries had no way to post their data themselves. In 2020, the World Bank, UNDP, and UNICEF were recognized as "supporting agencies," and the World Bank Data Group kindly created a database for national MPI reporting that drew on the structure used for national monetary poverty data (SDG 1.2.1) and included verification by other supporting agencies.

What is reported? The metadata rightly stresses that national MPIs cannot be compared but are similar in structure:

> The design of a measure of multidimensional poverty is different in each country, but regardless of the exact methodology selected, it still follows a similar process to define the features of the measure, which include: i) the purpose of the measure; ii) the unit of identification (most frequently either the household or the individuals); iii) the dimensions and respective indicators that delimit which deprivations should be measured; iv) the methodology for developing the measure (including deprivation cut-offs, weights, and poverty cut-offs). (p. 2)

It also observes that the 1.2.2 indicators use a counting approach—often using the adjusted headcount ratio methodology:

> The most commonly used method is the Alkire Foster (AF) methodology which identifies dimensions, typically health, education and living standards and several indicators in

> each dimension. The unit of analysis could be either the individual or the household. The individuals or households are considered as multidimensionally poor if they are deprived in multiple dimensions, exceeding certain thresholds.
>
> (p. 2)

The process outlined on the website, which occurs under the oversight of the World Bank Data Group,[14] is as follows:

> The data has been validated by a three-stage approach to assure its accuracy. First, the data is entered by World Bank staff assigned to each country, typically in consultation with the country NSO and/or country official documents. That data is sent to UNICEF and UNDP country officers for the validation. After integrating inputs from these three agencies, the data is sent to the SDGs focal point for each country for their final approval. For countries where the World Bank does not have any country offices, such as for OECD and EU countries, the World Bank collected the information based on data source available online, and sent it directly to the official counterparts of each country for verification.[15]

In 2024, Eurostat data are uploaded by the World Bank for 34 OECD and EU countries, while the remaining 46 countries follow the preceding procedure. In Eurostat countries, the 1.2.2 measure is the proportion of the population who are deprived in at least one of three indicators: material deprivation, quasi-joblessness, or the monetary at-risk-of-poverty indicator.

The SDG database contains the following five categories: (1) incidence and (2) intensity in the population, (3) incidence and (4) intensity by households, and (5) child-specific incidence. As the 1.2.2 metadata document explains:[16]

> The following five series are used to monitor the SDG 1.2.2.
>
> 1. Official multidimensional poverty headcount, by sex, and age (% of population)
>   - The percentage of people who are multidimensionally poor
> 2. Average share of weighted deprivations (intensity) for total population
>   - The average share of weighted dimensions in which poor people are deprived among total population
> 3. Official multidimensional poverty headcount (% of total households)
>   - The percentage of households who are multidimensionally poor
> 4. Average share of weighted deprivations (intensity) for total households
>   - The average share of weighted dimensions in which poor people are deprived among total households
> 5. Multidimensional deprivation for children (% of population under 18)
>   - The percentage of children who are simultaneously deprived in multiple material dimensions

Thus, the Sustainable Development Goals Indicator 1.2.2 reports countries' national multidimensional poverty metrics (the IAEG on SDG indicators procedure in which the child category was adopted is not documented). But what does it contain?

In 2024, SDG Indicator 1.2.2 is reported for 79 countries. As we share in related work,[17] 34 countries, home to 543 million people, use the Eurostat at risk of poverty and exclusion measure; 44 countries, home to 2.9 billion people, use a national MPI based on the AF method; and 2 countries (Burundi and Tonga) use a different methodology. In addition, five countries that do not have a national MPI report a multidimensional child poverty measure.[18]

We also analyze the dimension and indicator composition of these 44 countries, as well as the trends over time for the 62 countries. To summarize, in breve, whether absolute or relative changes are analyzed, only one-third to one-half of countries seem to be "on track" to halve the incidence of SDG 1.2.2 by 2030—half to two-thirds are off track according to the most recent data posted, which may be pre-pandemic. Still, this is a leap forward in terms of reporting—and half again as many countries are currently working to launch their national AF MPI by 2025.

2.4.3 Policy Uses of National MPIs

2.4.3.1 Atkinson: Measuring Poverty Around the World *(2019)*

The reason that national MPIs are launched as official poverty statistics is to complement monetary poverty statistics in monitoring poverty levels and trends, and to inform anti-poverty policies. Again, Atkinson's last book, *Measuring Poverty Around the World*, published posthumously in 2019, gets it exactly right. The book opens by stressing the action orientation of poverty measures:

> Learning about the extent of poverty is important . . . but it is the link with action that marks out this issue from many other subjects of study in the social sciences. Poverty statistics matter because they motivate people to tackle a key challenge.
>
> (2019: 1)

In the first page of the book, Atkinson recognizes the link between measurement and the SDGs:

> Poverty is one of the two great challenges facing the world as a whole today, along with climate change, with whose consequences poverty is intimately connected. Achieving the ambitious Sustainable Development Goals to which world leaders committed themselves in 2015 is challenging in the extreme, and the book . . . seeks to provide the evidence about the extent and nature of poverty that is necessary to spur action and to design effective policies.
>
> (2019: 1)

Building on the earlier observation about an essential link between poverty measurement and action, Atkinson recognizes that the "role of performance monitoring has acquired much greater salience with the adoption . . . of targets for the reduction or elimination of poverty." And because performance monitoring affects political careers, this gives poverty measures an additional import: "Statistical evidence about the extent and nature of poverty has been a major factor influencing political action" (2019: 30 both quotes).

2.4.3.2 OPHI: Building a National MPI *(2019)*

How does this work in practice? In 2019, OPHI also published with UNDP a handbook titled *Building a National MPI.* The first half provided information for technical teams on good practices for building a technically rigorous and robust poverty statistic. The second half described country experiences and procedures for consulting about, then institutionalizing, an official permanent national multidimensional poverty statistic, and using it for policies such as geographic targeting, budget allocation, household-level targeting, policy design, results-based management frameworks, policy communication, and coordination across sectors and across levels of government and with the wider public, civil society groups, and the private sector. This volume was a first step in an ongoing process of developing case studies for different audiences, from very detailed technical teams to their managers to high-level policy leaders. It is complemented by a magazine, *Dimensions*, that publishes 800- to 1,000-word overviews of policy innovations to catalyze exchanges across the MPPN. And it gave rise to a massive open online course (MOOC) by UNDP and OPHI that is still running, as well as intensive executive education courses for policy actors.

It is very difficult to distill the many case studies and cameos into a few pages, but a stellar documentation of policies that are informed by the MPI has been the case of Colombia, and that case is now available in detail from the view of a policy leader.

2.4.3.3 Santos: The Battle Against Poverty *(2023)*

In 2010, when Juan Manuel Santos became the president of the Republic of Colombia, he observed that he could not build peace if he could not visibly reduce the deprivations that impoverish many Colombians. By 2016, the historic Havana Peace Accord was signed, and he was awarded the Nobel Peace Prize. Published by Oxford University Press, *The Battle Against Poverty* tells Colombia's story of poverty reduction from 2010 to 2018.

In that period, the incidence of Colombia's national MPI fell from 29.7% in 2010 to 19.1% in 2018. That is, it fell by over one-third, by over 10 percentage points, in eight years. Similarly, monetary poverty fell from 40.3% in 2009 to 27.0% in 2018. The book details the policy actions—and surrounding discussions and institutional structures—that drove change, and how these were implemented despite the internal armed conflict and during complex peace negotiations. It also traces Colombia's international footprint as a significant contributor to Rio+20, Agenda 2030, and the SDGs. The book unfurls this story, accompanied by many anecdotes which highlight the pivotal role that a plethora of people and their insights, imagination, and actions, at many different levels and positions, played in driving change.

What emerges in Colombia as in so many other countries has been a vital role played by leadership who personally commit to doing what it takes to reduce poverty, as well as the grassroots workers who actually caused change and whose efforts became recognized. The author writes, "The fight against poverty is not just a technical issue but one of leadership."

The official adoption of the multidimensional poverty index made it possible for the Colombian government to understand and integrate different dimensions of poverty into its decisions; respond with innovative programs to warnings, indicating where timely state intervention was needed; and optimize public entities' investments and social actions by directing the budget and programs to alleviate the most pressing needs of the most vulnerable population segments.

Notes

1 Section 5.1, paragraph 135.
2 Addis Ababa Accord 2015, paragraph 119.
3 All quotes from the in-person launch of the report, November 4, 2016.
4 United Nations (2018: 2).
5 United Nations (2018: 13–14).
6 United Nations (2018: 13–14).
7 United Nations (2018: 15).
8 United Nations (2018: 15).
9 Alkire et al. (2023: table 6).
10 Plurinational State of Bolivia (2003–2008, 2008–2016), Cambodia (2010–2014, 2014–2021/2022), Democratic Republic of the Congo (2007–2013/2014, 2013/2014–2017/2018), Dominican Republic (2007–2014, 2014–2019), Ethiopia (2011–2016, 2016–2019), Gambia (2005/2006–2013, 2013–2018), Honduras (2005/2006–2011/2012, 2011/2012–2019), India (2005/2006–2015/2016, 2015/2016–2019/2021), Kyrgyzstan (2005/2006–2014, 2014–2018), Lesotho (2009–2014, 2014–2018), Liberia (2007–2013, 2013–2019/2020), Mali (2006–2015, 2015–2018), Mexico (2012–2016, 2016–2020), Mongolia (2010–2013, 2013–2018), Nepal (2011–2016, 2016–2019), North Macedonia (2005/2006–2011, 2011–2018/2019), Peru (2012–2018, 2019–2021), Rwanda (2010–2014/2015, 2014/2015–2019/2020), Sao Tome and Principe (2008/2009–2014, 2014–2019), Sierra Leone (2013–2017, 2017–2019), Suriname (2006–2010, 2010–2018), Thailand (2012–2015/2016, 2015/2016–2019), Zambia (2007–2013/2014, 2013/2014–2018), and Zimbabwe (2010/2011–2015, 2015–2019).
11 The countries that had no significant change: Armenia (2010–2015/2016), Benin (2014–2017/2018), Burkina Faso (2006–2010), Cameroon (2011–2014, 2014–2018), Guinea-Bissau (2014–2018/2019), Jordan (2012–2017/2018), Montenegro (2013–2018), State of Palestine (2010–2014, 2014–2019/2020), and Ukraine (2007–2012) (these listings are quoted from the 2023 UNDP–OPHI global MPI report).
12 Refer to the paper with Nicolai and Ricardo and the 2020 UNDP–OPHI report showing the trends.
13 Ranked from poorest in the first time period considered: Niger, Chad, Burkina Faso, Central African Republic, Ethiopia, Mali, Burundi, Mozambique, Madagascar, Congo, Democratic Republic of the, Guinea, Benin, Guinea-Bissau, Gambia, Liberia, Mauritania, Togo, Sierra Leone, Tanzania, Senegal, Rwanda, Uganda, Sudan, Zambia, Malawi, Cameroon, Côte d'Ivoire, Timor-Leste, Nigeria, Pakistan, Haiti, Kenya, Cambodia, Namibia, Zimbabwe, Ghana, and Lesotho.
14 Their contact details are available here: https://unstats.un.org/sdgs/dataContacts/.
15 https://unstats.un.org/sdgs/dataCalendar/1.2.2_World%20Bank.xlsx accessed June 1, 2024.
16 https://unstats.un.org/sdgs/metadata/files/Metadata-01-02-02.docx accessed June 1, 2024.
17 Alkire and Dirksen (mimeo).
18 Draft analysis by Sabina Alkire and Jakob Dirksen using 2023 UN DESA population division estimates (May 2024).

References

Alkire, S., and Fortacz, A. (2022). "Insights for policymaking from the Multidimensional Poverty Index", in S. Waglé and K. Wignaraja (eds.), *The Great Upheaval: Resetting Development Policy and Institutions for the Decade of Action in the Asia-Pacific*, Cambridge: Cambridge University Press.

Alkire, S., and Foster, J. (2011). "Counting and multidimensional poverty measurement", *Journal of Public Economics*, 95(7–8), pp. 476–487.

Alkire, S., Foster, J., Seth, S., Santos, M.E., Roche, J.M., and Ballon, P. (2015). *Multidimensional Poverty Measurement and Analysis*, Oxford: Oxford University Press.

Alkire, S., and Jahan, S. (2018). "The new global MPI 2018: Aligning with the Sustainable Development Goals", *OPHI Working Paper 121*, University of Oxford.

Alkire, S., and Kanagaratnam, U. (2021). "Revisions of the global multidimensional poverty index: Indicator options and their empirical assessment", *Oxford Development Studies*, 49(2), pp. 169–183.

Alkire, S., Kanagaratnam, U., Nogales, R., and Suppa, N. (2022). "Revising the global Multidimensional Poverty Index: Empirical insights and robustness", *Review of Income and Wealth*, 68(S2), pp. S347–S384.

Alkire, S., Kanagaratnam, U., and Suppa, N. (2023a). "The global Multidimensional Poverty Index (MPI) 2023: Country results and methodological note", *OPHI MPI Methodological Notes* **55**, University of Oxford.
Alkire, S., Kanagaratnam, U., and Suppa, N. (2023b). 'The global Multidimensional Poverty Index (MPI) 2023: Disaggregation results and methodological note', *OPHI MPI Methodological Notes 56*, University of Oxford.
Alkire, S., Kanagaratnam, U., and Suppa, N. (2023c). "A methodological note on the global Multidimensional Poverty Index (MPI) 2023: Changes over time results for 84 countries", *OPHI MPI Methodological Notes 57*, University of Oxford.
Alkire, S., and Robson, M. (2022). "On international household survey data availability for assessing pre-pandemic monetary and multidimensional poverty in developing countries", *Development Studies Research*, 9(1), pp. 277–295.
Alkire, S., and Santos, M.E. (2014). "Measuring acute poverty in the developing world: Robustness and scope of the Multidimensional Poverty Index", *World Development*, 59, pp. 251–274.
Alkire, S., and Ul Haq, R. (2023). "Analyzing individual deprivations alongside household poverty: Possibilities for gendered, intrahousehold and multidimensional analyses", *OPHI Research in Progress No. 65a*, University of Oxford.
Atkinson, A. (2017). *Monitoring Global Poverty: Report of the Commission on Global Poverty*. Washington, DC: World Bank.
Atkinson, A.B. (2019). *Measuring Poverty Around the World*, Princeton, NJ: Princeton University Press.
Independent Group of Scientists Appointed by the Secretary-General. (2019). *Global Sustainable Development Report 2019: The Future Is Now—Science for Achieving Sustainable Development*. https://sustainable development.un.org/gsdr2019
Narayan, D., Patel, R., Schafft, K., Rademacher, A., and Koch-Schulte, S. (1999). *Voices of the Poor: Can Anyone Hear Us? Voices from 47 Countries*. Washington, DC: The World Bank.
Santos, J.M. (2023). *The Battle Against Poverty ~ Colombia: A Case of Leadership*, Oxford: Oxford University Press.
Sen, Amartya. K. (2009). *The Idea of Justice*. Harmondsworth: Penguin.
UN. (2018). 'Implementation of the third United Nations decade for the eradication of poverty (2018–2027),' *UN.Doc A/73/298* (6 August). https://undocs.org/A/73/298.
UNDP and OPHI. (2019). *How to Build a National Multidimensional Poverty Index (MPI): Using the MPI to Inform the SDGs*, New York: UNDP.
UNDP and OPHI. (2021). *Global Multidimensional Poverty Index 2021: Unmasking Disparities by Ethnicity, Caste and Gender*, UNDP and OPHI.
UNDP and OPHI. (2022). *Global Multidimensional Poverty Index 2022: Unpacking Deprivation Bundles to Reduce Multidimensional Poverty*, UNDP and OPHI.
UNDP and OPHI. (2023). *Global Multidimensional Poverty Index 2023: Unstacking Global Poverty: Data for High Impact Action*, UNDP and OPHI.
United Nations (UN). (2015). "Transforming our world: The 2030 agenda for sustainable development", *A/RES/70/1*, New York: UN.
Wolff, J., and De-Shalit, A. (2007). *Disadvantage*, Oxford: Oxford University Press.

3

EXPLORING AGENDA 2030 THROUGH THE 5P LENS

Energy Beyond Prosperity

Matheus Gobbato Leichtweis and Raiana Schirmer Soares

3.1 Introduction

Comprising and cutting across different dimensions of human life, Agenda 2030 (2015) provided a comprehensive and unprecedented tool to quantify, measure, monitor, and report progress toward sustainable development. Containing "a plan of action for people, planet and prosperity" with the objective "to take the bold and transformative steps which are urgently needed to shift the world onto a sustainable and resilient path," *Transforming Our World* (2015) presented 17 Sustainable Development Goals (SDGs) and 169 targets, which were deemed as "integrated," "indivisible," and balancing the "three dimensions of sustainable development: the economic, social and environmental" (*Transforming Our World: The 2030 Agenda for Sustainable Development*, 2015). However, despite efforts to harmonize the goals, lack of integration remained one of the main obstacles for implementation, allowing for weak, cherry-picked, and politically innocuous strategies toward this objective so far.

As pointed out by Trucco et al. (2021), lack of SDGs integration has been singled out as one of the main challenges in their implementation (Trucco et al. 2021). As noted by Bleischwitz et al. (2018), "integrated approaches are still the exception rather than the rule." Indeed, in the search to understand and overcome fragmentation, and in view of the pressing need for decisive action to implement the 2030 Agenda ("Decade of Action," "Decade of Delivery") (2020), the scientific and international communities have been putting efforts to rethink the implementation of Agenda 2030 (United Nations 2023).

This chapter questions the role of the 5Ps ("people," "planet," "prosperity," "peace," and "partnership") in Agenda 2030, and their relationship with the SDGs. It is a critical and interdisciplinary endeavor that problematizes mainstream interpretations of this relationship, seeking to understand how the 5Ps can be better used to guide action toward the implementation of the goals, overcome fragmentation, strengthen the link between the SDGs, and better detect synergies and trade-offs between them.

The chapter is structured into four sections. The first one illuminates the historical process through which the 5Ps have been defined and added to Agenda 2030. The second part calls attention to the problem of fragmentation and lack of integrity between the 5Ps, under their mainstream recognition, and the SDGs, indicating some of the consequences this segmentation might have altogether. The third part, by its turn, explores the 5Ps as reflections of the core values of Agenda 2030, presenting the 5Ps as a meaningful tool for a multidimensional, harmonized interpretation

DOI: 10.4324/9781003285472-4

of the goals. It discusses the importance of values in development, law, and policymaking, arguing for an integrated approach to the SDGs through the 5Ps. Finally, the fourth part presents a concrete application of the 5Ps integrated lens, using SDG 7 and "prosperity" as a model, or case study, to illustrate the argument. Crucially, the section "Energy Beyond Prosperity" problematizes mainstream understandings of the relationship between the 5Ps and the SDGs that subsume each SDG into a single *P*, shadowing interconnections between them. Specifically, in the case at hand, these mainstream understandings tend to subsume SDG 7 exclusively as a matter of "prosperity," thus excluding energy from "people," "planet," "peace," and "partnership" concerns. Departing from this view, the study demonstrates that reducing the issue of energy exclusively to "prosperity" implies a series of misconceptions about the multidimensional nature of energy, as well as in relation to a harmonized understanding of the Agenda itself. Thus, by spotting these "distortions," or "blind spots," in the example of "energy" and "prosperity," the study presents a holistic lens that may not only throw light on the "dark side" of SDG 7 but also set cornerstones for a value-based, holistic, integrated, and multidimensional approach to Agenda 2030.

3.2 The SDGs and the Problem of Fragmentation in Agenda 2030

In the preamble of the *Transforming Our World* declaration, Agenda 2030 emphasized the interdependence between the SDGs: "[T]hey are integrated and indivisible and balance the three dimensions of sustainable development: the economic, social and environmental." Specifically, in the section "Partnership," the agenda reads: "The interlinkages and integrated nature of the Sustainable Development Goals are of critical importance to ensure that the purpose of the new Agenda is realised." However, in the years that followed, multiple studies recognized fragmentation as a problem of the Agenda, and the need for integrated, comprehensive approaches to the SDGs.[1]

Indeed, although the negotiation process and the unfolding of the 2030 Agenda have shown awareness regarding lack of integration among goals, fragmentation remained an issue, leading to contradictory, weak, and politically innocuous sustainable development decisions. For Kim (2016, p. 15), "the SDGs and targets reflect the fragmented structure of international law, and therefore would have limited utility for orchestration." As pointed out by Trucco et al. (2021),

> [a]lthough SDG 17 has been introduced to "harmonize" all the others, the adoption of integrated and systemic responses is still needed to identify synergies and trade-offs among SDGs and to promote the SDG as a package, rather than the cherry-picking approach.

Under this sense, considering that, for an adequate, orchestrated implementation of the Agenda, integrated and systemic approaches to the SDGs are needed, this chapter addresses the question of how the 5Ps can be used to better harmonize the SDGs in a principled manner. Through the so-called 5Ps lens, the study proposes to read the SDGs holistically and in accordance with an integrated and systemic interpretation, thus moving away from what it identifies as the fragmented, positivist approach.

The SDGs were originally supposed to counteract siloization. However, as shown by Kim (2016, p. 17) "the SDGs themselves are presented using a silo approach." According to Kim, illustrating the problems of such fragmented approach,

> [t]he drafters did not employ systems thinking when goal-setting and ended up forming a list of equally important global priorities. The non-hierarchical organization of the SDGs is problematic because the goals and targets interact. While some targets are interdependent or reinforce each other, some impose constraints on others.

More specifically, as this chapter demonstrates, the 5Ps and the specific SDGs that focus on "partnership" and implementation could be seen as attempts by the drafters to provide the Agenda with tools to overcome its own problem of fragmentation. However, this effort, alongside proclamations of interdependence and indivisibility among goals, did not seem enough to achieve such an objective. The following section problematizes what is hereby identified as the mainstream interpretation of the 5Ps and the SDGs and addresses some of its possible consequences.

3.3 The 5Ps: Origins and Content

The 5Ps appear in the preamble of *Transforming Our World*, presented as "areas of critical importance for humanity and the planet." Accordingly, it is toward these "areas of critical importance" that the goals and targets should stimulate action.

In their inside story of the SDGs, Kamau et al. (2018) observe that, during negotiations, six elements were initially proposed as "areas of critical importance": dignity, people, prosperity, planet, justice, and partnership. Brazil contended that not all these elements "adequately conveyed the priority and essence of th[e] agenda," and proposed "four Ps": people, prosperity, planet—which "capture[d] the three dimensions of sustainable development"—and partnership. According to Kamau et al. (2018), "many delegates from both North and South thought that a fifth 'P'—peace—needed to be added to highlight the relationship between peace and sustainable development" (Kamau et al. 2018, p. 300). Eventually, the Declaration (United Nations 2015) defined the 5Ps as:

People: "[W]e are determined to end poverty and hunger, in all their forms and dimensions, and to ensure that all human beings can fulfil their potential in dignity and equality and in a healthy environment."

Planet: "[W]e are determined to protect the planet from degradation, including through sustainable consumption and production, sustainably managing its natural resources and taking urgent action on climate change, so that it can support the needs of the present and future generations."

Prosperity: "[W]e are determined to ensure that all human beings can enjoy prosperous and fulfilling lives and that economic, social, and technological progress occurs in harmony with nature."

Partnership: "[W]e are determined to mobilize the means required to implement this Agenda through a revitalised Global Partnership for Sustainable Development, based on a spirit of strengthened global solidarity, focused on the needs of the poorest and most vulnerable and with the participation of all countries, all stakeholders and all people. The interlinkages and the integrated nature of the Sustainable Development Goals are of crucial importance in ensuring that the purpose of the new Agenda is realised. If we realise our ambition across the full extent of the Agenda, the lives of all will be profoundly improved and our world will be transformed for the better."

Peace: "[W]e are determined to foster peaceful, just and inclusive societies which are free from fear and violence. There can be no sustainable development without peace and no peace without sustainable development."

The definition of "people" as an area of critical importance for humanity and the planet reflects the development agenda's overriding commitment to eradicating extreme poverty and protecting the most vulnerable, a commitment that is encapsulated in the slogan "Leave No One Behind" and that is grounded in principles such as dignity and equality. As for "planet" and "prosperity," the priorities defined around these two areas generally express the main challenge of reconciling the preservation and conservation of the environment with the notion of technological progress and economic development. While "planet" seems to reinforce principles such as conservation, sustainability, rational management of natural resources, and intergenerational equality, "prosperity" gives a new layer to the development dimension of sustainable development, emphasizing the quest for "progress," but in harmony with nature.

Taken together, "people," "planet," and "prosperity" present in updated terms the most fundamental challenges of sustainable development. They can be seen as a continuation of the traditional three-dimensional understanding of *sustainable development* (the society–economy–environment tripod), as *society* translated to *people*, *environment* to *planet*, and *economy* to *prosperity*. One could question the extent to which this rewording has meant something more than just a rhetorical move or "rebranding" of the same old sustainable development challenges.

But it was the addition of two other dimensions ("peace" and "partnership") that has meant the most important improvement in the current understanding of *sustainable development* in the context of Agenda 2030. In comparison to "people," "planet," and "prosperity," the meaning and function of the Ps "peace" and "partnership" seem to be of a different nature. Whilst "people," "planet," and "prosperity" reflect core aspects of sustainable development, "peace" and "partnership" can be seen as evoking general principles of cooperation, multilateralism, and justice. Their call for a "global partnership for sustainable development," in a "spirit of strengthened global solidarity," together with the recognition of peace as essential for sustainable development, strengthens the principles of multilateralism, solidarity, integration, and inclusiveness.

"Peace" and "partnership" expand the concerns of the sustainable development agenda toward the broader areas of global governance—in other words, toward the institutional conditions, instances, and parameters of global sustainability governance and development. They seem to relate to the regulatory, normative, institutional, sociotechnical framework within which sustainable development is set to happen, pointing not only to the rule of law but also to cultural and contextual aspects of sustainable development that remain unaddressed by the traditional tripod.

Specifically, the addition of peace as an "area of critical importance" for the implementation of the SDGs was an important innovation. Acknowledging the fundamental interconnections between peace and sustainable development is crucial, as the violence of war, crime, and conflict causes damage and suffering to people and nature, seriously compromising the achievement of the SDG framework. Eventually, "peace" was addressed in the preamble and in SDG 16, which established important parameters of justice and protection from various forms of violence and crime.[2]

Partnership, by its turn, was not an innovation of Agenda 2030 either. However, the renewed emphasis on cooperation and multilateralism successfully raised the status of sustainable development within global governance structures, creating a new structure to discuss issues related to the SDGs and their implementation (The High-Level Political Forum on Sustainable Development) and new mechanisms for implementation, review, and follow-up.[3] Additionally, the recognition of partnership as an "area of critical importance" can be seen as an important step toward inclusiveness and integration of the most vulnerable, at least attempting to accommodate historical conflicting claims between developed and developing countries. In this sense, faithful to the commitment of "leaving no one behind," many SDGs do focus on issues such as support, access to resources, and participation of developing countries, including least developed countries and small island developing states. Finally, the new universal character of the 2030 Agenda (an improvement from the Millennium Development Goals, which were aimed only at developing countries) could also be seen as a sign of this improvement in terms of global partnership for sustainable development and cooperation, a milestone for global sustainability governance.

Partnership was also enshrined in SDG 17, with the purpose of "[s]trengthen[ing] the means of implementation and revitalise the global partnership for sustainable development." Thus, like a central axis, "peace" and "partnership" do provide an innovative, integrative function to the sustainable development agenda, acting transversally in relation to the other Ps. By doing so, that can be seen as "bridging the missing pillars" of sustainable development, adding important dimensions to the sustainable development debate by highlighting aspects of cooperation, multilateralism, international law, and policy. It is important to mention that "peace" and "partnership" should not be considered only at the "international" level but as cutting across distinct levels of governance and institutional layers, from global to local. Bleischwitz et al. (2018) mention Ostrom's (2010) concept of polycentric governance, which can be useful for this purpose, as it relates different institutional dimensions (Bleischwitz et al. 2018). An updated view of the global/local dimension of international law and development, as in Eslava (2015), might help address the challenges of orchestrating and implementing the SDGs across distinct levels of governance (Eslava 2015).

Also, by pointing to related implementation challenges, "peace" and "partnership" bring light to conjunctural aspects, historical and political challenges, social norms, forms, and institutions that may foster or hinder the achievement of the goals (Vvsg 2018). In this regard, Burford et al. (2016) hold a view that "peace" and "partnership" should imply the observation of local contexts, the respect for tradition, the inclusion of different values that are not captured by the Agenda's global universalist vision (Burford et al. 2016). A focus on "peace" and "partnership" on a local level shifts the gaze of international development from international contexts of war and conflict toward the structural violence that pervades countries, especially developing countries, both in cities and in the countryside, and which is related colonial and neocolonial structures of power and exploitation. In this sense, racial violence; the overexploitation of labor; and the lack of access to basic rights, which often condemns people to poverty, crime, and violence, must also be considered as vital in the struggle for peace and partnership in the context of the SDGs.

Considering this, the present study contends that the notion of partnership that should be supported by the global sustainable development agenda must be able to transcend the traditional dimension of states, international law, and international organizations, which purportedly has universal reach but is, in fact, Eurocentric. Conversely, only an integrated, bottom-up approach that discusses the meaning and significance of each of the Agenda's goals and priorities from the perspective of those "left behind" could guarantee the effective implementation of the goals in the years remaining until 2030.

This is the context in which the 5Ps emerged, with a specific purpose of integrating the multiple SDGs and addressing the problem of fragmentation and compartmentalization. Before discussing the role of the 5Ps as reflections of core values of Agenda 2030, the next section discusses the problem of its fragmentation more deeply.

3.3.1 The Mainstream Interpretation of the Relationship Between Ps and SDGs

Mainstream approaches for reading the 5Ps within Agenda 2030 tend to subsume SDGs 1, 2, 3, 4, and 5 exclusively under "people"; SDGs 6, 12, 13, 14, and 15 under "planet"; SDGs 7, 8, 9, 10, and 11 under "prosperity"; SDG 16 under "peace"; and finally, SDG 17, under "partnership" (Tremblay et al. 2020, p. 1586). Aside from the fact that the array of the goals within this clustered interpretation is rather unclear and questionable—given that organizations, authors, and countries tend to allocate and arrange them differently (Tremblay et al. 2020; Ki-Moon 2015)—the skepticism concerning this understanding comes from the fact that, since it considers that "[e]ach SDG

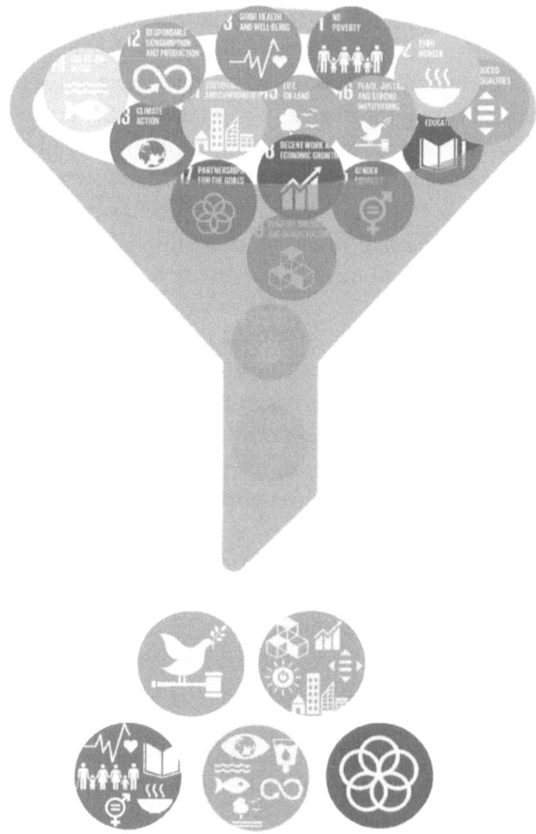

Figure 3.1 Mainstream approach—from 17 (isolated) goals to 5 (still isolated) clusters, or Ps.
Source: Authors.

can be categorised in one [exclusive] dimension" of the 5Ps, this clustered, siloed, segmented interpretation reinforces the very compartmentalization and fragmentation of the SDGs that it seeks to overcome, resulting in 5 still isolated Ps instead of the previous 17 even more isolated and fragmented goals (Figure 3.1). This would not only mean that the "areas of critical importance" of Agenda 2030 are unperturbed, unresponsive, and inert to each other's effect but also that each of the 17 goals will only be subjectively questioned by a singular P, interpreted and implemented via a singular dimension of the Agenda's values.

From a pragmatic perspective, adding synergic goals within the same "P," or cluster, might make it easier to understand their interactions—after all, it is easier to analyze interactions between 5 or 6 goals than between 17 (Chakravorti 2017). However, there seems to be no consensus concerning which of the Ps better adjust to each of the goals, and vice versa. As Tremblay et al. (2020) have demonstrated, organizations, authors, and countries tend to allocate them differently (Table 3.1) (Tremblay et al. 2020).

Although some SDGs are unanimously allocated into specific Ps, the disagreement regarding the allocation of some SDGs reveals that this dominant way of interpreting the relationship between the 5Ps and the SDGs is insufficient to account for complex nuances and interconnections between them. The challenge of implementation requires an understanding of this relationship, which is more attentive to such connections, one that is able to conceive the SDGs holistically and

Table 3.1 Examples of official classifications of the 17 Sustainable Development Goals (SDGs) in relation to the five pillars (5Ps)

	People	*Planet*	*Prosperity*	*Peace*	*Partnership*
United Nations in Rwanda (2016)	2, 3, 4	6, 12, 13, 14, 15	**7, 8, 9, 10, 11**	1, 5, 16	17
OECD (2017)	1, 2, 3, 4, 5	6, 12, 13, 14, 15	**7, 8, 9, 10, 11**	16	17
Stockholm Resilience Center (2016)	**1, 2, 3, 4, 5, 7, 11, 16**	6, 13, 14, 15	8, 9, 10, 12	—	17
France Diplomatie (2017)	1, 2, 3, 4, 5	**6, 7, 11, 13, 14, 15**	8, 9, 10, 12	16	17
UN System Staff College	1, 2, 3, 4, 5, 6, 10	**6, 7, 10, 13, 14, 15**	**1, 7, 8, 9, 10, 11, 12**	16	17
Palestine VNR (2018)	1, 2, 3, 4, 5, 10	**6, 7, 12, 13, 14, 15**	8, 9	11, 16	17

Source: Adapted from Tremblay et al. (2020). Cells that contain goal 7 are highlighted.

related to more than one "P" at the same time. It is in this sense, with the aim of providing normative guidance for the implementation of the Agenda, that the 5Ps lens will be further presented.

One of the practical risks associated with the selective and simplistic integration of the goals is allowing companies engaged in the Agenda to ignore the trade-offs of their business decisions while selling them as SDGs-oriented. This has been denounced as "SDG-washing," a phenomenon that reverberates problems such as greenwashing, weak interpretations of *sustainable development*, privatization, and the capture of environmental agendas by corporations. An example by Nieuwenkamp (2017, 1) illustrates this problem: "A car company may market their electric cars as saving the climate (SDG13 ↑). Yet, the cobalt in their batteries may be mined by five-year old kids in Congo (SDG8 ↓)." Under the mainstream approach, SDG 13 and SDG 8 are allocated into different "Ps" ("planet" and "prosperity," respectively), and surprisingly, no interconnection between them whatsoever was mentioned in the documents analyzed in Table 3.1. This example is illustrative of the problem implied in treating each "P" as isolated "areas of critical importance." If actors, especially companies, are free to choose a single "area" and fashion themselves as models of sustainable business, the Agenda's weakness for market interests' capture is then revealed. If, on the one hand, the growing electric car industry supposedly presents an alternative to carbon consumption and, in this sense, can be seen as "green" alternatives for transition, on the other hand, the demand they create for metals, such as lithium and cobalt, intensifies extractive exploitation in countries of the Global South, such as the Congo, but also in Argentina, Bolivia, and Chile, Latin American countries that are dependent on the exportation of these metals and, due to this dependence, are unable to break with the neocolonial cycle of inequality and exploitation that have been preventing the achievement of sustainable development in the region.

Trucco et al. (2021, p. 7) also address the risks of such an isolated approach:

> An isolated approach to the implementation of the SDGs could result in under-exploitation of the SDGs synergies in the best case, to conflicts for the prioritization of one SDG over another, or one (a group of) country over another, or even one generation over the next in the worst case.

Also, the mainstream approach does not consider the nexus between economic development, social well-being, and natural resources—in other words, the constant interplay between these

Table 3.2 Nexus × 5Ps mainstream approach

Nexus	*Main interlinkages between goals*	*SDG interlinkages between 5Ps under mainstream approach*
Land	SDG 2, SDG 7, SDG 11, SDG 13	People, prosperity, planet
Water	SDG 2, SDG 6, SDG 7, SDG 11, SDG 12	People, planet, prosperity
Energy	SDG 7, SDG 11, SDG 12, SDG 13	Prosperity, planet
Materials	SDG 2, SDG 11, SDG 12, SDG 13	People, prosperity, planet
Food	SDG 2, SDG 11, SDG 12	People, prosperity, planet

Source: Adapted from Bleischwitz et al. (2018).

dimensions. While defining the resource nexus as "a set of context-specific critical linkages between two or more natural resources used in delivery chains towards systems of provision for water, energy, food, land, and materials," Bleishwitz et al. (2018, 737–738) demonstrate, for instance, that relevant questions concerning human security necessarily correspond to a proper and integrated resource management. Yet as observed in Table 3.2, even though the main interlinkages between goals and nexus are mostly tridimensional (people, planet, prosperity), if synergies and trade-offs between goals were to be assessed through the 5Ps mainstream fragmented approach, these interlinkages would not be brought to light, remaining unaddressed as gray areas (blind spots) in between the clustered Ps.

Thus, given the risks associated with fragmented targeting and reporting of the SDGs and the incompleteness of the mainstream approach to identify synergies and trade-offs between them, the present study reiterates the 5Ps' role over this effort while suggesting an original approach to the subject matter.

3.4 The 5Ps as a "Lens": A Holistic, Integrated, Multidimensional View of the SDGs

This section presents the 5Ps as a lens through which the Agenda altogether and each of the 17 goals can be read from a holistic and integrated approach. Before delving into the lens metaphor, however, it might be useful to reflect briefly on the role of values for law and policymaking, as well as on the preambular, hermeneutic nature of the 5Ps within the framework of Agenda 2030.

3.4.1 The Role of Values in (International) Law, Policy, and Development

Debates regarding the implementation of Agenda 2030 revolve around the SDGs, their measurement and reporting, involving mostly empirical claims on data gathering, channeling resources, and improving mechanisms for reviewing progress toward the goals. The 5Ps, on the other hand, are not given the same priority, probably because, in contrast to the statistically modelled SDGs, they are expressed in rather subjective terms, such as "healthy environment," "protect the planet," "enjoy fulfilling lives," "harmony with the nature," "a spirit of strengthened global solidarity," and so on. As such, they cannot be measured, operationalized, and implemented in the same way as the SDGs, which are regarded as the most important contribution of the agenda.

Agenda 2030 is guided by the principles of universality, "leaving no one behind," and sustainable development, thus reflecting more general values, such as equity, fairness, and social justice.

The 17 SDGs also embody several values. For example, SDG 1 (No Poverty) and SDG 10 (Reduced Inequalities) are grounded in the value of equality and dignity. Similarly, SDG 5 embodies the value of gender equality, promoting social inclusion and eliminating gender-based discrimination.

A central contention of this chapter is that, as "areas of critical importance," the 5Ps reflect core values of Agenda 2030. The 5Ps of Agenda 2030 are not necessarily values in themselves, but they reflect broader values that underpin the SDG framework: sustainable development (and its three pillars) and cooperation.

The 5Ps are meant to guide the implementation of Agenda 2030 and thus ensure that development is people-centered, protects the planet, promotes prosperity, advances peace, and fosters partnership and collaboration. In this sense, as already discussed, they can be seen as providing or reflecting these values, with people, planet, and prosperity revamping the three-dimensional concept of sustainable development, with a specific focus on dignity and equality, and peace and partnership evoking the more general fundamental values of cooperation, inclusiveness, justice, and the rule of law.

However, the role of values in the Agenda and, more generally, in development, international law, and policymaking, has not been adequately appreciated. As Burford et al. (2016, 2) explain, "[v]alues have historically been excluded from the realm of indicator development because they are perceived as intangible and difficult to measure." In Dyer's (1996, 39) words, "[b]ecause intellectual tastes have tended to favour the analysis of quantifiable utilities, the analysis of values has been marginalised in social science."

Values do play a significant role in shaping international law, policy, and development, serving as the normative, moral, and ethical foundation for these domains and influencing decision-making processes and actions. Phillipe Sands (2003), for example, advocates that values are the basis for decision-making, alongside science, economics, and other social objectives. In the same sense, Louka (2006, 17) adds that "[o]nly by attaching values—a human artifact—can the course of action be determined what is appropriate for the protection of the environment and human health." Addressing values and their theoretical implications for the field of international relations, Dyer (1996, 39) draws attention to the increased interest in "a values-centred approach to public-policy analysis" and argues that, despite their subjectivity, values play a vital role in political actions. He presents an interesting point made by Di Bernardo (1988, 173, quoted in Dyer 1996, 35) that goes as follows:

> [I]f we employ subjective categories such as intentions, ends, rules, values, norms . . . [action] may be explained in terms of the contents of the consciousness of the agent which are linked with his vision of the world. The sense of his actions depends on these contents, and they contribute to the construction of the social world.

And then he expounds:

> [V]alues are the underlying substance of political systems and structures, and the appropriate objects of study. Any comprehension of interests or tangible assets in politics depends on a comprehension of the values at play, for it is they which endow political meaning.

In summary, values are essential for shaping international law, policy, and development, providing the normative framework for decision-making and actions, and helping ensure that they align with broader societal goals and aspirations. As such, they are critical for promoting a just and equitable world. Additionally, thinking in terms of values can also mean thinking critically about the political, ideological, and philosophical aspects of law, policy, and development. Crucially, this approach means a necessary move away from the technocratic, positivist, and empiricist approaches to international law and policy—such as the one crystallized around the Sustainable Development 2030 Agenda—toward a more contextual, value-based, culturally and historically sensitive approach to it.

In fact, one of the problems with the 2030 Agenda—that might explain its fragmented, insufficient approach—is precisely that, although it has tried to disentangle itself from this characteristic through the acknowledgment of the 5Ps, it still operates under a depoliticized and ahistorical paradigm of positivist science. The subjective character of the 5Ps has not been able to dismantle the empiricism and positivism of more traditional approaches to modern law and science.

According to Williams (1996, 59), "[p]ositivism separates subject from object, 'facts' from values, and presents a view of knowledge constructed around the concept of a knowable and objective external reality." By doing so, he observes, "[p]ositivist approaches prohibit critical self-reflexive theory." Indeed, the lack of political strength of the SDGs may lie in its focus on the empirical dimension, to the detriment of critical theory and self-reflection. At this point, the lens metaphor comes in handy, serving precisely as a point of critical reflection for the operationalization of the goals.

Williams (1996, 59) also notes that an "[a]nalysis of the complex interplay between social theory and social practice is an interpretative process, and it is [thus] impossible to produce critical theory without a recognition of the subjectivity of knowledge production." In this sense, it is crucial to mention that, despite having expanded, after 2015, its scope to the entire globe, the universalist global discourse of the Agenda is still far from adequately engaging with values (and their intersubjective nature), what in turn makes the Agenda incapable of addressing the regional (cultural) peculiarities and specific difficulties faced especially by developing countries in what comes to gathering resources and measuring and implementing the SDGs. However, in highlighting the 5Ps as interpretative tools of the Agenda and the SDGs (both together, in clusters, and each individually), recognizing the intersubjective dimension of the values it reflects, this chapter aims to bring together practice and social theory of world transformation.

Paterson (1996, 67) points in the same direction, criticizing the disregard for the question of values that results from strict positivism and empiricism. According to him, a "strict distinction between facts and values . . . creates an equivalent distinction between analytical, or empirical, theory and normative theory," privileging "the analytical, 'value-free' theory." As a result, he adds, "one ends up trying to theorise either about the way the world is or about how the world ought to be," and these activities are held to be irredeemably separate. In the case of the Agenda, this is revealed in the gulf that is perceived between the goals of transformation and the fragmented concrete data of the reality to be transformed. The separation between facts and values implies a disconnection from the concrete and complex reality to be transformed, segmenting interconnected dimensions of social reproduction, what eventually impedes the transformative power of the agenda as a whole.

The crux of the matter is realizing that there seems to be a mismatch between the empirical dimension of indicator development, data collection, measurement and progress in relation to the SDGs, and the comprehension of the constitutional values of the 2030 Agenda. This mismatch can be seen in the way the international community has failed to relate the challenge of implementation goals and the "areas of critical importance" for their achievement (the 5Ps). Despite the subjective character of the Ps, the very structure of the Agenda, divided into 17 non-hierarchical goals, confirms a reductionist way of fragmenting social reality: Although the 17 SDGs demonstrate a praiseworthy effort to operationalize transformation, they end up reducing the complexity and multidimensionality of global problems in the name of systematicity, oversimplifying them. As for the 5Ps, undertheorized and reproducing compartmentalization, the 5Ps have not been able to integrate the SDGs and orchestrate the SDGs' implementation. Committed to operationalization (to the detriment of critical thinking), the Agenda seems to be blind to important discussions regarding values, culture, history, and politics—its pragmatic impetus tends to circumvent such issues. As a result, the current interpretation and application of the agenda have been insufficient to generate the political energy necessary for the real transformation of our world—especially in developing countries.

The point is that the 5Ps may offer a unique opportunity for problematization and critical discussion about the meaning and direction of Agenda 2030, illuminating crucial questions, such as what kind of progress is being pursued and how exactly to perceive its points of harmony or disharmony with nature. What genuinely counts as the human needs, of both present and future generations? What does it mean to enjoy a prosperous and fulfilling life? What is *transformation*? Who are the winners and losers of the development story? As reflections of the core values of Agenda 2030, the 5Ps bring (inter)subjective aspects of sustainable development to discussion, illuminating yet other hidden aspects that fragmented, mainstream approaches fail to conceive.

3.4.2 The Soft Legal Nature of the SDGs and the 5Ps as Interpretation Tools of the Agenda

According to Kim (2016, 16), "[t]he SDGs themselves are political goals, not legal rules." More specifically, according to him, "[t]he SDGs and targets are . . . best conceptualized as a subset of existing intergovernmental commitments." Between politics, science, and law, the status of the SDGs is controversial, for if, on the one hand, the goals are presented as mainly political commitments, on the other the structure and form of their adoption and implementation carry traces of international law, which, in turn, is intertwined with the idea of development (Eslava 2015). Due to this hybrid character, which also presupposes a specific relationship between public and private entities, and several layers of governance, from global to local, the term *global sustainability governance* has been applied to study the SDGs.

Nevertheless, the SDGs can be classified as soft law or quasi-legal. In this respect, Kim (2016, 16) argues that "the substance that the SDGs reflect (some of which are international custom) and the process by, and form in, which they were adopted indicate that at least some SDGs or targets may qualify as soft law." It is important to acknowledge that, although it stands as a soft legal mechanism, the sustainable development agenda preserves some basic characteristics of contemporary international law, such as the centrality of the nation-state; the premise of equal formal sovereignty (the North/South divide); the elitist, Eurocentric character of international law; and also the fact that the production of knowledge is still concentrated in the institutions of the Global North. Regardless of the status of the Agenda, by themselves, such characteristics alone make it fair to criticize the instrument from the standpoint of critical international legal scholarship.

The point is that the SDGs find an important support point in international law, as Kim's study demonstrates. Notably, the study identifies the fragmentation problem of the agenda, searching in the principles of international law the function of integration between and among goals.

Certain principles of international law, such as the principle of integration, may provide normative guidance as to how the goals and targets should relate to each other, and thereby avoiding "a state of normative anarchy."

In this effort, Kim even finds this integrating function in the principle of sustainable development itself. The author identifies that *sustainable development* is increasingly understood as a general principle of international law, and notes that "[r]egardless of the exact legal status, it is reasonable to conclude that the concept of sustainable development has 'practical legal consequences,' one of these being the integrating function of the dimensions of development."

This chapter explores this integrative function of sustainable development as a principle of international law but chooses to do so in relation to the 5Ps, considered as the most sophisticated development of the concept of sustainable development, particularly regarding identifying the priorities for its implementation.

As already mentioned, the 5Ps are in the preamble of the Declaration. According to Jan Klabbers (2018, p. 172), typically, the preamble of a treaty is regarded as legally non-binding, that is, "as not giving rise to enforceable rights and obligations." Klabbers argues that

> [t]he preamble may enumerate some of the motives behind the conclusion of a treaty but is not considered to give rise to any practical legal effects: its role is generally considered to be largely metaphysical.

From this statement, one can understand that if the legal status of the SDGs is controversial, that of the 5Ps is even more obscure. Although the principle of sustainable development has gained practical legal effects in some historical decisions, its reformulation through the 5Ps has been little explored in this sense, and the effects of the 5Ps remain unaddressed by scholarship. Klabbers continues:

> And yet the preamble may come to influence the standard interpretation of a text and even come close to being enforceable. . . .Those who apply treaties tend to give meaning to the words used regardless of their place in the grander scheme of the treaty, and a promise made in the preamble may be as good as one made in the main body of text.
>
> (Klabbers 2018, p. 172)

Considering this, it is possible to identify a function, a role, for the Ps within the scope of the 2030 Agenda: a function which is not merely metaphysical, or programmatic, but building on the role of values as discussed earlier, which can be seen as hermeneutic, even constitutional. In a certain sense, the 5Ps, or five "areas of critical importance," can be seen as existential goals, "promised lands," that reflect the core values of Agenda 2030 (namely, sustainable development and its three pillars, as well as to "leave no one behind," which stands for dignity and equality, and also peace and cooperation). As such, they function like a multidimensional reading prism, or "lens," that can orient the interpretation of the agenda as a multidimensional whole and the implementation not of a single and isolated SDG but of all the SDGs, defining the horizon toward which implementation action must head.

As reflections of values, the 5Ps emphasize the "crucial" interconnectedness of the SDGs, cementing them into an integrated body. Furthermore, as promises, they can be seen as existential goals that should guide normative and institutional developments related to the implementation of the 2030 Agenda across different sectors of the economy, environmental regimes, and levels of governance.

In sum, exploring the Agenda through the 5Ps may provide a more integrated, value-based, and multidimensional reading of the Agenda and of *sustainable development*. The 5Ps "lens" may allow one to rethink not only the internal and external connections between the goals and the legal-political environment that structures them but also the cultural dimensions that drive and constrain actors, institutions, processes, and structures in the objective of transforming the world.

3.4.3 Visual Representations of "The Lens"

Through the 5P lens, each SDG can and should be understood simultaneously from the perspective of each "P," thereby overcoming the fragmentation that results from the mainstream approach to that relationship. As illustrated in Figure 3.2 and Figure 3.3, the integrated approach contrasts to

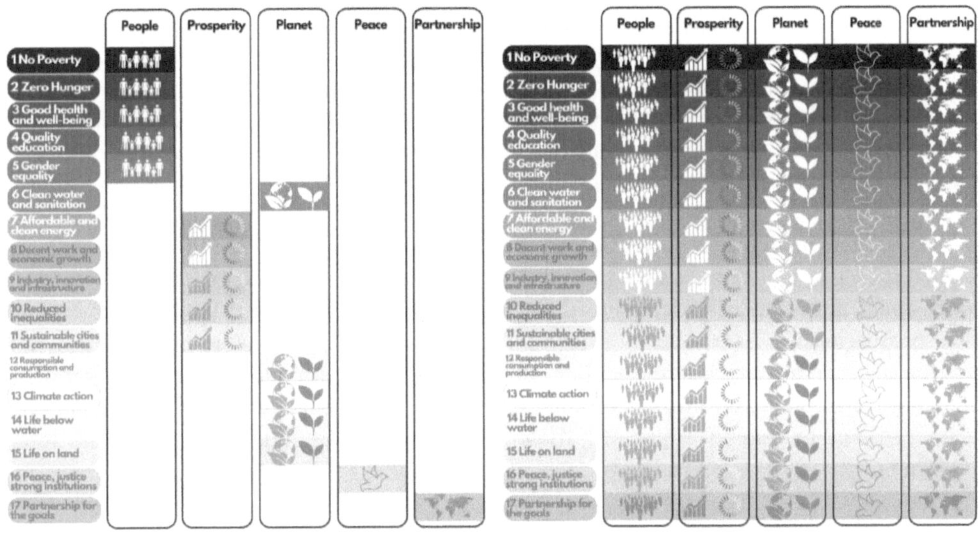

Figure 3.2 Comparison between the 5Ps mainstream approach and our integrated approach from the perspective of the SDGs.

Source: Authors.

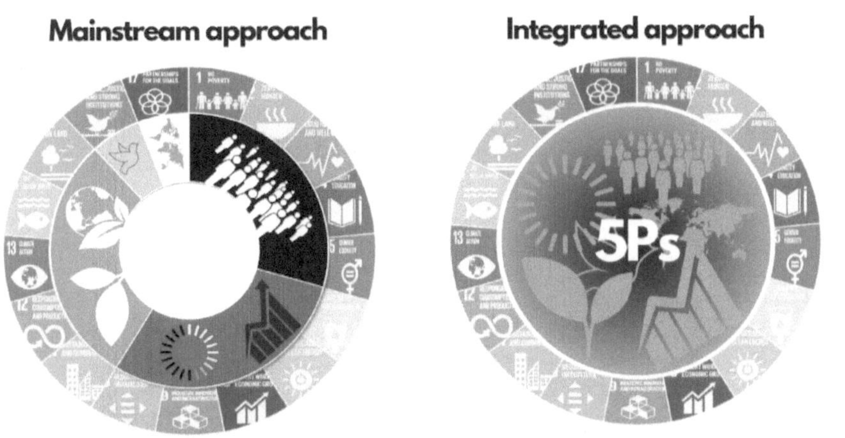

Figure 3.3 Comparison between the 5Ps mainstream approach and our integrated approach from the perspective of 5Ps.

Source: Authors.

the mainstream approach, as it provides a sense of integrity and complexity both from the perspective of the 5Ps and from the perspective of the SDGs. When presenting such figures in gray-black scale, it is important to make the following caveat: The visualization of these figures in multiple colors would allow one to see and feel the nuances and gradients generated by the integrated approach, as opposed to the fragmented and monochrome vision of the mainstream approach. The grayscale meets limits to demonstrate the complexity of the interconnection between the SDGs and the 5Ps when viewed through this innovative lens.

Contrarily to the clustered arrangement of 5Ps, this study suggests a lens to read 5Ps, though each goal, focusing on their interconnections. Thus, instead of five segregated lenses (each *P* as an independent lens), it suggests the use of a single integrated lens, whereby Ps themselves are deeply interconnected and the SDGs are multidimensional. The lens metaphor tools provide the opportunity for one to see each and every SDG, or the Agenda 2030 altogether, as a three-dimensional array in which society, environment, and economy are assessed as in relation to each particular conjuncture, with socio-political contexts also considered—as peace and partnership cut across sustainable development's three traditional pillars. Figure 3.4 and Figure 3.5 illustrate the proposition.

As illustrated in Figure 3.5, taking the 5Ps as reflections of the Agenda's core values, so as to better "read" the Agenda and its goals, the 5Ps lens should deliver a contextualized multidimensional and culturally sensitive understanding of the Agenda 2030, whereby people–planet–prosperity represents the substance of the Agenda and sustainable development, while the peace–partnership axis represents the transversal, cross-sectional elements and parameters of global governance that frame and regulate the actions and procedures of states, private actors, and stakeholders in general. Figure 3.5 illustrates that the same understanding applies to each SDG as a unity.

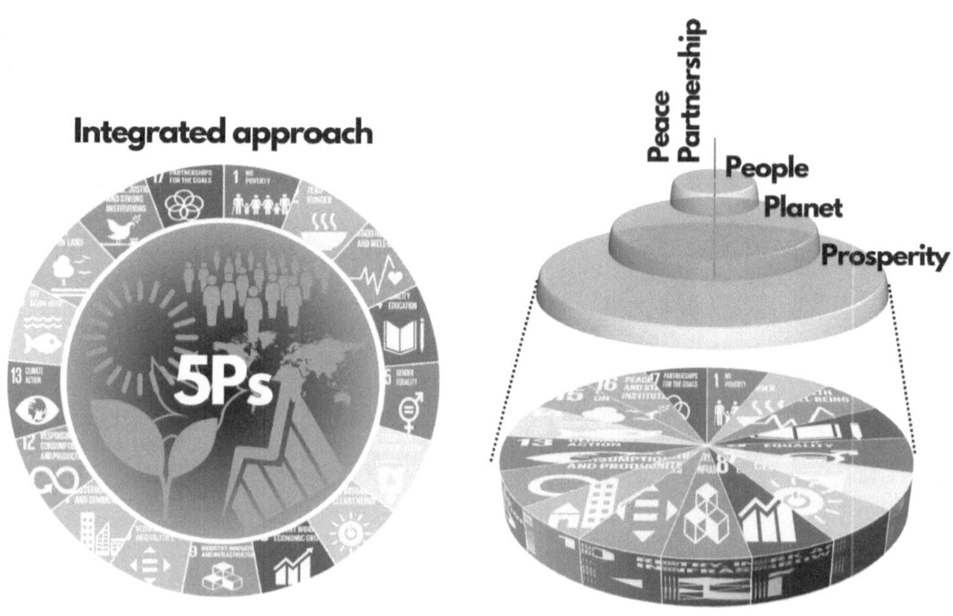

Figure 3.4 5Ps lens applied to Agenda 2030 as a whole.

Source: Authors.

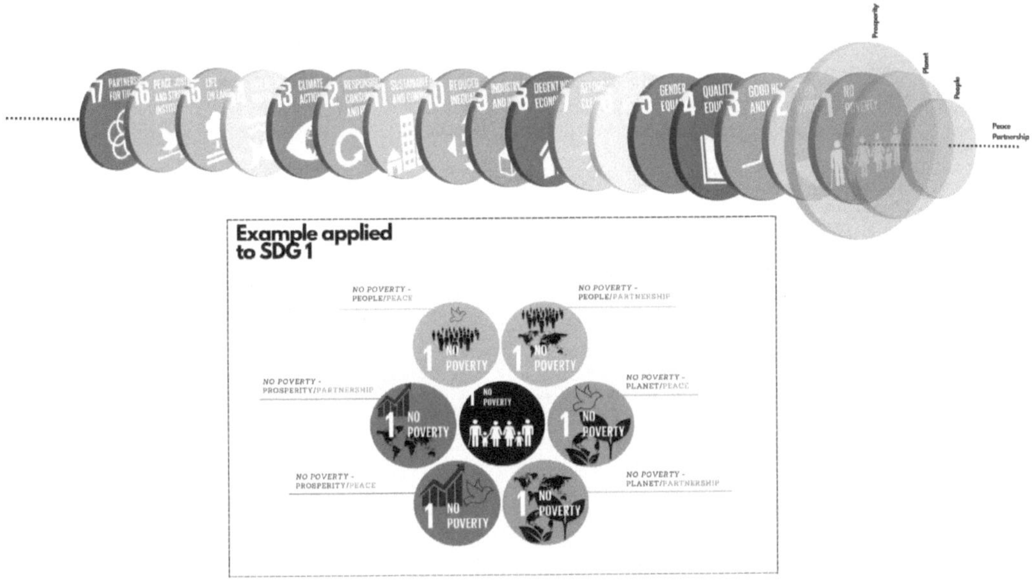

Figure 3.5 5Ps lens applied to each SDG.[4]

Source: Authors.

Ultimately, our "lens" is a critical exercise that allows us to rethink the internal and external connections between the goals and the legal-political environment that structures them. As far as the lens metaphor allows one to "see the unseen," to construct new narratives, the 5Ps lens intends to overcome the universalist and instrumental rationality of the Agenda, enlightening new reflections on the same object, whereby the point of view of the observers not only is permitted and incentivized but also recognized and expected. Such reflections are assumed to crosscut the boundaries between the goals, guiding a better comprehension of the synergies and trade-offs between them from a bottom-somewhere approach. Such reflections may also provide novel interpretations on the actual content of such values, favoring the view of those excluded from the development process or encouraging different approaches and epistemologies to it. This effort should be taken as a pedagogical exercise to explore the values and other subjectivities within Agenda 2030, allowing room for new interpretations concerning the issues it deals with and the scale assumptions on which it is based, also questioning the voices it amplifies and their discourses.

3.5 Energy Beyond Prosperity: SDG 7 Through the 5Ps Lens

This section presents the 5Ps lens as applied to a concrete object, a specific goal of the 2030 Agenda. SDG 7 was chosen as a model to illustrate the application of the lens as a tool to read the interconnectedness between goals and the "areas of critical importance" for their implementation.

As energy increasingly came to be recognized as a basic need, it was agreed that the new universal agenda should address energy specifically. Thus, expanding on the Millenium Development Goals (2000), SDG 7 was agreed upon in 2015, promising to "ensur[e] access to affordable, reliable, sustainable and modern energy for all." Figure 3.6 presents SDG 7's targets and indicators (United Nations 2000).

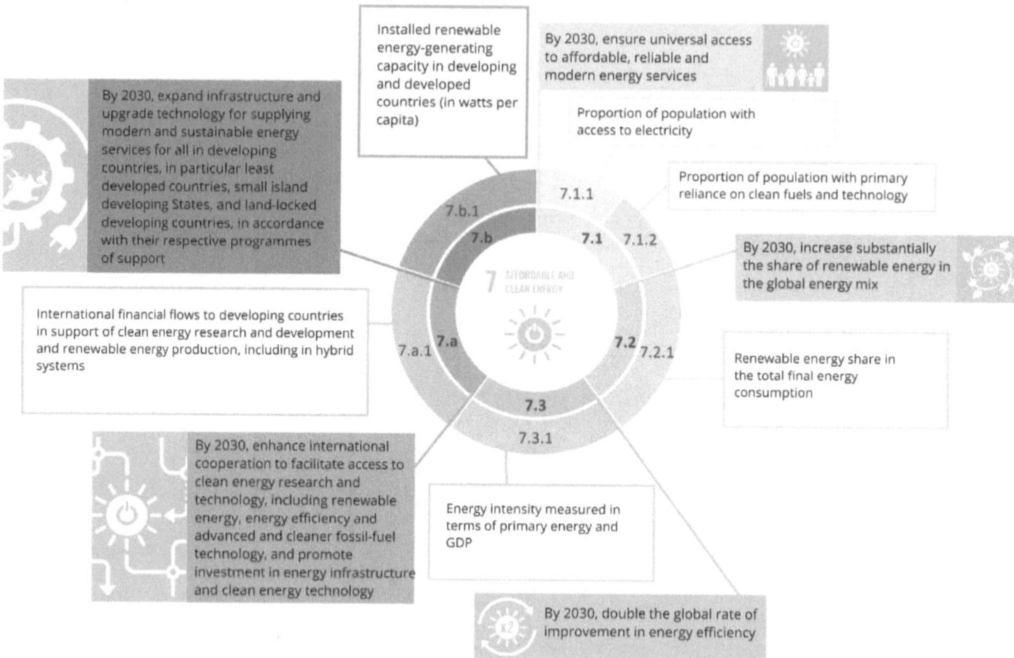

Figure 3.6 SDG 7's targets and indicators (United Nations 2000).
Source: Authors.

3.5.1 SDG 7: Ambiguities and Incongruences Between the Goal and the Targets

This section first looks at the internal contradictions between SDG 7 and its targets/indicators. Two important critiques are directed from this perspective, the first being the fact that SDG 7 is one of the 17 goals with the lowest number of targets, as mentioned by McCollum et al. (McCollum et al. 2017). Such a fact should not mislead to the conclusion that the energy issue is simple and easy to resolve. In fact, energy is foundational not only to economic development and progress (mainstream concerns, which respond to "prosperity") but also to human development ("people") and environment protection ("planet") (Nilsson et al. 2013). As the recent conflict between Russia and Ukraine has shown, energy is also crucial to maintaining peace.

Drawing from Villavicencio Calzadilla and Maugner (2018), the second critique concerns the fact that terms such as "access to," "modern," "substantially," and even "sustainable" are broad, are difficult to delimit, and may lead to important ambiguities if not well defined (Kroll et al. 2019). Here appears once again the importance of looking at the intersubjective and relative character of the goals and its underpinning values (reflected by the 5Ps). Accessibility, modernity, and substantiality may vary according to cultural and historical contexts, as well as according to socio-economic conditions, and only a broad and multidimensional interpretation of the instrument and its implementation can adequately define these terms. The measurement effort, which is essential for monitoring progress, may be severely hampered if these definitions are not established in advance.

Combined, both critiques reveal an internal contradiction of SDG 7 which is particularly hard to solve: On the one hand, the goal sustains broad and ambitious terms (relying on broad values for its achievement); on the other, it is sustained by no more than a few targets and only six indicators, what may cause the impression that their achievement is relatively easy compared to other goals.

Besides that, as the "affordability" aspect of the goal seems to have been completely ignored from the indicators' perspective (and this omission can be seen as yet another blind spot of the goal), "access" to energy can only be assessed in terms of infrastructure and technology.[5] Consequently, there is the distortion of the energy problem, given that it would be misleading to conclude that access to affordable, reliable, sustainable, and modern energy is really ensured for all in countries with a high score in progress toward SDG 7.

Moreover, intriguingly, only six indicators are presented as a means for SDG 7 to be achieved, three being measured in very economic terms: GDP, international financial flow, and investments. This study contends that this focus is a harmful consequence of the mainstream approach's subsumption of SDG 7 exclusively under "prosperity." The result is a notion of progress that privileges economic growth over social justice and nature conservation, revealing a bias in favor of the economy over society and the environment and the Agenda's commitment to an unsustainable capitalist notion of development and progress.

3.5.2 SDG 7: Connection With Other Goals

Different authors have been working on the interconnectedness of the SDGs from the targets and indicators perspective. Fuso Nerini et al. (2018) mapped the synergies and trade-offs between SDG 7 and other goals through a comprehensive, consensus-based expert elicitation process, demonstrating that, out of the 169 targets of the SDGs, 143 relate synergically with the pursuit of SDG 7, while 65 relate to it as trade-offs (Fuso Nerini et al. 2018). McCollum et al. (2017), by their turn, demonstrate that each SDG is, in a way, related to SDG 7.

The interconnections of SDG 7 with other SDGs and their targets illuminate the multiple sides of the energy issue and the difficulty in translating and reaching solutions through fragmented targets and isolated "Ps"—in the case at hand, solely through the "P" "prosperity." Stafford-Smith et al. (2017) explain the risks associated with uncoordinated actions in what comes to their interconnections, using poverty as an example: While increasing consumption may be taken as a means to alleviate poverty, it may also burden the environmental system. As for SDG 7, they note that

> [u]ncoordinated action may create internal conflicts, such as subsidies for both renewable and non-renewable fuel sources, or missed synergies, for example, where appropriately targeted investment in renewable energy reduces emissions, but it could also reduce pollution, improve human health, and increase equality.
>
> (Stafford-Smith et al. 2017, 912)

Moreover, thinking in terms of poverty, the concept of energy poverty allows us to identify aspects of SDG 7 which have been overlooked by mainstream readings. Under an indicator's perspective, SDG 7 seems to overlook important social justice concerns: the distributive aspect concerning "energy for all," for example. This means, in a direct sense, that, while its targets may point to a cleaner and more modern future, problems such as energy poverty may rest unaddressed.

Even though energy poverty also relates to other targets (SDG 1, SDG 2, and SDG 4, for example), the very reason for "energy poverty" existing as a specific concept should be reason enough for SDG 7 to engage in its measurement and eradication. Distancing SDG 7 from a social perspective and leaving for other goals the responsibility to deal with energy poverty–related issues could distort the ultimate goal of transforming energy systems for all. Besides that, given that SDG 1 treats poverty in an econometric, unidimensional manner, as Alston demonstrates (United Nations 2020), it is therefore unable to conceive of the energy dimension of poverty. In sum, given the trade-offs between the goals, if not manifestly imperative in SDG 7, energy poverty may not be addressed by any goal whatsoever.

Finally, the single fact that SDG 7 alone relates to other SDGs (Fuso Nerini et al. 2018; Fonseca et al. 2020; McCollum et al. 2017) demonstrates that SDG 7 is, in itself, multidimensional. If the mainstream rationale implies that each SDG corresponds to one single dimension (*P*), this means, in a direct sense, that each goal has overlooked dimensions, blind spots that could be better read through a holistic, multidimensional lens that integrates different Ps, given that targets and goals are mutually affecting each other.

3.5.3 Energy Beyond Prosperity

The mainstream approach to the 5Ps bluntly presents SDG 7 exclusively as a matter of "prosperity." However, this section shows that "prosperity" is only one among the many dimensions of the energy issue, and that interpreting it solely as a matter of prosperity risks detaching energy from nature ("planet") and society ("people"), as well as from central questions concerning "peace" and "partnership."

Regarding "planet," this detachment seems even more evident, as energy systems are responsible for around 60% of anthropogenic emissions of greenhouse gases, and thus, any solution to the climate issue necessarily involves SDG 7 (Fuso Nerini et al. 2018). In fact, Tremblay et al. (2020) demonstrated that more people linked SDG 7 to the "planet" than to the "prosperity" dimension (Tremblay et al. 2020). Still, on "planet," there is the already-mentioned mainstream's lack of attention to "resource nexuses," and the fact that, from production/generation to consumption, energy often uses minerals and land in controversial ways, reproducing violent colonial/imperialist and extractivist forms of violence that generate significant impacts on nature and people. Addressing the issue, but from a social-ecological perspective, Eisenmenger et al. (2020) show that the SDGs prioritize economic growth over sustainable resource use (Eisenmenger et al. 2020). This criticism applies to SDG 7 as far as the goal and its targets prioritize technological issues over ecological and social ones.

As for the blind spot "people," building on the mentioned mainstream's disregard for questions of social justice, such as energy poverty, the study stresses the importance of promoting a conception of energy as a basic and universal human right and not as a commodity so as to guarantee a real accessibility/affordability to energy services. In this sense, this chapter echoes Third World calls for a "people-centered development agenda" (Third World Network 2013) and emphasizes the fact that energy consumption must be seen related to a different set of human necessities and, therefore, should be seen mainly as a human right (Löfquist 2020; Walker 2015). Moreover, besides promoting access to all, it is also important that the integrated lens addresses the trade-offs concerning energy production and social justice, as the idea of energy as a resource nexus corresponds to the notion that energy production impacts other resources and, thus, all people that are, under any sense, related to it—whether from a production/consumption subsistence perspective (Biggs et al. 2015) or from an existential one (Dunlap 2017).

Finally, in what refers to "peace" and "partnership," their relationship with the energy issue manifests in different ways, from the global to the local arena. First, there is the fact that the energy transition to a low-carbon economy will require a level of multilateral cooperation and partnership not yet seen, so that a regulatory, institutional, normative framework favorable to the transition can be put in place.[6] In addition, this cooperation must also consider those aspects related to human resources development, education, technology transfer, sociotechnical capacity building, and democratic participation of grassroots social movements; face the challenges of the crisis of multilateralism; and going further, focus on transforming more radically the socio-economic relations that underlie the energy transformation agenda. Also, the question of imperialism and dependency needs to be better addressed by the SDGs, as well as geopolitical instability related to energy resources and the great disparity in energy consumption between the Global North and Global South countries (Foster 2008; Musso and Crouzet 2019; Hamouchene 2020). In this sense, studies on the relationship between SDG 7 and resources, sovereignty and political instability, armed conflicts, proxy wars, humanitarian interventions, coups d'état, etc. should be stimulated.

Finally, an integrated, multidimensional, values-based approach could introduce to the SDG 7 debate some perspectives of those "left out" by the "locomotive of progress." For example, SDG 7's exclusive focus on prosperity means that renewable energy production will always be a commendable effort, as it is obviously pointed at progress. However, is there a consensus about what *progress* may actually mean to everyone? Or even *prosperity*? These questions may seem a little too philosophical in relation to the propositional targets of the Agenda; however, they address a hidden problem, since, oftentimes, the progress for some trespasses the planet of other people.

For example, the Brazilian Indigenous leader, environmentalist, and writer Ailton Krenak (2019) described the construction of the hydroelectric power plant Aymorés in the beginning of the 2000s in Brazil as an "unbearable hug of progress," given that, as he explains, the construction damaged the whole ecosystem of the community (Krenak 2019). Dunlap (2017), in the same sense, argued that, instead of being a simple and obvious "solution," the development of wind energy production in the coastal Isthmus of Tehuantepec region of Oaxaca, Mexico, became, in fact, a "problem" to the traditional communities set in the land, reproducing colonial destructive patterns that point to what he calls the genocide–ecocide nexus. In what comes to these impacts, it is of major importance to acknowledge that they cannot be simply translated to those who do not share the same community values. What for some might represent a sad loss to be compensated in financial terms for others may represent the end (of peace, of an identity, of life), a war.

3.6 Conclusion

This study concerned the problem of fragmentation, finding limits in mainstream interpretations and discussing new possibilities for interpreting Agenda 2030 and the SDGs. It argued that mainstream ways of conceiving the relationship between the 5Ps and the SDGs continued to reproduce fragmented and isolated readings, creating distortions by interpreting each SDG as unidimensional, from the perspective of a single "P" at a time, consequently failing to deliver an integrated implementation of the Agenda as a whole (Fonseca et al. 2020).

Alternatively, the chapter argued that the relationship between the 5Ps and the SDGs should be conceived according to a holistic, integrated, multidimensional, critical, contextual, and human rights–based approach to poverty and sustainable development, rather than the segmented, sectorial, positivist, and empiricist approach put forth by mainstream interpretations. Our proposed "holistic lens" construed the 5Ps as core values (or reflections of the Agenda's core values) that potentially endow the Agenda, its goals, and its targets with clearer political meaning and purpose and, as such, provide for a more substantial and meaningful definition of the goals, illuminating their interconnections, synergies, and trade-offs. As reflection of values, the 5Ps may be used to better guide the interpretation and implementation of the SDGs, from the local to the global level, across multiple sectors and layers of governance, in a dynamic and dialectic interplay between different dimensions of sustainable development. In sum, together, the 5Ps potentially allow for an integrated and multidimensional interpretation of each of the SDGs and of Agenda 2030.

To illustrate the argument, the lens was applied to SDG 7, bringing the energy issue into "sharp focus." The exercise found that interpreting "energy" solely as a matter of prosperity contradicts the efforts other authors have put to analyze synergies and trade-offs between SDG 7 and other goals, which are, in fact, surprisingly high.

Besides that, it was found that under the mainstream interpretation, the nexus relation is precluded and, because of that, the dependencies between goals related to resources are not brought to light. This simplistic interpretation may also reveal SDG 7's bias and compromise to capitalist imperatives of growth and accumulation. Also, it may be revealing of an overreliance on technology, to the detriment of social and environmental concerns, when it comes to the point of what kind of transition is envisioned for a sustainable future. Conversely, through the holistic lens of the 5Ps, it was able to assess dimensions of the energy issue which are hidden from its targets and, therefore, of the Agenda altogether.

The aim of this study was to demonstrate that illuminating the 5Ps within each goal is a feasible means to ensure SDGs can read transformative capabilities through different social, economic, and environmental systems. It was not meant, however, to undermine the efforts to better understand and measure synergies and trade-offs between the goals from targets' perspective. On the contrary, it is understood that this conceptual tool, as demonstrated with the energy case, would reveal even more synergies and trade-offs between them.

Finally, this brief dive into the hermeneutics of Agenda 2030 is of paramount importance for a future research agenda: Firstly, because a more refined understanding of the values and the process of interpreting the SDGs as an integrated whole is what makes it possible to understand the operation of the lens in movement—the 5Ps as reflecting core values for a dynamic process of reinterpretation, application, and critique of the SDGs. Secondly, because this comprehension of values and their role can illuminate and guide action in international relations, lawmaking, development, and policy planning—understanding how values work in relation to interests, goals, principles, and norms can help weigh different (and sometimes contrasting) values and inform decision-making and problem-solving in the face of crises and impasses. Thirdly, because the critical thinking implied in this "holistic" interpretation proposal is crucial to avoid a blind and automatic (mechanical) application of the targets (positivist approach).

Notes

1 For important studies on measurement, integration, nexus, synergies, and trade-offs between the SDGs and the indicators, see Nilsson et al. (2013), Norström et al. (2014), Costanza et al. (2016), Fukuda-Parr (2016), Koehler (2016), Burford et al. (2016), Stafford-Smith et al. (2017), Fuso Nerini et al. (2018), Bleischwitz et al. (2018), Kroll et al. (2019), Bhatt (2020), Tremblay et al. (2020), Coscieme et al. (2020), Shachi and Ram (2021), and Trucco et al. (2021), among others.

2 A critical commentary by Liora Lazarus raises concern about the securitization of sustainable development present in SDG 16. Lazarus argues that SDG 16 is reflective of a broader shift toward the securitization of the rule of law and human rights, meaning, the enhancement of the coercive function of the state. See Lazarus, L. 2020. "Securitizing Sustainable Development? The Coercive Sting in SDG 16."

In *Sustainable Development Goals and Human Rights*, edited by M. Kaltenborn, M. Krajewski, and H. Kuhn. *Interdisciplinary Studies in Human Rights*, vol. 5. Cham: Springer. https://doi.org/10.1007/978-3-030-30469-0_9.

3 The idea of partnership reflects the principles of cooperation and multilateralism that have characterized the international legal order at least since the post-war period. Since then, and especially after the end of the Cold War, discourses of global partnership for development have gained the upper hand, guided by principles of liberal internationalism linked to the market forces of neoliberal globalization, thus deepening the idea of partnership based on ordo-globalist terms, entailing new relations between the public and the private.

4 Notice that "peace" and "partnership" function as an axis, equally framing each and every SDG.

5 Even from a technological perspective, the indicators neglect a number of technological assets necessary for human development.

6 "Improving global energy supply security through clean technology will require unprecedented international cooperation and considerable financial investment. International law can facilitate this process. However, as argued, the current international legal architecture is insufficient to achieve these goals." See: Bruce and Stuart 2013, *Melbourne Journal of International Law*, 14, 1, June 2013, pp. 1–36.

References

Biggs, Eloise M., Eleanor Bruce, Bryan Boruff, John M.A. Duncan, Julia Horsley, Natasha Pauli, Kellie McNeill, et al. 2015. "Sustainable Development and the Water-Energy-Food Nexus: A Perspective on Livelihoods." *Environmental Science and Policy* 54: 389–397. https://doi.org/10.1016/j.envsci.2015.08.002.

Bleischwitz, Raimund, Catalina Spataru, Stacy D. VanDeveer, Michael Obersteiner, Ester van der Voet, Corey Johnson, Philip Andrews-Speed, Tim Boersma, Holger Hoff, and Detlef P. van Vuuren. 2018. "Resource Nexus Perspectives Towards the United Nations Sustainable Development Goals." *Nature Sustainability*. Nature Publishing Group. https://doi.org/10.1038/s41893-018-0173-2.

Burford, Gemma, Peter Tamás, and Marie K. Harder. 2016. "Can We Improve Indicator Design for Complex Sustainable Development Goals? A Comparison of a Values-Based and Conventional Approach." *Sustainability (Switzerland)* 8 (9). https://doi.org/10.3390/su8090861.

Chakravorti, Bhaskar. 2017. "How Companies Can Champion Sustainable Development." https://hbr.org/2017/03/how-companies-can-champion-sustainable-development.

Di Bernardo, Giouliano (ed.). 1988. *Normative Structures of the Social World*. Amsterdam: Editions Rodopi.

Dunlap, Alexander. 2017. "The 'Solution' Is Now the 'Problem': Wind Energy, Colonisation and the 'Genocide-Ecocide Nexus' in the Isthmus of Tehuantepec, Oaxaca." *The International Journal of Human Rights*.

Dyer, Hugh C. 1996. "Environmental Security as a Universal Value." In *The Environment and International Relations*, edited by Mark Imber and John Vogler, vol. 1, pp. 1–248. New York: Routledge.

Eisenmenger, Nina, Melanie Pichler, Nora Krenmayr, Dominik Noll, Barbara Plank, Ekaterina Schalmann, Marie Theres Wandl, and Simone Gingrich. 2020. "The Sustainable Development Goals Prioritize Economic Growth Over Sustainable Resource Use: A Critical Reflection on the SDGs from a Socio-Ecological Perspective." *Sustainability Science* 15 (4): 1101–1110. https://doi.org/10.1007/s11625-020-00813-x.

Eslava, Luis. 2015. *Local Space, Global Life: The Everyday Operation of International Law and Development*, vol. 68, 351 p. Cambridge University Press, 2015. https://doi.org/10.17103/redi.68.2.2016.4.06.

Fonseca, Luis Miguel, José Pedro Domingues, and Alina Mihaela Dima. 2020. "Mapping the Sustainable Development Goals Relationships." *Sustainability (Switzerland)* 12 (8). https://doi.org/10.3390/SU12083359.

Foster, John Bellamy. 2008. "Peak Oil and Energy Imperialism." https://monthlyreview.org/2008/07/01/peak-oil-and-energy-imperialism/.

Friedmann, Wolfgang. 1964. *The Changing Structure of International Law*. 1st ed., vol. 1. Columbia: University Presses of California.

Fuso Nerini, Francesco, Julia Tomei, Long Seng To, Iwona Bisaga, Priti Parikh, Mairi Black, Aiduan Borrion, et al. 2018. "Mapping Synergies and Trade-Offs between Energy and the Sustainable Development Goals." *Nature Energy* 3 (1): 10–15. https://doi.org/10.1038/s41560-017-0036-5.

Hamouchene, Hamza. 2020. "Energy Transitions and Colonialism." *Committee for the Abolition of Illegitimate Debt* (2 November 2020). https://www.cadtm.org/Energy-transitions-and-colonialism.

Kamau, Macharia, Pamela Chasek, and David O'Connor. 2018. *Transforming Multilateral Diplomacy: The Inside Story of the Sustainable Development Goals*. Routledge.
Ki-Moon, Ban. 2015. "The 5Ps of the Sustainable Development Goals." Accessed February 2, 2023. www.unescwa.org/sites/default/files/inline-files/the_5ps_of_the_sustainable_development_goals.pdf.
Kim, Rakhyun E. 2016. "The Nexus between International Law and the Sustainable Development Goals." *RECIEL (Review of European, Comparative & International Environmental Law)* 25 (1): 15–26.
Klabbers, Jan. 2018. "7 Treaties and Their Preambles." In *Conceptual and Contextual Perspectives on the Modern Law of Treaties*, edited by Michael J. Bowman and Dino Krisiotis, pp. 172–200.
Krenak, Ailton. 2019. *O Insustentável Abraço Do Progresso Ou Era Uma Vez Uma Floresta No Rio Doce*. Edited by Ângela Domingues, Maria Leônia Chaves de Resende, and Pedro Cardim. 1st ed., vol. 1. Lisboa: Centro de História.
Kroll, Christian, Anne Warchold, and Prajal Pradhan. 2019. "Sustainable Development Goals (SDGs): Are We Successful in Turning Trade-Offs into Synergies?" *Palgrave Communications* 5 (1). https://doi.org/10.1057/s41599-019-0335-5.
Löfquist, Lars. 2020. "Is There a Universal Human Right to Electricity?" *International Journal of Human Rights* 24 (6): 711–723. https://doi.org/10.1080/13642987.2019.1671355.
Louka, Elli. 2006. *International Environmental Law: Fairness, Effectiveness, and World Order*. Cambridge: Cambridge University Press.
McCollum, David, Luis Gomes Echeverri, Keywan Riahi, and Simon Parkinson. 2017. "Ensure Access to Affordable, Reliable, Sustainable and Modern Energy for All." In *A Guide to SDG Interactions: From Science to Implementation*, vol. 1, pp. 127–169. International Council for Science.
Musso, Marta, and Guillemette Crouzet. 2019. "Energy Imperialism? Introduction to the Special Issue." *Journal of Energy History Revue d'Historie d l'Énergie*.
Nilsson, Måns, Paul Lucas, and Tetsuro Yoshida. 2013. "Towards an Integrated Framework for Sdgs: Ultimate and Enabling Goals for the Case of Energy." *Sustainability (Switzerland)* 5 (10): 4124–4151. https://doi.org/10.3390/su5104124.
Ostrom, Ellinor. 2010. "Beyond Markets and States: Polycentric Governance of Complex Economic Systems". *American Econonomy Review* 100: 641–672.
Paterson, Matthew. 1996. "IR Theory: Neorealism, Neoinstitutionalism and the Climate Change Convention." In *The Environment and International Relations*, edited by John Vogler and Mark F. Imber, vol. 1, pp. 59–76. New York: Routledge.
Roel Nieuwenkamp. 2017. "Ever Heard of SDG Washing? The Urgency of SDG Due Diligence | Development Matters," no. March: 2021. https://oecd-development-matters.org/2017/09/25/ever-heard-of-sdg-washing-the-urgency-of-sdg-due-diligence/.
Sands, Philippe. 2003. *The Environment and International Relations*. 2nd ed. Cambridge: Cambridge University Press.
Solomon, Margot E. 2013. "From Nieo to Now and the Unfinishable Story of Economic Justice." *International & Comparative Law Quarterly* 62 (1): 55–95.
Stafford-Smith, Mark, David Griggs, Owen Gaffney, Farooq Ullah, Belinda Reyers, Norichika Kanie, Bjorn Stigson, Paul Shrivastava, Melissa Leach, and Deborah O'Connell. 2017. "Integration: The Key to Implementing the Sustainable Development Goals." *Sustainability Science* 12 (6): 911–919. https://doi.org/10.1007/s11625-016-0383-3.
Third World Network. 2013. "G77: Key Components of a 'People-Centred Development Agenda'." www.twn.my/twnf/2013/3975.htm.
Tremblay, David, François Fortier, Jean François Boucher, Olivier Riffon, and Claude Villeneuve. 2020. "Sustainable Development Goal Interactions: An Analysis Based on the Five Pillars of the 2030 Agenda." *Sustainable Development* 28 (6): 1584–1596. https://doi.org/10.1002/sd.2107.
Trucco, Sara, Maria Chiara Demartini, and Valentina Beretta. 2021. "The Reporting of Sustainable Development Goals: Is the Integrated Approach the Missing Link?" *SN Business & Economics* 1 (2): 1–13. https://doi.org/10.1007/s43546-021-00046-9.
United Nations. 2000. "Millennium Declaration: General Assembly Resolution 55/2."
United Nations. 2015. "Transforming Our World: The 2030 Agenda for Sustainable Development (A/RES/70/1)."
United Nations. 2018. "Transforming Our World: The 2030 Agenda for Sustainable Development." *A New Era in Global Health*. https://doi.org/10.1891/9780826190123.ap02.
United Nations. 2020. "A/HRC/44/40: The Parlous State of Poverty Eradication—Report of the Special Rapporteur on Extreme Poverty and Human Rights."

United Nations. 2023. "Decade of Action, Ten Years to Change Our World." *Decades of Action: Ten Years to Transform Our World.*

Villavicencio Calzadilla, Paola, and Romain Mauger. 2018. "The UN's New Sustainable Development Agenda and Renewable Energy: The Challenge to Reach SDG7 While Achieving Energy Justice." *Journal of Energy and Natural Resources Law* 36 (2): 233–254. https://doi.org/10.1080/02646811.2017.1377951.

Vvsg. 2018. "Integrating the SDGS into Your Context Analysis: How to Start?" https://www.local2030.org/library/431/Integrating-the-SDGs-into-policy-planning-context-analysis.pdf

Walker, Gordon. 2015. "The Right to Energy: An Analysis of Meaning and Implications." doi:10.3917/eufor.378.0026

Williams, Marc. 1996. "International Political Economy and Global Environmental Change." In *The Environment and International Relations*, edited by John Vogler and Mark F. Imber, vol. 1, p. 41. New York: Routledge.

4

CHALLENGES AND THE WAY FORWARD IN A DIFFICULT DEVELOPMENT LANDSCAPE

Achieving a Sustainable Future in Africa and Its Sahel Region

Mohammad H. Sepahvand

4.1 Introduction

Africa's Sahel region is potentially one of the richest regions in the world, with abundant human and natural resources. However, it faces many complex and interconnected challenges. Decades of conflict and war in the region have eroded the national government's capacity to deal with crisis. They are less able to cope due to pervasive liquidity constraints and lack of effective countercyclical policy tools. This has transformed Sahel into one of the world's most neglected and conflict-ridden regions, with more than four million displaced people across the Sahel (UNHCR, 2023a).

With up to 80% of the population living on less than 2 dollars a day, poverty is more widespread in the Sahel than in most other parts of Africa (World Bank, 2020a). This, combined with one of the world's highest population growth rates—consisting of overwhelmingly young people (USAID, 2023)—makes the situation in Sahel highly unmanageable for its weak national and regional institutions.

The aforementioned challenges combined impose a significant impact on the Sahel region's ability to reach the targets set in the Sustainable Development Goals (SDGs), which will put the region on a dangerous path to harming future generations. The challenges are of such magnitude that, according to the World Bank, we can only accomplish the mission of ending extreme poverty in Africa by focusing and prioritizing the Sahel region (World Bank, 2020a).

The broad aim of this chapter is to discuss the main development patterns and challenges that lie ahead in reaching the targets set in the Sustainable Development Goals (SDGs), with a particular focus on SDGs 1 (No Poverty), 2 (Zero Hunger), and 5 (Gender Equality), and how they are impacted by economic and societal factors caused by crises and shocks, such as political instability, violent conflict, and the COVID-19 pandemic in the Sahel region.

I start by exploring Sahel-wide patterns and then focus the analysis on Burkina Faso, a least-developed and landlocked country particularly hard-hit by climate change, conflict, poverty, and food insecurity, and for which I have access to detailed data. While the former analysis offers a bird's-eye view within as well as across country borders, the latter provides in-depth knowledge on how different factors influence the achievement of the SDGs.

DOI: 10.4324/9781003285472-5

4.2 Impacts From Conflict and Climate Change

The occurrence of armed violence in the region is a major obstacle for achieving the Sustainable Development Goals (SDGs) by 2030, as it was for the Millennium Development Goals (MDGs) by 2015 (UN Chronicle, 2018).

The Central Sahel countries Burkina Faso, Mali, and Niger are experiencing a major humanitarian crisis due to the unstable security situation with armed violence and conflicts, which is limiting the access to basic social services. In 2023, more than 10 million people were in need of emergency assistance in the Central Sahelian countries due to the expansion of conflicts, food insecurity, and infectious diseases (IRC, 2023). A number which has more than doubled since 2020 (UNOCHA, 2021).

In Burkina Faso, where non-state armed actors, such as Islamists, and local and ethnic homegrown militia groups have established a foothold (Benedikter and Ouedraogo, 2017), putting the country at the epicenter of violent conflict in the Sahel (ACLED, 2022), an estimated 3.4 million people are severely food-insecure (UN, 2023), and nearly 2 million were displaced by the end of 2023 (NRC, 2023), projected to increase during 2024 (UNHCR, 2023b). The population displacement, in turn, fuels crisis, leading to reduced access to essential services and increased vulnerability (UNICEF, 2023).

The Sahel is also highly exposed to the impacts of climate change, with temperatures climbing at a rate 1.5 times faster on average than in other parts of the world (UN, 2021). Heat, drought, water shortages, and extreme weather events cause dwindling economic opportunities in the region, amplifying vulnerabilities (World Bank, 2023a). Climate impacts are threatening the habitability of entire areas. Millions of people within the region have either been forcibly displaced or migrated as a coping strategy. Governments in the region have limited capacity to respond to natural and human calamities due to a wide range of vulnerabilities that stem from political instability, weak governance, violent conflict, and high level of poverty (Vigil, 2017).

Future climate extremes are likely to lead to more armed conflicts (Von Uexkull and Buhaug, 2021). In Niger alone, up to 100,000 hectares of arable land are lost yearly to soil erosion and desertification (IRC, 2023), despite innovative community plantation programs (World Bank, 2020b). This loss in arable land is likely to increase competition for natural resources and intensify the challenges of a sustainable development in terms of reducing poverty and food insecurity.

4.3 The Way Forward

With less than a decade away from the United Nations Agenda 2030, reaching many of the targets of the Sustainable Development Goals (SDGs) is growing more difficult in Sahel.

According to several international actors, the year 2023 and 2024 are defining years: (1) marking the midpoint for the implementation of the United Nations Agenda 2030, (2) the evaluation of the first decade since the implementation of the African Union's Agenda 2063 (UNDP, 2023), and (3) the 2024 UN Summit of the Future to help advance the SDGs by 2030 (IISD, 2023). There is a unison consensus that the SDGs and all other targets are occurring in a challenging development landscape.

Because, even though these two agendas are complementary, with a common framework and objective to promote sustainable development, laying aside compatibility and measurability issues, they are being implemented in a landscape characterized by numerous large shocks and crises on several fronts – extreme weather events, the COVID-19 pandemic, and violent conflicts causing national, regional, and worldwide uncertainty – this brings slower economic growth and increased vulnerability, which is set to have severe consequences for reaching targets set in these agendas, unless the pressure and pace are drastically increased.

So what is the way forward? Is it possible to address these challenges and reach the SDGs in the midst of geopolitical uncertainty, or are the targets too utopian for the countries in the region to reach? One thing we know for sure based on previous empirical research is that achieving other SDGs in the region cannot be accomplished without a productive and sustainable agricultural sector (e.g., Nhemachena et al., 2018). At least 8 out of the 17 SDGs are, in one way or another, related to agriculture, signifying the multidimensional status of the sector (SDSN, 2015; FAO, 2015). The importance of the sector is also acknowledged through Aspiration 1 of the African Union's Agenda 2063, emphasizing the agricultural sector's role in economic and sustainable development through a modernized agriculture for increased production and productivity (AU, 2023).

However, before delving into the role of agriculture in the region as a possible "silver bullet" to reach the SDGs, let us start with discussing the absence of data in many Sahelian countries, which plays a fundamental role in allowing to identify priority areas in need of investment in order to achieve the Goals by 2030.

4.4 Absence of Data—Impediment to Measuring and Achieving the SDGs

The implementation, monitoring, and evaluation of regional and continental commitments in Sahel and Africa toward the SDGs are critical for the achievement of the targets at the national and local levels. Many countries in the region, such as Chad and Sierra Leone, have prioritized SDGs, their targets, and their indictors as part of their national development plans, which is also aligned to the African Union's Agenda 2063. Others have gone even further and included the SDGs as part of their national budget, such as Ghana, implementing it through a decentralized planning system that allows for stakeholder participation (IISD, 2019; HLPF, 2022).

However, to be able to implement and monitor the strategic areas of the Sustainable Development Goals, there is a need for statistics and a sustainable statistical system.

The official statistics in many sub-Saharan African and Sahelian countries have historically been limited and not reliable because of small sample size, reduced geographic variation, and lack of panel structure (Sepahvand, 2019). This is despite the fact that the region, in order to address areas that the Millennium Development Goals could not achieve and to implement the SDGs, has made improvements in the quantity and quality, as well as the use, of data to support policy, planning, and development processes (GGA, 2022).

One of the reasons for this lack of reliable data is the underdevelopment of national institutions, caused in part by unpredictable or volatile aid flows. Many donors have inconsistent requirements, and aid is often fragmented into many small projects. International institutions could also have the effect of crowding out or undermining local efforts to collect national statistics, or the donors and recipients of national official statistics have different goals (Sepahvand, 2019).

However, even if donors manage to reduce the volatility in aid flows, there could still be diminishing returns to foreign aid in the region, implying that aid can instead undermine development (McGillivray, 2016). There are several explanations as to why this might occur. One such is related to the fact that there are institutional constraints to using aid effectively, since high levels of aid place a significant administrative burden on recipients, with public sector officials facing excessive management and reporting requirements. This could be relevant for the Sahelian setting, with fragile states, where these capacities are especially weak (Caselli and Presbitero, 2021).

As a consequence, during voluntary national reviews (i.e., the VNRs) of the SDGs, which aim to facilitate the sharing of national experiences in order to accelerate SDG implementation, many countries in the region emphasize the need to reinforce national statistical capacity technically, but mainly financially (IISD, 2019). The focus on the financial part of it is related to the fact that there is a lack of development of statistical data because of limited data production, a lack of archiving, and the security policy of data.

So any efforts to improve the official national statistics need to focus on improving and increasing the capacity and ownership of the national statistical systems in the region. The argument is that the availability of official statistics is vital for the development of these Sahelian countries in monitoring economic and sustainable development. One such example is the Swedish financed technical assistance project in statistics in Burkina Faso, with the overall aim to strengthen the capacity of the national statistical system (for an overview, see Sepahvand, 2019).

A considerable part of this assistance was dedicated to the development of a large nationally representative household survey dataset with a panel structure to ensure a sustainable statistical production system. The main objective of this household survey was (and still is) to support and provide data to the overall majority of the indicators on the poverty level and other factors needed to monitor whether Burkina Faso is achieving the goals of the development initiatives, such as previously the MDGs, and now the SDGs.

However, the challenge in creating and implementing a survey to get access to reliable data in the region is the cost of the survey. Burkina Faso is no exception. For example, a national survey could cost somewhere around US$1.5–2 million and only cover a short period of data collection, which is a couple of weeks.

The reason that the cost of collecting data becomes an issue is that it is related to the core question when it comes to aid and the sustainability of development. If the costs are not low enough, the national statistical office will never be able to implement the survey itself and instead be dependent on donors and the international community. Given the existence of volatilities in aid flows and other aspects of why aid may not function well, it becomes difficult, if not impossible, to have a longer perspective in terms of data collection.

These limitations in the data collection process create institutions that are forced to reinvent the wheel for each survey, which has a devastating effect on the national statistical process and its ability to provide data needed to monitor the implementation of development programs, such as the SDGs.

One way to deal with this within the technical assistance project in statistics in Burkina Faso was to contribute to a sustainable statistical system. The idea was to implement a survey that would be continuous and implemented with a cost low enough so that the national statistical office could bear it.

Now that we have discussed the challenges imposed on reaching the SDGs given absence of data, let us delve into what role agriculture could have for a sustainable development and achieving the SDGs.

4.5 The Role of Agriculture in Reaching the SDGs

Growth in agricultural productivity has been identified as a key driver of poverty reduction and increased food security in sub-Saharan Africa, including the Sahel (e.g., Irz et al., 2001; Ligon and Sadoulet, 2008). The United Nations Agenda 2030 formalizes these targets as Sustainable Development Goals (SDGs) 1 (No Poverty) and 2 (Zero Hunger).

Across the region, agriculture has been considered the main sector in the economies of most countries, accounting for between 30 and 40% of gross domestic product, and over two-thirds of the workforce are within the sector (World Bank, 2013). This implies that when there is growth in agriculture, due to the sector's links with off-farm stages of the agri-food system and non-farm sectors, it leads to expanding employment in the rest of the economy (e.g., Jayne et al., 2020; Fuglie et al., 2020).

During the past two decades, the region has achieved the highest rate of growth in agricultural production compared to the rest of the world (World Bank, 2023b). However, most of this growth came from low levels and from the expansion of area under cultivation rather than higher crop yield, that is, productivity (Jayne and Sanchez, 2021), explaining why the rates of extreme poverty across the region are still very high (Castaneda et al., 2016; Our World in Data, 2022).

Agricultural growth simply through area expansion is not an effective and sustainable option in dealing with food insecurity in the region, as it affects biodiversity and adds to the destruction of natural vegetation (Jayne and Sanchez, 2021). The feasible and more cost-effective way would instead be through productivity improvements on existing farmland (e.g., van Ittersum et al., 2016; Sanchez, 2019). This implies that there remains great unmet potential for crop yield improvement.

4.5.1 Gender Differences in Agricultural Productivity

The Food and Agriculture Organization of the United Nations (FAO) estimates that female farmers can achieve 20 to 30% higher crop yields just by having the same access to productive resources as men. If female farmer productivity could be increased to the same level as that of males, the total agricultural output in developing countries could potentially grow by 2.5–4.0%, lifting 100 to 150 million people out of hunger (FAO, 2011).

Thus, implying that the alleviation of gender differences in agricultural productivity could be considered as one of the key drivers of agricultural growth in order to reduce poverty and increase food security, while simultaneously aiming for individuals' livelihoods and promoting gender equality (Quisumbing et al., 2014).

This also prominently figures in United Nations 2030 Agenda of Sustainable Development, as SDG 5 (Gender Equality), which has led to an increased attention to gender analysis in agricultural research in the past decade (Doss and Quisumbing, 2020).

These gender analyses have identified certain structural barriers preventing the reduction of gender differences in sub-Saharan Africa and the Sahel (Quisumbing, 1996; Udry, 1996). Barriers such as how women in Western Kenya (Moock, 1976) and Ghana (Doss, 2001) are having less access to agricultural extension services and credit. Or as in the case of Ethiopian, Nigerian, and Ugandan women cultivating on plots with less soil quality and access to water sources (Tiruneh et al., 2001; Nkedi-Kizza et al., 2002; Peterman et al., 2011). This could be due to legal norms that impose restrictions on females' property rights to have access to high-quality land or soil (Cotula, 2006). Cultural norms and intra-household bargaining could be other important reasons behind these differences in productivity between women and men as they influence crop choices (Doss, 2002) and available labor endowment for agriculture (Udry et al., 1995).

It has recently been argued that the gender gap disappears when women express a higher willingness to take risk (e.g., Averett et al., 2018). Consequently, it is argued that lower female productivity, or the inability of female farmers to respond to economic incentives, is due to their unwillingness to take higher risk. Therefore, through this line of reasoning, female farmers will experience a higher level of productivity if they take higher risk. If this will be the case, understanding the determinants of women's risk-taking and promoting riskier behavior can give a push to female productivity, hence increasing agricultural production and possibly reducing poverty and food insecurity.

Empirical findings from the region indicate that this seems not to be the case: Female farmers do not increase their productivity, even when they are willing to take more risks (Sepahvand, 2022). In other words, productivity is not higher among women who take higher risks. On the contrary, it is lower. Among the men, there are no clear differences between those who take more or less risk; instead, their productivity seems to increase somewhat with the level of risk. The differences depend, among other things, on the type of crop cultivated.[1] This indicates that cultural norms could play an important role for understanding the persistent disadvantage among female farmers as they influence crop choices (e.g., Doss, 2002).

However, one of the potential mechanisms that may drive the effect behind the gender gap is cooperation, or the lack of it, between female farmers (Sepahvand, 2022). Being part of farmer associations (which allows for cooperation) seems to make female farmers, on average, more productive. Further, gender differences seem to be lower in polygynous households compared to monogamous households, where cooperation between co-wives could be a way to manage demands on their time and labor.

4.5.2 Gender Equality, the Way Toward a Sustainable Growth in Agriculture

As already mentioned earlier, all literature dealing with gender differences in agricultural productivity indicates that gender is an important consideration to take into account when designing agricultural policy interventions, as the results from previous studies point that women are less productive than men.

Therefore, in order to increase total agricultural output in Sahelian countries, such as Burkina Faso, but also across sub-Saharan Africa, policies should support female farmers by focusing on strengthening local networks of female farmers and encouraging more opportunities to get access to credit, as it allows females to share experience, get support, and learn from each other, but also be able to adopt new technology, buy better fertilizers, and hire more labor. These, alongside what crops are grown, access to better soil quality, and improved seeds that can attain high crop yields, are important factors which could explain why female and male farmers have different agricultural production and define areas where there is a need to be gender-sensitized in policies that support growth in agricultural productivity.

Thus, gender, and by that SDG 5, should be viewed as a key target to take into account when designing policies that support growth in agricultural productivity, with the aim to deal with poverty and food insecurity.

4.6 New Challenges Within the Agricultural Sector

Estimates from the African Development Bank (AfDB) show that the challenges ahead are becoming more related to the availability and access to nutritious food and not only to feeding all the people in the region. According to the AfDB, over 200 million children in sub-Saharan Africa suffer from malnutrition, with it being the second leading cause of death among children, after malaria. The situation is especially challenging in the Sahel region (AfDB, 2023).

However, although there are differences among sub-regions, countries, and rural and urban areas in the progress toward the SDGs, national governments, together with development partners, are stepping up initiatives to increase Africans' capacity to combat malnutrition by implementing specific regional programs on food security and nutrition. These range from projects that will benefit people directly, such as school meal projects in Burkina Faso that aim to reinforce the Burkinabè government's capacity to ensure that every child in primary school has at least one balanced meal a day (AfDB, 2022), to programs that have a more indirect effect, such as those with the aim of mitigating the effects of climate change by increasing productivity from agrarian, pastoral, fisheries, and forestry value chains in Guinea, Niger, and Togo (AfDB, 2023).

On a regional level, the African Union's Agenda 2063 (complementary with and mutually reinforcing the United Nations Agenda 2030) has a unified view on transforming Africa's food system, mainly on resilience in the face of the growing vulnerability and shocks caused from conflicts and crisis, such as the current one between Russia and Ukraine, which are among the main exporters of cereal crops to the continent. The rapid expansion in the continent's agricultural productivity and incorporation of specific food crops into countries' agriculture value chain have been identified as key in preventing future disruptions in the supply chain for cereal crops.

Thus, relating it back to the alleviation of gender differences and a sustainable growth in agricultural productivity, understanding what factors influence differences in productivity could ease the identification of possible policy recommendations to reduce reliance on cereal imports from outside of Africa and promote intra-African trade and sustainable agribusiness sectors by an increase in agricultural productivity.

But despite these initiatives, governments and regional and international actors on the continent are faced with a race against time: The targets in the United Nations SDGs should be reached by 2030. So the question arises if these initiatives would be enough to reach many of the SDG targets in Africa and its Sahel region. In particular, those of utmost importance for the Sahel: ending poverty and hunger. Especially as their implementation is occurring in a challenging development landscape.

4.7 A Challenging Development Landscape

Agriculture is a risky activity, and farmers in the region operate under uncertain situations (Ellis, 1998; Di Falco and Chavas, 2009). Natural disasters and other external shocks could affect farm production, investment, and management decisions as they influence farmers in avoiding higher-payoff activities and investments (Alderman, 2008) or adopting new technology (Liu, 2013).

In advance of the UN Climate Change Conference in Dubai, United Arab Emirates, a report was launched in October 2023 as part of a global effort to measure progress toward the SDGs and a more sustainable future. This report provided an assessment on how disasters, defined as "serious disruptions of the functioning of society," have increased from 100 events yearly in the 1970s to 400 events per year in the past 20 years. It also assessed how disasters are affecting the agriculture sector and food security. Estimates from the report show that agricultural losses caused by disasters are relatively higher in Africa, and in particular in the Sahel region, than the rest of the world (FAO, 2023).

One of the main amplifiers of disasters in the region has been caused by climate change, armed conflict, and pandemics. This adds to the clear picture provided by all the voluntary national reviews (VNRs) in Africa during the United Nations 2022 High-Level Political Forum on Sustainable Development: Countries described how the effects of the COVID-19 pandemic and the onset of multiple conflicts have caused setbacks in progress toward the SDGs (HLPF, 2022).

4.8 COVID-19 and Civil Conflict, a Barrier for Reaching the SDGs

Meeting the targets of the Sustainable Development Goals was a great challenge for the countries in the region right from their adoption in 2015. However, since March of 2020, the implementation challenge has been made even more difficult by the COVID-19 pandemic. COVID-related effects associated with government policies to slow the spread of the virus through, for example, lockdowns resulted in negative or lowered economic growth, offsetting some of the gains made toward achieving the SDGs.

These so-called second-order effects that stem from human responses to the pandemic, rather than from the virus itself, have had a particularly negative effect on the possibilities to achieve SDG 1 (No Poverty) and SDG 2 (Zero Hunger). They have resulted in reduced government revenues, reduced household income, and increases in extreme poverty for many Sahelian countries. Already in 2021, one year after the start of the COVID-19 pandemic, the reductions in GDP growth ranged from −2.6 to −10.6% across the region (UNDP, 2021). In addition, a worldwide increase in inflation began in mid-2021, with many countries seeing their highest inflation rates in decades (World Bank, 2022).

In addition, the COVID-19 pandemic underscored the fragility and imposed new development challenges for the sub-Saharan African and Sahelian countries. For instance, the World Bank (2020c) noticed in the early months of the COVID-19 pandemic, with lockdowns and travel restrictions, a historical decline in remittances, implying increased insecurity and volatility in income. This resulted in an increased probability of recruitment of joining armed groups to compensate for the loss of livelihood. A high level of disease prevalence and loss in life expectancy could increase the risky behavior of individuals and lower the threshold of joining armed groups (Kustra, 2017).

A possible explanation to these negative patterns is linked to state capacity. States with strong institutions deal with the impacts of a pandemic, such as COVID-19, by extending their social security net (Sobek, 2010). However, in states with weak and underdeveloped institutions, like those in the Sahel, the pandemic exposed the limitations of state capacities during the lockdowns, with loss in much-needed revenue and aid (Ataguba, 2020), hence resulting in slowed economic growth.

Moreover, the COVID-19 pandemic, in the midst of political instability and violence, has amplified these negative COVID-19-related second-order effects in the fragile, conflict-affected Sahelian countries. Decades of conflicts and war in the region have eroded the national governments' public health capacity. For instance, in Burkina Faso, by mid 2023, over 500 local health facilities were forced to shut down or be used as temporary shelter for internally displaced people due to violence in the country, implying that over 2 million individuals had restricted access to health care (ECHO, 2023).

These described patterns suggest that the combination of the ongoing humanitarian and security crises with a health crisis imposed by the COVID-19 pandemic has a major impact on the stability of the region: Recent empirical findings show that poverty, conflict, and COVID-19 form a deadly triangle, implying that the fight against the spread of COVID-19 is thus entangled with the fight against poverty and violence (Sepahvand and Verwimp, 2023). This imposes significant barriers to achieving SDG 1 (No Poverty) and SDG 2 (Zero Hunger).

And finally, the impact of COVID-19 differs by gender in the region, which affects the possibilities to achieve SDG 5 (Gender Equality). It involves an extra burden for women (Bundervoet et al., 2022), including caring for family members, childcare, and homeschooling. These time-consuming activities could affect their work and, hence, is a barrier to growth in the much-needed agricultural productivity, as discussed earlier.

4.8.1 Low Vaccine Uptake, Still a Major Challenge

According to the World Health Organization (2021), preventive measures against the spread of COVID-19 are still the main tools for the national governments in the Sahel to deal with the pandemic, as the region has the lowest vaccine uptake in the world. By way of illustration, Burkina Faso has a vaccine rate of 3 million out of 13 million individuals (African CDC, December 28, 2023; WHO data on vaccination rate as of November 26, 2023).

Recent empirical findings offer several insights on the dynamics of violent conflicts and their effect on the adoption of preventive measures to prevent infection by COVID-19 (Sepahvand and Verwimp, 2023). Based on a unique nationally representative panel survey collected during the first six months of the pandemic in Burkina Faso, the results point to a higher infection rate and a lower rate of adoption of COVID-19 prevention measures as violence increases.

These findings are especially relevant from an Agenda 2030 perspective. According to experts from the WHO, the more time it takes to tackle the COVID-19 pandemic in the region, the higher the risk of mutations of the virus, with continued severe consequences for public health and for already-poor economies that already struggle with extreme poverty and access to food, medicine, and education, thus implying more barriers to reach the SDGs for sub-Saharan African and Sahelian countries.

4.9 Concluding Remarks: A Race Against Time

So reaching the end of this chapter, the question that is still pending is whether Africa and its Sahel region will be able to end poverty and hunger as part of the Sustainable Development Goals and if the international community can live up to the promise to "leave no one behind," especially as the world's most neglected crises are all in Africa (NRC, 2022).

Based on what has been described in this chapter, it is possible for the region to reach the goals (in whole or in part) if it is given more time. However, it is not only the region that needs more time but also the international community.

The all-absorbing war in Europe's Ukraine, with its effect on global energy, fertilizer, and food prices, has further exacerbated the situation of reaching SDGs 1 and 2. The diversion of funding, media attention, and international political initiatives has pushed the region's needs at the bottom of the international agenda instead of addressing the ongoing conflicts and human suffering on this continent.

Against this background, and given these difficulties, a likely scenario could be that countries in the region might not be able to achieve all the SDGs and instead will focus on a much smaller, more manageable set of targets, possibly those that overlap with the African Union's Agenda 2063, such as SDGs 1 (No Poverty) and 2 (Zero Hunger), through, as described earlier, a sustainable growth in agricultural productivity.

Therefore, the international community should, in the case of Africa, maintain and enhance a cooperative approach to the provision of global public goods (e.g., public health, education, and security) and support the region's most vulnerable climate- and conflict-affected states, those in the Sahel.

Acknowledgments

I gratefully acknowledge the National Institute of Statistics and Demographics (INSD, Institut National de la Statistique et de la Démographie) in Burkina Faso for collecting the data used in this chapter, and Statistics Sweden and the World Bank, for technical assistance. All remaining errors are my own.

Note

1 These findings are from a large national representative survey of farmers in Burkina Faso. The risk measurement is based on a self-reported risk attitude question similar to the German Socio-Economic Panel (SOEP), assessing the farmers' willingness to take risks in general.

References

ACLED. (2022). *10 Conflicts to Worry About in 2022: The Sahel*. Retrieved from https://acleddata.com/10-conflicts-to-worry-about-in-2022/sahel/. Accessed 2022-02-08.

AfDB. (2022). *Burkina Faso: A Food and Nutrition Project Has Been a 'Godsend' for Students at Village Schools*. Retrieved from www.afdb.org/en/success-stories/burkina-faso-food-and-nutrition-project-has-been-godsend-students-village-schools-52382. Accessed 2022-06-30.

AfDB. (2023). *Child Malnutrition: The African Development Bank Steps Up Action to Save Thousands of African Children at Risk of Dying Before Their Fifth Birthday*. Retrieved from www.afdb.org/en/news-and-events/child-malnutrition-african-development-bank-steps-action-save-thousands-african-children-risk-dying-their-fifth-birthday-62273. Accessed 2023-07-13.

Africa CDC COVID-19 Vaccine Dashboard. *Africa Centers for Disease Control and Prevention*. Retrieved from https://africacdc.org/covid-19-vaccination/. Accessed 2023-12-28.

Alderman, H. (2008). *Managing Risk to Increase Efficiency and Reduce Poverty*. World Development Report 2008. Washington, DC: World Bank.

Ataguba, J. E. (2020). COVID-19 pandemic, a war to be won: Understanding its economic implications for Africa. *Applied Health Economics and Health Policy*, 18(3), 325–328.

AU. (2023). *Agenda 2063: Our Aspirations for the Africa We Want*. Retrieved from https://au.int/en/agenda2063/aspirations. Accessed 2023-11-23.

Averett, S. L., Argys, L. M., & Hoffman, S. D. (Eds.). (2018). *The Oxford Handbook of Women and the Economy*. Oxford: Oxford University Press.

Benedikter, R., & Ouedraogo, I. (2017). Violent extremism takes root in Burkina Faso. *Foreign Affairs*, April 25.

Bundervoet, T., Dávalos, M. E., & Garcia, N. (2022). The short-term impacts of COVID-19 on households in developing countries: An overview based on a harmonized dataset of high-frequency surveys. *World Development*, 153, 105844.

Caselli, F. G., & Presbitero, A. F. (2021). Aid effectiveness in fragile states. In: Chami, R., Espinoza, R., & Montiel, P. J. (Eds.), *Macroeconomic Policy in Fragile States*, vol. 16, pp. 439–520. Oxford: Oxford University Press.

Castaneda, R., Doan, D., Newhouse, D. L., Nguyen, M., Uematsu, H., & Azevedo, J. P. (2016). *Who Are the Poor in the Developing World*? World Bank Policy Research Working Paper (7844).

Cotula, L. (2006). *Gender and Law: Women's Rights in Agriculture*. Food & Agriculture Org.

Di Falco, S., & Chavas, J. P. (2009). On crop biodiversity, risk exposure, and food security in the highlands of Ethiopia. *American Journal of Agricultural Economics*, 91(3), 599–611.

Doss, C. R. (2001). Designing agricultural technology for African women farmers: Lessons from 25 years of experience. *World Development*, 29(12), 2075–2092.

Doss, C. R. (2002). Men's crops? Women's crops? The gender patterns of cropping in Ghana. *World Development*, 30(11), 1987–2000.

Doss, C. R., & Quisumbing, A. R. (2020). Understanding rural household behavior: Beyond Boserup and Becker. *Agricultural Economics*, 51(1), 47–58.

ECHO. (2023). *European Civil Protection and Humanitarian Aid Operations: "He Who Has Health Has Hope": Emergency Medical Care for IDPs in Burkina Faso*. Retrieved from https://civil-protection-humanitarian-aid.ec.europa.eu/news-stories/stories/he-who-has-health-has-hope-emergency-medical-care-idps-burkina-faso_en. Accessed 2023-12-19.

Ellis, F. (1998). Household strategies and rural livelihood diversification. *The Journal of Development Studies*, 35(1), 1–38.

FAO. (2011). *The State of Food and Agriculture: Women in Agriculture—Closing the Gender Gap for Development*. Rome, Italy: FAO. Retrieved from www.fao.org/docrep/013/i2050e/i2050e.pdf. Accessed 2018-02-10.

FAO. (2015). *FAO and the 17 Sustainable Development Goals*. Rome, Italy: Food and Agriculture Organisation of the United Nations.

FAO. (2023). *The Impact of Disasters on Agriculture and Food Security 2023—Avoiding and Reducing Losses Through Investment in Resilience*. Rome. https://doi.org/10.4060/cc7900en

Fuglie, K., Gautam, M., Goyal, A., & Maloney, W. F. (2020). *Harvesting Prosperity: Technology and Productivity Growth in Agriculture*. Washington, DC: World Bank Group.

GGA. (2022). *Harnessing the Data Revolution for Sustainable Development in Africa*. Retrieved from https://gga.org/harnessing-the-data-revolution-for-sustainable-development-in-africa/. Accessed 2023-11-05.

HLPF. (2022). *United Nation's High-Level Political Forum 2022, Voluntary National Reviews: Synthesis of Main Messages*. Retrieved from https://hlpf.un.org/sites/default/files/inlinefiles/Synthesis%20of%20VNR%20main%20messages%204%20July.pdf. Accessed 2023-11-10.

IISD. (2019). *African Countries Stress Need for Data, Financing in VNR Main Messages*. Retrieved from https://sdg.iisd.org/news/african-countries-stress-need-for-data-financing-in-vnr-main-messages/. Accessed 2023-11-20.

IISD. (2023). *Accelerating the SDGs Through the 2024 Summit of the Future*. Retrieved from https://sdg.iisd.org/commentary/guest-articles/accelerating-the-sdgs-through-the-2024-summit-of-the-future/. Accessed 2023-10-03.

IRC. (2023). *The Central Sahel: How Conflict and Climate Change Drive Crisis*. Retrieved from www.rescue.org/article/central-sahel-how-conflict-and-climate-change-drive-crisis. Accessed 2023-10-27.

Irz, X., Lin, L., Thirtle, C., & Wiggins, S. (2001). Agricultural productivity growth and poverty alleviation. *Development Policy Review*, 19(4), 449–466.

Jayne, T. S., Fox, L., Fuglie, K., & Adelaja, S. (2020). Agricultural productivity growth, resilience, and economic transformation in sub-Saharan Africa: Implications for USAID. *Bureau for International Food and Agricultural Development*. Retrieved from www.usaid.gov/sites/default/files/documents/BIFAD_Agricultural_Productivity_Growth_Resilience_and_Economic_Transformation_in_SSA_Final_Report_4.20.21_2_2.pdf. Accessed 2023-11-27.

Jayne, T. S., & Sanchez, P. A. (2021). Agricultural productivity must improve in sub-Saharan Africa. *Science*, 372(6546), 1045–1047.

Kustra, T. (2017). HIV/AIDS, life expectancy, and the opportunity cost model of civil war. *Journal of Conflict Resolution*, 61(10), 2130–2157.

Ligon, E. A., & Sadoulet, E. (2008). Estimating the effects of aggregate agricultural growth on the distribution of expenditures.

Liu, Elaine M. (2013). Time to change what to sow: Risk preferences and technology adoption decisions of cotton farmers in China. *Review of Economics and Statistics*, 95(4), 1386–1403.

McGillivray, M. (2016). State capacity and aid effectiveness in weak states in the greater middle east. In: Kamrava, M. (Ed.), *Fragile Politics: Weak States in the Greater Middle East* (2016 online edition). Oxford Academic.

Moock, P. R. (1976). The efficiency of women as farm managers: Kenya. *American Journal of Agricultural Economics*, 58(5), 831–835.

Nhemachena, C., Matchaya, G., Nhemachena, C. R., Karuaihe, S., Muchara, B., & Nhlengethwa, S. (2018). Measuring baseline agriculture-related sustainable development goals index for Southern Africa. *Sustainability*, 10(3), 849.

Nkedi-Kizza, P., Aniku, J., Awuma, K., & Gladwin, C. H. (2002). Gender and soil fertility in Uganda: A comparison of soil fertility indicators for women and men's agricultural plots. *African Studies Quarterly*, 6(1), 27–43.

NRC. (2022). *The World's Ten Most Neglected Crises Are All in Africa*. Retrieved from www.nrc.no/news/2022/june/the-worlds-ten-most-neglected-crises-are-all-in-africa/. Accessed 2022-11-10.

NRC. (2023). *Burkina Faso Is the World's Most Neglected Crisis*. Retrieved from www.nrc.no/news/2023/june/burkina-faso-is-the-worlds-most-neglected-crisis/. Accessed 2023-01-06.

OurWorld in Data. (2022). *Increasing Agricultural Productivity Across Sub-Saharan Africa Is One of the Most Important Problems This Century*. Retrieved from https://ourworldindata.org/africa-yields-problem. Accessed 2023-06-10.

Peterman, A., Quisumbing, A., Behrman, J., & Nkonya, E. (2011). Understanding the complexities surrounding gender differences in agricultural productivity in Nigeria and Uganda. *Journal of Development Studies*, 47(10), 1482–1509.

Quisumbing, A. R. (1996). Male-female differences in agricultural productivity: Methodological issues and empirical evidence. *World Development*, 24(10), 1579–1595.

Quisumbing, A. R., Meinzen-Dick, R., Raney, T. L., Croppenstedt, A., Behrman, J. A., & Peterman, A. (2014). Gender in agriculture. *Springer*, 102072, 444.

Sanchez, P. A. (2019). *Properties and Management of Soils in the Tropics*. Cambridge University Press.

SDSN. (2015). Indicators and a monitoring framework for the sustainable development goals: Launching a data revolution for the sdgs. In *A Report to the Secretary-General of the United Nations by the Leadership Council of the Sustainable Development Solutions Network*. Paris, France and New York: Sustainable Development Solutions Network.

Sepahvand, M. H. (2019). *Essays on Risk Attitudes in Sub-Saharan Africa*. Economic Studies, ISSN 0283-7668; no. 178, Uppsala University, Department of Economics, p. 215.

Sepahvand, M. H. (2022). *Agricultural Productivity in Burkina Faso: The Role of Gender and Risk Attitudes*. Working Paper 2022:19. Department of Economics, Lund University.

Sepahvand, M. H., & Verwimp, P. (2023). Fighting Covid-19 amidst civil conflict: Micro-level evidence from Burkina Faso. *World Development*, 170, 106343.

Sobek, D. (2010). Masters of their domains: The role of state capacity in civil wars. *Journal of Peace Research*, 47(3), 267–271.

Tiruneh, A., Tesfaye, T., Mwangi, W., & Verkuijl, H. (2001). *Gender Differentials in Agricultural Production and Decision-Making Among Smallholders in Ada, Lume, and Gimbichu Woredas of the Central Highlands of Ethiopia*. Cimmyt.

Udry, C. (1996). Gender, agricultural production, and the theory of the household. *Journal of Political Economy*, 104(5), 1010–1046.

Udry, C., Hoddinott, J., Alderman, H., & Haddad, L. (1995). Gender differentials in farm productivity: Implications for household efficiency and agricultural policy. *Food Policy*, 20(5), 407–423.

UN. (2021). *Human Rights Climate Change and Migration in the Sahel*. Retrieved from www.ohchr.org/sites/default/files/2021-11/HR-climate-change-migration-Sahel.pdf. Accessed 2023-04-18.

UN Chronicle. (2018). *The Sustainable Development Goals and a Substantial Reduction in Illicit Arms Flows*. UN Chronicle, United Nations. Retrieved from www.un.org/en/chronicle/article/sustainable-development-goals-and-substantial-reduction-illicit-arms-flows. Accessed 2023-11-20.

UNDP. (2023). *Africa Sustainable Development Report: Accelerating the Recovery from the Coronavirus Disease (COVID-19) and the Full Implementation of the 2030 Agenda for Sustainable Development and African Union Agenda 2063 at All Levels*. AU/UNECA/AFDB/UNDP. Addis Ababa, Ethiopia.

UNHCR. (2023a). *Emergencies: Sahel Refugee Crisis*. Retrieved from www.unrefugees.org/emergencies/sahel-crisis/. Accessed 2023-11-20.

UNHCR. (2023b). *Operational Data Portal: Sahel Crisis*. Retrieved from https://data.unhcr.org/en/situations/sahelcrisis. Accessed 2023-11-30

UNICEF. (2023). *Burkina Faso: New Academic Year Starts with One Million Children Out of School Due to Ongoing Violence and Insecurity*. Retrieved from www.unicef.org/press-releases/burkina-faso-new-academic-year-starts-one-million-children-out-school-due-ongoing. Accessed 2023-10-15.

UNOCHA. (2021). *Global Humanitarian Overview 2021*. Retrieved from https://cerf.un.org/sites/default/files/resources/Session%202%20%20Global%20Humanitarian%20Overview%202021%20.pdf. Accessed 2022-01-27.

USAID. (2023). *Sahel Regional: Country Overview*. Retrieved from www.usaid.gov/sahel-regional. Accessed 2023-11-24.

Van Ittersum, M. K., Van Bussel, L. G., Wolf, J., Grassini, P., Van Wart, J., Guilpart, N., . . . & Cassman, K. G. (2016). Can sub-Saharan Africa feed itself? *Proceedings of the National Academy of Sciences*, 113(52), 14964–14969.

Vigil, S. (2017). Climate change and migration: Insights from the sahel. In Carbone, G. (Ed.), *Out of Africa: Why People Migrate*. Milano: LediPublishing.

Von Uexkull, N., & Buhaug, H. (2021). Security implications of climate change: A decade of scientific progress. *Journal of Peace Research*, 58(1), 3–17.

WHO. (2021). *Opening Statement, End-of-Year Press Conference*, December 14.

WHO. (2023). *WHO Data on Vaccination Rate: Africa COVID-19 Dashboard*. Retrieved from https://data.who.int/dashboards/covid19/vaccines?n=c. Accessed 2023-11-26.

World Bank. (2013). *Unlocking Africa's Agricultural Potential*. Retrieved from https://openknowledge.worldbank.org/server/api/core/bitstreams/3cbc1ec4-8400-5971-8b0b-6ada9679cec5/content. Accessed 2014-02-25.

World Bank. (2020a). *The World Bank Can Only Accomplish Its Mission of Ending Extreme Poverty in Africa by Prioritizing the Sahel Region*. Retrieved from www.worldbank.org/en/news/opinion/2020/12/16/

the-world-bank-can-only-accomplish-its-mission-of-ending-extreme-poverty-in-africa-by-prioritizing-the-sahel-region. Accessed 2021-02-21.

World Bank. (2020b). *In Niger, a 'Crazy Idea' to Restore Degraded Land Pays Off*. Retrieved from https://blogs.worldbank.org/nasikiliza/niger-crazy-idea-restore-degraded-land-pays.Accessed 2021-02-21.

World Bank. (2020c). *World Bank Predicts Sharpest Decline of Remittances in Recent History*. Retrieved from www.worldbank.org/en/news/press-release/2020/04/22/world-bank-predicts-sharpest-decline-of-remittances-in-recent-history. Accessed 2020-10-06.

World Bank. (2022). *Inflation, Consumer Prices (Annual %): 1960–2022*. Retrieved from https://data.worldbank.org/indicator/FP.CPI.TOTL.ZG. Accessed 2023-12-03.

World Bank. (2023a). *Results That Change Lives in the Sahel*. Retrieved from www.worldbank.org/en/country/mali/results-sahel. Accessed 2023-12-01.

World Bank. (2023b). *World Development Indicators: Sub-Saharan Africa*. Retrieved from https://data.worldbank.org/region/sub-saharan-africa. Accessed 2023-10-03.

5

SUSTAINABILITY AS A DEFINING MOMENT FOR CHANGING (M)EATING BEHAVIOR?

Comparing Eastern and Western Food Choices, Moral Evaluations, and Harmonization Strategies in Consumption and Gaps in SDG 12

Franzisca Weder, Stella Lemke, and Thomas Fenzl

5.1 Introduction

Communication is fundamental for sustainable development (Robertson, 2018; Weder et al., 2021; Newig et al., 2013), established as institutionalized framework for corporate and organizational action (UN, 2021); 17 Sustainable Development Goals define transformation processes mainly in those areas that are affected by the climate crisis or are related to an imbalanced human–nature relationship, a market-driven economy, and related resource exploitation. However, from a terminological and philosophical perspective, sustainability is still highly complex and offers many ways of interpretation, and therefore many ways for being used and (ab)used in corporate and political communication, by framing organizational and institutional behavior as future-oriented, transformative, and innovative (Weder, 2021; Baker & Weder, 2021).

Furthermore, from an individual perspective, the term "sustainability" itself brings in a certain degree of normativity, and with that morality, in everyday life choices regarding transportation and mobility, retail, or food and nutrition. From a theoretical as well as empirical perspective, we are interested in the *degree of morality in sustainability communication* in general and in food-related consumption practices, as well as the specific choices that are made and their interpretations, in particular. This is an important puzzle piece to further elaborate our overarching research interest in how far sustainability is already *normalized* in the sense of being an established, "new" societal norm and not only "label" for certain behavior or on certain products. The study at hand, and the theoretical concepts presented, will also help critically reflect on the SDGs as normative framework built on sustainability as principle of action, however a framework which is perceived as rather inconsistent and hard to implement and which still needs more detailed elaboration regarding its measurability (Weder, 2022).

DOI: 10.4324/9781003285472-6

Based on the existing literature in the areas of sustainable consumption (Fischer et al., 2021; Cohen, 2005; Béné et al., 2019) and sustainability communication (Weder et al., 2019), our assumption is that, still, sustainability is mostly used as an "empty" term, "feel-good" label, or "trend" (Shrivastava et al., 2020), with the tendency to overmoralization. To explore this assumption, we work with an innovative framework for capturing the *degree of morality in food-related conversations* to trace back moralization and demoralization in individual evaluations of eating behavior and food choices.

Demoralization and moralization, which potentially lead to moral convictions and ideologization (Nezlek & Forestell, 2020; Skitka et al., 2021), are discussed in social psychology. We define *morality* from an ethical perspective as different to norms, which are principles of action. Morality encompasses the principles concerning the distinction between right and wrong or good and bad behavior. Therefore, morality and the process of moralization/demoralization, as it will be talked about in this chapter, refer to a particular set of values and principles of conduct—and sustainability is one of those principles of conduct or behavior (Weder & Eriksen, 2023). In this chapter, we want to particularly explore the *defining communicative moments* which influence the processes of (de)moralization—again in relation to sustainability as normative principle of action; this is why we bring in communication theories and learnings from existing studies in environmental communication (Kristiansen et al., 2021; Adams, 2021; Moore & Yang, 2020; Urbanovich & Bevan, 2020; Buttlar et al., 2021; Stea & Pickering, 2019), and particularly food communication (Larson & Story, 2009; Vandenbroele et al., 2020; Azzurra et al., 2019; Graham & Abrahamse, 2017; Sanchez-Sabate & Sabate, 2019), talking about morality in decision-making processes. With this communication perspective on (de)moralization, we can reconceptualize and study those defining moments that change food choices and eating behavior and answer the question of how (far) sustainability as a normative principle overmoralizes food-related decision-making and behavior.

In this research chapter, we draw on a series of explorative, guideline-based interviews, conducted in three different socio-cultural and political settings (2019–2020: Austria, $n = 25$; Australia, $n = 5$; Indonesia, $n = 5$), and the overarching interview questions addressed the following dimensions: The focus of the interviews was on exploring the narratives around certain food choices, including reasoning for sustainable food choices and on identifying the influences (encounters, communicative moments). We analyzed the semi-structured food-related conversations applying inductive category formation (Mayring & Fenzl, 2019; Mayring, 2014) using the open-access analysis web tool QCAmap (Fenzl & Mayring, 2017; Mayring & Fenzl, 2014).

The findings show that sustainability is not (yet) a moral imperative or defining communicative moment for reflections on and changes of eating behavior and food choices (from meat to no meat, for example). People rather tend to demoralize (harmonize) a certain fact (animal cruelty, CO_2 emissions of a product, plastic packaging, long transportation) or behavior (buying avocados or beef) because "sustainability" seems too complex, intangible, and "blurry" to actually work as a principle of action or norm. Instead, a certain behavior of food choice is mostly influenced by interpersonal encounters and related to life-changing events, like moving out or having a new partner. Information in the media and other mediated encounters (influencer, documentaries, etc.) are not as influential as assumed based on existing studies and an environmental communication perspective.

These findings from the exploratory pilot study and the apparent lack of a deeper discursive construction of sustainability as new norm and normality in mediated public discourses need to be further explored by larger comparative and rather quantitative studies, which will be part of the recommendations at the end of the discussion of this chapter, including comments on the limitations of the study.

5.2 External and Internal Defining Communicative Moments for (Me)ating Behavior

Food choices cannot be viewed isolated from global developments. For example, the second goal of the UN Sustainable Development Goals (#2) is to end hunger, achieve food security, and improve nutrition. This goal is related to goal #12, which defines *sustainable consumption* as "responsible," meaning, that buyers and citizen consumers must make their choices taking into account environmental impacts at all stages of the product life cycle (UN, 2022).

The cultural and social roles of food must always be seen in the context of access, financial and informational resources, as well as social condition. Marsden and Morley (2014) argue that, rather than thinking in terms of concrete definitions of what is and is not sustainable, understanding which elements of social regulation are involved as *external moment* of change in conditioning developments in an unsustainable way may enable us to condition a more sustainable future. This means that the movement toward greater sustainability is only possible with the joint effort of consumers, producers, and regulators; however, the role and impact of the consumer are vital for a positive change (Vermeir & Verbeke, 2006). And indeed, literature shows that increasing information in the media, and particularly in the social media sphere, influences individual habits (Bellotti & Panzone, 2016) and more sustainable consumption (Fischer et al., 2021). More recently, Moran et al. (2020) estimated that through the realistic implementation of already-known changes in consumer behavior, the European Union (EU) could reduce its carbon footprint by approximately 25%. Although there is much evidence to suggest that dietary changes to reduce meat consumption are a key factor for tackling climate change (Gerber et al., 2013; Machovina et al., 2015; Clune et al., 2017), some still argue in favor of expanding animal farming to feed the population. Studies on consumer willingness to cut down on carbon-intensive food products, such as meat, showed that climate-sensitive labeling and communication were effective for and influenced the food-related decision-making processes (Beattie & McGuire, 2020; Thøgersen, 2021), followed by organizations increasingly displaying more information on sustainability as a core value in their production and processes (Wognum et al., 2011).

This can also be supported by studies with a stronger marketing focus which discuss the effects of labeling a certain product (avocado) or eating behavior (organic) as more or less sustainable (Swenson & Olsen, 2018; Kronthal-Sacco et al., 2020; Asioli et al., 2020). Next to labels, reports, and mediated information, critical issue representations, like documentaries, on mass meat production, animal cruelty, methane/CO_2 problems, climate change, etc., are *external defining moments of change*, which are considered as being influential for individual decisions as well (Gottwald & Weder, 2021). Current environmental communication studies also state that media play a strong role in defining more- or less-sustainable food choices (Kristiansen et al., 2021; Moore & Yang, 2020). Assuming that new norms are negotiated specifically in the media and related public discourses (Fairclough, 1992; Krzyżanowski, 2020) and media are used in order to create a common understanding about societal values on sustainability, as well as to set some concrete objectives (Newig et al., 2013), those external moments potentially create conflicts and cognitive dissonances for individuals, which will be further explained later.

However, if sustainability and its related issues (energy, natural hazards, food security, social justice, environmental care, etc.) are typically characterized by a high level of complexity and uncertainty, much more tangible than information presented in and received through the media are interpersonal conversations within and stories from our direct social networks (family and friends). Previous studies show that influences about food choices cannot be divorced from media communication on the one hand and activist networks people participate in, and their family and friends, on the other (Stephens Griffin, 2017; Rodan & Mummery, 2016; Buddle et al., 2018).

As outlined previously, interpersonal communication and, rather, *internal defining moments*, like emotions, beliefs, and values initiating and perpetuating communication, play a major role in sense- and meaning-making processes (Weick, 1987), where values and norms are "restoried" (Weder et al., 2019). Individual perspectives, standpoints, and orientations around eating behavior and food choices are those *internal moments* which are (communicatively) exchanged on an inter- and intrapersonal level (Fischer et al., 2021): Eating habits, traditions, and behaviors which have been internalized through education or childhood experiences are, thus, shaped by social and cultural contexts (Bourdieu, 1984; Gallimore & Lopez, 2002). Individuals internalize and enact social structures and cultures in which they grew up and had lived. Thus, eating habits and traditions may develop from rituals for cultural or religious celebrations or as a daily routine that offers structure and comfort through predictability. Additionally, personal food choices are further shaped and internalized throughout our childhood and adolescence, and they provide a sense of identity (Sobal & Bisogni, 2009; Jastran et al., 2009). So family and social networks are key social environments and settings for the development of eating patterns and food preferences (Campbell & Crawford, 2001; Delormier et al., 2009), which may even influence people's health (Christakis & Fowler, 2007). On the other hand, health awareness and health issues may crucially influence eating habits and behavior (Fox & Ward, 2008; Hoek et al., 2004; Hoffman et al., 2013).

Change and a transition or shift in a person's life course and, thus, in the individual social situation can influence changes in food choices (Wethington, 2005), for example, changing residence due to attending university, or a new partnership. This turning point could lead to having to cook for oneself for the first time, or to a change in eating routines.

Eating as social practice also comes with certain demands; it can be a constant battle between personal desire and social acceptance. Therefore, specific food choices—consciously or subconsciously—can vary according to our social surroundings and expectations (Feunekes et al., 1998; Jastran et al., 2009), and they can be led by social norms (Farrow et al., 2017; Salmivaara et al., 2021). The most prominent example to discuss the internal defining communicative moments is veganism; here, the most commonly stated personal motivation to adopt and follow a vegan diet is a concern about the ethics of raising and slaughtering non-human animals (Fox & Ward, 2008; Hussar & Harris, 2009; Ruby, 2012), next to personal health concerns (Hoffman et al., 2013).

Moral and normative contemplations and values as motivators for environmentally significant behavior are being discussed in different contexts (Bieling et al., 2020). Meat consumption leads to negative affective states for meat-eaters if they are confronted with a—dissonant—view of themselves that is unfavorable (Bastian & Loughnan, 2017; Aquino & Reed, 2002).

If we want to identify the defining communicative moments of change mainly influencing the shift from a meat to a plant-based diet, we must consider dissonances and morality. Thus, we assume that external and internal defining communicative moments can influence either an increased moralization of eating behavior or harmonization strategies and, therefore, a demoralization of (m)eating behavior, which is captured in Figure 5.1 and further elaborated in the following.

5.3 Scenarios of Moralization and Demoralization of (M)eating Behavior

Seeking to understand *moralization* and *demoralization* and what the communicative moments are that influence moralization or demoralization of individual eating behavior and food-related decision-making processes, we complement the following two concepts: the psychological framework, to understand moral convictions (Skitka et al., 2021; Haidt, 2013) that describe moral judgments as accompanied by strong emotions (Skitka et al., 2005), and strong links between anger and disgust (Clifford, 2019; Wisneski & Skitka, 2017). This is complemented by the existing SDH (*sustainability dissonance harmonization*) framework introduced by Weder et al. (2019), an innovative model that describes coping strategies with cognitive and moral dissonance.

The SDH model is derived from Festinger's (1962) cognitive dissonance model, where two conflicting ideas must be harmonized and belief adjustment and behavioral justification are described as coping strategies. Adding Lowell's (2012) moral dissonance model, the SDH model covers unsustainable actions which would harm others in the long run and would be therefore unsustainable and immoral. Finally, Gardiner's (2013, pp. 301–321) descriptions of *justification* and *demoralization* were added from his "moral corruption model." Thus, *sustainability dissonance harmonization* (SDG) suggests that when it comes to sustainability-related internal dissonance, which would occur by doing unsustainable actions, individuals will try to harmonize their conflicting thoughts by justifying their immoral (unsustainable) behavior using logic which is inclined toward convenience in their lifestyle, further described as "demoralization," as a way where individuals try to deconstruct sustainability to make it applicable in their behavior or cope with dissonance appearing with unsustainable behavior. Moralization, possibly leading to moral conviction, and demoralization, captured as "harmonization" strategies, are processes that are, from our point of view, stimulated and perpetuated by *defining communicative moments* on an individual (internal) and discursive media level (external).

This feeds into *four types* or variations of sustainability influencing (m)eating behavior that we seek to identify in the interview data (see Table 5.1).

Table 5.1 Typology of defining communicative moments driving (de)moralization and changes of eating behavior

	External defining moments (documentary, media report, SDG, fair trade label)	*Internal defining moments (life change/moving out, tradition, new flatmate, conversation)*
Demoralization/sustainability dissonance harmonization	Type 1	Type 2
Moralization, ideologization	Type 3	Type 4

The four types are differentiated related to internal and external defining moments and how these moments stimulate and direct demoralization and moralization.

- The first scenario would be that sustainability is already a social norm, established in public discourses. If individuals refer to external defining moments changing and directing their food choices and eating habits, then they could either feel *supported*, through a documentary on animal cruelty (moralization, moral conviction) (type 3), or *disrupted*, which would lead to dissonances and related harmonization strategies (type 1). A low consciousness about internal communicative moments would support this. This also implies that individuals use mediated communication to accommodate their own food choices and eating behavior, by "linking" it to sustainability. Examples would be a product label ("fair" or "organic"), doing something "trendy," or the feeling of being part of a trend (following a sustainable fashion influencer on social media).
- The second scenario would be that sustainability is not a social norm (yet) but much more a belief or possible ideology, employed to *evaluate* something as "bad" or "good" (veganism vs. meat-eating). This relates to the thinking that internal defining moments of change are related to values lying in the specific worth of objects, their quality, or their behaviors (Leiserowitz, 2007); they have the potential to invoke strong feelings and, as mentioned, are used as standard to judge (meat-eating) behavior or the quality of an object (beef). An either strong moralization and confirmation of already-established individual eating habits and food choices (type 4) or reflections on and "harmonization" of conflictual communicative moments (*experiencing* animal cruelty, natural hazards, conversations, and individual encounters, *living* traditions, "family rules," etc.) (type 2) would support this scenario.

With the dataset of internationally comparative interviews, we will further explore this typology.

5.4 Methodology

With the qualitative empirical study at hand, we seek to explore respondents' views and thus get a deeper understanding of the degree of morality that sustainability brings along as principle, and how far sustainability is already "normalized" and acts as a moral compass, influencing eating behavior and mainly meat or plant-based food choices. Research questions that guided the empirical research presented in this chapter are if sustainability is

1. a norm that is referred to in moral evaluations and justifications of individual food choices and eating behavior, or even
2. a defining communicative moment for change in eating behavior, with a specific focus on the shift to a plant-based diet—which would be an indicator for a moral evaluation and a relatively high degree of morality—or
3. a "label" and buzzword that is not referred to in communicative encounters and does not influence individual behavior.

In three phases, we conducted guideline-based interviews in, firstly, face-to-face settings (2019), then in telephone interviews and online sessions, depending on the then-current COVID-19 restrictions (2020). The sample consisted of individuals aged between 20 and 40 years in Austria (n = 25) (2019), Australia (n = 5), and Indonesia (n = 5) (2020), where the sample from Austria included different nationalities. The overarching interview questions addressed the following dimensions: (1) beliefs and values about food and nutrition,

(2) influences on individuals' food habits (internal/external), (3) central factors making people change their eating behavior (internal/external), and (4) attitudes toward meat consumption and the connection to environmental issues, climate change, and sustainability. Following several introductory questions regarding the meaning of food in the respondents' lives and their typical dieting, the following specific questions were asked in the interviews, addressing the aforementioned topics:

- When thinking about your personal choice of food, what are central factors that you take into consideration when choosing food?
- When you go shopping for groceries, how do you choose the foods you purchase?
- When you are "out and about," how do you choose what to eat?
- If you think back to your eating habits a few years ago, how have they changed over the time?
- What do you think about food trends?
- When thinking about meat as part of your regular diet, what are your thoughts?
- What influences your choice to eat/not eat meat?
- What could be arguments that consuming meat is a good/bad thing?
- If you think back to your meat consumption a few years ago, how has it changed over the time?
- Food choices and climate change are being linked in current public debates. What are your thoughts in this respect?
- Which factors would/could influence society/others to change food choices?
- Which factors would/could influence you to change your food choices?

The recordings of the interviews were literally transcribed (Mayring, 2002) and analyzed using qualitative content analysis (Mayring, 2014; Mayring & Fenzl, 2019). Based on the data at hand, the content analytical technique of inductive category building (multiple coding of categories counted) was applied. Text analysis was performed with the online tool www.QCAmap.org (Fenzl & Mayring, 2019; Mayring & Fenzl, 2014). The result of the analysis is the category system with the inductively developed categories and the corresponding frequencies.

5.5 Findings

Participants in this study provided rich data on eating habits, influences on, and reasons for changes in food choices. Furthermore, information on the role of morality and communication in the context of changing one's eating behavior could be analyzed. The following sections report the category systems and exemplifications of the found inductive categories based on the statements of the interview partners as the key findings on the different topics addresses in the interviews.

5.5.1 Beliefs About and Influences on Individual Food Choices

When looking at the portrayal of their associations with food and the main influences on their individual habits, participants described (1) social factors, (2) emotional associations, and (3) health-related topics as recurring topics. The participants attributed social factors as the greatest influence on their individual food habits (25%) and as one of the main functions/roles of food (25%). Food consumption was closely linked to being *part of the family connection* (40%), and the respondents stated that the specific choice is primarily *influenced by family and societal/cultural views* (76%; 44%). For example, participants stated that whole family gatherings were organized around food and shaped their personal eating routines.

> "My parents used to cook meals when I was young and we gathered together and had like dinner on Sundays maybe, or and on Saturdays as well. So, it was like a joint time spent together with each other." (BELIEFS)
>
> "[F]ood gathers my family. . . . [W]e usually have dinner together." (BELIEFS)

It can be seen as a "social glue" connecting people, supporting cultural representation, social identity, and belonging. One participant explained food as "essential when [they] celebrate something." "I learnt that food plays [an] important role in making connection with the neighbors. We exchange dishes, and that makes us feel closer to each other, even though we rarely meet." This brings in sustainability as core value and norm for social interaction, without explicitly labeling it.

Overall, food choices have increasingly become a domain within which people express their ideals and identities. Participants described the role emotional associations had on their food choices (36%), and also the role the *quality of food* played (12%) when choosing it. Again, communication within the family was mentioned as a central factor when learning about the value of food. One interviewee explained:

> "[O]ur parents have always taught us the meaning of good quality food so they have always paid attention and they always taught us to pay attention to the quality of food we eat."

Table 5.2 Beliefs and values about food, found in 25 guideline-based interviews (inductive category system)

Category name	*Abs. frequency*	*Rel. frequency*
Social function		
Food is part of the family connection.	10%	40%
Food can be a celebration.	2%	8%
Existential/health function		
Food is necessary for survival.	9%	36%
Food as facilitator for good health.	4%	16%
Quality of food is essential.	3%	12%
Food is a chore.	3%	12%
Balance in consumption of food.	2%	8%
Emotional function		
Positive emotional connection toward food.	3%	12%
Food is enjoyable.	3%	12%
Food/eating is appreciated and loved.	2%	8%
Other		
Food is nothing special.	2%	8%

Table 5.3 Influences on individual food habits, found in 25 guideline-based interviews (inductive category system)

Category name	*Abs. frequency*	*Rel. frequency*
Social system		
Family impacts food choices.	19%	76%
Cultural/societal views have impacts.	11%	44%
Important others impact personal choices.	9%	36
Traditional occasions.	6%	24%
Advice from competent people.	2%	8%
Experience and knowledge		
Health and nutrients of food.	18%	72%
Source/quality of food.	9%	36%
Experiences from different countries.	2%	8%
Previous experience.	2%	8%
Media		
Internet/(social) media has an impact.	14%	56%
TV shows impact choices.	6%	24%
Morals and values		
Body awareness.	10%	40%
Religious associations.	4%	16%
Emotion		
Feelings and emotions impact choice.	9%	36%
Food preferences.	7%	28%
Food has to be tasty/delicious.	4%	16%
Resources and skills		
Financial ability.	9%	36%
Cost and pricing.	6%	24%
Cooking own food.	5%	20%
Time and effort for preparation.	4%	16%
Availability of food.	4%	16%
Place of living.	4%	16%

5.5.1.1 How Do Eating Habits Change?

When considering the central influences on and defining moments of changing one's eating behavior, (1) information, (2) health concerns, (3) external changes, and (4) internal changes were mentioned as dominant motives. The role of (social) media (56%) as well as television content and advertisement (24%) as sources of information was listed as the main stimuli for behavior change. An Australian interviewee described this influence as follows:

> "Yes, I think that media and people following other people [on social media] definitely have a big influence on the change of lifestyles for people." (CHANGE)

Also, in the area of health, the aspects of body consciousness (32%), thoughts on nutrition (32%), and age (16%) had an impact on changing eating habits. Participants indicated that a deeper awareness and engagement with the effects different foods could have on their own bodies had an impact on their choices, especially with advancing age and experience:

> "[I]t wasn't that important when I was a kid, but it kept getting more important when I was a teenager. When I started to really understand what food does to your body. And in the last few years, it really changed my way of thinking." (CHANGE)

Thus, it can be seen that external influences shaped eating habits in many ways. A change of the living environment (36%) and the personal network (36%) had an effect on individual eating behaviors:

> "I moved from my hometown to another city for studying. It was [a] turning point of my life, and my preferences were changed. . . . My social circle has changed, and I live among people who really care about their health." (CHANGE)

For some participants, external influences also contributed to changing their meat consumption. Friends and family (32%) and thoughts on the environmental aspects of meat production (24%) were reported as factors. The analysis of beliefs about and influences on eating habits is crucial when examining the relationships between food choices and motivations to change eating behavior. Also, taking into consideration the connection and impact of one's food choices on the environment. Particularly in relation to meat consumption. An Australian participant highlighted the connection between environmental concerns and food choices:

> "There's the mental element, too, as I am deeply aware of the environmental impacts of intensive meat production here in Australia." (CHANGE MEAT)

Table 5.4 Factors influencing a change in eating habits, found in 25 guideline-based interviews (inductive category system)

Category name	*Abs. frequency*	*Rel. frequency*
Information		
(Social) media/internet	14%	56%
Television content/advertisement	6%	24%
(Knowledge about) environmental aspects	5%	20%
Research/educational sources	5%	20%
Introduction to (meat-free) alternatives	3%	12%
Health		
Body consciousness	8%	32%
Nutrition/health	8%	32%
Age	4%	16%
External changes		
Influence by personal network	9%	36%
Change of living environment	9%	36%
Cultural influence	3%	12%
Constitutional changes/ban	3%	12%
Internal changes		
Financial resources	6%	24%
Ethical problems with meat production	4%	16%
Change in teste	2%	8%

5.5.1.2 Moralization of Meat Consumption

Looking at the (potential) change in meat consumption, the influences started to change predominantly toward aspects that were communicated by the personal network (internal defining moments, however type 2/strong harmonization). (1) External influences, (2) accessibility, and (3) internal influences were mentioned as motives for changing one's meat consumption, but the role of and conversations with family and friends (32%) were mentioned as main stimuli for a behavior change. The need for harmonization and demoralization became obvious when interviewees framed the influence as "bad" influence, making it more difficult to stick to one's new habit. Three participants explained pointedly the difference:

> "I haven't eaten meat since I came to university. . . . I went into the meat section of the supermarket the other day. I couldn't handle it. So I don't think I could eat any meat again. . . . Yeah. As long as I'm away from my family. . . . Being with my family, and my mom next to me, when we're eating, she would give me some of her chicken or something to eat." (DISRUPTIVE)
>
> "When I became vegetarian in my early 20s, my family [who were chicken farmers] didn't think much of it. I had to argue with my brothers a lot about the [in]efficiencies of meat production in Western Europe." (DISRUPTIVE)
>
> "I am really interested in becoming vegan. However, I found the options are very limited, and even my family and friends hardly support me on this. They would think that I am too much of an 'environmental maniac.'" (DISRUPTIVE)

It needs to be pointed out that even if individual beliefs, conversations, and considerations can be seen as defining communicative moments, certain context factors are considered as well: Especially the role of pricing dictates choice and is seen as a crucial factor to influence the amount of meat consumption.

> "The factor of money, the price of this kind of vegan alternatives, plays a big role." (PRICE)
>
> "I think, for a lot of people, it's always the price, because, um, many people think it is less, um, less expensive to just buy meat, um, and they also think that veggies and fruit are expensive." (PRICE)

However, the individual reflections and value-related considerations were the dominant communicative moments that we identified in the interview data:

> "I think we should educate ourselves more about the results and circumstances of our meat consumption. . . . We should experience the troubles, so we can actually learn and understand."

Table 5.5 Factors influencing a change in meat-eating habits, found in 25 guideline-based interviews (inductive category system)

Category name	*Abs. frequency*	*Rel. frequency*
External influences		
Impact of family and friends on meat consumption	8%	32%
Impact of environmental factors on consuming meat	6%	24%
Impact of ethical factors on consuming meat	4%	16%

(*Continued*)

Table 5.5 (Continued)

Category name	*Abs. frequency*	*Rel. frequency*
Early introduction of vegetarian meal options	4%	16%
Amendments in laws and policies	4%	16%
Representation in (social) media	3%	12%
Sustainability concerns of meat consumption	2%	8%
Accessibility		
Pricing/cost	8%	32%
Alternatives/substitutes to meat	5%	20%
Quality of meat	2%	8%
Internal influences		
Exposure to information toward meat consumption	7%	28%
Cooking for oneself	4%	16%
Health reasons	4%	16%
Change in living conditions	3%	12%
Accessibility of alternatives	3%	12%
Political/governmental influence	2%	8%
Negative feelings toward consuming meat	2%	8%
Exposure to other diets not involving meat	2%	8%

5.5.1.3 Morality of Food Choices

Taking a closer look at the explanations given for the process of deciding and reconsidering specific food choices, aspects of moralization and demoralization strategies come into place. People's strong ideas about what was "right" shaped their food choices. Recurring topics in the moralization of food choice decisions were the global responsibility for climate (20%) and animal welfare (12%). Participants described these responsibilities in the following way:

> "I think that the link between eating meat and the environment is the most present and the most important for me, and maybe also the most influencing. When I look at my friends, for example, and my social surroundings, I can see there is a trend, that is, to eating rather local, and also thinking more about that."

While illustrating the decision to eat meat, health concerns (20%) were a strong point of reflection—more than environmental issues, climate change, or sustainability. Harmonization and demoralization could be identified in comments like the following:

> "I think that people who don't eat meat and make a trend of it are fools, because they think that the other humans that eat meat are killers and other stuff that harasses me. This is why I don't like these people."

Overall, study participants reported many different kinds of beliefs and influences on their food choices that could be analyzed and characterized in different ways. On the one hand, food-related decision-making processes were described as being part of a multifaceted and complex framework including social, individual, and emotional processes. On the other hand, the food choice decisions

Table 5.6 Role of morality in food choices, found in 25 guideline-based interviews (inductive category system)

Category name	*Abs. frequency*	*Rel. frequency*
Moralization		
Global responsibility for climate	5%	20%
Animal welfare	3%	12%
Questioning of necessity	2%	8%
Demoralization		
Necessity of meat for health	5%	20%
Morality of forced choice	2%	8%
Overmoralization of eating meat	2%	8%

were rather situational and less "justified" or reflected than we assumed. Thus, the interviews show, firstly, that sustainability apparently is less established as a social norm and framework that individuals refer to in their individual decisions leading to a specific eating behavior. This is mainly seen in only few sustainability-related dissonances. Secondly, harmonization strategies were applied, and explanations and justifications of certain behavior could be identified in the interviews. Eating and meat-eating in particular were less (over)moralized as we assumed.

While previous studies show that, mainly in Western society, a conscious decision is made to convert from a meat-eating to a plant-based diet (Beardsworth & Keil, 1992; Rosenfeld & Burrow, 2017; Lindeman & Sirelius, 2001; Jabs et al., 1998), indeed, those changes were only detected in the Austrian and Australian sample.

Generally, there were less differences to be detected when we compared the interviews in a Western socio-political and socio-cultural context. Interviewees from Indonesia valued their family background, traditions, and habits in the same way the interviewees in Austria did; similarly, meat-eating was less moralized in Austria and Australia rather than related to health reasons. Media and related external defining communicative moments were less influential in all three samples—at least less than we assumed.

Participants reported that they changed or modified their eating habits when circumstances required an adjustment; rational decisions were more frequent than emotional or ideological, and aspects of moralization in the decision-making process of food choices and the topic of sustainability played a rather minor role.

5.6 Summary and Discussion

The research knowledge of this study is particularly interesting, because it allows a deeper understanding of influences and beliefs on individual food habits and the potentials for behavior adjustments to changing environmental demands and improving sustainability.

The two major findings of our study showed that (1) sustainability is less established as a social norm, is not yet "normalized," and apparently plays a minor role in food choices; furthermore, even if climate change is perceived as a threat, thinking about the horrors of industrial livestock farming, animal rights, pollution, and waste (and their representation in the media as "external defining communicative moment") does not directly influence people to change from a meat to a plant-based diet. (2) Individual food choices and changes from meat consumption to veganism, or becoming vegetarian, are mainly influenced by being exposed to new ideas within the closer network of and related conversations with family and friends rather than media or related mediatized key events (documentaries). Even more interestingly, accessibility as well as general cooking and eating practices (food is celebration vs. food is necessary for survival) influence meat- or plant-based food choices.

Therefore, referring to Table 5.1, type 2 seems to be dominant—at least in the samples at hand (Table 5.1). Related to the dimensions we conceptualized in the theoretical part of the chapter, type 2 is the slightly reflective individual, confronted with new thoughts and behavior, but rather demoralizing eating behavior in general. Overall, it can be stated that, considering the interconnection of sustainability, food security, and environmental care, there still seems to be a huge gap between theoretical understanding of and "knowing about sustainability" and personal action, especially when looking at the aspect of meat consumption. Further research and representative samples are required to further elaborate the typology and create more generalizable results.

5.7 Implications and Applications

Most of the sustainability (communication) literature assumes that for sustainable development (and a realization of the 17 goals, UN, 2021), there is the need for change in human values, beliefs, attitudes, and behaviors in order to achieve a sustainability transition that will meet human needs (Leiserowitz, 2007). The study presented in this chapter supports this by showing that the greatest impact on food choices have internal moments and moments where food connects people to communities of individuals who share similar values and beliefs. Already being part of a family and their values is the first connection; thereby, linking food choices to both personal identity and social identity seems important. When thinking about impacts and opportunities of external defining communicative moments, our results implicate a rather-limited external (media) communication influence and, instead, a great impact on family and peers. Therefore, the role of communication in influencing sustainable consumption must be re-evaluated—at least in the dimension of food and eating behavior.

When looking at the influence of cultural background, Béné and his colleagues (2019) argue that defining a culturally acceptable diet is complicated because it always depends on the individuals, families, and communities concerned. Producing or consuming food that is culturally acceptable does not necessarily mean that it is healthy, or even sustainable. The aspect of social and cultural values significantly complicates this discussion. It opens up another aspect of the challenge in communicating the interrelation of sustainability and food choices. Interventions designed to enhance (food system) sustainability may work at one level but might fail to improve sustainability at other levels (Eakin et al., 2017). On another level, if meat consumption is moralized, it rather leads to a defensiveness (harmonization strategies), which deepens the rifts and does not hold out the prospect of a desired change in behavior.

Appeals based on social norms could be a promising approach to foster sustainable food choices and their related eating behavior, as the compliance of most people with social norms has been shown to be conditional upon the compliance of others (Fehr et al., 2002; Fehr and Gächter, 2002; Fischbacher et al., 2001). Hence, the challenge is to make norms of desired behavior more salient, for example, by providing information on desirable eating behavior from a sustainability perspective, or on how other people take environmentally beneficial food choices in certain situations.

Overall, it has to be stated that ethical reasons and sustainability as moral compass play a smaller role than we assumed and apparently are rather insignificant when people consider their eating behavior. Where Chuck et al. (2016) and colleagues found "spillover effects" in which people's food choices affected their broader philosophical views, which in turn had an influence on their activism and moral views, we rather found the self-concept to be stable due to family values and influences from a narrow circle of people.[1]

Understanding the beliefs and motives associated with food choices and the often-missing association with morality is important when considering the scientific foundation on the interconnection of meat consumption and climate change. Still, very little is known on how food choices work, and only partial knowledge is available to help decision-makers influence the system to work toward a more sustainable and a higher quality in choice and availability. When looking at the systematic review the research group around Tobi et al. (2019) conducted on food labeling, they found a preference for environmental and social responsibility product attributes. Consumers evaluated environmental and social responsibility labeling information more favorably than nutrition labeling information. Maybe this passive approach of fact-based information, in a non-reactive situation, could be a strategy to support conscious (and sustainable) food choices—which needs to be further explored in experiment-based studies.

5.8 Limitations of the Study

The findings of the present study are liable to potential limitations. There was a certain selection bias within our sample, as only interviewees who were interested in participating were included. Due to the method of recruitment, it is not traceable how many potential participants were contacted and how many of them declined during the process. Moreover, the sample characteristics did not consider specific diets, social backgrounds, and professional backgrounds. These are aspects which should be considered in further studies. Even though generalizing from the present sample to other populations regarding the prevalence of various environmental beliefs would be inappropriate, this qualitative exploratory pilot study on eating habits and potential changes of behavior is giving a deeper insight into the interconnection of decision-making processes with personal and societal influences. The findings need to be extended and elaborated upon with further studies, focusing on different sampling strategies, for example, a purposive (heterogenous) sample of vegan and meat-eating participants, with a variety of characteristics. It would be interesting how these aspects contribute to the development of sustainable behavior as one of the major challenges for modern society. Decision-makers are challenged to define a specific intervention or even a particular vision of the future as "sustainable"; still, there seems less consensus on the appropriate means to achieve it.

Ethics approval: University of Queensland, 2020/HE001141.

Note

1 We must keep in consideration that our sample was not restricted to a specific diet, whereas other studies specifically interview vegetarians and vegans.

References

Adams, M. (2021). Communicating vegan utopias: The counterfactual construction of human-animal futures. *Environmental Communication*, 1–14.

Aquino, K. F., & Reed, A., II. (2002). The self-importance of moral identity. *Journal of Personality and Social Psychology*, *83*, 1423–1440.

Asioli, D., Aschemann-Witzel, J., & Nayga Jr, R. M. (2020). Sustainability-related food labels. *Annual Review of Resource Economics*, *12*, 171–185.

Azzurra, A., Massimiliano, A., & Angela, M. (2019). Measuring sustainable food consumption: A case study on organic food. *Sustainable Production and Consumption*, *17*, 95–107.

Baker, T., & Weder, F. (2021). The evolution of the sustainability story in corporate communication. *IAMCR Proceedings/Chapter*.

Bastian, B., & Loughnan, S. (2017). Resolving the meat-paradox: A motivational account of morally troublesome behavior and its maintenance. *Personality and Social Psychology Review*, *21*(3), 278–299.

Beardsworth, A., & Keil, T. (1992). The vegetarian option. Varieties, conversions, motives and careers. *The Sociological Review*, *40*(2), 253–293.

Beattie, G., & McGuire, L. (2020). The modifiability of implicit attitudes to carbon footprint and its implications for carbon choice. *Environment and Behavior*, *52*, 467–494.

Bellotti, E., & Panzone, L. (2016). Media effects on sustainable food consumption. How newspaper coverage relates to supermarket expenditures. *International Journal of Consumer Studies*, *40*(2), 186–200.

Béné, C., Oosterveer, P., Lamotte, L., Brouwer, I. D., de Haan, S., Prager, S. D., Talsma, E. F., & Khoury, C. K. (2019). When food systems meet sustainability—Current narratives and implications for actions. *World Development*, *113*, 116–130. https://doi.org/10.1016/j.worlddev.2018.08.011

Bieling, C., Eser, U., & Plieninger, T. (2020). Towards a better understanding of values in sustainability transformations: Ethical perspectives on landscape stewardship. *Ecosystems and People*, *16*(1), 188–196.

Bourdieu, P. (1984). *Distinction: A Social Critique of the Judgment of Taste*. London: Routledge & Kegan Paul.

Bucchi, M., & Trench, B. (2021). Rethinking science communication as the social conversation around science. *Journal of Science Communication*, *20*(3), Y01.

Buddle, E. A., Bray, H. J., & Ankeny, R. A. (2018). Why would we believe them? Meat consumers' reactions to online farm animal welfare activism in Australia. *Communication Research and Practice*, *4*(3), 246–260.

Buttlar, B., Rothe, A., Kleinert, S., Hahn, L., & Walther, E. (2021). Food for thought: Investigating communication strategies to counteract moral disengagement regarding meat consumption. *Environmental Communication*, *15*(1), 55–68.

Campbell, K., & Crawford, D. (2001). Family food environments as determinants of preschool-aged children's eating behaviours: Implications for obesity prevention policy. A review, Australian *Journal of Nutrition and Dietetics*, *58*(1), 19–25.

Christakis, N. A., & Fowler, J. H. (2007). The spread of obesity in a large social network over 32 years. *The New England Journal of Medicine*, *357*, 370–379. https://doi.org/10.1056/NEJMsa066082.

Chuck, C., Fernandes, S. A., & Hyers, L. L. (2016). Awakening to the politics of food: Politicized diet as social identity. *Appetite*, *107*, 425–436. https://doi.org/10.1016/j.appet.2016.08.106

Clifford, S. (2019). How emotional frames moralize and polarize political attitudes. *Political Psychology*, *40*(1), 75–91.

Clune, S., Crossin, E., & Verghese, K. (2017). Systematic review of greenhouse gas emissions for different fresh food categories. *Journal of Cleaner Production*, *140*, 766–783.

Cohen, M. J. (2005). Sustainable consumption in national context: An introduction to the special issue. *Sustainability: Science, Practice and Policy*, *1*(1), 22–28.

Crouch, D. (2018). The Swedish 15-year-old who's cutting class to fight the climate crisis. *The Guardian*, September 1. www.theguardian.com/science/2018/sep/01/swedish-15-year-old-cutting-class-to-fight-the-climate-crisis

Delormier, T., Frohlich, K. L., & Potvin, L. (2009). Food and eating as social practice—understanding eating patterns as social phenomena and implications for public health. *Sociology of Health & Illness*, *31*(2), 215–228. https://doi.org/10.1111/j.1467-9566.2008.01128.x

Eakin, H., Connors, J. P., Wharton, C., et al. (2017). Identifying attributes of food system sustainability: Emerging themes and consensus. *Agriculture and Human Values*, *34*, 757–773. https://doi.org/10.1007/s10460-016-9754-8

Fairclough, N. (1992). *Discourse and Social Change*. Cambridge: Polity Press.

Farrow, K., Grolleau, G., & Ibanez, L. (2017). Social norms and pro-environmental behavior: A review of the evidence. *Ecological Economics*, *140*, 1–13. https://doi.org/10.1016/j.ecolecon.2017.04.017

Fehr, E., Fischbacher, U., & Gächter, S. (2002). Strong reciprocity, human cooperation and the enforcement of social norms. *Human Nature*, *13*, 1–25. https://doi.org/10.1007/s12110-002-1012-7

Fehr, E., & Gächter, S. (2002). Altruistic punishment in humans. *Nature*, *415*(10), 137–140. https://doi.org/10.1038/415137a

Festinger, L. (1962). Cognitive dissonance. *Scientific American*, *207*(4), 93–106.
Feunekes, G. I., de Graaf, C., Meyboom, S., & van Staveren, W. A. (1998). Food choice and fat intake of adolescents and adults: Associations of intakes within social networks. *Preventive Medicine*, *27*(5 Pt 1), 645–656. https://doi.org/10.1006/pmed.1998.0341
Fenzl, T., & Mayring, P. (2017). QCAmap: eine interaktive Webapplikation für Qualitative Inhaltsanalyse. *Zeitschrift für Soziologie der Erziehung und Sozialisation ZSE*, *37*(3), 333–340. www.researchgate.net/publication/327262349_QCAmap_eine_interaktive_Webapplikation_fur_Qualitative_Inhaltsanalyse
Fischbacher, U., Gächter, S., & Fehr, E. (2001). Are people conditionally cooperative? Evidence from a public goods experiment. *Economics Letters*, *71*(3), 397–404. https://doi.org/10.1016/S0165-1765(01)00394-9
Fischer, D., Reinermann, J. L., Mandujano, G. G., DesRoches, C. T., Diddi, S., & Vergragt, P. J. (2021). Sustainable consumption communication: A review of an emerging field of research. *Journal of Cleaner Production*, *300*, 126880.
Fox, N., & Ward, K. (2008). Health, ethics and environment. A qualitative study of vegetarian motivations. *Appetite*, *50*, 422–429.
Furst, T., Connors, M., Bisogni, C. A., Sobal, J., & Falk, L. W. (1996). Food choice: A conceptual model of the process. *Appetite*, *26*(3), 247–265. https://doi.org/10.1006/appe.1996.0019
Gallimore, R., & Lopez, E. M. (2002). Everyday routines, human agency, and ecocultural context: Construction and maintenance of individual habits. *OTJR: Occupation, Participation and Health*, *22*(Suppl 1), 70S–77S.
Gardiner, S. (2013). *Reflecting on a Perfect Moral Storm*.
Gerber, P. J., Steinfeld, H., Henderson, B., Mottet, A., Opio, C., Dijkman, J., Falcucci, A., & Tempio, G. (2013). *Tackling Climate Change Through Livestock. A Global Assessment of Emissions and Mitigation Opportunities*. Food and Agriculture Organization of the United States, Rome.
Gottwald, F.-T., & Weder, F. (2021). Documentaries and implicit morality. Chapter in Weder, F., Karmasin, M., & Krainer, L. (Eds.), *The Sustainability Communication Reader*. A Reflective Compendium. Heidelberg, Berlin et al.: Springer VS.
Graham, T., & Abrahamse, W. (2017). Communicating the climate impacts of meat consumption: The effect of values and message framing. *Global Environmental Change*, *44*, 98–108.
Haidt, J. (2013). Moral psychology for the twenty-first century. *Journal of Moral Education*, *42*(3), 281–297.
Hoek, A. C., Luning, P. A., Stafleu, A., & de Graaf, C. (2004). Food-related lifestyle and health attitudes of Dutch vegetarians, non-vegetarian consumers of meat substitutes, and meat consumers. *Appetite*, *42*(3), 265–272.
Hoffman, S. R., Stallings, S. F., Bessinger, R. C., & Brooks, G. T. (2013). Differences between health and ethical vegetarians. Strength of conviction, nutrition knowledge, dietary restriction, and duration of adherence. *Appetite*, *65*, 139–144.
Hussar, K. M., & Harris, P. L. (2009). Children who choose not to eat meat. A study of early moral decision-making. *Social Development*, *19*(3), 627–641.
Jabs, J., Devine, C. M., & Sobal, J. (1998). Model of the process of adopting vegetarian diets. Health vegetarians and ethical vegetarians. *Journal of Nutrition Education*, *30*(4), 196–203.
Jackson, T. (2005). Motivating sustainable consumption: A review of evidence on consumer behaviour and behavioural change. *Sustainable Development Research Network*, *1*, 30–40. https://timjackson.org.uk/wp-content/uploads/2018/04/Jackson.-2005.-Motivating-Sustainable-Consumption.pdf (Accessed 12 July 2021).
Jastran, M. M., Bisogni, C. A., Sobal, J., Blake, C., & Devine, C. M. (2009). Eating routines. Embedded, value based, modifiable, and reflective. *Appetite*, *52*(1), 127–136. https://doi.org/10.1016/j.appet.2008.09.003
Kearney, J. (2010). Food consumption trends and drivers. *Philosophical Transactions of the Royal Society B: Biological Sciences*, *365*(1554), 2793–2807. https://doi.org/10.1098/rstb.2010.0149
Kristiansen, S., Painter, J., & Shea, M. (2021). Animal agriculture and climate change in the US and UK elite media: Volume, responsibilities, causes and solutions. *Environmental Communication*, *15*(2), 153–172.
Kronthal-Sacco, R., Van Holt, T., Atz, U., & Whelan, T. (2020). Sustainable purchasing patterns and consumer responsiveness to sustainability marketing messages. *Journal of Sustainability Research*, *2*(2).
Krzyżanowski, M. (2020). Normalization and the discursive construction of "new" norms and "new" normality: Discourse in the paradoxes of populism and neoliberalism. *Social Semiotics*, *30*(4), 431–448. https://doi.org/10.1080/10350330.2020.1766193
Lally, P., Bartle, N., & Wardle, J. (2011). Social norms and diet in adolescents. *Appetite*, *57*(3), 623–627. https://doi.org/10.1016/j.appet.2011.07.015

Larson, N., & Story, M. (2009). A review of environmental influences on food choices. *Annals of Behavioral Medicine*, *38*(1), 56–73.
Leiserowitz, A. (2007). International public opinion, perception, and understanding of global climate change. *Human Development Report*, *2008*(2007), 31.
Lindeman, M., & Sirelius, M. (2001). Food choice ideologies: The modern manifestations of normative and humanist views of the world. *Appetite*, *37*(3), 175–184. https://doi.org/10.1006/appe.2001.0437
Lowell, J. (2012). Managers and moral dissonance: Self justification as a big threat to ethical management? *Journal of Business Ethics*, *105*(1), 17–25.
Machovina, B., Feeley, K. J., & Ripple, W. J. (2015). Biodiversity conservation: The key is reducing meat consumption. *Science of the Total Environment*, *536*, 419–431. https://doi.org/10.1016/j.scitotenv.2015.07.022
Marsden, T., Hebinck, P., & Mathijs, E. (2018). Re-building food systems: Embedding assemblages, infrastructures and reflexive governance for food systems transformations in Europe. *Food Security*, *10*, 1301–1309. https://doi.org/10.1007/s12571-018-0870-8
Marsden, T., & Morley, A. (Eds.). (2014). *Sustainable Food Systems: Building a New Paradigm*. Routledge.
Mayring, P. (2002). *Einführung in die Qualitative Sozialforschung: Eine Anleitung zu qualitativem Denken*. Weinheim: Beltz.
Mayring, P. (2014). *Qualitative Content Analysis: Theoretical Foundation, Basic Procedures and Software Solution*. www.ssoar.info/ssoar/handle/document/39517 (Accessed 3 October 2019).
Mayring, P., & Fenzl, T. (2014). *QCAmap. A Software for Qualitative Content Analysis*. www.qcamap.org (Accessed 3 October 2019).
Mayring, P., & Fenzl, T. (2019). Qualitative inhaltsanalyse. In Baur, N., & Blasius, J. (Hrsg.), *Handbuch Methoden der empirischen Sozialforschung*. Wiesbaden: Springer VS Verlag, 633–648. https://doi.org/10.1007/978-3-658-21308-4_42.
Moore, M. M., & Yang, J. Z. (2020). Using eco-guilt to motivate environmental behavior change. *Environmental Communication*, *14*(4), 522–536.
Moran, D., Wood, R., Hertwich, E., Mattson, K., Rodriguez, JFD, Schanes, K., & Barrett, J. (2020). Quantifying the potential for consumer-oriented policy to reduce European and foreign carbon emissions *Climate Policy*, *20*, S28–S38.
Newig, J., Schulz, D., Fischer, D., Hetze, K., Laws, N., Lüdecke, G., & Rieckmann, M. (2013). Communication regarding sustainability: Conceptual perspectives and exploration of societal subsystems. *Sustainability*, *5*(7), 2976–2990.
Nezlek, J. B., & Forestell, C. A. (2020). Vegetarianism as a social identity. *Current Opinion in Food Science*, *33*, 45–51. https://doi.org/10.1016/j.cofs.2019.12.005.
Robertson, M. (2018). *Communicating Sustainability*. Routledge.
Rodan, D., & Mummery, J. (2016). Doing animal welfare activism everyday: Questions of identity. *Continuum: Journal of Media & Cultural Studies*, *30*(4), 381–396.
Rosenfeld, D. L., & Burrow, A. L. (2017). The unified model of vegetarian identity: A conceptual framework for understanding plant-based food choices. *Appetite*, *112*, 78–95.
Ruby, M. B. (2012). Vegetarianism. A blossoming field of study. *Appetite*, *58*, 141–150.
Salmivaara, L., Lombardini, C., & Lankoski, L. (2021). Examining social norms among other motives for sustainable food choice: The promise of descriptive norms. *Journal of Cleaner Production*, *311*, 127508.
Sanchez-Sabate, R., & Sabaté, J. (2019). Consumer attitudes towards environmental concerns of meat consumption: A systematic review. *International Journal of Environmental Research and Public Health*, *16*(7), 1220.
Schwartz, S. (1968). Awareness of consequences and the influence of moral norms on interpersonal behavior. *Sociometry*, *31*, 355–369.
Shrivastava, P., Stafford Smith, M., O'Brien, K., & Zsolnai, L. (2020). Transforming sustainability science to generate positive social and environmental change globally. *One Earth*, *2*(4), 329–340. https://doi.org/10.1016/j.oneear.2020.04.010
Skitka, L. J., Bauman, C. W., & Sargis, E. G. (2005). Moral conviction: Another contributor to attitude strength or something more? *Journal of Personality and Social Psychology*, *88*(6), 895.
Skitka, L. J., Hanson, B. E., Morgan, G. S., & Wisneski, D. C. (2021). The psychology of moral conviction. *Annual Review of Psychology*, *72*(1), 347–366. https://doi.org/10.1146/annurev-psych-063020-030612
Sobal, J., & Bisogni, C. A. (2009). Constructing food choice decisions. *Annals of Behavioral Medicine*, *38*(Suppl 1), S37–46. https://doi.org/10.1007/s12160-009-9124-5
Stea, S., & Pickering, G. J. (2019). Optimizing messaging to reduce red meat consumption. *Environmental Communication*, *13*(5), 633–648.

Stephens Griffin, N. (2017). *Understanding Veganism: Biography and Identity*. Cham: Springer International.

Stoll-Kleemann, S., & Schmidt, U. J. (2017). Reducing meat consumption in developed and transition countries to counter climate change and biodiversity loss: A review of influence factors. *Regional Environmental Change*, *17*, 1261–1277. https://doi.org/10.1007/s10113-016-1057-5

Swenson, R., & Olsen, N. (2018). Food for thought: Audience engagement with sustainability messages in branded content. *Environmental Communication*, *12*(7), 973–988. https://doi.org/10.1080/17524032.2017.1279202

Thøgersen, J. (2021). Consumer behavior and climate change: Consumers need considerable assistance. *Current Opinion in Behavioral Sciences*, *42*, 9–14.

Tobi, R. C. A., Harris, F., Rana, R., Brown, K. A., Quaife, M., & Green, R. (2019). Sustainable diet dimensions. Comparing consumer preference for nutrition, environmental and social responsibility food labelling: A systematic review. *Sustainability*, *11*(23), 6575. https://doi.org/10.3390/su11236575

Urbanovich, T., & Bevan, J. L. (2020). Promoting environmental behaviors: Applying the health belief model to diet change. *Environmental Communication*, *14*(5), 657–671.

Vandenbroele, J., Vermeir, I., Geuens, M., Slabbinck, H., & Van Kerckhove, A. (2020). Nudging to get our food choices on a sustainable track. *Proceedings of the Nutrition Society*, *79*(1), 133–146.

Vermeir, I., & Verbeke, W. (2006). Sustainable food consumption: Exploring the consumer "attitude–behavioral intention" gap. *Journal of Agricultural and Environmental Ethics*, *19*, 169–194.

Weder, F. (2021). Sustainability as master frame of the future? Potency and limits of sustainability as normative framework in corporate, political and NGO communication. In Weder, F., Karmasin, M., & Krainer, L. (Eds.), *The Sustainability Communication Reader. A Reflective Compendium*. Heidelberg, Berlin, et al.: Springer VS, 103–119.

Weder, F. (2022). Nachhaltigkeit kultivieren. Öffentliche Kommunikation über Umwelt, Klima, nachhaltige Entwicklung und Transformation. *Communicatio Socialis (ComSoc)*, *55*(2), 146–159.

Weder, F., & Erikson, M. B. (2023). *CSR Communication and Cultures of Sustainability*. Brisbane: UQ Pressbook.

Weder, F., Karmasin, M., Krainer, L., & Voci, D. (2021). Sustainability communication as critical perspective in media and communication studies—an introduction. In Weder, F., Karmasin, M., & Krainer, L. (Eds.), *The Sustainability Communication Reader. A Reflective Compendium*. Heidelberg, Berlin, et al.: Springer VS, 1–12.

Weder, F., Lemke, S., & Tungarat, A. (2019). (Re)storying sustainability: The use of story cubes in narrative inquiries to understand individual perceptions of sustainability. *Sustainability*, *11*(19), 5264.

Weick, K. E. (1987). Organizational culture as a source of high reliability. *California Management Review*, *29*(2), 112–127.

Wethington, E. (2005). An overview of the life course perspective: Implications for health and nutrition. *Journal of Nutrition Education and Behavior*, *37*(3), 115–120. https://doi.org/10.1016/s1499-4046(06)60265-0

Wisneski, D. C., & Skitka, L. J. (2017). Moralization through moral shock: Exploring emotional antecedents to moral conviction. *Personality and Social Psychology Bulletin*, *43*(2), 139–150.

Wognum, P. N., Bremmers, H., Trienekens, J. H., Van Der Vorst, J. G., & Bloemhof, J. M. (2011). Systems for sustainability and transparency of food supply chains–Current status and challenges. *Advanced Engineering Informatics*, *25*(1), 65–76.

6

ENERGY POVERTY AND EDUCATION

Evidence From Indian Households

Manasi B. and Jyoti Prasad Mukhopadhyay

6.1 Introduction

Energy poverty is a basic deprivation experienced by millions globally, especially by those belonging to emerging economies. Lack of access to basic energy services has a detrimental effect on the quality of life of energy-poor households (Phoumin & Kimura, 2019). Embracing energy poverty as a critical issue, the United Nations has included universal access to energy as Sustainable Development Goal 7 (SDG 7). The goal aims to "ensure universal access to affordable, reliable, sustainable, and modern energy by 2030." According to the IEA 2023 Report, there are still around 2.3 billion people in the world who do not have access to clean cooking fuel, and 675 million lack access to electricity services. Though energy poverty is a global challenge, its impact and measurement differ significantly between developed and emerging economies. Developed countries consider the inability to afford a reliable supply of energy services for their cooling and heating purposes as energy poverty (Middlemiss et al., 2019; Thomson et al., 2019). On the other hand, in a developing economy, the lack of access to or inability to afford basic energy services, like electricity and clean cooking fuel for basic livelihood needs, is considered energy deprivation (Barnes et al., 2011; Abbas et al., 2020; Jayasinghe et al., 2021).[1]

As a developing country, India faces the issue of lack of adequate and affordable supply of basic energy services for daily household needs (Sadath & Acharya, 2017). Energy poverty does not occur in isolation. It has several direct and indirect implications for socio-economic and development indicators, such as income, health, education, employment, and overall well-being (Koomson & Awaworyi Churchill, 2022; Nguyen & Su, 2021; Omar & Hasanujzaman, 2021; Ahmad et al., 2014). Income and energy poverty are closely related such that energy poverty acts as both the means and an end to income poverty (Nguyen & Nasir, 2021). The detrimental impact of energy poverty on health outcomes is direct and well researched. Exposure to solid fuel burning in the household causes indoor air pollution, where women, children, and elderly members become the most vulnerable to health problems, such as respiratory diseases, cardiovascular problems, and even premature death of infants (Ang'u et al., 2023; Barron & Torero, 2017; Boy et al., 2000; Teariki et al., 2020).

Similarly, being energy-deprived has indirect and persistent impact on education. United Nations SDG 4 focuses on quality education, where it aims to "ensure inclusive and equitable

DOI: 10.4324/9781003285472-7

quality education and promote lifelong learning opportunities for all." Thus, it is imperative to understand the relation between energy poverty and education for improving the quality of human lives in a sustainable way. There is a dearth of studies that explore the relation between energy poverty and education in developing countries. A handful of studies document a negative relation between energy poverty and educational outcomes (Omar & Hasanujzaman, 2021; Oum, 2019; Rafi et al., 2021). However, the evidence is scant and not well explored in the context of India. This chapter fills this gap in the related literature. In this chapter, we examine the impact of energy poverty on the educational outcomes of children from Indian households.

6.2 Empirical Evidence of Energy Poverty and Education

Energy deprivation is a crucial factor that stifles the process of human and economic advancement. Figure 6.1 explains the relation between energy poverty and educational outcomes of children.

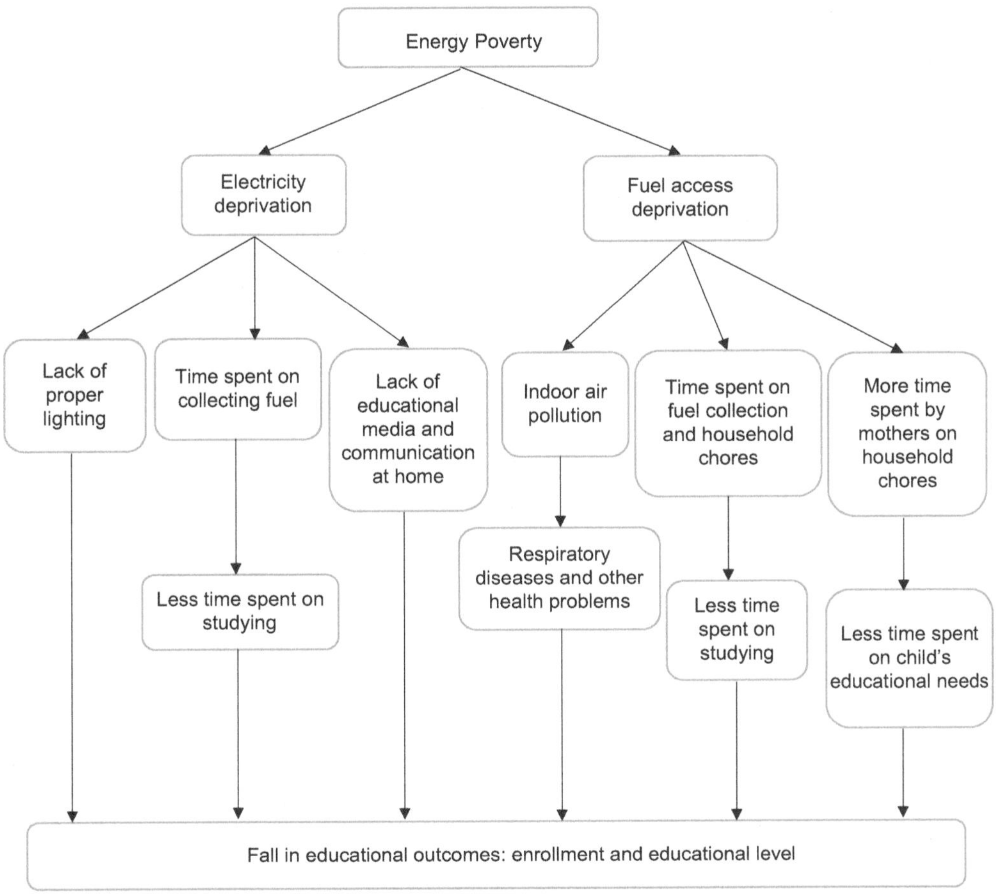

Figure 6.1 Conceptual framework of the effect of energy poverty on the educational outcomes of children.
Source: Authors' own creation.

The two major energy poverty components are electricity deprivation and clean fuel deprivation. Lack of basic energy services in the form of electricity and clean cooking fuel severely impacts the educational outcomes of children in such energy-deprived households. Lack of electricity can imply poor lighting in the house. Electricity access leads to quality lighting, which can improve the time children expend on learning, especially during nighttime. Electrification at home can facilitate access to educational media and thereby facilitate distance learning (Karekezi et al., 2012). The school enrolment rates and the completed schooling outcomes tend to be better for households with electricity (Oum, 2019). Children in electrified households spend more time studying and doing other educational activities, especially girls, compared to non-electrified households. The time spent by children from electrified households in fuel collection has also decreased, which implies a shift in household preferences in allotting more time for education rather than fuel collection (Khandker et al., 2014).

Solid fuel usage and educational outcomes are potentially linked in multiple ways. First, those households that do not have access to modern cooking fuels resort to traditional fuels, like firewood, cow dung, or crop residue, for their cooking needs. Persistent use of solid fuel in the household causes indoor air pollution, which can adversely impact health and, thereby, the enrollment of children in schools (Barron & Torero, 2017; Rafi et al., 2021). Second, school-going children who are required to help their mothers with household chores lose the time they intend to spend on studying. The children are required to spend a significant amount of time collecting traditional fuels, like firewood, which could have been invested in studying. This affects their school performance, and more so for girls, since fuel collection is predominantly a female-centric activity (Biswas & Das, 2022; Levison et al., 2018). Households that are more inclined toward investing in children's education tend to reduce other family expenses and purchase clean fuels. Third, households that use traditional fuels leave mothers with less time to concentrate on their child's educational needs, as household chores, including the collection of traditional fuel resources, consume most of their time. The lack of supervision by the mother affects the learning outcomes of children (Choudhuri & Desai, 2021).

6.3 Data

This study employs the fifth round of the National Family Health Survey (NFHS henceforth) conducted during 2019–2021. The survey is conducted by the International Institute of Population Sciences (IIPS), Mumbai, and is part of the multi-country Demographic Health Surveys (DHS). The fifth round of the survey is a nationally representative survey consisting of details on sample households' household-specific, socio-economic, demographic, health, gender-specific, and sanitation-related information. The survey consists of four questionnaires: household, woman's, man's, and biomarker. The survey period was interrupted by the COVID-19 pandemic. Hence, we have limited our sample by excluding the observations beyond March 2020. This is done to avoid the disruptions and shocks caused by the pandemic, where the country went into a nationwide lockdown in March 2020 during the first wave of the pandemic. The survey was then restarted in October 2020, but those observations are excluded from this sample. The unit of analysis is children aged between 6 and 14 years.

6.3.1 Educational Outcomes

In this chapter, we use different indicators of educational outcomes as dependent variables in regression analysis. The first indicator is the enrollment status of children aged between 6 and

14 years. It is a binary variable that takes the value 1 if the child is currently enrolled and 0 if the child is not currently enrolled. We consider the enrollment status for the age group based on Article 21-A of the Indian Constitution, mandating free and compulsory education for all children between 6 and 14 years of age. In a developing country like India, mere school enrollment does not guarantee completion of education or learning outcomes (Chatterjee et al., 2018). Therefore, we consider the current educational level of children aged between 6 and 14 years as the second indicator of educational outcome. The different levels of education are no education, primary, and secondary. *Primary* refers to grades 1 to 5, and *secondary* refers to grades 6 to 12.[2]

We measure energy poverty in terms of the multidimensional energy poverty index (MEPI henceforth) score. We have calculated the MEPI scores following the framework proposed by Nussbaumer et al. (2012). This framework consists of five dimensions: cooking, lighting, services from household appliances, services from education appliances, and telecommunication services. The MEPI scores are calculated based on the weights assigned by Nussbaumer et al. (2012). The methodology adopted for calculating MEPI is based on the method developed by the Oxford Poverty and Human Development Initiative (OPHI) (Alkire & Foster, 2011). The MEPI score ranges from 0 to 1, with 0 indicating the least energy-poor and 1 being the most energy-poor. We find that the MEPI score of Indian households is 0.29, which indicates its moderately energy-poor status (Manasi & Mukhopadhyay, 2024).

The MEPI is an all-encompassing index and, hence, provides the gross effect of energy poverty on educational outcomes. To deepen our analysis, we further use two of its constituent components of energy poverty, and thus, we disentangle the impact of its individual components on the educational outcomes. We consider electricity and fuel deprivation as the components of energy poverty, since these two are the most significant contributor to MEPI and are assigned the highest weights in the computation of the said index. Electricity deprivation is measured using a dummy (takes the value 1 for yes, and 0 otherwise) that denotes whether the household to which the child belongs has access to electricity or not. The other component of energy poverty is fuel deprivation, which is captured by a dummy variable (takes the value 1 for yes, and 0 otherwise) indicating whether the child's household has access to clean cooking fuel (LPG, biogas, electricity).

The descriptive statistics of the key variables are given in Table 6.1.

Table 6.1 Descriptive statistics of the key variables

	Observations	*Mean*	*Std dev*
Outcome variables			
Enrollment status			
Enrolled	3,00,495	0.94	
Educational level			
Primary	1,86,113	0.58	
Secondary	1,14,382	0.36	
Explanatory variables			
MEPI score	3,18,386	0.37	0.29
Electricity deprivation (1 = yes, 0 = no)	14, 032	0.44	
Fuel deprivation (1 = yes, 0 = no)	1,65,963	0.52	

Source: Authors' own calculations.

Table 6.2 Definition of the variables included in the analysis

Dependent variable	*Description*
Enrollment status, educational level	
Control variables	
Residence (0 = urban, 1 = rural)	Type of place of residence (urban or rural)
Child gender (0 = male, 1 = female)	Gender of the child
Household head gender (0 = male, 1 = female)	Gender of the household head
Household size	Number of members in the household of the child
Religion (0 = Hindu, 1 = Muslim, 2 = Christian, 3 = Sikh, 4 = Buddhist, 5 = Others)	Religion followed by household head
Caste (0 = ST, 1 = SC, 2 = OBC, 3 = others)	Caste of the household head
Household head education (0 = no education, 1 = primary, 2 = secondary, 3 = higher)	Education completed by household head
Age	Age of the child
BPL status (0 = no, 1 = yes)	Whether the household holds a BPL ration card or not
House ownership (0 = no, 1 = yes)	Whether the household owns this or any other house

We use several household-specific control variables used in the study, which are provided in Table 6.2.

6.4 Methodology

Table 6.3 presents the five dimensions, their corresponding indicators, and the weights used for the computation of the MEPI scores following the framework developed by Nussbaumer et al. (2012).

We estimate the following empirical model to analyze the effects of energy poverty on educational outcomes:

$$Y_{ijd} = \alpha_0 + \beta_1 Energy\, Deprivation_i + \beta_2 X_j + \beta_4 \eta_d + \varepsilon_{ijd} \quad (1)$$

Here, Y is the educational outcome of the ith individual from household j in district d. The educational outcomes are *enrollment status* and *educational level* of children in the age group 6–14 years. We use three proxy measures of energy poverty: MEPI score, electricity deprivation dummy, and fuel deprivation dummy. MEPI scores is a continuous variable which ranges from 0 to 1. Electricity deprivation and fuel deprivation are dummy variables, as explained in the previous section. X denotes the vector of control variables, which are provided in Table 6.2, and η is the district fixed effects. ε is the error term. We use the linear probability model (LPM) as a baseline regression for the enrollment model. We then employ probit regression to examine the impact of energy deprivation on the current enrolment status of the children, since enrollment is a binary variable. We use LPM as the baseline regression, followed by ordered logit regression for the current education level, since education level is an ordered categorical variable: 0 = no education, 1 = primary, 2 = secondary.

There is a concern that energy poverty can be potentially endogenous. The endogeneity can occur due to reverse causality running from educational outcomes to energy deprivation, unobserved heterogeneity, or omitted variables that are closely associated with energy poverty and

Table 6.3 Dimensions, respective indicators of multidimensional energy poverty index

Dimension	*Indicator (weights)*	*Variable*	*Energy-poor if . . .*
Cooking	Using modern cooking fuel *(0.205)*	Type of cooking fuel	Any fuel besides LPG, biogas, electricity
	Indoor pollution *(0.205)*	Food cooked on stove or open fire (no hood/chimney), indoor, if using any fuel besides electricity, LPG, biogas	True
Lighting	Electricity access *(0.2)*	Has access to electricity	False
Services from household appliances	Household appliance ownership *(0.13)*	Has a fridge	False
Entertainment	Entertainment appliance ownership *(0.13)*	Has a radio OR television	False
Communication services	Telecommunication means *(0.13)*	Has a phone landline OR mobile phone	False

Source: Reproduced from Nussbaumer et al. (2012), Table 6.2, p. 235.

also impact educational outcomes. As a result, LPM coefficients estimated using OLS become biased and inconsistent. To account for endogeneity concerns, we adopt the two-stage least squares (2SLS) method and employ an instrument variable (*IV*) for our estimation. Following Omar and Hasanujzaman (2021), we use the presence of sanitation facilities as an IV for energy poverty.

The two-stage least squares model is as follows:

$$Y_{ijd} = \mu_0 + \mu_1 Energy\ deprivation_i + \mu_2 X_j + \mu_3 \eta_d + \sigma_{ijd} \tag{2}$$

The instrument variable (*IV*) employed is "sanitation facilities," which is a dummy that takes the value of 1 if there is sanitation facility in the house of the child i and 0 if there is no sanitation facility at the house of the child. Y denotes the educational outcomes of child i at household j in district d. The educational outcomes are enrollment status and education levels attained by children aged 6–14 years.

For robustness check, we measure MEPI scores using alternate weights, such as equal weights and principal component analysis (PCA) adjusted weights. We measure the influence of MEPI scores on enrollment status and education level using these alternate MEPI scores.

6.5 Results

Table 6.4 provides the baseline LPM and probit regression results for the effects of energy poverty on the enrollment status of children aged between 6 and 14 years. MEPI score, electricity deprivation, and fuel deprivation are the explanatory variables in panels A, B, and C, respectively. Column 1 presents LPM regression results, and column 2 shows the marginal effects of energy deprivation on the current enrollment status of the children estimated from probit regression. The sign and significance of energy deprivation are consistent across all the three panels, that is, for MEPI scores, electricity deprivation, and fuel deprivation. We find that energy deprivation significantly reduces the likelihood of a child being enrolled in school. The estimated

Table 6.4 Energy poverty and enrollment status of the children

	LPM (1)	*Probit* (2)
Panel A: Explanatory variable—MEPI score		
MEPI score	−0.0783***	−0.0759***
	(0.00383)	(0.0295)
Constant	0.966***	
	(0.00814)	
Household controls	YES	YES
District FE	YES	YES
N	299,211	296,176
R-squared	0.068	
Pseudo R-squared		0.14
Panel B: Explanatory variable—electricity deprivation		
Electricity deprivation	−0.0629***	−0.0413***
	(0.00465)	(0.00327)
Constant	0.949***	
	(0.00806)	
Household controls	YES	YES
District fixed effects	YES	YES
N	299,211	296,176
R-squared	0.066	
Pseudo R-squared		0.13
Panel C: Explanatory variable—fuel deprivation		
Fuel deprivation	−0.0262***	−0.02547***
	(0.00174)	(0.00137)
Constant	0.951***	
	(0.00797)	
Household controls	YES	YES
District fixed effects	YES	YES
N	299,211	296,176
R-squared	0,065	
Pseudo R-squared		0.13

Note: The dependent variable is the current enrollment status of children between the ages of 6 and 14 years. Details of the household characteristics are given in Table 4.2. Column 1 denotes the OLS regression coefficients, and column 2 denotes the margin effects of probit regression. Standard errors are clustered at district level and are reported in parentheses. *** $p < 0.01$; ** $p < 0.05$; * $p < 0.1$.

coefficient of −0.0759 for the MEPI score suggests that, on average, one SD increase in the MEPI scores leads to a decrease in likelihood of a child being enrolled in school by 2.17 percentage points (−0.0759 × 0.286).[3]

We present Table 6.5 with the results on the effect of energy deprivation on the current educational level of children aged 6–14 years. Our coefficients of interest are the coefficients of MEPI score, electricity deprivation, and fuel deprivation in panels A, B, and C, respectively. Columns 1 and 2 present the LPM and ordered logit regression results. We find that children belonging to

Table 6.5 Energy poverty and educational level of the children

	LPM (1)	*Ordered logit* (2)
Panel A: Explanatory variable—MEPI score		
MEPI score	−0.136***	−0.641***
	(0.00645)	(0.0294)
Constant	−0.0470**	
	(0.0183)	
Household controls	YES	YES
District fixed effects	YES	YES
N	299,211	299,211
R-squared	0.426	
Pseudo R-squared		0.35
Panel B: Explanatory variable—electricity deprivation		
Electricity deprivation	−0.106***	−0.562***
	(0.00695)	(0.0431)
Constant	−0.0763***	
	(0.0179)	
Household controls	YES	YES
District fixed effects	YES	YES
N	299,211	299,211
R-squared	0.424	
Pseudo R-squared		0.35
Panel C: Explanatory variable—fuel deprivation		
Fuel deprivation	−0.0452***	−0.208***
	(0.00298)	(0.0146)
Constant	−0.0729***	
	(0.0178)	
Household controls	YES	YES
District fixed effects	YES	YES
N	299,211	299,211
R-squared	0.424	
Pseudo R-squared		0.35

Note: The dependent variable is the current educational level of children between the ages of 6 and 14 years. Details of the household characteristics are given in Table 4.2. Column 1 denotes the OLS regression coefficients, and column 2 denotes the coefficients of ordered logit estimation. Standard errors are clustered at district level and are reported in parentheses. *** $p < 0.01$; ** $p < 0.05$; * $p < 0.1$.

energy-poor households are more likely to be non-educated or attain primary level of education but less likely to attain secondary level of education. The results remain consistent in significance and direction across all three forms of energy deprivation across panels A, B, and C in Table 6.5.

We present the marginal effects of energy deprivation on education level in Table 6.6. Here, we find that an increase in energy poverty increases the probability of a child to belong to the categories of either no education or primary level of education. In contrast, an increase in energy poverty decreases the probability of a child to attain the secondary level of education. Coming to

Table 6.6 Energy poverty and educational level of the children (marginal effects)

	No education (1)	*Primary* (2)	*Secondary* (3)
Panel A: Explanatory variable—MEPI score			
MEPI score	0.027***	0.053***	−0.081***
	(0.001)	(0.002)	(0.004)
Household controls	YES	YES	YES
District fixed effects	YES	YES	YES
N	299,211	299,211	299,211
Panel B: Explanatory variable—electricity deprivation			
Electricity deprivation	0.025***	0.037***	−0.063***
	(0.002)	(0.002)	(0.004)
Household controls	YES	YES	YES
District fixed effects	YES	YES	YES
N	299,211	299,211	299,211
Panel C: Explanatory variable—fuel deprivation			
Fuel deprivation	0.009***	0.017***	−0.027***
	(0.0006)	(0.001)	(0.001)
Household controls	YES	YES	YES
District fixed effects	YES	YES	YES
N	299,211	299,211	299,211

Note: The dependent variable is the current educational level of children between the ages of 6 and 14 years. Details of the household characteristics are given in Table 4.2. The results show the marginal effects of logit regression on educational levels. Column 1 denotes no education level, column 2 denotes primary level, and column 3 denotes secondary level of education. Standard errors are clustered at district level and are reported in parentheses. *** $p < 0.01$; ** $p < 0.05$; * $p < 0.1$.

the secondary level of education, an increase in MEPI score by 1 point (implying more energy-poor households) reduces the likelihood of a child to attain secondary level of education by 8 percentage points. In other words, there is an inverse association between energy poverty and the attainment of secondary level of education. The direction and significance of the relation between levels of education and energy deprivation remain qualitatively similar across MEPI scores, electricity deprivation, and fuel deprivation across panels A, B, and C, respectively.

6.5.1 Robustness Tests

6.5.1.1 Addressing Endogeneity

As discussed earlier, there is a concern for energy poverty variables to be potentially endogenous, rendering our previous results biased and inconsistent. We allay such concern by using the IV 2SLS model, where the instrumental variable is the access to having sanitation facilities at home. We report the IV estimation results in Table 6.7. MEPI score, electricity deprivation, and fuel deprivation are the explanatory variables in panels A, B, and C, respectively. Column 1 reports the effect of energy deprivation on current enrolment status, and column 2 reports the effects on educational level. We find that energy deprivation affects both enrollment status and educational

Table 6.7 Energy poverty and educational outcomes of the children (IV estimates)

	Enrollment status	*Education level*
	(1)	(2)
Panel A: Explanatory variable—MEPI score		
MEPI score	−0.302***	−0.853***
	(0.0182)	(0.0827)
Constant	1.030***	1.189***
	(0.0090)	(0.0256)
Household controls	YES	YES
District fixed effects	YES	YES
N	299,211	299,211
First-stage F stat	2,121.28	179.64
Kleibergen–Paap rk LM	278.41	109.48
(p-value)	(0.00)	(0.00)
Panel B: Explanatory variable—electricity deprivation		
Electricity deprivation	−0.762***	−1.069***
	(0.0590)	(0.118)
Constant	1.013***	1.041***
	(0.0103)	(0.0163)
Household characteristics	YES	YES
District FE	YES	YES
N	299,211	299,211
First-stage F stat	254.74	80.32
Kleibergen–Paap rk LM	130.00	80.32
(p-value)	(0.00)	(0.00)
Panel C: Explanatory variable—fuel deprivation		
Fuel deprivation	−0.220***	−1.313***
	(0.0134)	(0.272)
Constant	1.008***	1.328***
	(0.00855)	(0.0814)
Household FE	YES	YES
District FE	YES	YES
N	299,211	299,211
First-stage F stat	1,439.9	25.07
Kleibergen–Paap rk LM	253.09	23.45
(p-value)	(0.00)	(0.00)

Note: The instrumental variable is "household sanitation facilities." Details of the household characteristics are given in Table 4.2. Column 1 denotes enrollment status, and 2 denotes education level. Standard errors are clustered at the district level and are reported in parentheses. P-values are also reported in parenthesis. *** $p < 0.01$; ** $p < 0.05$, * $p < 0.1$.

levels negatively and statistically significantly, consistent with our previous results. The results are statistically similar across all the models. The first-stage F statistics and Kleibergen–Paap rk LM statistics in all panels A, B, and C show that our IV does not suffer from weak IV problem and, hence, further adds credence to our IV results.

Table 6.8 Robustness test: alternate measurements of MEPI score

	Enrollment status *(1)*	*Education level* *(2)*
Panel A: Explanatory variable—MEPI score (equal weights)		
MEPI score	−0.0873***	−0.151***
	(0.00405)	(0.00685)
Constant	0.970**	−0.0397***
	(0.00817)	(0.0183)
Household controls	YES	YES
District fixed effects	YES	YES
N	299,211	299,211
Panel B: Explanatory variable—MEPI score (PCA-generated weights)		
MEPI score	−0.0721***	−0.125***
	(0.00353)	(0.00601)
Constant	0.964***	−0.0489***
	(0.00811)	(0.0182)
Household controls	YES	YES
District fixed effects	YES	YES
N	299,211	299,211

Note: Panel A comprises MEPI scores calculated using equal weights, and panel B comprises MEPI scores calculated using PCA-generated weights. Column 1 denotes enrollment status, and column 2 denotes the education level of children between the ages of 6 and 14 years. Details of control variables are given in Table 4.2. Standard errors are clustered at district level and are reported in parentheses. *** $p < 0.01$; ** $p < 0.05$; * $p < 0.1$.

6.5.1.2 Using Alternate Measures of Energy Poverty

We employ alternate measurements of MEPI scores by using different weighing schemes, such as equal weights and PCA-generated weights. The results remain qualitatively similar across all the measures. We find that the results reaffirm the negative and significant effect of energy deprivation on educational outcomes.

6.6 Conclusion

In this chapter, we explore the influence of energy poverty on one of the most important socio-economic outcomes: the educational outcomes of children in India. We find that energy poverty adversely impacts educational outcomes. It adversely affects the school enrolment status of children between 6 and 14 years of age. We then examine the educational level of children in the age group of 6–14 years. The results indicate that the higher the energy poverty, the lesser the likelihood of a child attaining higher education levels. An increase in energy poverty increases the probability of a child belonging to either no education level or primary level. On the other hand, a decrease in energy poverty increases the probability of a child belonging to a secondary-level education. We indicate energy poverty as the MEPI score and as its individual components, such as electricity and fuel deprivation. All forms of energy poverty seem to have a negative effect on educational outcomes. In sum, the results suggest that energy poverty dampens the country's efforts to improve the quality of its human capital. We incorporate MEPI scores into the analysis of this study.

This study faces limitations in including alternate learning outcome variables. The data source offers information about the reading and writing abilities of women alone and not children. Therefore, such learning outcomes of children are not included in the analysis. We also face data limitations concerning the time children spend studying, which limits the scope of this study. This poses a challenge in delving deeper into the channels between energy poverty and educational outcomes.

The empirical results suggest that energy poverty is deeply interconnected with other socio-economic indicators. The mere presence of energy poverty has further effects on socio-economic outcomes, such as education. The dependence on solid fuels and the lack of access to modern electricity services most affect vulnerable sections, like children. Sustainable Development Goals 4 and 7 focus on ensuring quality education and universal energy access. Though the Government of India has implemented several programs, like Pradhan Mantri Ujjwala Yojana (PMUY), and rural electrification programs, like the Saubhagya scheme, energy poverty still persists in the country.

It is possible to achieve the goal of quality education by effectively tackling energy poverty. There should be an affordable and reliable power and fuel supply for households to boost the process of creating human capital. Special efforts are needed in providing essential energy services to the marginalized sections of society, as the children belonging to those sections suffer more in terms of their educational outcomes. Lack of ownership of household appliances and communication services, which enable education and facilitate distance learning, is detrimental to effective learning. The subsidized provision of such appliances can improve the overall educational outcomes of children. Electricity and LPG access in households can also help bridge the gap in the enrollment rate between boys and girls, since girls tend to spend more time on household chores, sacrificing their study time.

Beyond financial support, it is important to create awareness of the harmful health effects of solid fuels and associated indoor air pollution. Use of solid fuels can cause respiratory ailments, and even deaths, and the effects are more pronounced for children. There should be a shift in the focus of government policies, where people should be encouraged to rely on renewable sources, like biogas, electric stoves, etc. Subsidies for the usage of such renewables can go a long way in their acceptance among households. A shift to clean energy can improve the children's health and, in turn, enhance their school attendance and enrollment. The better the energy status, the higher the educational level attained. A household with access to basic energy services, essential household appliances, and communication services is able to help their children achieve higher levels of education.

The nexus between energy poverty and educational outcomes can be better understood by exploring their channels. The time spent by children of energy-deprived households in different activities is one such major link to energy poverty. The scope of this study can thus be further extended to include the time spent by children in different activities (studying, household chores, fuel collection). An important direction for future research will be to examine these potential pathways through which they inform the understanding of energy poverty in India.

Notes

1 We use *energy poverty* and *energy deprivation* interchangeably in this study.

2 DHS follows the International Standard Classification of Education (ISCED) for classifying the education level attained. According to ISCED, grades 1 to 5 belong to the primary level, grades 6 to 10 in the lower secondary level, grades 11 to 12 in an upper secondary level, and education after 12 grades as higher. We use secondary level for grades 6 to 12 here.

3 SD of enrollment = 0.286.

References

Abbas, K., Li, S., Xu, D., Baz, K., & Rakhmetova, A. (2020). Do socioeconomic factors determine household multidimensional energy poverty? Empirical evidence from South Asia. *Energy Policy*, *146*, 111754. https://doi.org/10.1016/j.enpol.2020.111754

Ahmad, S., Mathai, M. V., & Parayil, G. (2014). Household electricity access, availability and human well-being: Evidence from India. *Energy Policy*, *69*, 308–315. https://doi.org/10.1016/j.enpol.2014.02.004

Alkire, S., & Foster, J. (2011). Counting and multidimensional poverty measurement. *Journal of Public Economics*, *95*(7), 476–487. https://doi.org/10.1016/j.jpubeco.2010.11.006

Ang'u, C., Muthama, N. J., Mutuku, M. A., & M'IKiugu, M. H. (2023). Analysis of energy poverty in Kenya and its implications for human health. *Energy Policy*, *176*, 113506. https://doi.org/10.1016/j.enpol.2023.113506

Barnes, D. F., Khandker, S. R., & Samad, H. A. (2011). Energy poverty in rural Bangladesh. *Special Section on Offshore Wind Power Planning, Economics and Environment*, *39*(2), 894–904. https://doi.org/10.1016/j.enpol.2010.11.014

Barron, M., & Torero, M. (2017). Household electrification and indoor air pollution. *Special Issue on Environmental Economics in Developing Countries*, *86*, 81–92. https://doi.org/10.1016/j.jeem.2017.07.007

Biswas, S., & Das, U. (2022). Adding fuel to human capital: Exploring the educational effects of cooking fuel choice from rural India. *Energy Economics*, *105*, 105744. https://doi.org/10.1016/j.eneco.2021.105744

Boy, E., Bruce, N., Smith, K. R., & Hernandez, R. (2000). Fuel efficiency of an improved wood-burning stove in rural Guatemala: Implications for health, environment and development. *Energy for Sustainable Development*, *4*(2), 23–31. https://doi.org/10.1016/S0973-0826(08)60239-2

Chatterjee, I., Li, I., & Robitaille, M.-C. (2018). An overview of India's primary school education policies and outcomes 2005–2011. *World Development*, *106*, 99–110. https://doi.org/10.1016/j.worlddev.2018.01.016

Choudhuri, P., & Desai, S. (2021). Lack of access to clean fuel and piped water and children's educational outcomes in rural India. *World Development*, *145*, 105535. https://doi.org/10.1016/j.worlddev.2021.105535

Jayasinghe, M., Selvanathan, E. A., & Selvanathan, S. (2021). Energy poverty in Sri Lanka. *Energy Economics*, *101*, 105450. https://doi.org/10.1016/j.eneco.2021.105450

Karekezi, S., McDade, S., Boardman, B., Kimani, J., & Lustig, N. (2012). Energy, poverty, and development. In *Global Energy Assessment Towards a Sustainable Future* (pp. 151–190). Cambridge University Press. https://doi.org/10.1017/cbo9780511793677.008

Khandker, S. R., Samad, H. A., Ali, R., & Barnes, D. F. (2014). Who benefits most from rural electrification? Evidence in India. *The Energy Journal*, *35*(2). https://doi.org/10.5547/01956574.35.2.4

Koomson, I., & Awaworyi Churchill, S. (2022). Employment precarity and energy poverty in post-apartheid South Africa: Exploring the racial and ethnic dimensions. *Energy Economics*, *110*, 106026. https://doi.org/10.1016/j.eneco.2022.106026

Levison, D., DeGraff, D. S., & Dungumaro, E. W. (2018). Implications of environmental chores for schooling: Children's time fetching water and firewood in Tanzania. *The European Journal of Development Research*, *30*(2), 217–234. https://doi.org/10.1057/s41287-017-0079-2

Manasi, B., & Mukhopadhyay, J. P. (2024). Definition, measurement and determinants of energy poverty: Empirical evidence from Indian households. *Energy for Sustainable Development*, *79*, 101383. https://doi.org/10.1016/j.esd.2024.101383

Middlemiss, L., Ambrosio-Albalá, P., Emmel, N., Gillard, R., Gilbertson, J., Hargreaves, T., Mullen, C., Ryan, T., Snell, C., & Tod, A. (2019). Energy poverty and social relations: A capabilities approach. *Energy Research & Social Science*, *55*, 227–235. https://doi.org/10.1016/j.erss.2019.05.002

Nguyen, C. P., & Nasir, M. A. (2021). An inquiry into the nexus between energy poverty and income inequality in the light of global evidence. *Energy Economics*, *99*, 105289. https://doi.org/10.1016/j.eneco.2021.105289

Nguyen, C. P., & Su, T. D. (2021). Does energy poverty matter for gender inequality? Global evidence. *Energy for Sustainable Development*, *64*, 35–45. https://doi.org/10.1016/j.esd.2021.07.003

Nussbaumer, P., Bazilian, M., & Modi, V. (2012). Measuring energy poverty: Focusing on what matters. *Renewable and Sustainable Energy Reviews*, *16*(1), 231–243. https://doi.org/10.1016/j.rser.2011.07.150

Nussbaumer, P., Fuso Nerini, F., Onyeji, I., & Howells, M. (2013). Global insights based on the multidimensional energy poverty index (MEPI). *Sustainability*, *5*(5), 2060–2076.

Omar, M. A., & Hasanujzaman, M. (2021). Multidimensional energy poverty in Bangladesh and its effect on health and education: A multilevel analysis based on household survey data. *Energy Policy*, *158*, 112579. https://doi.org/10.1016/j.enpol.2021.112579

Oum, S. (2019). Energy poverty in the Lao PDR and its impacts on education and health. *Energy Policy*, *132*, 247–253. https://doi.org/10.1016/j.enpol.2019.05.030

Phoumin, H., & Kimura, F. (2019). Cambodia's energy poverty and its effects on social wellbeing: Empirical evidence and policy implications. *Energy Policy*, *132*, 283–289. https://doi.org/10.1016/j.enpol.2019.05.032

Rafi, M., Naseef, M., & Prasad, S. (2021). Multidimensional energy poverty and human capital development: Empirical evidence from India. *Energy Economics*, *101*, 105427. https://doi.org/10.1016/j.eneco.2021.105427

Sadath, A. C., & Acharya, R. H. (2017). Assessing the extent and intensity of energy poverty using Multidimensional Energy Poverty Index: Empirical evidence from households in India. *Energy Policy*, *102*, 540–550. https://doi.org/10.1016/j.enpol.2016.12.056

Teariki, M. A., Tiatia, R., O'Sullivan, K., Puloka, V., Signal, L., Shearer, I., & Howden-Chapman, P. (2020). Beyond home: Exploring energy poverty among youth in four diverse Pacific island states. *Energy Research & Social Science*, *70*, 101638. https://doi.org/10.1016/j.erss.2020.101638

Thomson, H., Simcock, N., Bouzarovski, S., & Petrova, S. (2019). Energy poverty and indoor cooling: An overlooked issue in Europe. *Energy and Buildings*, *196*, 21–29. https://doi.org/10.1016/j.enbuild.2019.05.014

7

EDUCATION, EMPLOYMENT, AND GENDER EQUITY

Interactions and Interlinkages Between SDGs 4, 5, and 8

M. Niaz Asadullah, Kazi Mukitul Islam, Uma Kambhampati, and Reem Ashraf

7.1 Introduction

Education has long been seen as key to development outcomes. Not surprisingly, therefore, it was a Millennium Development Goal (MDG 2) and is now a Sustainable Development Goal (SDG 4). Education enters as an input into economic growth as well as into progress on many fronts—social, economic, political, and health. While MDG 2 highlighted the quantity of education (the proportion of children enrolled in schooling), SDG 4 has progressed beyond this to consider the quality of education, including the identification of relevant skills for education, decreasing gender inequalities and literacy outcomes (rather than just enrolment), increasing qualified teachers, etc. It is clear from the sub-indicators within SDG 4 that there will be interlinkages from education to gender equality (SDG 5), and also from education to employment (SDG 8[1]). It is these interlinkages that we will consider in this chapter.

SDGs are more interdependent than MDGs, both across and within goals and targets, and these interlinkages are critical for the attainment of the 2030 Agenda (Swain and Ranganathan 2020). Although there is a consensus about the synergies between the goals, there is no consensus regarding the SDGs with the strongest connections or even the direction of the synergies (Le Blanc 2015; Pradhan et al. 2017).[2] While some scholars (Le Blanc 2015) have found SDG 12 to have the highest connections and SDG 14 to have the lowest connections, others (Wu et al. 2022) have observed that SDGs 3, 9, 16, and 6 are the most positively connected.

Amongst the three SDGs we are concerned with in this chapter—SDGs 4, 5, and 8—these interlinkages are crucial because they could presage virtuous circles from education to gender equality to decent work for women to economic growth or, at the microeconomic level, from education to employment to increased incomes and, therefore, reduced inequalities or reduced poverty (see Figure 7.1). Numerous policy documents acknowledge the critical nature of SDG 4 for the process of human development (e.g., United Nations (2022). Despite this, interlinkages between human development goals such as SDG 4 (Quality Education), SDG 5 (Gender Equality), and SDG 8 (Decent Work and Economic Growth) remain under-researched in the literature on global goals.

A recent shift in emphasis has been from the quantitative to the qualitative dimension of education in global policy frameworks. At least three out of ten targets in SDG 4 stress quality education,

DOI: 10.4324/9781003285472-8

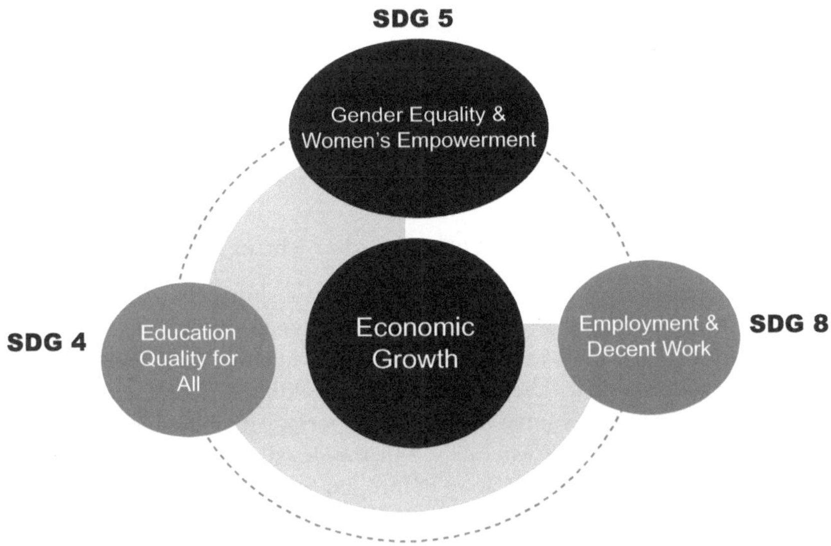

Figure 7.1 Interlinkages and Virtuous Cycle.

including access to affordable and equal tertiary education (Target 4.3), among others. Compared to the MDG goals, which only stressed school enrolment, SDG 4 emphasizes skills for jobs and work (Sayed and Moriarty 2020, p. 203). The recent literature has therefore rightly focused on the quality deficit and the phenomenon of schooling without learning (Asadullah et al. 2019; Pritchett and Sandefur 2020) and how that, in turn, may undermine the agenda of women's progress through education (see for examples Kaffenberger and Pritchett 2020, 2021). The quality of education in the form of inputs (e.g., student–staff ratios, separate classrooms, better-trained teachers) as well as the content of the education is key to educational outcomes and to reinforcing these virtuous interlinkages. For instance, if gender equality (SDG 5) is to be progressed through education, then this has to be reflected in the presence of female teachers, the representation of women in textbooks, and so on. In contrast, hidden gender inequalities in SDG 4, defined as the low presence of women in textbooks and educational leadership, will undermine progress in SDG 5.[3]

Using the aforementioned concept of hidden gender inequalities in education quality, this chapter assesses the interactions between SDGs (SDGs 4, 5, and 8.5) for the case of women's economic empowerment in South Asia. We argue that progress in women's empowerment requires accounting for the interactions between these SDGs, for example, combining gender-sensitive curriculum and women's leadership within education systems is conducive for simultaneously meeting SDGs 4 and 5 as well as other SDGs (e.g., SDG 13, Climate Action; for example, see Angrist et al. 2023).

The rest of the chapter is organized as follows. Section 2 describes the study context by briefly summarizing the somewhat-contested relationship between female education and labor market outcomes in the academic literature, particularly in South Asia. Here we also summarize the broad trends in women's education and labor market participation, over time and across regions. Section 3 then builds on this and presents a review of the literature on SDGs 4 and 5. Section 4 explains some data and methodological issues. Section 5 presents the main argument in terms of descriptive evidence on the broken and intact interlinkages between SDGs 4 and 5.6. Section 6 offers an analysis of policy provisions on gender mainstreaming, while Section 6 offers some discussion and recommendations for addressing the broken linkages and to strengthen the positive interactions between these two goals. Section 8 concludes with the message that if gender inequality within the production of education is overlooked (e.g., gender-biased textbooks and gender-biased school leadership), simply improving girls' access to education will not be enough for the economic empowerment of women and will only reinforce the broken links between SDGs 4 and 5.

7.2 Women's Education and Labor Market Outcomes

Policymakers have generally expected a positive relationship between women's education and employment. More educated women, it is thought, would wish to use their education in the labor market to improve their incomes. They would also be more likely to rebel against patriarchal norms, which aim to keep them secluded within the household and to restrict their mobility or reinforce the "women homemaker" norm. This positive correlation between education and employment is well established in the development and education literature. One of the earliest studies on the topic, long before the MDG campaign, confirmed a positive relationship between education and female labor force participation (FLFP) using data from 136 countries (Psacharopoulos and Tzannatos 1987). Literate women were 54% more likely to participate in the labor force than illiterate women, while an extra year of schooling was associated with two more years of labor force participation. Another study on 17 high- and middle-income countries also noted that women with higher education were more likely to be employed (England et al. 2012).

For non-OECD countries, Bussemakers et al. (2017) reported robust evidence that one additional year of schooling increases the odds of women's employment by 14%. Nguyen (2015), analyzing 40 developing countries, finds that education (years of schooling) increases women's employment chances by 2.74%.[4]

Research has also investigated the link by levels of education. In one study on five Asian countries, tertiary education appeared to have the strongest effect on FLFP (Cameron et al. 2001). Kanjila-Bhaduri and Pastore (2018) found a U-shaped relation between female education and FLFP in India where labor market returns are insignificant for lower levels of education. Klasen and Pieters (2012) also observed a positive relation between education and employment among very well-educated women in India. Higher female education has a positive effect on better job positions in Iran (Yousefy and Beratali 2011). In Pakistan, empirical studies found a stronger relation between education and employment among highly educated women beyond secondary education (Batool et al. 2019; Aslam et al. 2008). Most of the existing research has continued to analyze the impact of enrolment in education rather than quality of education.

However, despite these studies, there is an imperfect relationship between progress in female schooling and employment across developing regions. In some low-income countries, female labor force participation has stagnated or remained relatively low despite high female schooling enrolment (Klasen and Pieters 2015; Lahoti and Swaminathan 2016; Ilkkaracan 2012; Assad et al. 2020). In India, FLFP has stagnated since the late 1980s at 18%, despite a decline in fertility and rising wages and school enrolment (Klasen and Pieters 2015). A similar trend is also observed in Turkey (Ilkkaracan 2012) and in the countries of the MENA region (Assad et al. 2020). In the MENA region, female enrolment in secondary education increased from around 50% to 80% between 1985 and 2022, whereas female labor force participation increased by less than 2 percentage points (from 18.2% to 20%).

7.3 Female Education and the Role of Gender Norms

The imperfect link between female education and employment might be related to labor market conditions, socio-cultural norms, and institutional weakness (Psacharopoulos and Tzannatos 1987; Dildar 2015; Ejaz 2007; Moghadam 2015; Bussemakers et al. 2017). To start with, it is clear that the presence of jobs is a necessary condition for female education to result in FLFP (Baird et al. 2016). Bussemakers et al. (2017) found a stronger relationship between education and employment where service sector jobs are scarce (higher educational levels push lower-educated girls out of employment). In India, strong family income partly explains this paradox, but further analysis finds the importance of the lack of suitable employment for moderately educated women (Chatterjee et al. 2018). In particular, the lack of expansion of sectors that traditionally draw female workers and as well as low labor market returns to female education in both rural and urban areas are key reasons for this stagnation in the country (Kanjila-Bhaduri and Pastore 2018; Klasen and Pieters 2015).

Likewise, cultural norms and social attitudes negatively affect the direction and strength of the linkage. Using World Values Survey data for 74 countries (between 1995 and 2014), Bussemakers et al. (2017) found that women, in general, are less employed in countries with conservative gender norms, but they gain more from higher education. This means, in a conservative setting, lower-educated women bear the most burden (Bussemakers et al. 2017, p. 35). In contrast, Cameron et al. (2001) noted that in Southeast Asian countries such as Thailand and the Philippines, where women's gender roles are not rigidly defined, there is a stronger relationship between education and FLFP. Countries in the region have some of the highest female labor participation rates in the world (The Economist 2019).

The contrast between South and Southeast Asia suggests that the difference could be primarily a matter of economic development. However, support for this hypothesis remains contested (Gaddis and Klasen 2014). Even within a group of countries at the same level of development, there are large differences in female employment, and variation in cultural norms could help explain this difference (see Jayachandran 2015, 2021). Moreover, a review of the more recent literature on how social norms limit women's employment in developing countries, particularly causal evidence, suggests that poverty reduction alone is not the answer. In sum, economic disparity, institutional factors, and socio-cultural norms could all undermine the empowering effect of education and may therefore attenuate the positive role of education. We will revisit this conclusion when discussing our study implications for policy later in Section 6.

Women in South Asia face a variety of barriers to entering the formal labor force due to cultural norms that prioritize their role within the household (Moghadam 2015; Kabeer 2003). Even within the household, physical abuse is a common phenomenon for women who try to bargain with patriarchy (Chaudhuri et al. 2014). They face restrictions on their mobility and need permission or the company of a male member (mahram) to travel or work. For instance, in Afghanistan, women are compelled to stay at home, and those who want to go outside are required to not only follow a strict dress code but also need male mahram to travel more than 75 kilometers (UNHCR 2023). Women's participation is low not only in the labor force but also in other socio-political spheres, including decision-making autonomy within families. Women have limited rights to inheritance and property (Gonzales et al. 2015; Agarwal 1994). In South Asia, around half of the health-care decisions in the household are made without the participation of women (Senarath and Gunawardena 2009). A qualitative study among adolescent girls found very little decision-making autonomy in Bangladesh (Shahabuddin et al. 2016). In some countries, girls' education and employment are allowed, but under restrictive conditions. For instance, in Afghanistan, less parental support was found for high school female students, and they only accepted teaching and medical profession for females (Karlsson and Mansory 2007). However, it is expected that many of these issues would also be positively impacted by women's education, especially education which helps establish progressive social norms.

7.4 Methodology and Data

SDG indicators work through complex networks of interaction and are difficult to measure and monitor (Swain and Ranganathan 2020). Therefore, quantitative researchers use network analysis, text mining, qualitative visualization, and mathematical inference to identify synergies and trade-offs across the SDGs (Pradhan et al. 2017; Pham-Truffert et al. 2020; Alcamo et al. 2020; Le Blanc 2015; Wu et al. 2022).[5] In general, most studies observe more synergies than trade-offs (Le Blanc 2015; Nerini et al. 2018; Pradhan et al. 2017; Nerini et al. 2018; Swain and Ranganathan 2020; Dawes 2022).

Since we are particularly concerned with the impact of education on female labor force participation, we first consider whether there is a relationship between enrolment and female labor force participation. We will then consider why such a relationship is not as strong as we might expect. To consider the latter, we concentrate on two main factors—the content of education and establishing gender-equal norms, for instance, through textbooks and the role of female leadership. We adopt a descriptive qualitative methodological approach throughout this chapter.

7.5 Visualizing Hidden Inequality in Education and Interlinkages Between SDGs 4 and 5

Figures 7.2 and 7.4 indicate that school enrolment, both at secondary and tertiary levels, for girls has been increasing in all three regions (South Asia, Southeast Asia, and the OECD) since the pre-MDG era. The rate of increase in secondary enrolment is higher for South Asia and East Asia than for the OECD, since they started at a lower base. Tertiary education has shown a

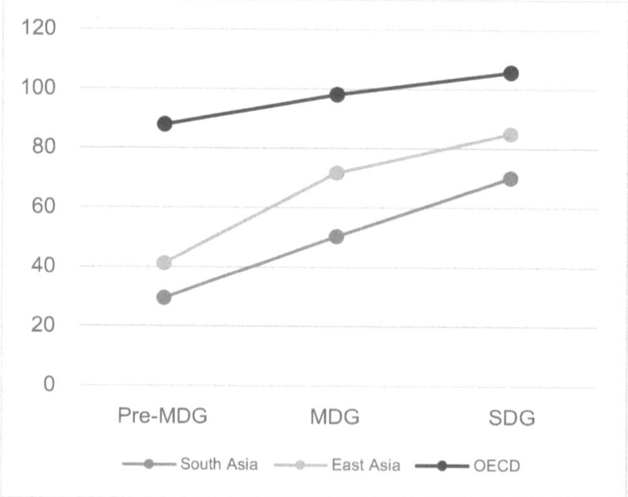

Figure 7.2 School enrollment, secondary, female (% gross).

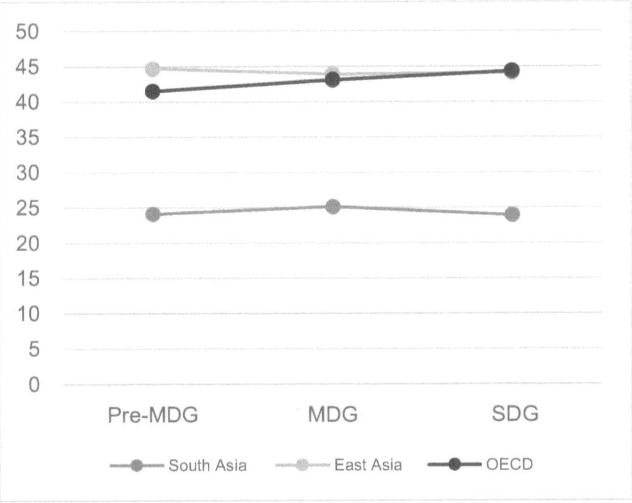

Figure 7.3 Labor force, female (% of total labor force).

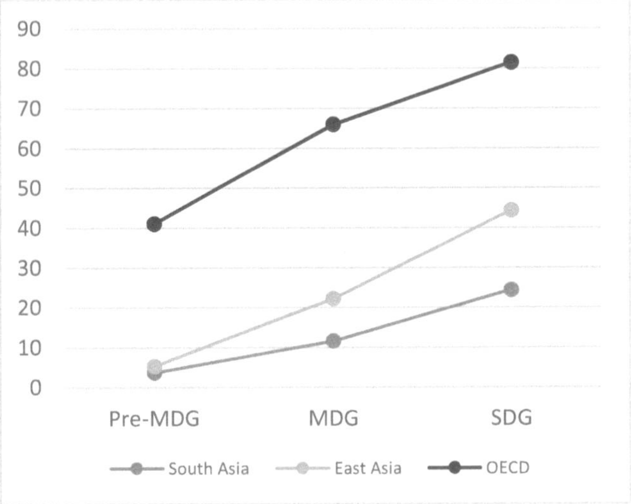

Figure 7.4 School enrollment, tertiary, female (% gross).

significant increase from 40% to 80% in the OECD countries too. However, when we look at the female labor force (Figure 7.3), we see a decrease in women as a percentage of the total labor force in South Asia between the pre-MDG period and the SDG period. This is a small decrease, but it is nevertheless surprising. Such a decline exists in East Asia, too, albeit at a higher level of employment.

Turning more specifically to employment in services, we see that this has increased in all three regions between the pre-MDG and the SDG eras, with the increase being smallest in the case of South Asia. Thus, the trend in South Asia seems to contradict expectations. It is this puzzle that we

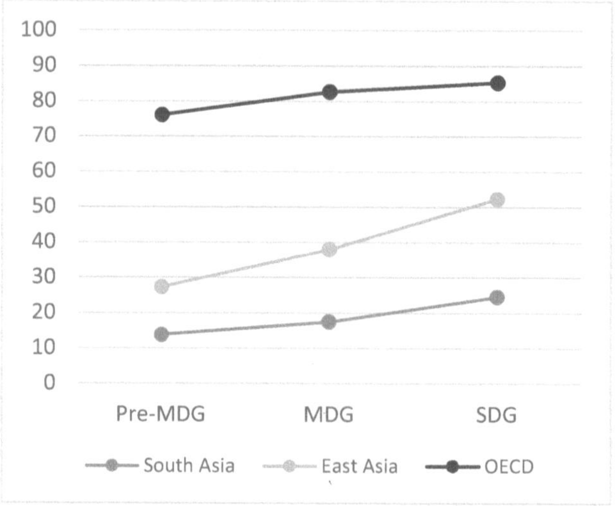

Figure 7.5 Employment in services, female (% of female employment). Note: *Pre-MDG* refers to average values for the period 1985–1999; *MDG* spans 2000–2014, and *SDG* spans 2015–2022.

Source: Authors, based on World Bank DataBank.

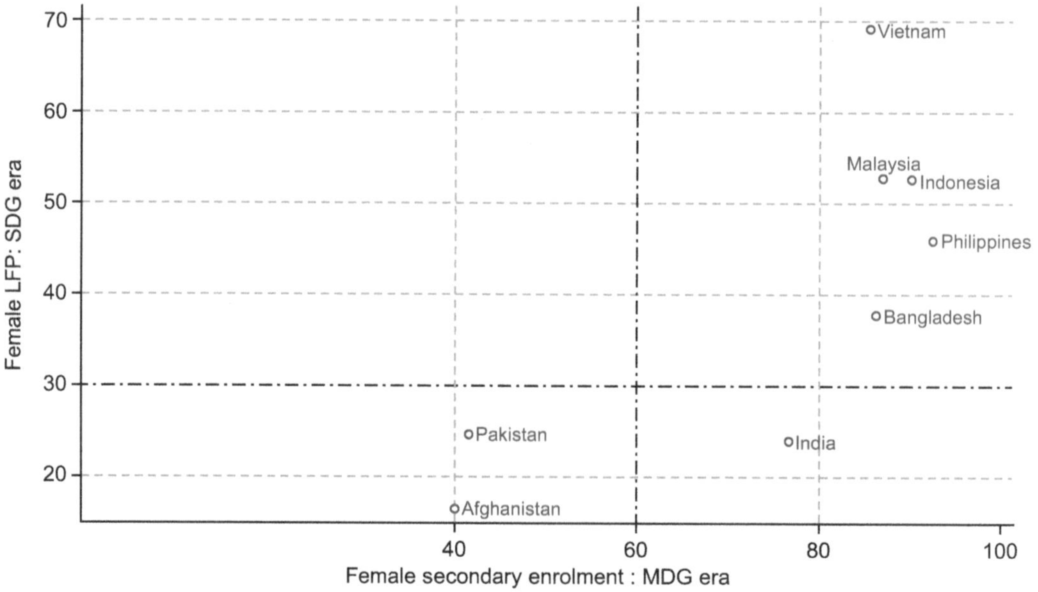

Figure 7.6 Female LFP and secondary school enrolment.

consider in more detail in the next section, identifying one possible factor that might influence the outcome—the content of textbooks being used.

Figures 7.6 and 7.7 indicate the association between school enrolment and FLFP. Figure 7.6 indicates that female secondary enrolment is between 75 and 85% in all countries in our sample,

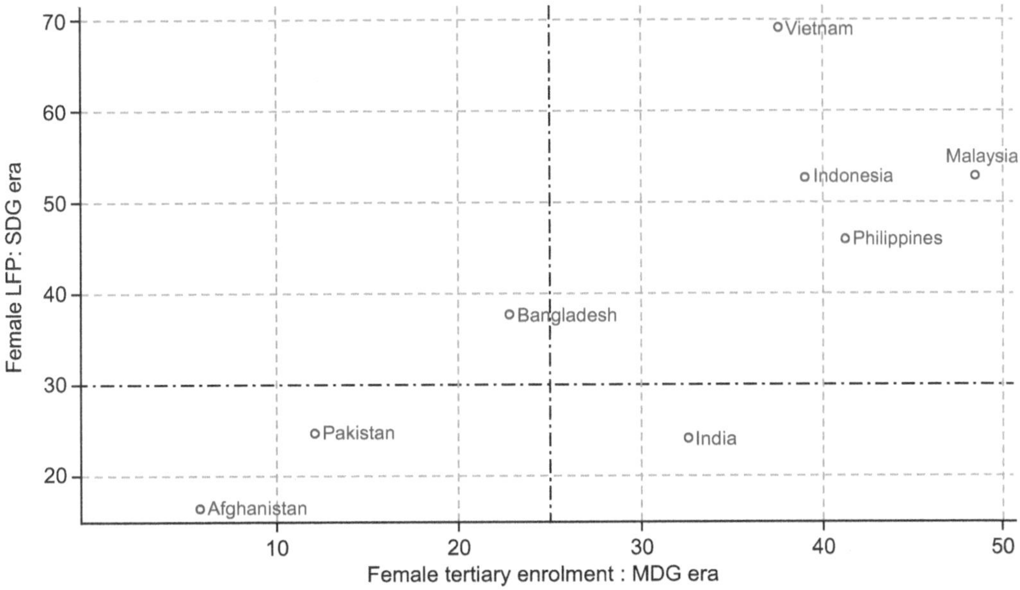

Figure 7.7 Female LFP and tertiary enrolment.

Source: Authors.

except Afghanistan and Pakistan (where it was approximately 40%). However, despite similar levels of secondary enrolment, the labor market outcomes are very different, ranging from approximately 23% in India to approximately 70% in Vietnam.

Turning to consider tertiary enrolment, we see that it ranges from below 10% in Afghanistan to approximately 50% for women in Malaysia. The graph indicates that, despite having relatively high levels of female tertiary involvement, women in India had very low FLFP, while Bangladesh, with lower tertiary enrolment, had higher FLFP, as did the Southeast Asian countries, which had higher tertiary enrolment and higher FLFP. Putting all the graphs together, we see the following patterns:

1. Increasing female education across regions
2. Stagnant or worsening FLFP in South Asia
3. India, in particular, occupying an anomalous position, with high enrolment but poor FLFP
4. An improvement in enrolment across South Asia but continuing poor FLFP in all countries, with Bangladesh being highest at 32%

These results cause us to consider why education does not seem to be helping, and in what follows, we focus on the quality of education and whether it is able to transfer progressive gender norms as one possible reason for this pattern.

7.5.1 Hidden Measures of Quality of Education

Our main indicator for quality of education is the content of textbooks used in schools. In particular, we are concerned with the representation of women in these textbooks, given that this may influence how young men and women see their own roles in the future.

To do this, we combine quantitative indicators on schooling and labor market with qualitative proxies of female representation in textbook. Data on the latter is not officially available for all countries. Hence, it is not feasible to undertake the conventional quantitative analysis on SDG interlinkages for a larger cross section of countries. We therefore limit our case study analysis to eight countries, for which we compile data on female representation in textbooks from individual country studies. In other words, to facilitate analysis across the MDG and the SDG eras, data is compiled to capture variation over time and across country in indicators of hidden gender inequality in education and women's labor force participation. We use these to help visualize the interactions and interlinkages. We focus on four countries from SA (Afghanistan, Bangladesh, Pakistan, and India), and four from SE Asia (Philippines, Vietnam, Indonesia, and Malaysia). The following table summarizes figures on female representation in textbooks for our eight case study countries.

Table 7.1 presents the findings of the recent literature on gender stereotypes in textbooks in South Asia and East Asia. We find women's visibility ranges from 13% to 44% in the textbooks (see Figure 7.6 and Figure 7.7). The highest female presence is noticed in Malaysian textbooks, and the lowest in Afghanistan textbooks. No country in our sample countries has more than 50% female visibility in the textbooks.

Figure 7.8 indicates that even the highest scorers on female representation in textbooks in the MDG era fall below the 50% line. Four Asian countries—Afghanistan (20%), Pakistan (20%), Vietnam (20%), and India (below 10%)—all indicate very poor representation of women in textbooks. Turning to Figures 7.8 and 7.9, we find a higher female representation in Southeast Asian countries than in Asia. Barring Vietnam, all these countries are to the right of Figure 7.8, indicating that female representation in textbooks was between 40 and 50%. Bangladesh is close to 40%. India, Pakistan, and Afghanistan are all in the lower left-hand corner, with female representation being low: below 10% in India and around 20% in Pakistan and Bangladesh. We might therefore conclude that education in South Asia is unlikely to challenge existing patriarchal norms. It is therefore not surprising that this translates to very low FLFP in all three countries (below 25%), while it is relatively high (between 30 and 40%) in Bangladesh, Malaysia, Indonesia, Philippines, and Malaysia.

Table 7.1 Female representation in the textbooks in selected South and Southeast Asian countries

Country/region	*Papers*	*Female representation (%)*
South Asia		
Afghanistan	Orfan (2022), Khattak and Akhtar (2017)	13
Bangladesh	Suchana (2020), Islam et al. (2018)	40
India	Kosir and Lakshminarayanan (2023), Kuruvilla and Thasniya (2015)	25.5
Pakistan	Mahmood et al. (2021), Islam et al. (2018), Agha et al. (2018)	25
East Asia		
Malaysia	Islam et al. (2018)	44.4
Indonesia	Islam et al. (2018), Dewi et al. (2022)	40
Philippines	Curaming and Curaming (2020)	34.4
Vietnam	Vu and Pham (2023)	37.8

Note: Figures on female representation correspond to percentage textbook content (words, sentences, etc.) being female. The table shows average figures by source. If we have the figure from more than one study for a given country, we use the country average. We only include studies that report quantitative figures. We have prioritized the most recent and comprehensive study.

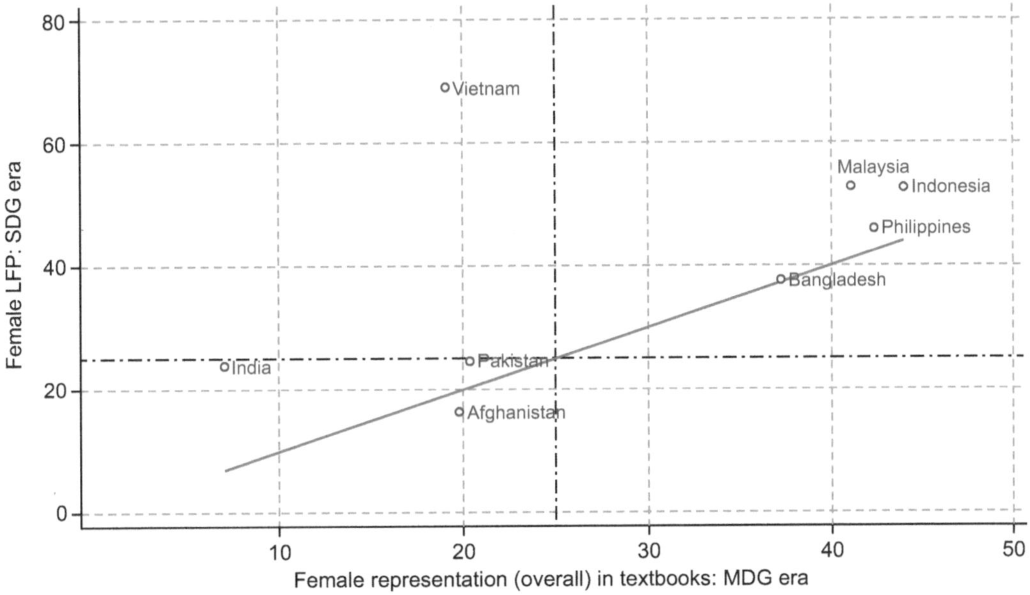

Figure 7.8 Female LFP and textbook representation.

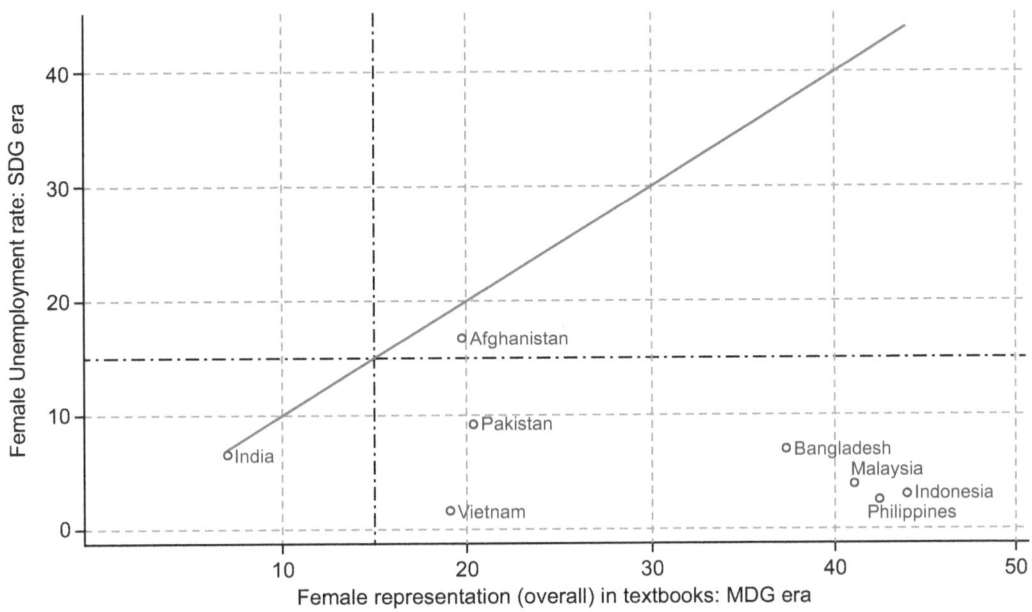

Figure 7.9 Female unemployment and textbook representation.

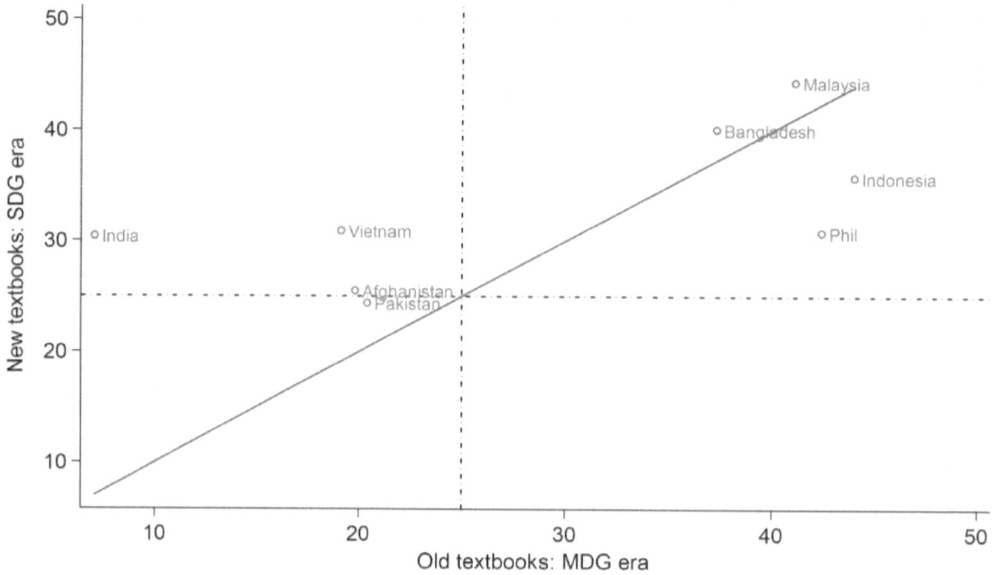

Figure 7.10 Persistence in textbook gender representation.

An immediate conclusion of the preceding figures (and a review of the evidence on gender stereotypes and female employment) is that improving textbooks could be of both intrinsic value for SDG 4 and instrumental value for SDG 5 and SDG 8. Yet our analysis also shows that progress in textbook improvement is slow. Figure 7.10, which follows, plots data on old and new textbooks for study countries. While there is improvement for almost all countries, female underrepresentation is still a serious issue in India, Pakistan, and Afghanistan. Alongside pursuing textbook reforms, a complementary approach is to change gender attitudes in school through direct interventions targeting female and male adolescent learners (Dhar et al. 2020).

The scientific support for the SDG 5 consequences of the hypothesis of hidden gender inequality in education is indirect. It primarily comes from empirical research that suggests a strong association between gender stereotypes and women's self-image, school performance, interest, and future career choices (Nohemi 2015; Kim et al. 2018; De la Torre-Sierra and Guichot-Reina 2022; De Gioannis et al. 2023; Bian et al. 2017). Within schools, gender discriminatory treatment by teachers and gender-stereotyped curriculum may lead to internalization of gender stereotypes over time (Blumberg 2008; Starr and Simpkins 2021; McKellar et al. 2019). Gender discrimination within the school is found to be linked with high depression among girls (Bell and Juvonen 2020). This can weaken the motivation of female students and push them into activities in gender-typed domains (Rogers et al. 2021, p. 407). Several studies also show that perceived gendered differential treatment, including discouraging comments by teachers, leads to female students' lower valuing of math and math score achievement (Brown and Leaper 2010; McKellar et al. 2019). Gender-stereotyped occupations in primary school textbooks also limit children's educational and professional identity (Nohemi 2015). Classroom instructors claim that textbook gender stereotypes in early grades influence career choices in later life (Teliousi et al. 2020).

Additional insights are available from the literature that employs experimental research design to study the potential impact of exposure to stereotypes or how to reverse the adverse effect. A review of the literature by Olsson and Martiny (2018) found that exposure to counter-stereotypical role models can change stereotypical beliefs about girls and women. Another study, on the effects of gender stereotypic and counter-stereotypic textbooks on science performance (Good et al. 2010), found that female students had higher comprehension after viewing counter-stereotypic images, whereas it was the opposite for male students. Chhaochharia and Niessen-Ruenzi (2022) found a positive relationship between admiring counter-stereotypical female role models and more women working in the labor market and male-dominated industries. An experimental intervention study in Zimbabwe found that students who were exposed to non-traditional careers for women over a three-year period in the learning material expressed more aspiration for non-traditional career choices. A similar result was found in another experimental research with ninth-grade students in the United States (Savenye 1990). An experimental study with two different primary school students in Israel found that children who were exposed to gender-fair textbooks attributed more activities that are appropriate for women than those who were exposed to stereotyped textbooks (Karniol and Gal-Disegni 2009). Another study—Bursztyn et al. (2020)—shows how husbands' perception of social attitude toward employed women can reduce their actual support for wives' employment. Their experimental design randomly alters male perception and attitudes toward support for women working outside the home in Saudi Arabia. This is found to not only increase men's willingness to help their wives look for jobs but also led to more job applications and interviews for outside jobs.

7.6 Analysis of Policy Provisions on Gender Mainstreaming

Our review of existing gender mainstreaming policies and curriculum frameworks guiding the gender mainstreaming in education does not indicate across-the-board progress. As for gender mainstreaming in education through gender-sensitive curriculum and textbooks, in Southeast Asia, we found evidence of early recognition of gender bias in textbooks and the need for gender mainstreaming in curriculum during the MDG period which continued into the SDG era. In the context of Vietnam, for example, Article 29 of the Education Law 2005 recognized the need to address existing gender biases in Vietnamese textbooks (MoET 2013). Also, Article 32 of the Education Law 2019 states that the content of the general education and the textbooks should be free from biases or prejudices related to ethnicity, religion, occupation, gender, or social status (Socialist Republic of Vietnam 2019). In the Philippines, basic education curriculum was directed to be gender- and culture-responsive, as per Republic Act 10533 (Republic of Philippines 2013). Persistently, during the SDG phase, the Philippines' Gender-Responsive Basic Education Policy called attention to the need for gender-sensitive curriculum, learning materials, and use of an inclusive language which is gender-fair (DepEd 2017). A review found that at the start of the MDG phase, Indonesia seemed to focus on reviewing curriculum and textbooks to improve education quality (Ministry of National Education 2005); soon we see policy focus on gender mainstreaming in education within the timeline of the MDG (BAPPENAS 2013).[6] In Malaysia, an exception in Southeast Asia, the review was, however, unable to find policy evidence on gender mainstreaming through curriculum or textbooks. In fact, even the education blueprint guiding the education sector priorities until 2025 revealed reforms for curriculum and assessments but no specific mention with respect to gender sensitivity (Ministry of Education, Malaysia 2013).

In the case of South Asia, the review on the same lines as the previous one indicates that gender sensitivity in policy mandates has been more recent and less systematic. For instance, the review could locate provisions to revise the curriculum and learning material at the primary level to include positive and progressive images of women, induct life stories of inspirational and influential women, and include essays written by women and the concept of equal rights for women (Ministry of Finance 2017) for Bangladesh only in the SDG phase. During the same phase, the review found Pakistan to introduce the Single National Curriculum, which focuses on gender mainstreaming and has issued a gender-sensitive checklist for textbook writers, teachers, and researchers to account for gender biases (National Curriculum Council 2020). In India, however, the review found gender mainstreaming in the curriculum to figure in both phases. The National Curriculum Framework (NCF) 2005 recognized the existing gender biases in the textbooks and called for the need to make curriculum gender-responsive (NCERT 2005), and the recent NCF (2023) specifies the need for gender-balanced representation in the reading material (NCERT 2023). The only country for which the review found gender mainstreaming a policy priority during the MDG but no specific mention of the same in the SDG phase is Afghanistan. During the MDG phase, the unified curriculum for all public and private schools called for textbooks to be gender-sensitive (UNESCO 2021). However, during the SDG era, the focus was largely on the availability of textbooks and language of publication and new books (Ministry of Education 2016). Our review also finds that for South Asia, the framework and the guidelines for textbook writers have not been very specific and systematic with reference to gender sensitivity or producing a gender-balanced content for India, Bangladesh, and Pakistan[7] (CBSE 2013; Sedere 2011; Paul 2014).

7.7 Discussion and Policy Implications

Numerous studies on the social returns to education provide support to interlinkages between SDG 4 and other SDGs (Canton 2007; Sianesi and Reenen 2003; Grossman 2006). Similar research linking SDG 4 and SDG 8 has been reviewed in Section 3. SDG 4 and SDG 5 are critical for the realization of other goals, both individually as well as in conjunction. A review of 37 UN flagship reports finds links between SDG 4 and all other SDGs, except SDG 14 (Vladimirova and Le Blanc 2016). Likewise, Rosengren et al. (2023) found that SDG 5 (Gender Equality) had both the widest and deepest interlinkages with other SDGs in terms of affecting the system. If the economic goal (SDG 8) is conserved as a means, not an end goal, of development, SDG 1 (No Poverty) and SDG 4 (Quality Education) become more central and require prioritization in global partnership (Laumann et al. 2022). In contrast, negative loops can be found between other goals and education. For instance, high poverty (SDG 1) is associated with a higher school dropout rate (SDG 4) and a decrease in access to basic sanitation (SDG 6) (Antoniades et al. 2020).

Several studies show a close connection between SDG 4 and SDG 5 (Le Blanc 2015; Pradhan et al. 2017; Vladimirova and Le Blanc 2016; Manandhar et al. 2018; Fiho et al. 2022). These synergies were observed both ways. Vladimirova and Le Blanc (2016) found a close connection between SDGs 4.1, 4.2, 4.3, 4.5, 4.6, and 4.7 and SDG 5 (Gender Equality). Likewise, SDG 5.6 was connected to SDG 4 (Quality Education). Leal-Fiho et al. (2022) found that seven of the ten SDG 4 targets are impacted by gender inequality. Vladimirova and Le Blanc (2016) identified a reciprocal relationship between the two goals in their review of global reports. SDG 4 influences SDG 5 when education expands opportunities for young women, including raising aspirations for work outside the home. On the other hand, education is influenced by gender outcomes (SDG 4), including early marriage, which curtails educational opportunities. Lastly, using micro data, a number of studies have established the causal link between SDG 4 and SDG 5 in a variety of settings and in different directions, such as the benefit of mothers' education on children's health (Keats 2018; Wu 2022).

In sum, our review of the literature suggests very close connections between SDGs 4 and 5, with reciprocal relationships (synergies) between the two goals. Yet these synergies are far from being fully utilized. In Bangladesh, for instance, girl-child marriage remains a norm despite a steady increase in female schooling (Amin et al. 2017). Within this context, the attainment of the SDG 5 goal will largely depend on leveraging synergies and negotiating trade-offs with mutually reinforcing actions (Pradhan et al. 2017; Pham-Truffert et al. 2020; Alcamo et al. 2020; Nilsson et al. 2016).

By emphasizing the fragile nature of empirical interlinkages, we have established that the attainment of individual SDGs does not guarantee that synergies will be fully leveraged. Goals and target-setting may have shifted the focus away from process- and capacity-related factors. For illustration purposes, we singled out a specific hypothesis: Hidden gender inequality in learning materials, including textbooks, plays an important role in shaping and reshaping gender stereotypes among learners.

Lastly, our analysis also has added implications for policy as it highlights another important gap in the SDG framework, that is, poor coverage of gender-related data in the SDG monitoring framework. One recent review confirms that only a third of gender-related SDG data is available (Beegle et al. 2023). To our knowledge, there is no global project under consideration to compile data on women's underrepresentation in textbook and educational leadership positions. Overcoming this data gap is a necessary first step to address hidden gender inequality in education that we have argued as critical for realizing the synergy between SDGs 4 and 5. Another related potential constraint includes state capacity. For instance, research reporting that synergy is 40% larger than the trade-off also confirms that countries with stronger and more capable government (e.g., Japan, Germany, and Finland) have the highest synergy (Pradhan et al. 2017). This could also be true for our case study countries, whereby Southeast Asian nations have a long legacy of benefiting from developmental states compared to South Asia. This possibility is left out for future research.

7.8 Conclusion

While some SDGs are more connected than others, with fewer trade-offs, research on interlinkages is lacking on SDG 4 (Quality Education) and SDG 5 (Gender Equality). Progress in these two goals not only depends on one another but can also potentially influence the achievement of overall SDG Agenda 2030 goals. It is in this context that this chapter has presented illustrative evidence on the issue, using South Asia and Southeast Asia as comparative case studies. These two regions were selected given their contrasting experience in women's labor force participation.

We have conjectured that the answer to the broken linkage between education and labor force participation in South Asia and the strong link in Southeast Asia can be found within the education system itself. By way of reviewing the evidence on social/cultural barriers to women's education and employment as well as gender stereotypes in textbook, we have singled out a particular form of *hidden gender inequality within education* (gender bias in learning materials) that may potentially undermine interlinkages between SDGs 4 and 5. We present descriptive evidence based on country-level secondary data from South Asia, two regions with broken interlinkage between female schooling and women's market outcomes. As part of the country case studies, we examine national policies for SDGs 4 and 5 in SA and contrast this with the same on selected countries from Southeast Asia (including China).

We have argued that one key factor, often ignored in policy discussions, is the role of gender inequality within education, which may undermine positive synergies and increase trade-offs in SDG interlinkage. Improving textbooks can be of both intrinsic value for SDG 4 and instrumental value for SDG 5 and SDG 8. Progress in women's empowerment (SDG 5) requires accounting for gender-sensitive curriculum and women's leadership within the education system. Co-benefits and synergies coming out of this linkage are conducive for simultaneously meeting SDG 4 and SDG 5 as well as other SDGs, especially for regions where the link is broken or relatively weaker: South Asia. On the other hand, Southeast Asia's progress in attracting more women into the labor force was not simply about increasing access to education and creating jobs; it also coincided with learning materials that are more gender-balanced. This may have helped transmit progressive social norms related to outside employment to the next generation, that is, gender-sensitive textbooks may have also facilitated women's labor market entry, while in South Asia schools continue to remain a site of social reproduction. At the same time, based on the evidence presented, progress in textbook improvement has been slow during the second half of the SDG campaign.

Our conclusion is that, if gender inequality within the production of education is overlooked, simply improving girls' access to education, a constituent target of SDG 4, will not be enough for the economic empowerment of women and may even reinforce the broken links between SDGs 4 and 5.

Acknowledgments

We thank the reviewer and participants in the October 2023 handbook webinar for comments on an earlier version of the chapter.

Notes

1 This particularly emphasizes equality of productive employment, decent work, and pay for all women and men.
2 One key explanation of high synergy is the fact that a similar indicator was used for multiple SDGs.
3 We use the term "hidden" because of the fact that none of the SDG 4 indicators focus on "gender inequality in educational production." In SDG 4 targets, *gender gaps* refer to disparities in skills and opportunities, whilst "effective learning outcomes" has no reference to gender-mainstreamed learning materials. Moreover, Target 4.7 lumps gender together with numerous other transformative themes: "Target 4.7: By 2030, ensure that all learners acquire the knowledge and skills needed to promote sustainable development, including, among others, through education for sustainable development and sustainable lifestyles, human rights, gender equality, promotion of a culture of peace and non-violence, global citizenship and appreciation of cultural diversity and of culture's contribution to sustainable development." For the full list of indicators, see https://tcg.uis.unesco.org/wp-content/uploads/sites/4/2020/09/SDG4_indicator_list.pdf.
4 Other country-specific studies show similar positive associations across LDCs (Yousef and Beratali 2011; Ince 2010; Aslam 2008; Batool et al. 2019; Klasen and Pieters 2012).
5 For instance, Le Blanc (2015) found an average of six to eight connections of each SDG to other goals using network analysis. A review of Global Sustainable Development Report (GSDR 2019) data by Dawes (2022) finds a total of 5,758 interactions, and this could be doubled, or even more, if three or more ways of interactions are considered across the 169 SDG targets (Alcamo et al. 2020).
6 By Presidential Decree (Inpres) No. 9/2000 on Gender Mainstreaming in National Development Planning and Programming and Permendiknas No. 84/2008 states guidelines for implementing gender mainstreaming in education (9).
7 For Pakistan, the review found guidelines for the textbook writers to be less specific with respect to gender sensitivity until the introduction of the Single National Curriculum in 2020.

References

Agarwal, B. (1994). *A Field of One's Own: Gender and Land Rights in South Asia*. Cambridge: Cambridge University Press.

Agha, N., Syed, G. K., & Mirani, D. A. (2018, January). Exploring the representation of gender and identity: Patriarchal and citizenship perspectives from the primary level Sindhi textbooks in Pakistan. In *Women's Studies International Forum* (Vol. 66, pp. 17–24). Pergamon.

Alcamo, J., Thompson, J., Alexander, A., Antoniades, A., Delabre, I., Dolley, J., . . . & Scharlemann, J. P. (2020). Analysing interactions among the sustainable development goals: Findings and emerging issues from local and global studies. *Sustainability Science*, *15*, 1561–1572.

Amin, S., Asadullah, M. N., Hossain, S., & Wahhaj, Z. (2017). Eradicating child marriage in the commonwealth: Is investment in girls' education sufficient? *The Round Table*, *106*(2), 221–223.

Angrist, N., Winseck, K., Patrinos, H. A., & Graff Zivin, J. S. (2023). Human capital and climate change. *World Bank*. https://cepr.org/voxeu/columns/untapped-potential-education-battle-against-climate-change

Antoniades, A., Widiarto, I., & Antonarakis, A. S. (2020). Financial crises and the attainment of the SDGs: An adjusted multidimensional poverty approach. *Sustainability Science*, *15*, 1683–1698.

Asadullah, M. N. (2006). Returns to education in Bangladesh. *Education Economics*, *14*(4), 453–468.

Asadullah, M. N., Alim, M. A., & Anowar, H. M. (2019). Enrolling girls without learning: Evidence from public schools in Afghanistan. *Development Policy Review*, *37*(4), 486–503.

Asadullah, M. N., Amin, S., & Chaudhury, N. (2019a). Support for gender stereotypes: Does madrasah education matter? *The Journal of Development Studies*, *55*(1), 39–56.

Asadullah, M. N., Mukitul Islam, K. M., & Wahhaj, Z. (2018). Gender bias in Bangladeshi school textbooks: Not just a matter of politics or growing influence of Islamists. *The Review of Faith & International Affairs*, *16*(2), 84–89. https://doi.org/10.1080/15570274.2018.1469821

Asadullah, M. N., Savoia, A., & Sen, K. (2020). Will South Asia achieve the sustainable development goals by 2030? Learning from the MDGs experience. *Social Indicators Research*, *152*(1), 165–189.

Aslam, M., Kingdon, G., & Söderbom, M. (2008). Is female education a pathway to gender equality in the labor market? Some evidence from Pakistan. In Tembon, M. M., & Fort, L. (Eds.), *Girl's Education in the 21st Century: Gender Equality, Empowerment and Growth*. The World Bank.

Baird, S., McIntosh, C., & Ozler, B. (2016). When the money runs out: Do cash transfers have sustained effects on human capital accumulation? *Mimeo*.

BAPPENAS. (2013). Review of a decade of gender mainstreaming in education in Indonesia. In *Ministry of National Development Planning/National Development Planning Agency (BAPPENAS)*.

Barekzai. (2014). *Women's Participation in School Management Shuras in Afghanistan: Obstacles and Opportunities for Women's Involvement in Government Schools in Kabul City*. Retrieved November 9, 2023, from www.diva-portal.org/smash/get/diva2:813316/FULLTEXT01.pdf

Batool, S. A., Tabassum, A., & Saghir, S. (2019). Dynamics of female labor force participation in Pakistan. *Pakistan Journal of Social Sciences*, *39*(3), 751–762.

Beegle, K. G., Serajuddin, U., Stacy, B. W., & Wadhwa, D. (2023). *Missing SDG Gender Indicators (English)*. Policy Research Working Paper; No. WPS 10544. Washington, DC: World Bank Group. http://documents.worldbank.org/curated/en/099401008102333778/IDU0948ba62905408048c609d430b3c1e874dc00s

Bell, A., & Juvonen, J. (2020). Gender discrimination, perceived school unfairness, depressive symptoms, and sleep duration among middle school girls. *Child Development*, *91*(6), 1865–1876.

Bian, L., Leslie, S. J., & Cimpian, A. (2017). Gender stereotypes about intellectual ability emerge early and influence children's interests. *Science*, *355*(6323), 389–391.

Blumberg, R. L. (2008). The invisible obstacle to educational equality: Gender bias in textbooks. *Prospects*, *38*, 345–361.

Brown, C. S., & Leaper, C. (2010). Latina and European American girls' experiences with academic sexism and their self-concepts in mathematics and science during adolescence. *Sex Roles*, *63*(11– 12), 860–870. https://doi.org/10.1007/s11199-010-9856-5

Bruce, R., Cavgias, A., Meloni, L., & Remígio, M. (2022). Under pressure: Women's leadership during the COVID-19 crisis. *Journal of Development Economics*, *154*(C).

Bursztyn, L., González, A. L., & Yanagizawa-Drott, D. (2020). Misperceived social norms: Women working outside the home in Saudi Arabia. *American Economic Review*, *110*, 2997–3029.

Bussemakers, C., van Oosterhout, K., Kraaykamp, G., & Spierings, N. (2017). Women's worldwide education–employment connection: A multilevel analysis of the moderating impact of economic, political, and cultural contexts. *World Development*, *99*, 28–41.

Cameron, L. A., Malcolm Dowling, J., & Worswick, C. (2001). Education and labor market participation of women in Asia: Evidence from five countries. *Economic Development and Cultural Change*, *49*(3), 459–477.

Canton, E. (2007). Social returns to education: Macro-evidence. *De Economist*, *155*, 449–468. https://doi.org/10.1007/s10645-007-9072-z

CBSE. (2013, May 14). *Guidelines to Schools for the Selection of Books for Classes I–VIII*. Central Board of Secondary Education. Retrieved November 29, 2023, from https://cbseacademic.nic.in/web_material/Circulars/201

Chatterjee, E., Desai, S., & Vanneman, R. (2018). Indian paradox: Rising education, declining women's employment. *Demographic Research, 38*, 855.

Chaudhuri, S., Morash, M., & Yingling, J. (2014). Marriage migration, patriarchal bargains, and wife abuse: A study of South Asian women. *Violence Against Women, 20*(2), 141–161.

Chhaochharia, V., Du, M., & Niessen-Ruenzi, A. (2022). Counter-stereotypical female role models and women's occupational choices. *Journal of Economic Behavior & Organization, 196*, 501–523.

Curaming, E. M., & Curaming, R. A. (2020). Gender (in) equality in English textbooks in the Philippines: A critical discourse analysis. *Sexuality & Culture, 24*(4), 1167–1188.

Dawes, J. H. P. (2022). SDG interlinkage networks: Analysis, robustness, sensitivities, and hierarchies. *World Development, 149*, 105693.

De Gioannis, E., Bianchi, F., & Squazzoni, F. (2023). Gender bias in the classroom: A network study on self and peer ability attribution. *Social Networks, 72*, 44–51.

De la Torre-Sierra, A. M., & Guichot-Reina, V. (2022). The influence of school textbooks on the configuration of gender identity: A study on the unequal representation of women and men in the school discourse during the Spanish democracy. *Teaching and Teacher Education, 117*, 103810.

DepEd. (2017). *Gender-Responsive Basic Education Policy*. The Department of Education (DepEd). Retrieved November 8, 2023, from www.deped.gov.ph/wp-content/uploads/2017/06/DO_s2017_032.pdf

Dewi, P., Rizki Yuliatin, R., & Eka Sari, D. (2022). The representation of gender roles in Indonesian national English textbooks for senior high schools. *Language, Discourse & Society, 10*(2), 35–46.

Dhar, D., Jain, T., & Jayachandran, S. (2020). *Reshaping Adolescents' Gender Attitudes: Evidence from a School-Based Experiment in India*. Working paper. Northwestern University.

Dildar, Y. (2015). Patriarchal norms, religion, and female labor supply: Evidence from Turkey. *World Development, 76*, 40–61.

Dollar, D., & Gatti, R. (1999). *Gender Inequality, Income, and Growth: Are Good Times Good for Women?* (Vol. 1). Washington, DC: Development Research Group, The World Bank.

The Economist. (2019). *Vietnam Has One of the Highest Shares of Women in Work in the World 8 June*.

Ejaz, M. (2007). Determinants of female labor force participation in Pakistan an empirical analysis of PSLM (2004–05) micro data. *Lahore Journal of Economics, 12*(Special Edition), 204–235.

England, P., Gornick, J., & Shafer, E. F. (2012). Women's employment, education, and the gender gap in 17 countries. *Monthly Labor Review, 135*, 3.

Filho, W. L., Kovaleva, M., Tsani, S., Ţîrcă, D., Shiel, C., Dinis, M. a. P., Nicolau, M., Sima, M., Fritzen, B., Salvia, A. L., Minhas, A., Kozlova, V., Doni, F., Spiteri, J., Gupta, T., Wakunuma, K., Sharma, M., Barbir, J., Shulla, K., . . . Tripathi, S. (2022). Promoting gender equality across the sustainable development goals. *Environment Development and Sustainability, 25*(12), 14177–14198. https://doi.org/10.1007/s10668-022-02656-1

Gaddis, I., & Klasen, S. (2014). Economic development, structural change, and women's labor force participation. *Journal of Population Economics, 27*, 639–681.

Garikipati, S., & Kambhampati, U. (2021). Leading the fight against the pandemic: Does gender really matter? *Feminist Economics, 27*(1–2), 401–418.

Goedderz, A., & Calanchini, J. (2023). Regional traditional gender stereotypes predict the representation of women in the workforce in 35 countries across five continents. *Current Research in Ecological and Social Psychology, 5*, 100138.

Gonzales, M. C., Jain-Chandra, M. S., Kochhar, M. K., & Newiak, M. M. (2015). *Fair play: More equal laws boost female labor force participation*. International Monetary Fund.

Good, J. J., Woodzicka, J. A., & Wingfield, L. C. (2010). The effects of gender stereotypic and counter-stereotypic textbook images on science performance. *The Journal of Social Psychology, 150*(2), 132–147.

Government of Pakistan. (2012). *The Gazette of Pakistan, Extraordinary Published by Authority—Acts, Ordinances, President's Order and Regulations*.

Govinda, R., & Bandyopadhyay, M. (2010, May). Changing framework of local governance and community participation in elementary education in India. *Consortium for Research on Educational Access, Transitions and Equity (CREATE), Research Monograph No. 3*.

Grossman, M. (2006). Education and nonmarket outcomes. *Handbook of the Economics of Education, 1*, 577–633.

Ilkkaracan, I. (2012). Why so few women in the labor market in Turkey? *Feminist Economics, 18*(1), 1–37. https://doi.org/10.1080/13545701.2011.649358

ILO (2023). *Employment in Afghanistan in 2022: A Rapid Impact Assessment*. www.ilo.org/asia/publications/issue-briefs/WCMS_869949/lang-en/index.htm

Ince, M. (2010). How the education affects female labor force? Empirical evidence from Turkey. *Procedia-Social and Behavioral Sciences*, *2*(2), 634–639.

Islam, K. M. M., & Asadullah, M. N. (2018). Gender stereotypes and education: A comparative content analysis of Malaysian, Indonesian, Pakistani and Bangladeshi school textbooks. *PLoS ONE*, *13*(1), e0190807.

Jayachandran, S. (2015). The roots of gender inequality in developing countries. *Annual Review of Economics*, *7*(1), 63–88.

Jayachandran, S. (2021). Social norms as a barrier to women's employment in developing countries. *IMF Economic Review*, Springer.

Kabeer, N. (2003). Gender equality, poverty eradication and the millennium development goals: Promoting women's capabilities and participation. *Women in Development Discussion Paper Series*, *1*, 26.

Kaffenberger, M., & Pritchett, L. (2020). Aiming higher: Learning profiles and gender equality in 10 low- and middle-income countries. *International Journal of Educational Development*, *79*(C).

Kaffenberger, M., & Pritchett, L. (2021). Effective investment in women's futures: Schooling with learning. *International Journal of Educational Development*, *86*(C).

Karlsson, P., & Mansory, A. (2007). An Afghan dilemma: Education, gender and globalisation in an Islamic context. *Studies in International and Comparative Education*, *72*. https://doi.org/10.25656/01:22648

Karniol, R., & Gal-Disegni, M. (2009). The impact of gender-fair versus gender-stereotyped basal readers on 1st-grade children's gender stereotypes: A natural experiment. *Journal of Research in Childhood Education*, *23*(4), 411–420.

Keats, A. (2018). Women's schooling, fertility, and child health outcomes: Evidence from Uganda's free primary education program. *Journal of Development Economics*, *135*, 142–159.

Khalid, Z., & Ghania, O. (2019). Gender positioning in the visual discourse of Algerian secondary education EFL textbooks: Critical image analysis vs teachers' perceptions. *Journal of Language and Linguistic Studies*, *15*(3), 773–793.

Khattak, S., & Akhtar, H. (2017). Gender mainstreaming in textbooks discourse via the metaphorical account of Malalai of Maiwand, Afghanistan. In *Globalisation and Historiography of National Leaders: Symbolic Representations in School Textbooks* (pp. 229–243).

Kim, A. Y., Sinatra, G. M., & Seyranian, V. (2018). Developing a STEM identity among young women: A social identity perspective. *Review of Educational Research*, *88*(4), 589–625.

Klasen, S., & Pieters, J. (2012). *Push or Pull? Drivers of Female Labor Force Participation During India's Economic Boom*. IZA Discussion Paper 6398.

Klasen, S., & Pieters, J. (2015). What explains the stagnation of female labor force participation in Urban India? *The World Bank Economic Review*, *29*(3), 449–478.

Kuruvilla, M., & Thasniya, K. T. (2015). Gender stereotypism in the pictorial depictions of primary school text books. *IOSR Journal of Humanities and Social Science*, *20*(6), 16–26.

Lahoti, R., & Swaminathan, H. (2016). Economic development and women's labor force participation in India. *Feminist Economics*, *22*(2), 168–195.

Laumann, F., von Kügelgen, J., Uehara, T. H. K., & Barahona, M. (2022). Complex interlinkages, key objectives, and nexuses among the Sustainable Development Goals and climate change: A network analysis. *The Lancet Planetary Health*, *6*(5), e422–e430.

Le Blanc, D. (2015). Towards integration at last? The sustainable development goals as a network of targets. *Sustainable Development*, *23*(3), 176–187.

Llego. (n.d.). *DepEd Guidelines on the Establishment of School Governance Council (SGC)*. TEACHERPH. Retrieved November 12, 2023, from www.teacherph.com/deped-guidelines-school-governance-council-sgc/

Mahmood, S., Kaur, S., & Daghigh, A. J. (2021). Role of gender in Pakistani ELT textbooks: A critical discourse analysis approach. *Journal of Cultura and Lingua*, *2*(1), 9–14.

Manandhar, M., Hawkes, S., Buse, K., Nosrati, E., & Magar, V. (2018). Gender, health and the 2030 agenda for sustainable development. *Bulletin of the World Health Organization*, *96*(9), 644.

McKellar, S. E., Marchand, A. D., Diemer, M. A., Malanchuk, O., & Eccles, J. S. (2019). Threats and supports to female students' math beliefs and achievement. *Journal of Research on Adolescence*, *29*(2), 449–465. https://doi.org/10.1111/jora.12384

Ministry of Education. (2016). National education strategic plan 2017–2021. In *Government of Islamic Republic of Afghanistan*.

Ministry of Education, Malaysia. (2013). *Malaysia Education Blueprint 2013–2025*. Retrieved November 29, 2023, from www.moe.gov.my/menumedia/media-cetak/penerbitan/dasar/1207-malaysia-education-blueprint-2013-2025/file

Ministry of Finance. (2017). Gender budgeting report 2017–18. In *Ministry of Finance*. Ministry of Finance, Government of the People's Republic of Bangladesh. Retrieved November 29, 2023, from https://mof.portal.gov.bd/site/page/3bb14732-b5b1-44df-9921-efedf1496295

Ministry of National Education. (2005). *Strategic Plan 2005–2009*. Ministry of National Education, the Republic of Indonesia. Retrieved November 8, 2023, from https://planipolis.iiep.unesco.org/sites/default/files/ressources/indonesia_educationplan2005-2009.pdf

MOET. (2013). Vietnam EFA Action Plan 2003–2015. In *Review and Update 2012*. Ministry of Education and Training (MOET), Vietnam. Retrieved November 8, 2023, from https://planipolis.iiep.unesco.org/2012/vietnam-efa-action-plan-2003-2015-review-and-update-2012-5589

Nandwani, B., & Jain, C. (2022). Female representation in school management and school quality. *Indira Gandhi Institute of Development Research, Working Paper, 2022–002*.

National Curriculum Council. (2020). Single national curriculum. National curriculum. In *Ministry of Federal Education and Professional Training, Government of Pakistan*.

NCERT. (2005). National curriculum framework 2005. In *National Council of Educational Research and Training (NCERT)*. Retrieved November 3, 2023, from https://ncert.nic.in/pdf/nc-framework/nf2005-english.pdf

NCERT. (2023, August). National curriculum framework for school education 2023. In *National Council for Educational Research and Training (NCERT)*.

Nerini, F., Tomei, J., To, L. S., Bisaga, I., Parikh, P., Black, M., . . . & Mulugetta, Y. (2018). Mapping synergies and trade-offs between energy and the Sustainable Development Goals. *Nature Energy*, *3*(1), 10–15.

Nilsson, M., Griggs, D., & Visbeck, M. (2016). Policy: Map the interactions between Sustainable Development Goals. *Nature*, *534*(7607), 320–322.

Nohemi, H. P. (2015). Can gender-stereotyped depictions of occupations in primary school textbooks help shape students' career choice in the Dominican Republic? *CUNY Academic Works*. https://academicworks.cuny.edu/cgi/viewcontent.cgi?article=1530&context=cc_etds_theses

Olsson, M., & Martiny, S. E. (2018). Does exposure to counterstereotypical role models influence girls' and women's gender stereotypes and career choices? A review of social psychological research. *Frontiers in Psychology*, *9*, 2264.

Orfan, S. N. (2022). Gender voices in Afghanistan primary and secondary school English textbooks. *The Curriculum Journal*, 1–23. https://doi.org/10.1002/curj.178

Paul. (2014, October). *General Standard of Pakistani Textbooks*. Pakistan Minorities Teachers' Association. Retrieved November 29, 2023, from www.itacec.org/document/learning_resources/2014/GENERAL%20STANDARD%20OF%20PAKISTANI%20TEXTBOOKS.pdf

Pham-Truffert, M., Metz, F., Fischer, M., Rueff, H., & Messerli, P. (2020). Interactions among sustainable development goals: Knowledge for identifying multipliers and virtuous cycles. *Sustainable Development*, *28*(5), 1236–1250. https://doi.org/10.1002/sd.2073

Pradhan, P., Costa, L., Rybski, D., Lucht, W., & Kropp, J. P. (2017). A systematic study of sustainable development goal (SDG) interactions. *Earth's Future*, *5*, 1169–1179.

Pritchett, L., & Sandefur, J. (2020). Girls' schooling and women's literacy: Schooling targets alone won't reach learning goals. *International Journal of Educational Development*, *78*(C).

Psacharopoulos, G., & Patrinos, H. A. (2004). Returns to investment in education: A further update. *Education Economics*, *12*(2), 111–134.

Psacharopoulos, G., & Tzannatos, Z. (1987). *Education and Female Labor Force Participation*. Paper presented at the Sixth World Congress of Comparative Education, Rio de Janeiro, July 6–10.

Republic of Philippines. (2013). *Implementing Rules and Regulations of the Enhanced Basic Education Act of 2013 (Republic Act No. 10533)*. Retrieved September 2023, from www.officialgazette.gov.ph/2013/09/04/irr-republic-act-no-10533/

Rogers, A. A., Boyack, M., Cook, R. E., & Allen, E. (2021). School connectedness and STEM orientation in adolescent girls: The role of perceived gender discrimination and implicit gender-science stereotypes. *Sex Roles*, *85*, 405–421.

Rosengren, L. M., Schinko, T., Sendzimir, J., Mohammed, A. R., Buwah, R., Vihinen, H., & Raymond, C. M. (2023). Interlinkages between leverage points for strengthening adaptive capacity to climate change. *Sustainability Science*, 1–20.

Sarvarzade, S., & Wotipka, C. M. (2017). The rise, removal, and return of women: Gender representations in primary-level textbooks in Afghanistan, 1980–2010. *Comparative Education*, *53*(4), 578–599. www.tandfonline.com/doi/abs/10.1080/03050068.2017.1348021

Savenye, W. C. (1990). Role models and student attitudes toward nontraditional careers. *Educational Technology Research and Development*, *38*, 5–13. https://doi.org/10.1007/BF0229817

Sayed, Y., & Moriarty, K. (2020). SDG 4 and the 'education quality turn': Prospects, possibilities, and problems. In Wulff, A. (Ed.), *Grading goal four: Tensions, threats, and opportunities in the Sustainable Development Goal on quality education Grading goal four* (pp. 194–213). Brill. https://library.oapen.org/bitstream/handle/20.500.12657/53875/9789004430365.pdf?sequence=1#page=221

Sedere. (2011, March 16). *Guidelines for Curriculum Review and Development National Curriculum and Textbook Board—Bangladesh*. Institute of Educational Development BRAC University. Retrieved November 29, 2023, from https://files.eric.ed.gov/fulltext/ED533200.pdf

Senarath, U., & Gunawardena, N. S. (2009). Women's autonomy in decision making for health care in South Asia. *Asia Pacific Journal of Public Health*, *21*(2), 137–143.

Shahabuddin, A. S. M., Nöstlinger, C., Delvaux, T., Sarker, M., Bardají, A., De Brouwere, V., & Broerse, J. E. W. (2016). What influences adolescent girls' decision-making regarding contraceptive methods use and childbearing? A qualitative exploratory study in Rangpur District, Bangladesh. *PLoS ONE*, *11*(6), e0157664. https://doi.org/10.1371/journal.pone.0157664

Sianesi, B., & Reenen, J. V. (2003), The returns to education: Macroeconomics. *Journal of Economic Surveys*, *17*, 157–200. https://doi.org/10.1111/1467-6419.00192

Starr, C. R., & Simpkins, S. D. (2021). High school students' math and science gender stereotypes: Relations with their STEM outcomes and socializers' stereotypes. *Social Psychology of Education*, *24*(1), 273–298. https://doi.org/10.1007/s11218-021-09611-4

Suchana, A. A. (2020). Investigating gender equity in a primary level English language textbook in Bangladesh. In *The Routledge Handbook of English Language Education in Bangladesh* (pp. 298–311). Routledge.

Swain, R. B., & Ranganathan, S. (2021). Modeling interlinkages between sustainable development goals using network analysis. *World Development*, *138*, 105136.

Teliousi, V., Zafiri, M., & Pliogou, V. (2020). Occupation and gender stereotypes in primary school: The case of the English language coursebooks in Greek primary schools. *Universal Journal of Educational Research*, *8*(4), 1135–1148.

UNDP (United Nations Development Programme) (2022). *Human Development Report 2021–22: Uncertain Times, Unsettled Lives: Shaping Our Future in a Transforming World*. New York.

UNESCO. (2021). *The Right to Education: What's at Stake in Afghanistan?* Retrieved November 7, 2023, from https://en.unesco.org/sites/default/files/afghanistan_v11.pdf

UNHCR. (2023). *Afghanistan: UN Experts Say 20 Years of Progress for Women and Girls' Rights Erased Since Taliban Takeover*. www.ohchr.org/en/press-releases/2023/03/afghanistan-un-experts-say-20-years-progress-women-and-girls-rights-erased

Vladimirova, K., & Le Blanc, D. (2016). Exploring links between education and sustainable development goals through the lens of UN flagship reports. *Sustainable Development*, *24*(4), 254–271.

Vu, M. T., & Pham, T. T. T. (2023). Still in the shadow of Confucianism? Gender bias in contemporary English textbooks in Vietnam. *Pedagogy, Culture & Society*, *31*(3), 477–497.

Wafa, D. (2021). Reinforced stereotypes: A case study on school textbooks in Egypt. *Journal of International Women's Studies*, *22*(1), 374–385.

Wang, M., Janssen, A. B. G., Bazin, J., et al. (2022). Accounting for interactions between Sustainable Development Goals is essential for water pollution control in China. *Nature Communications*, *13*, 730. https://doi.org/10.1038/s41467-022-28351-3

Wijetunga, L. M. S. (2017). *Sexism in ESL Textbooks in Sri Lanka: A Case Study of the GCE O/L Textbooks*.

Wu, H. (2022). The effect of maternal education on child mortality in Bangladesh. *Population and Development Review*. *Women in Politics in Vietnam*. (2006). Women's Environment & Development Organisation. Retrieved November 8, 2023, from https://wedo.org/tag/women-leaders/

Wu, X., Fu, B., Wang, S., Song, S., Li, Y., Xu, Z., . . . & Liu, J. (2022). Decoupling of SDGs followed by re-coupling as sustainable development progresses. *Nature Sustainability*, *5*(5), 452–459.

Yousefy, A., & Baratali, M. (2011). Women, employment and higher education schoolings. *Procedia-Social and Behavioral Sciences*, *15*, 3861–3869. https://doi.org/10.1016/j.sbspro.2011.04.386

8

MEASURING THE IMPACT OF SDG 5 ON EMPOWERMENT IN INDIA

Pritha Dev, Ananya Acharya, and Astha Kushwaha

8.1 Introduction

It was not too long ago that the discussion of women's empowerment entered global agendas for international organizations. While the UN Charter of Human Rights gave equal rights to men and women in 1975, it was only in the 1995 Beijing Declaration and Platform for Action which critical areas for women's empowerment were emphasized. The next wheel that set this global goal was the Millennium Development Goals (MDG). With the mixed performance of the MDGs, the Sustainable Development Goals (SDG) were launched, with a target of eliminating 17 crucial socio-economic problems by 2030. The SDGs were drawn from the lessons taken from the MDGs and are a much-refined version of its predecessor. For example, one of the significant downsides of the MDGs was that they focused primarily on developing countries (assuming the developed would fund the developing countries) and on economic goals (Kumar et al., 2016). The SDGs emphasize focus on all countries and about equal importance on all three pillars: economic, social, and environmental goals. While the methodology of determining targets for the MDGs and their outcomes can be contested, MDGs acted as the first platform that gave world gender equality a separate target. With a little more than five years remaining to achieve the Sustainable Development Goals (SDGs), it is crucial to look back at its initial promise.

Across the vast literature that attempts to define and measure empowerment, Kabeer (1999) brings up the essence of empowerment: "In fact, human agency is indeterminate and hence unpredictable in a way that is antithetical to requirements of measurement." However, some definitions have tried to encompass or at least compartmentalize the vastness of empowerment, enhancing the measurement's robustness. Duflo (2012) emphasizes that supporting either gender equality or economic development in isolation will not be able to push the other up consistently. Particularly for the SDGs, unless progress on gender equality is accelerated, the global community will not only fail to achieve SDG 5 but will also forgo the catalytic effect that gender equality can have for achieving the 2030 Agenda more broadly (UN, 2018).

8.2 Interlinkages of SDG 5 With the Other SDGs

SDG 5 is interdisciplinary, as gender equity cannot occur in isolation. It is connected with SDG 1, which aims at no poverty and equal rights for men and women. It is related to SDG 2, which

DOI: 10.4324/9781003285472-9

doubles agricultural productivity, particularly of women farmers. It applies to SDG 3, which seeks to provide better maternal care, and SDG 4, which aims at providing equitable quality education. SDG 6 strives to improve sanitation for women and girls, and SDG 7 ensures access to clean fuel for enhancing women's health. SDG 8, SDG 10, SDG 13, SDG 15, and SDG 16 are also aligned with SDG 5 since they aim to improve women's quality of life.

8.3 SDG 5 in India

While India joined 193 other countries at the United Nations in September 2015 to commit to the 17 SDGs, the baseline for impact evaluation of the initiatives undertaken to achieve the SDGs is 2018. As of 2018, India got an overall score of 57 across all the SDGs. Over the last five years, India has risen to 66. In terms of SDG 5, the growth has been relatively slow. The SDG 5 rank in India was at 36, which, as of 2022, is at 48, the latest score stagnant over the last three years. State-wise, Kerala, Himachal Pradesh, Tamil Nadu, Andhra Pradesh, and Goa are the forerunners of SDG 5. However, in terms of maximum progress over the last five years, Bihar, Uttar Pradesh (UP), Madhya Pradesh, Goa, Himachal Pradesh, and Tamil Nadu have made an enormous difference in jumping a minimum of 20 ranks. At the same time, some states have had negative changes or drops in their ranks. Telangana and Assam did poorly over the last five years under SDG 5.

8.4 Policies Undertaken by the Center and State to Strengthen Gender Equality

The central government, in line with the principle of cooperative federalism and larger downstream channelization of funds to the states, emphasized the localization of the SDGs, thereby making the states the primary stakeholder in managing and implementing policies and, thereby, taking responsibility for making improvements in line with the SDGs in their states. NITI Aayog has been entrusted with conducting reviews and audits periodically to note the states' progress and assist them with challenges. However, some flagship programs of the center were launched earlier or in line with the SDG timeline and do work toward addressing the challenges targeted by the SDGs, such as *Poshan Abhiyaan (National Nutrition Mission)*, which is a multi-ministerial convergence mission whose goal is to achieve a malnutrition-free India by 2022. The mission's objectives are to reduce stunting in the highest-malnutrition districts in India by enhancing the quality of Anganwadi service delivery and enhancing the utilization of essential services; additionally, the mission aims to ensure the holistic development and adequate nutrition of pregnant women, mothers, and children. Under the *Ayushman Bharat (National Health Protection)* scheme, the government guarantees that over ten crores of impoverished and vulnerable families will be covered, including secondary and tertiary hospitalization up to ₹5 lakh per family annually. The *Beti Bachao Beti Padhao (Care for the Girl Child)* initiative seeks to alter national attitudes while addressing the societal inequalities that girls experience. The program aims to tackle many concerns, such as lowering the child sex ratio (CSR), promoting women's empowerment, eradicating gender-based inequality, and protecting girls from harm. Additional program goals include educating girls and encouraging their participation in society.

To address gender equality, the center has been committed to launching several redistributive welfare policies, increasing fiscal expenditure, enabling financial incentives and smooth disbursal of credit to women, to name a few themes. The government has launched schemes like *Mahila Shakti Kendra (MSK)*, which was established under the Pradhan *Mantri Mahila Shashaktikaran Yojana (PMMSY)* umbrella program, which gives rural women a forum to contact the government to assert their rights and become more empowered through capacity building, dissemination of knowledge, and training. The goal of *Women Working Hostel (WWH)* is to encourage the availability of secure and convenient housing for working women, along with childcare services when feasible, in urban, semi-urban, and even rural regions where there are opportunities for employment for women. The program supports initiatives for the construction of new hostels, the enlargement of current hostels, and the construction of hostels in rented spaces. *Mahila Police Volunteers (MPV)* is a central sector scheme that intends to serve as a public policy interface to help women in distress and enable police outreach on gender issues; this plan aims to establish a connection between the local community and the police authorities. *Rashtriya Mahila Kosh (RMK)* provides women working in the unorganized sector with hassle-free, client-friendly microcredit for income-generating activities without the need for collateral. To promote the ideas of microfinance, enterprise development, thrift, and credit, and the establishment and bolstering of women SHGs through intermediate organizations, under the *Pradhan Mantri Matru Vandana Yojna (PMMVY)* scheme, for the first living child in the family, a monetary incentive of approximately $5,000 is sent directly to the bank or post office account of pregnant women and lactating mothers, provided that they meet specific requirements related to maternal and child health. *Pradhan Mantri Kaushal Vikas Yojana (PMKVY)* encourages a significant number of Indian adolescents to enroll in industry-relevant skill training that helps them secure better employment opportunities; the fees for training and assessments are fully covered by the government under this scheme. *Sukanya Samriddhi Yojana (SSY)* is a government savings program designed to assist girls as part of the *Beti Bachao, Beti Padhao* effort. Under this program, a 10-year-old or younger girl-kid's parent or guardian may open an account. This plan offers multiple tax benefits at a higher interest rate. *Pradhan Mantri Mudra Yojana* is a program for providing micro and small businesses with financing at reasonable rates. PM Mudra plans aim to "fund the unfunded" or integrate businesses into the established financial system. Loans under the PM Mudra lending plan are offered to micro, small, and non-farm businesses that generate revenue through trading, manufacturing, and services.

The NITI Aayog has published two reports, in 2019 and 2022, respectively, addressing schemes or programs undertaken by the states to address the 17 SDGs. For the upliftment of the groups in situations of vulnerability, the state of Bihar has launched several schemes, like the *Mukhya Mantri Balika Cycle Yojana*, which ensures that every participating girl in grades 9 or above receives funding under the scheme to purchase a bicycle for use as a form of transportation to school. This program's primary goal is to prevent any girl-child from quitting school due to the distance to the school. *Mukhya Mantri Balika Poshak Yojana* started drawing girls and their guardians to educational institutions. Girls who consistently study in classes 1–8 are given a set amount of 600 rupees for classes I to II, 700 rupees for classes III to V, and 1,000 rupees for classes VI to VIII to purchase school outfits. Meanwhile, the objective of the *Mukhya Mantri Nari Shakti Yojana* scheme is integrated work related to social empowerment, cultural empowerment, and economic empowerment associated with the development of women. *Akshar Aanchal Yojana*'s goal is to connect the underprivileged children of the Mahadalit, Alpsankhyak, and "extremely backward class" categories to schools and to give elementary education to women. *Hunar*'s scheme is for minority women to take up the skill development of Bihari girls from marginalized and socially vulnerable groups. The intention is to integrate the girls into society and guarantee their socio-economic and educational empowerment, by encouraging self-employment and employment. Goa, another forerunner, has launched the *Grah Aadhar* scheme for women to address the problem of spiraling prices and to provide support to housewives/homemakers from the middle, lower-middle, and poor sections of society in maintaining a reasonable standard of living for their families, and for the *Ladli Laxmi* scheme, the state government provides a gift of Rs. 1 lakh to every village girl on the occasion of their marriage. The state has also worked toward legislation increasing women's representation in local bodies. Himachal Pradesh, in fact, has launched *Swaran Jayanti Poshahar Yojna* to eliminate malnutrition in women and children, and *Wo Din Yojna* to raise awareness about menstruation among women and adolescent girls. The state of Kerala has held high rankings since the baseline report in SDG 5 and has made consistent efforts to increase its score further by launching 259 women-related programs under the Kerala Institute of Local Administration: *Jaagratha Samithi* is a vigilant group of citizens that plays a vital role in preventing crimes against women. The Gender Resource Centre coordinates the activities for gender development and women empowerment under the local government. Under the *Nirbhaya* policy, the Government of India set up a dedicated fund, called the *Nirbhaya Fund*, to pursue measures targeted at improving the safety and security of women in the country and for the implementation of initiatives aimed at enhancing women's safety and its flagship program, *Kudumbashree*, for women empowerment. Kudumbashree works toward improving the standard of living of poor women in rural and urban areas by setting up microcredit facilities and productive enterprises. It has also launched Pink Patrol for the safety of women and children.

Maharashtra has had a shaky trajectory toward improving SDG 5. It scored 43 at the baseline, followed by 36 in 2019, and has been 51 since then. To provide direct financial support to the poorest women in rural areas, the state government launched a scheme under the *Tejashree Financial Services* through various measures, such as an ultrapoor women's loan scheme, financial assistance to women to repay debts, and working capital for social enterprises. The state government has published the Gender Budget and Child Budget for 2020–2021, in collaboration with UNICEF, and designed programs, such as the Livelihood Programme for Rural Women, to reach the most marginalized, to be implemented by *Mahila Arthik Vikas Mahamandal (MAVIM)* PSU under the Human Development Mission for 125 blocks in 23 districts from 2019–2020 to 2021–2022. This provides financial services to about 1,12,900 women, directly contributing to SDGs 1, 5, 8, and 10.

An interesting initiative launched by the state of Jharkhand is thrust on the *POSHAN Abhiyaan*, which is on the first "golden" 1,000 days of the child and preventive care for adolescent girls and women. Poshan Abhiyan focuses on the thematic convergence of nutrition with hygiene, water and sanitation, growth monitoring, complementary feeding, antenatal checkups, anemia, ECCE, and food fortification and micronutrients. To improve child sex ratio at birth, the government of Haryana has strictly implemented the PC-PNDT Act by involving the Women and Child Development and Health Departments. A few initiatives have also been taken by Arunachal Pradesh, such as incentivizing girls' education and operationalizing women police stations to promote safety and security among women, and Punjab launched *Mata Tripta Mahila Yojna*, for the empowerment of women-headed households. Uttarakhand has been active in nearly 30,874s, such as *Sankalp*, for skill development training for women and candidates of weaker sections. The *Sarva Utthan Sarva Samridhhi* campaign has been launched by the chief minister to address the socio-economic issues of people in situations of vulnerability.

However, it must be noted that these policies or increasing fiscal expenditures for gender equality do not always improve empowerment measures. For example, Bihar, as stated earlier, is a forerunner in the drastic improvement it has received in the last five years. Despite all the policy efforts and nearly 30,874 crores allocated from the budget for gender, as of 2019, Bihar falls short of the institutional facility to implement the funds received for improving gender equality. Nearly 46% of women in Bihar are illiterate, increasing their chances of being married early (2020). This will not be reflected in the SDG, as the SDG needs to account for education. While it accounts for domestic violence, it does not account for the age of marriage and autonomy of decision-making for strategic life decisions. Likewise, it lacks policies that effectively address gendered issues, such as piped water and sanitation, safety in public spaces, and gender-sensitized infrastructure development.

Its neighboring state, Uttar Pradesh, has also done exceptionally well in SDG 5. With the latest NFHS-5 numbers, it has nearly reached the ideal sex ratio at birth with 945, which is close to the norm of 950. It also has reduced domestic violence from 36% to 34.8%. While this is commendable, a contention against the state is that it is still a record holder for the highest cases of "missing child" at birth. While violence has reduced, it does not imply a safer state for women living in UP, as police reported nearly 50,000 violence cases against women. Moreover, the state has been declared to still engage in shoddy paperwork post-reporting cases of rape and has reported instances of activists calling out victim-blaming by state authorities.

These cases suggest that while the states may be improving extensively per the SDG parameters, much more must be done to achieve true gender equity.

The Niti Aayog uses various indicators for measuring the targets specified by SDG 5. For measuring Target 1 (95–100), the sex ratio at birth (female per 1,000 male) and the average female-to-male ratio of average salaries/wages received per day by regular salaried employees aged 15 to 59 for rural and urban areas are used. These mainly account for male-child preference and the gender pay gap to some degree (1). Target 2 (0) is measured using the percentage of married women between the ages of 15 and 49 who have ever experienced domestic violence. The percentage of women who have won general elections to the state legislative assembly and the ratio of female labor participation to male labor force participation are utilized to measure Target 5.1. Target 6.100 is accounted for using the percentage of women between the ages of 15 and 49 using modern family planning methods.

8.5 Empowerment and Outcomes

We now empirically evaluate how the SDG 5 measures are linked over time in India. We evaluate how past performance of an SDG 5 target impacts current achievements toward gender equality. We narrow our focus on two aspects of empowerment as measured by the impact of decision-making and freedom of movement. We then evaluate how past levels of the two impact outcomes for women regarding family planning, years of education, and present employment status.

8.6 Empowerment in Literature

"Empowerment" is vast and means different things to different people in other contexts. Mosedale (2005) presents criteria to ascertain empowerment. First, one must have been disempowered in order to be empowered. Second, a third party cannot grant empowerment. Third, the ability to make and carry out decisions about crucial aspects of one's life is typically cited in definitions of empowerment. Finally, empowerment is a continuous process as opposed to a finished good. Batliwala (1993) cites an NGO activist who likes the term *empowerment* "because no one has defined it yet; so it gives us breathing space to work it out into action terms before we have to pin ourselves down to what it means." Kabeer (1999) states that for some feminists, the value of the concept lies in its fuzziness. However, the importance of this variable commands for a method of defining and measuring empowerment. The chapter builds empowerment on three interrelated dimensions of choice: resources, agency, and achievements. Having access to resources is a precondition for having the ability to exercise a choice. *Resources* refer to control over physical, financial, human, and intellectual resources (Kabeer, 1994), while *agency* implies having the capability and freedom to make individual life choices (Desai, 2010; Sen, 1992, 1999, 2009). Although it is not an absolute assurance, controlling resources can act as a "catalyst for empowerment." Resources cannot lead to empowerment if women cannot, individually or collectively, identify and use resources in their interests (Malhotra et al., 2002). *Agency* is the process of defining one's goals and attaining them, while *achievements* are the outcomes achieved by exercising the choice. One of the definitions of *women's empowerment* is "an expansion in the range of potential choices available to women so that actual outcomes reflect the particular set of choices which the women value" (Kabeer, 2008). This rationale is similar to Sen's capability approach, which builds on the foundation of certain doings and beings that people have a reason to value. Mahmud et al. (2012) emphasizes three considerations one must undertake while measuring empowerment. Firstly, empowerment is not directly measurable but approximated using various indicators. Secondly, empowerment is a multidimensional process. Gender inequality is not one homogenous phenomenon but a bunch of disparate and interlinked problems (Sen, 2001). Thus, the pathways through which resources convert into agency can be varied: cognitive, relational, perceptual, and material (Chen & Mahmud, 1995). Thirdly, context is pertinent to measuring women's empowerment. The pathways of change vary between contexts, and even within the same context, all women may not experience empowerment within the exact dimensions. A woman may be empowered in one area or aspect of life but not in another (Kishor, 1995, 2000). Nussbaum (2000) argues that confusing cultural relativism with respect for diversity ignores the fact that most cultures have exhibited considerable intolerance of diversity and prevents us from adopting any more general norm of tolerance that could help us limit the intolerance of cultures. This is particularly applicable to patriarchal societies in developing countries. Nussbaum continues to say:

> If divorce and career difficulties are painful, as they surely are, they are a lot less painful than being unable to work when one is starving because one will be beaten if one goes outdoors, or being unable to leave an abusive marriage because of illiteracy and lack of employable skills.

Mahmud et al. (2012) reiterates that in a patriarchal society, where women are customarily housebound, an indicator of freedom of mobility, for instance, is far more pertinent than in a Western society.

The majority of scholars in the field concur that there are a few crucial variables that determine or affect empowerment. Age, gender, marital status, citizenship, social position, economic activity, intra-household distribution, decision-making, and health are a few of these (Sen, 1992, 1999, 2009; Kabeer, 1994; Trommlerová et al., 2015). The cultivation of women's empowerment and agency through such measures as women's education and gainful employment has shown to be particularly helpful in eliminating gender inequities' harshest consequences, especially mortality rates. According to Eyben et al. (2008), when women are economically empowered, they have more access to economic opportunities and resources. Women's empowerment involves women taking proactive steps to challenge and change the established societal norms and cultural values that may have restricted their well-being and opportunities. Financial empowerment helps women challenge traditional gender roles and improve their overall well-being, contributing to their empowerment (Swain & Wallentin, 2009). It is vital to appreciate the perspective taken by SDG 5, beyond the typical MDG focus on primary schools, and completion of secondary education and beyond has consistently extensive positive connections, suggesting there may be significant threshold benefits for education (Hanmer & Klugman, 2016). However, according to Drydyk (2013), empowerment is fleeting when there is a transient improvement in one's ability to make decisions and exert influence over how their life is shaped. However, the benefit is insufficient to endure strict societal and institutional opposition. Thus, he suggests the idea of "durable empowerment," which has two elements. The first is to increase one's ability to affect one's own life more, and the second is to maintain one's ability to influence over the long term. Therefore, empowerment that contains the first and lacks the second components is less valuable and less durable.

In addition, the position of a woman within the household structure is vital for her empowerment in terms of her participation in the household's decision-making (Ayevbuomwan, 2016). In some regions of sub-Saharan Africa (SSA), patriarchal authority makes it difficult for women to exercise their fundamental rights to reproductive health (Greno, 2021; Haider, 2013). In addition, some family planning programs in SSA failed to meet their goals because they ignored the power dynamics between women and their partners.

However, as Duflo (2012) points out, improving women's active agency can significantly improve the lives of men, women, children, and adults (Sen, 2001). According to Pratley P. and Sandberg, J. F. (2018), one social factor influencing maternal and child health is women's empowerment. In low- and middle-income countries (LMICs), it has a favorable correlation with women's use of contraception (James-Hawkins et al., 2018). Although there seem to be beneficial connections between reducing poverty and promoting economic growth, more is needed (Hanmer & Klugman, 2016).

8.7 Methodology

We use data from the National Family Health Survey (NFHS) Round 4 and Round 5, conducted during 2015–2016 and 2019–2021, respectively, for our study. The National Family Health Survey refers to India's Demographic and Health Survey (DHS) data. It is a multi-round survey that provides nationally representative information over a large sample of the country's population. For instance, NFHS-4, conducted in 2015–2016, surveyed 601,509 households across India, with individual-level information on 699,686 women. The NFHS-5 sample comprises 636,699 households surveyed, with 724,115 qualified women aged 15 to 49 participating in individual women's interviews. We first use NFHS-4 to calculate the level of empowerment at the district level.

One of our key independent variables is the decision-making capability of the women in the district using NFHS-4. Decision-making for an individual woman is measured by calculating the total number of decisions made from strategic decisions on health care, household purchases, visiting family, and expanding the husband's income. Our second independent variable is freedom of movement, which is calculated at the individual level using the total mobility of a woman when visiting the market, medical facility, and outside the village. We aggregate these variables to the district level to understand how much say women of the district have in decision-making and their ability to move freely.

The dependent variables used for this study are current employment status, years of education, and family planning (using contraceptive methods). We aim to see how individual-level values of these dependent variables are impacted by past levels of empowerment in the district as measured by the past district-level averages for decision-making and freedom of movement.

The summary statistics of the variables used are presented in Table 8.1.

We focus on understanding how past levels of empowerment in a district impact future outcome for women. Our outcomes of interest are women's education and employment. These variables were chosen based on the available data and the variables' relevance for SDG 5. Additionally, decision-making and freedom of movement are critical to achieving the SDGs, and previous research suggests using these variables to study the relationship (Miedema et al., 2018; Phan, 2016).

We have regressed education, employment, and family planning from NFHS-5 on previous NFHS rounds' "decision-making" and "freedom of movement" to understand the relationship between the said variables. Our regression model is as follows:

$$Y_{id,5} = B_o + B_1 DM_{d,4} + B_2 FoM_{d,4} + X_{id,5} + State\ FE + \epsilon_{id,5}$$

Where $Y_{id,5}$ captures the dependent variable of female respondent i from district d, which was covered in NFHS-5. We use three dependent variables capturing employment status, years of

Table 8.1 Summary statistics of NFHS-4 and NFHS-5

Variable	*Definition*	*Mean (std dev)*
Decision-making		
A person who usually decides on the respondent's health care	Coded as 3 if the respondent takes a decision alone, coded as 2 if the decision is taken jointly with the partner/husband, and coded as 1 otherwise.	2.418 (1.062)
A person who usually decides on large household purchases		2.513 (1.104)
A person who usually decides on visits to family or relatives		2.471 (1.071)
A person who usually decides what to do with money their husband earns		2.599 (1.194)
Freedom of movement		
Usually allowed to go to the market	They are coded as 3 if the respondent is allowed to go alone, 2 if allowed to go with someone else only, and 1 if not allowed to go at all.	1.274 (0.617)
Usually allowed to go to the health facility		1.382 (0.606)
Usually allowed to go to places outside this village		1.345 (0.636)
Currently employed	Coded as 0 if currently unemployed, and coded as 1 if presently employed.	0.234 (0.423)
Years of education	Number of years of education, ranging from 1 to 20 years.	6.729 (5.19)
Contraceptive method	Coded as 0 if not using any contraceptive methods, and 1 otherwise.	0.484 (0.5)

Note: Responses for the decision-making variables have been collected only from currently married women (during the survey). The summary statistics encompass information on decision-making and freedom of movement sourced from NFHS-4 and data on current employment, years of education, and contraceptive methods from NFHS-5.

education, and if the woman is employing any family planning method. $DM_{d,4}$ captures the average level of decision-making in the district *d* in NFHS-4, and *FoMd,4* captures the intermediate level of "freedom of movement" in the district *d* in NFHS-4. $X_{id,5}$ captures the individual-level control, such as age, marital status, religion, caste, and area of residence (urban/rural). State-fixed effects are included to control for state-level variations.

8.8 Data Analysis

We first present a graphical representation of the data to see how district-level average of the dependent variables in NFHS-5 correlates with previous (NFHS-4) district-level averages of the independent variables.

In Figure 8.1(a), we present a scatterplot to explore the correlation between the average district-level decision-making scores in NFHS-4 and the level of current employment in NFHS-5 at the district level. Similarly, in Figure 8.1(b), we depict a scatterplot to examine the relationship between the district-level freedom of movement scores in NFHS-4 and the district-level employment figures in NFHS-5. Both figures indicate a positive, linear association between these

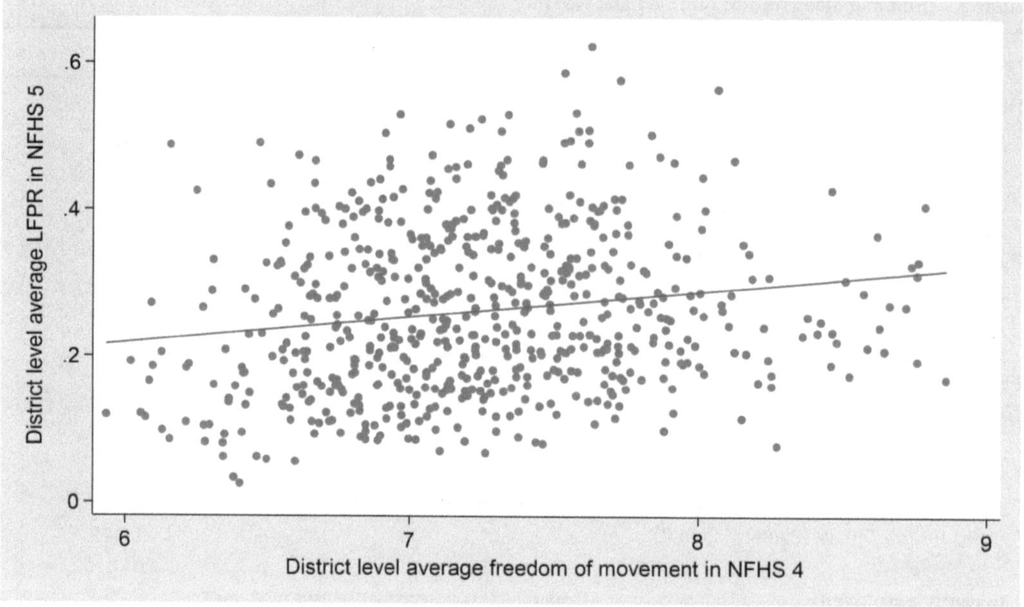

Figure 8.1a Women's current employment as a function of past decision-making.

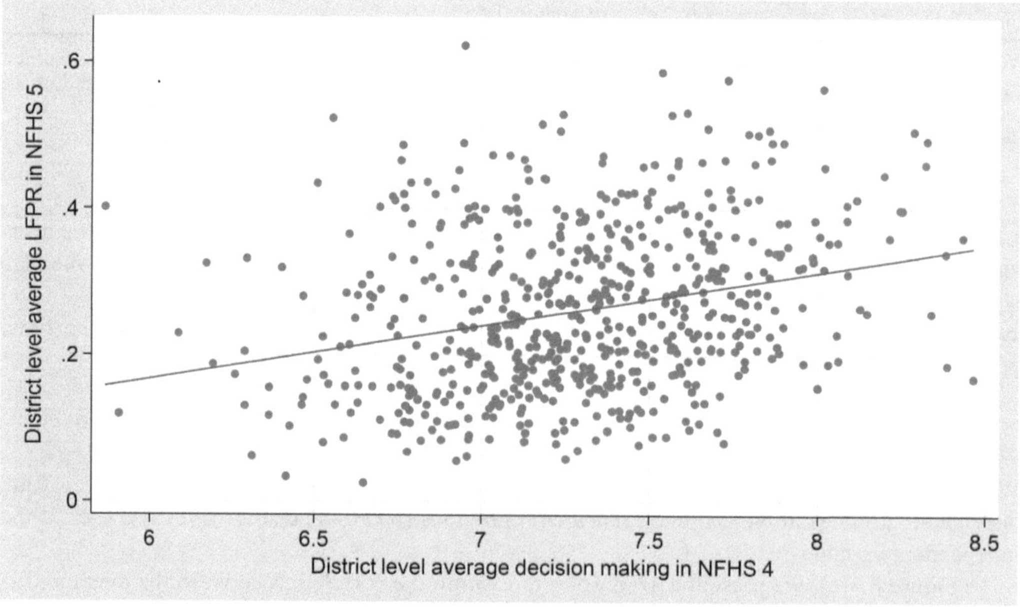

Figure 8.1b Women's current employment as a function of past freedom of movement.

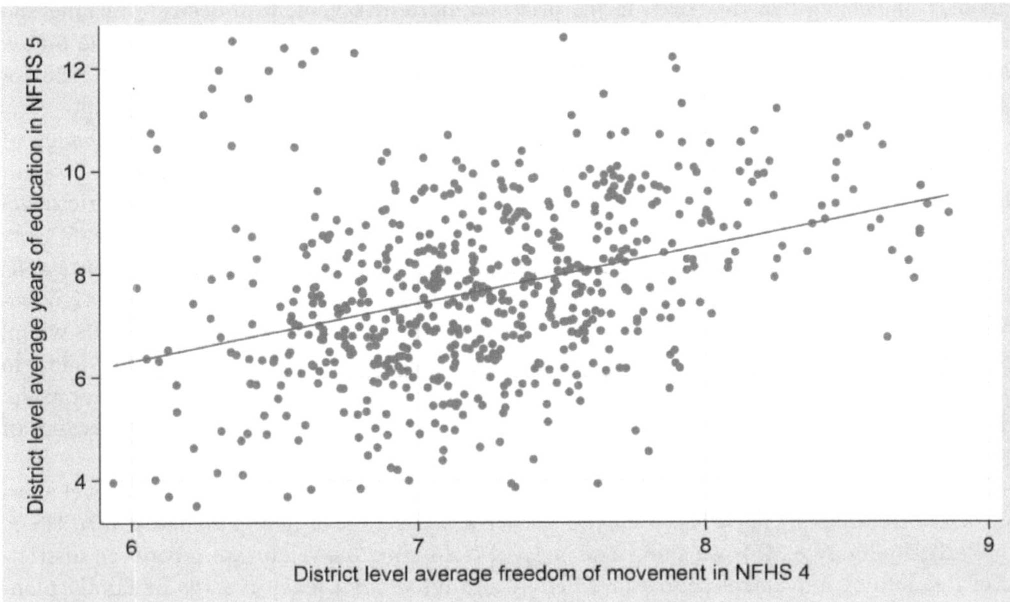

Figure 8.2.a Women's years of education as a function of past decision-making.

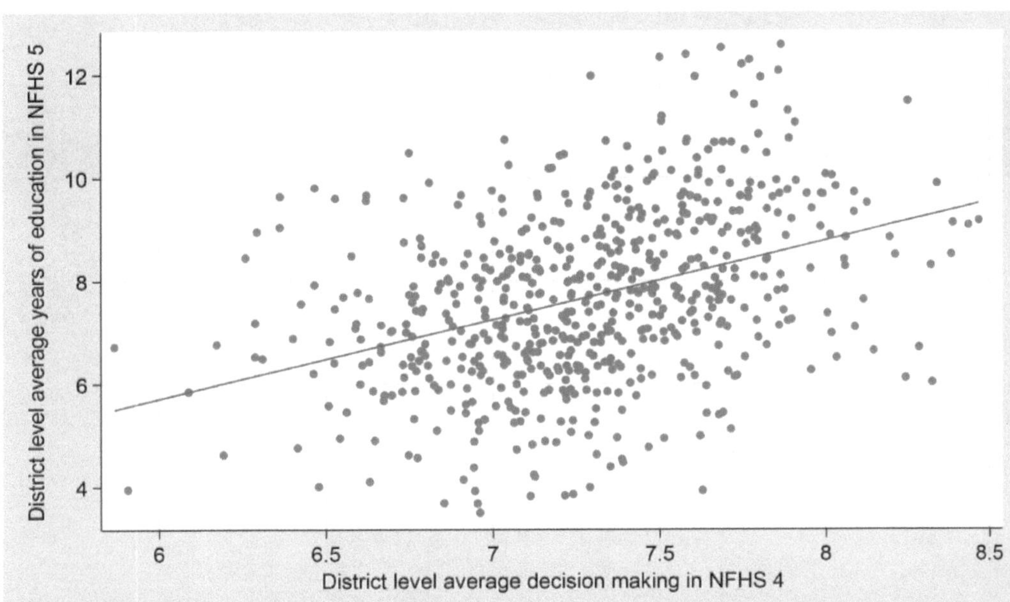

Figure 8.2.b Women's years of education as a function of past freedom of movement.

variables, indicating that if women in the past had more involvement in decision-making and greater freedom of movement, they tend to show higher female employment rates in the future. Data mainly falls within the 6.5–8 range for both variables, with a correlation of 0.2788 for decision-making and employment and 0.1987 for freedom of movement and employment.

In Figure 8.1(a), we display scatterplots illustrating the relationship between the average of district-level decision-making in NFHS-4 and district-level years of education in NFHS-5. Similarly, in Figure 8.1(b), we depict a scatterplot to examine the relationship between the district-level freedom of movement scores in NFHS-4 and district-level years of education in NFHS-5. Both graphs display a positive and linear relationship between these variables, suggesting that regions where women historically participated more in decision-making and had greater freedom of movement tend to exhibit higher levels of education in the future. The data predominantly falls within the 6.8–7.8 range for both factors, with values clustering around the trendline between 7 and 8 in the decision-making and years of education chart, emphasizing the strength of the relationship. The correlation stands at 0.4101 for decision-making, and education and 0.3432 for freedom of movement and education.

Figure 8.3(a) presents scatterplots illustrating the correlation between the district-level average of decision-making in NFHS-4 and the district-level average of family planning in NFHS-5. Similarly, in Figure 8.3(b), we show a scatterplot examining the relationship between district-level freedom of movement scores in NFHS-4 and the district-level average of family planning in NFHS-5. Figure 8.3(a) reveals a negative correlation between the variables, suggesting that the higher the participation of women in decision-making in the past, the lower the level of family planning in the future. In contrast, Figure 8.3(b) displays a positive linear relationship, indicating that the greater the freedom of movement in the past for women, the higher the level of family planning in the future. The data mostly falls within the range of 7–7.7 for

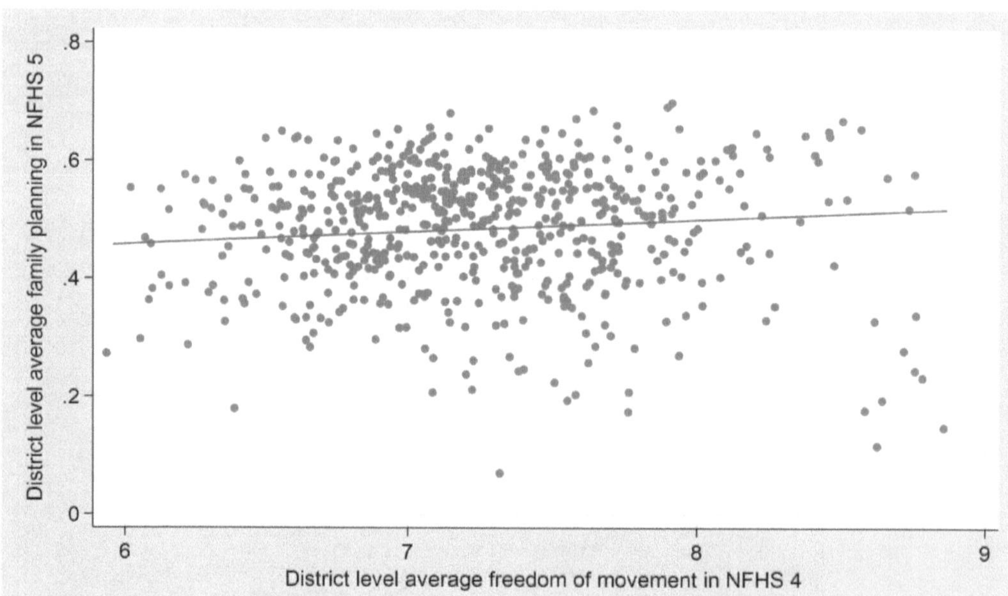

Figure 8.3.a Women's family planning as a function of past decision-making.

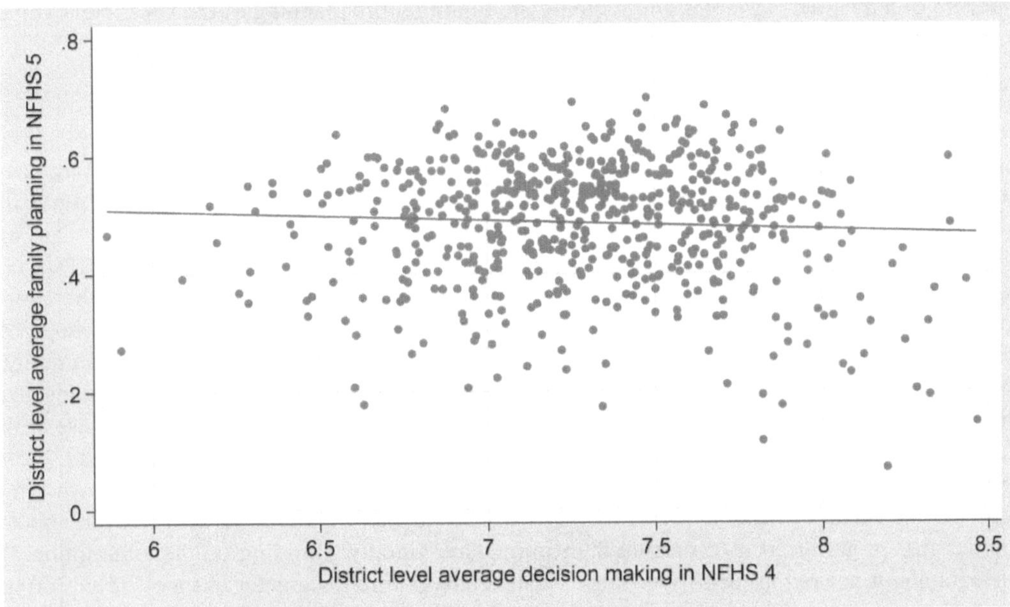

Figure 8.3.b Women's family planning as a function of past freedom of movement.

decision-making factors, and within the range of 6.8–7.8 for freedom of movement. The correlation is −0.0976 for decision-making and family planning, and 0.0460 for freedom of movement and family planning.

The regression results show that freedom of movement is crucial in determining employment in the current setting. The *p*-value for freedom of movement in this variable is below 0.01. The coefficients associated with the independent variables are positive; therefore, the relationship of the variable with current employment status is direct and positive. This indicates that an increase of one unit in freedom of movement corresponds to an increase in the likelihood of the future current working status.

Regressing years of education against decision-making and freedom of movement from NFHS-4 led to reaffirming results. Decision-making and freedom of movement play a statistically significant role in determining the years of education. As the coefficient attached to the variables decision-making and freedom of movement is positive, it indicates that a unit increment in these variables will increase the number of years of education in later years. When women have the freedom to move without restrictions, they can choose safer and more convenient transportation options. Freedom of movement allows women to choose educational institutions and programs that align with their career aspirations. They can explore a wider range of educational opportunities, including higher education institutions farther from their homes. This can alleviate safety concerns that may have previously deterred them from attending school. This has been extended by policies such as the *Mukhya Mantri Balika Cycle Yojana* in Bihar.

Regressing family planning against decision-making and freedom of movement from previous NFHS-4 rounds gave significant results. Earlier rounds of decision-making and freedom of movement play a statistically significant role in determining the latest years of family planning. As the coefficient attached to the variables is positive, a unit increment in decision-making and freedom of movement variables will increase the future family planning by the coefficient value in a later year.

8.9 Discussion

Employment, education, and decisions related to family planning are crucial to drafting sustainable policies on women's empowerment. Using NFHS-4 and 5 data, we establish empirical interlinkages in the measures over time by considering the impact of district-level averages of empowerment measures related to decision-making and freedom of movement from NFHS-4 on individual-level outcomes related to employment, education, and family planning in the later-round NFHS-5. Our results show that as the freedom of movement increases, the likelihood of present employment increases among women. As women attain more power to their autonomy, the chances of employment increase. Gupta and Yesudian (2006) concurs with this and adds that their study shows that the empowerment of Indian women does not necessarily increase with their engagement in the labor field. However, those who work and make a living are more likely to have great home autonomy; their odds ratio is more than three times that of non-working women. Greater freedom of movement allows women to explore a broader range of job opportunities that might be located outside their immediate vicinity. Building on the assumption of the data about not requiring permission to visit relatives implies greater chances of exploring employment opportunities.

The regression in measuring empowerment via total years of education depicts that as decision-making and freedom of movement increase, years of education also increase. When women are involved in making choices within their households, they can promote the allocation of increased resources to education, which includes securing funds for school fees, textbooks, and other educational necessities. Women with the capacity to influence household decisions may be able to postpone or manage family responsibilities, such as marriage and child-rearing, as suggested, as we have seen in the preceding results, that as decision-making increases, the decisions related to family planning increase. This flexibility enables them to concentrate on their education without being hindered by additional obligations that frequently disrupt their schooling.

The regression in family planning methods concurs with the literature. As decision-making and freedom of movement increase, the usage of family planning methods increases. When women have a more significant say in family decisions, they are more likely to have access to information about various family planning methods, their benefits, and their potential health implications. This empowers women to take control of their reproductive health and be more likely aware of a broader range of family planning methods, including contraceptives and spacing methods. Women with greater freedom of movement have better access to information about various family planning methods, benefits, and potential health implications. They can gather information from health-care providers, community resources, and educational materials. Improved mobility enables women to access health-care facilities more efficiently and to travel to pharmacies, clinics, or health-care centers to obtain contraceptives of their choice. Women's choices about the type of contraception might be informed by their knowledge of side effects, effectiveness or, otherwise, availability, and accessibility (Adedini et al., 2019; Tibaijuka et al., 2017). Qurra et al. (2015) further add that more empowered women tend to have or be associated with smaller family sizes, especially when educated. This can be attributed to awareness.

8.10 Conclusion

In analyzing the impact of SDG 5 on the interplay of education, employment, and women's empowerment, both positive strides and potential shortcomings come to the fore. SDG 5's dedication to gender equality and empowerment has undeniably propelled progress across education, employment, and empowerment. By prioritizing dismantling gender-based barriers, SDG 5 has driven investments in girls' education, equipping them with skills that bolster employability and broaden horizons. As educational opportunities expand for women, their capacity to enter diverse sectors grows, rendering them competitive job candidates.

Moreover, SDG 5's emphasis on empowerment aligns harmoniously with findings that underscore the profound connection between decision-making autonomy, mobility, education, and women's active engagement in the workforce. As women have more freedom of mobility, their likelihood of gaining employment improves. This parallel underscores how empowering women at home influences their career paths and opportunities. SDG 5's emphasis on empowerment reinforces the importance of fostering women's agency, creating an atmosphere where their potential can flourish.

Nevertheless, amid these advancements, potential vulnerabilities and challenges surface. The intricate relationship between decision-making, mobility, and years of education prompts a need for deeper exploration.

Furthermore, as SDG 5 champions empowerment, it may encounter hurdles tied to deeply embedded cultural norms and systemic inequalities. Despite strides in educational and employment access, biases and structural barriers can impede women's agency. This highlights that true empowerment necessitates not only policy shifts but also shifts in societal perceptions and practices. Additionally, this study acknowledges a limitation—limited data on variables interacting with cultural norms. These variables play a pivotal role in shaping women's empowerment yet remain insufficiently explored and could uniquely influence the outcomes of this relationship.

In conclusion, SDG 5's impact on education, employment, and women's empowerment is multifaceted. The advancements through education and heightened agency are commendable, showcasing the potential for positive transformations. Nevertheless, the intricacies highlighted by complex dynamics, uncharted variables, and potential bias underscore the importance of comprehensive approaches addressing systemic inequalities, cultural barriers, and methodological

Table 8.2 Regressing currently employed, years of education, and family planning from NFHS-5 on decision-making and freedom of movement from NFHS-4 with individual-level controls, taking states as fixed effect

Variables	*Currently employed* (1)	*Years of education* (2)	*Family planning* (3)
Decision-making	0.00615 (0.00511)	0.303*** (0.0197)	0.0120*** (0.00186)
Freedom of movement	0.0289*** (0.00477)	0.662*** (0.0184)	0.0275*** (0.00174)
Individual-level controls	Yes	Yes	Yes
Fixed effects	Yes	Yes	Yes
Observations	86,796	576,812	576,809
R-squared	0.098	0.345	0.376

Note: *** $p < 0.01$, ** $p < 0.05$, * $p < 0.1$.

limitations. Recognizing the strengths and vulnerabilities inherent in this nexus, we can endeavor to foster a more equitable and empowering environment aligned with SDG 5's overarching aim, promoting inclusivity for all.

References

Adde, K. S., Ameyaw, E. K., Dickson, K. S., Paintsil, J. A., Oladimeji, O., & Yaya, S. (2022). Women's empowerment indicators and short- and long-acting contraceptive method use: Evidence from DHS from 11 countries. *Reproductive Health, 19*(1), 222. https://doi.org/10.1186/s12978-022-01532-5

Adedini, S. A., Omisakin, O. A., & Somefun, O. D. (2019). Trends, patterns and determinants of long-acting reversible methods of contraception among women in sub-Saharan Africa. *PLoS ONE, 14*(6), e0217574. https://doi.org/10.1371/journal.pone.0217574

Akter, S., & Chindarkar, N. (2020). An empirical examination of sustainability of women's empowerment using panel data from India. *The Journal of Development Studies, 56*(5), 890–906. https://doi.org/10.1080/00220388.2019.1605054

Alsop, R., & Heinsohn, N. (2005). Measuring empowerment in practice: Structuring analysis and framing indicators. *The World Bank*. https://doi.org/10.1596/1813-9450-3510

Analysis of Women Empowerment in Rural Nigeria: A Multidimensional Approach. (2016).

Anderson, S. (2022). Unbundling female empowerment. *Canadian Journal of Economics/Revue Canadienne d'économique, 55*(4), 1671–1701. https://doi.org/10.1111/caje.12628

Bali Swain, R., & Wallentin, F. Y. (2012). Factors empowering women in Indian self-help group programs. *International Review of Applied Economics, 26*(4), 425–444. https://doi.org/10.1080/02692171.2011.595398

Bali Swain, R., & Yang-Wallentin, F. (2020). Achieving sustainable development goals: Predicaments and strategies. *International Journal of Sustainable Development & World Ecology, 27*(2), 96–106. https://doi.org/10.1080/13504509.2019.1692316

Batliwala, S. (1993). *Empowerment of Women in South Asia: Concepts and Practices*. Asian-South Pacific Bureau of Adult Education; FAO's Freedom from Hunger Campaign—Action for Development.

Batliwala, S. (2007). Taking the power out of empowerment—an experiential account. *Development in Practice, 17*(4–5), 557–565. https://doi.org/10.1080/09614520701469559

Chen, M., & Mahmud, S. (1995). Assessing change in women's lives: A conceptual framework.

Desai, M. A. (2010, July 7). *Hope in Hard Times: Women's Empowerment and Human Development*. UNDP-HDRO Occasional Papers No. 2010/14, Available at SSRN: https://ssrn.com/abstract=2351509

Drydyk, J. (2013). Empowerment, agency, and power. *Journal of Global Ethics, 9*(3), 249–262. https://doi.org/10.1080/17449626.2013.818374

Duflo, E. (2012). Women empowerment and economic development. *Journal of Economic Literature, 50*(4), 1051–1079. https://doi.org/10.1257/jel.50.4.1051

Eyben, R., Kabeer, N., & Cornwall, A. (2008). Conceptualising empowerment and the implications for pro poor growth. *DAC Poverty Network by the Institute of Development Studies*, Brighton.

Goulart, C. M., Purewal, A., Nakhuda, H., Ampadu, A., Giancola, A., Kortenaar, J.-L., & Bassani, D. G. (2021). Tools for measuring gender equality and women's empowerment (GEWE) indicators in humanitarian settings. *Conflict and Health, 15*(1), 39. https://doi.org/10.1186/s13031-021-00373-6

Greno, S., & Saikia, U. (2021). Does patriarchy rule impede contraception use in most African communities? *IOSR Journal of Humanities and Social Science, 26*(5), 39–43.

Gupta, K., & Yesudian, P. P. (2006). Evidence of women's empowerment in India: A study of socio-spatial disparities. *GeoJournal, 65*(4), 365–380. https://doi.org/10.1007/s10708-006-7556-z

Haider, T. L., & Sharma, M. (2013). Barriers to family planning and contraception uptake in sub-Saharan Africa: A systematic review. *International Quarterly of Community Health Education, 33*(4), 403–413.

Hanmer, L., & Klugman, J. (2016). Exploring women's agency and empowerment in developing countries: Where do we stand? *Feminist Economics, 22*(1), 237–263. https://doi.org/10.1080/13545701.2015.1091087

Heckert, J., & Fabic, M. S. (2013). Improving data concerning women's empowerment in Sub-Saharan Africa. *Studies in Family Planning, 44*(3), 319–344. https://doi.org/10.1111/j.1728-4465.2013.00360.x

James-Hawkins, L., Peters, C., VanderEnde, K., Bardin, L., & Yount, K. M. (2018). Women's agency and its relationship to current contraceptive use in lower- and middle-income countries: A systematic review of the literature. *Global Public Health, 13*(7), 843–858. https://doi.org/10.1080/17441692.2016.1239270

Kabeer, N. (1994). *Reversed Realities: Gender Hierarchies in Development Thought*. Verso.
Kabeer, N. (1999). Resources, agency, achievements: Reflections on measuring women's empowerment. *Development and Change*, *30*(3), 435–464. https://doi.org/10.1111/1467-7660.00125
Kishor, S. (1995). Autonomy and Egyptian women: Findings from the 1988 Egypt demographic and health survey. Occasional Papers 2, *Macro International Inc*, Calverton, MD.
Kishor, S. (2000). Empowerment of women in Egypt and links to the -survival and health of their infants. In *Women's Empowerment and Demographic Processes: Beyond Cairo*, Oxford University Press.
Kumar, S., Kumar, N., & Vivekadhish, S. (2016). Millennium Development Goals (MDGs) to Sustainable Development Goals (SDGs): Addressing unfinished agenda and strengthening sustainable development and partnership. *Indian Journal of Community Medicine: Official Publication of Indian Association of Preventive & Social Medicine*, *41*(1), 1–4. https://doi.org/10.4103/0970-0218.170955
Lassi, Z. S., Ali, A., & Meherali, S. (2021). Women's participation in household decision making and justification of wife beating: A secondary data analysis from Pakistan's demographic and health survey. *International Journal of Environmental Research and Public Health*, *18*(19), 10011. https://doi.org/10.3390/ijerph181910011
Mahmud, S., Shah, N. M., & Becker, S. (2012). Measurement of women's empowerment in rural Bangladesh. *World Development*, *40*(3), 610–619. https://doi.org/10.1016/j.worlddev.2011.08.003
Malhotra, A., Schuler, S. R., & Boender, C. (2002, June). Measuring women's empowerment as a variable in international development. In *Background Paper Prepared for the World Bank Workshop on Poverty and Gender: New Perspectives* (Vol. 28, p. 58). Washington, DC: The World Bank.
Miedema, S. S., Haardörfer, R., Girard, A. W., & Yount, K. M. (2018). Women's empowerment in East Africa: Development of a cross-country comparable measure. *World Development*, *110*, 453–464. https://doi.org/10.1016/j.worlddev.2018.05.031
Mosedale, S. (2005). Assessing women's empowerment: Towards a conceptual framework. *Journal of International Development*, *17*(2), 243–257. https://doi.org/10.1002/jid.1212
Niti Aayog. (2018). *SDG India Index—Baseline Report 2018*. https://planning.up.nic.in/Go/SDG/Baseline%20 2018.pdf
Niti Aayog. (2019a). *Localising SDGs: Early Lessons from India 2019*. https://india.un.org/en/173169-localising-sdgs-early-lessons-india-2019
Niti Aayog. (2019b). *SDG India: Index and Dashboard (2019–20)*. https://pminewyork.gov.in/pdf/menu/SDGIndia_Dec2.pdf
Niti Aayog. (2020a). *India VNR 2020—Decade of Action Taking SDGs from Global to Local*. https://sustainabledevelopment.un.org/content/documents/26281VNR_2020_India_Report.pdf
Niti Aayog. (2020b). *SDG India Index and Dashboard 2020–21*. https://india.un.org/en/173167-sdg-india-index-and-dashboard-2020-21
Niti Aayog. (2021). *North Eastern Region District SDG Index Report*. www.undp.org/india/publications/north-eastern-region-district-sdg-index-report
Niti Aayog. (2022). *The Indian Model of SDG Localisation*. https://india.un.org/en/192868-indian-model-sdg-localisation
Niti Aayog. (2023). *National Multidimensional Poverty Index: A Progress Review*. www.undp.org/india/publications/national-multidimensional-poverty-index-progress-review-2023
Nussbaum, M. C. (2000). *Women and Human Development: The Capabilities Approach* (Vol. 3). Cambridge University Press.
O.S, A., O.A, P., & A.I, A. (2016). Analysis of women empowerment in Rural Nigeria: A multidimensional approach. *Global Journal of Human-Social Science Research*, 16.
Phan, L. (2016). Women's empowerment and fertility preferences in Southeast Asia.
Pratley, P., & Sandberg, J. F. (2018). Refining the conceptualization and measurement of women's empowerment in Sub-Saharan Africa using data from the 2013 Nigerian demographic and health survey. *Social Indicators Research*, *140*(2), 777–793. https://doi.org/10.1007/s11205-017-1811-1
Qurra et al. (2015). Measuring and modeling women's economic empowerment in Pakistan. *Journal of Asia Pacific Studies*, *34*.
Sen, A. K. (1992). *Inequality Reexamined*. Oxford: Clarendon Press.
Sen, A. K. (1999). *Development as Freedom*. Oxford: Oxford University Press.
Sen, A. (2001). Many faces of gender inequality. *Frontline*, *18*(22), 35–40. https://frontline.thehindu.com/cover-story/many-faces-of-gender-inequality/article64924736.ece
Sen, A. K. (2009). *The Idea of Justice*. Harvard University Press. https://doi.org/10.2307/j.ctvjnrv7n
Sen, A. K. (2014). The ends and means of sustainability. In *The Capability Approach and Sustainability*. Routledge.

Swain, R. B., & Wallentin, F. Y. (2009). Does microfinance empower women? Evidence from self-help groups in India. *International Review of Applied Economics*, *23*(5), 541–556.
Tibaijuka, L., Odongo, R., Welikhe, E. et al. (2017). Factors influencing use of long-acting versus short-acting contraceptive methods among reproductive-age women in a resource-limited setting. *BMC Women's Health*, *17*, 25. https://doi.org/10.1186/s12905-017-0382-2
Trommlerová, S. K., Klasen, S., & Leßmann, O. (2015). Determinants of empowerment in a capability-based poverty approach: Evidence from The Gambia. *World Development*, Elsevier, *66*(C), 1–15.
Upadhyay, U. D., & Karasek, D. (2012). Women's empowerment and ideal family size: An examination of DHS empowerment measures in Sub-Saharan Africa. *International Perspectives on Sexual and Reproductive Health*, *38*(2), 078–089. https://doi.org/10.1363/3807812
Yount, K. M., Peterman, A., & Cheong, Y. F. (2018). Measuring women's empowerment: A need for context and caution. *The Lancet Global Health*, *6*(1). https://doi.org/10.1016/S2214-109X(17)30459-X

9

BRIDGING RESOURCES, BRIDGING GOALS

Water-Energy-Food Nexus for Integrated SDG Advancement

Xin Zhou, Bijon Kumer Mitra, and Vibhas Sukhwani

9.1 Introduction

As we passed the midpoint toward the 2030 Agenda, a sobering assessment of global progress on the Sustainable Development Goals (SDGs) confronts us: Of the 140 measurable targets, merely 15% are on track, while a concerning 37% are stagnating or regressing, and the remaining majority demand accelerated efforts (UNDESA, 2023). As we explore ways to accelerate SDG progress, it becomes increasingly clear that a siloed approach to managing resources falls short of addressing the multifaceted challenges our world faces. This chapter introduces a perspective that reconceptualizes resource management: the interlinkages between the water–energy–food (WEF) nexus and the SDGs. This integrated framework serves as a tool for holistic thinking, guiding us toward concerted SDG advancement.

The WEF nexus transcends traditional resource management, advocating for an integrative approach that considers the synergies and trade-offs between water, energy, and food systems. This chapter reflects on the evolution of the WEF nexus concept, exploring its potential to reshape our strategies for sustainable development. A core part of our discussion revolves around the intimate interlinkages between the WEF nexus and the broader SDGs. Through systematic mapping of the interlinkages between the WEF nexus and the SDGs, we provide an overall picture of the WEF–SDG system, highlighted by key SDGs and their linkages. To ground our discourse in reality, we present a practical case examining the water–energy nexus within the Ganges River basin—a vital lifeline for millions.

Further, we enrich this chapter with illustrative examples from the research works of the Institute for Global Environmental Strategies (IGES) which demonstrate the tangible benefits of the WEF approach. These case studies exemplify how the integrated management of water, energy, and food resources can not only propel resource efficiency but also fortify other development imperatives.

The purpose of this chapter is not merely to inform but to equip policymakers, researchers, and practitioners with scientific evidence and practical insights. By showcasing the interlinkages of the WEF nexus with the SDGs and illustrating how this approach can act as a lever for integrated SDG advancement, we aim to provide a practical tool to guide actions toward a future where resources are managed with a deep sense of interconnectivity.

DOI: 10.4324/9781003285472-10

9.2 Understanding the WEF Nexus Approach

By 2050, the global demand for each of the WEF resources is estimated to rise by over 50% as compared with the 2015 level (Ferroukhi et al., 2015). Against the already-existing resource shortfalls, the optimal management and preservation of WEF resources has become one of the major challenges of the 21st century. Acknowledging the tight interconnections among the WEF systems, the nexus approach has emerged as a basis for integrated management and governance of WEF resources across sectors and different scales (Stephan et al., 2018). It provides an integrative strategy to balance and optimize the WEF sectors, which are competing but also complementary for the same scarce resources. For instance, water is required for energy production and supply, energy is required for water treatment and transporting, and both water and energy are vital for food production (Mohtar and Daher, 2012). While factors like changing climate, demographics, and economic growth continue to strain the limited WEF resources, the growing occurrences of drought situations, lack of fresh water, intensification of food crises, and shortages in energy supply pose a clear threat to social, economic, and environmental sustainability (de Andrade Guerra et al., 2021; Carmona-Moreno et al., 2021). The implicit interconnections between WEF resources further complicate the challenge of achieving resource security, as during any potential resource shortage situation, such intricate linkages can implicitly affect the whole resource supply nexus.

A number of studies have established the evolution of the WEF nexus approach, including Albrecht et al. (2018) and Sukhwani et al. (2020). After the initial attention gained through the United Nations University's (UNU) Food–Energy Nexus Programme, launched in 1983, the Bonn Nexus Conference in 2011 played a pivotal role in bringing the WEF nexus approach into the limelight of global policy and research (Hoff, 2011). Based on state-of-the-art review, Zhang et al. (2018) imply that the "nexus" concept is today gaining increasing attention, and it is explicit that ignorance of the WEF interlinkages may raise future uncertainties. Biofuel, for instance, was initially advocated as a good practice for mitigating climate change, but gradually its negative implications on biodiversity and land use are being realized. Thus, in regard to balancing different interests in WEF resource use, addressing conflicts, and capitalizing on the synergies, significant efforts are now being made to investigate the WEF nexus from different aspects, in terms of conceptual development, quantification of resource flows and their dependencies, as well as technology and policy application. Stephan et al. (2018) underscore that there is still a need to better quantify the WEF-related trade-offs for future planning and management.

In reference to the messages from the Bonn 2011 Nexus Conference, literature reviews, and expert knowledge, GIZ (2020) put forth ten core principles of the WEF nexus, namely: (1) equitable and balanced weighting of WEF sectors in decision-making; (2) leaving no one behind; (3) political will and commitment; (4) cross-departmental and multi-sectoral cooperation; (5) mechanisms and tools for data exchange and modelling; (6) capacity development of people, institutions, and organizations; (7) inclusive and participatory multistakeholder approach; (8) financing schemes and investments; (9) sustainable and efficient resource use and consumption; and (10) furthering peace and preventing conflict.

Clearly, the nexus approach promotes a rational, efficient, and balanced management of WEF resources, which widens its scope to also understand the interconnections between areas like health, gender, human mobility, and conflict management. In that manner, the WEF nexus provides a systemic and multi-sectoral perspective toward sustainable development and is vital for the integrated achievement of the SDGs, particularly in fast-growing developing countries (de

Andrade Guerra et al., 2021). Along with the direct connections to SDG 6 on water (Clean Water and Sanitation), SDG 7 on energy (Affordable and Clean Energy), and SDG 2 on food (Zero Hunger), the WEF nexus approach also links indirectly to several other SDGs, like SDG 10 (Reduced Inequalities), SDG 11 (Sustainable Cities and Communities), SDG 12 (Responsible Consumption and Production), SDG 13 (Climate Action), etc.

Through capitalizing on the synergies of multiple SDGs and addressing the trade-offs, the nexus approach lays emphasis on transitioning from a silo-based thinking to an integrated and transdisciplinary perspective (Ghodsvali et al., 2019). However, despite the growing dialogue, human societies are still confronted with the challenge of realizing an integrated WEF resource management for sustainable outcomes. A range of governance challenges restrain the localized implementation of the WEF nexus, like, for example, how majority of the policy decisions on WEF resource management continue to be made by separate institutions and departments, with minimal consideration to cross-sectoral impacts (Sukhwani et al., 2019). The practical implementation of WEF nexus solutions, even for the achievement of the SDGs and other global agendas on sustainability, has been limited. In regard to the integrated governance approach, Sukhwani and Shaw (2022) summarize the two key points that are consistent in the existing literature, namely, the fragmented policy approach on WEF nexus and the essence of achieving policy coherence by enhancing synergies and minimizing trade-offs.

9.3 WEF–SDG Interlinkages: Developing a Network Model

Beginning in the early 1980s with studies on the bilateral relationships among water, energy, and food systems, WEF nexus research has been actively developed since the early 1990s. It has since evolved to incorporate additional physical sectors—such as land, climate, and the environment—that are closely linked with the WEF systems. Nexus thinking has further expanded to cover broader development sectors, particularly the economic and social sectors, including economic growth, trade, employment, poverty, health, and justice. Just a few years before the adoption of the SDGs in 2015, the interlinkages between WEF and the SDGs began to attract attention. While research on WEF–SDG interlinkages has rapidly evolved, most studies have concentrated on the relationships between WEF and individual SDGs or a subset of the SDGs. There is a lack of scientific research mapping the interlinkages between WEF and all SDGs.

In this section, the interlinkages between the WEF nexus and the SDGs are mapped based on a systematic review. A network model is further developed to illustrate key SDG involved and their relationships within the WEF–SDG system.

9.3.1 A Systematic Mapping of WEF–SDG Interlinkages

A systematic review of the literature, selected from Scopus—a bibliographic database—was conducted to map the interlinkages between the WEF nexus and all the SDGs. Natural language processing (NLP) techniques were utilized to systematically process and synthesize the data following a transparent and replicable protocol.

A protocol with six processes was developed to guide an NLP-based systematic review and mapping of WEF–SDG interlinkages from selected literature (Figure 9.1). The methodology and processes are built upon existing studies (Zhou and Moinuddin, 2023; Zhou, Jain et al., 2022; Zhou, Moinuddin et al., 2022).

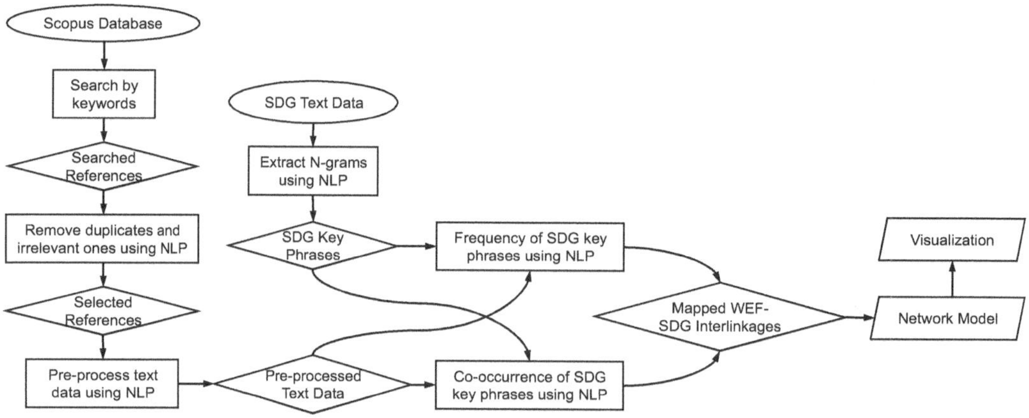

Figure 9.1 Workflow for NLP-based systematic review and mapping of WEF–SDG interlinkages.

Note: NLP = natural language processing; WEF = water–energy–food; SDG = Sustainable Development Goals.

9.3.1.1 Process 1: Selection of Literature

To conduct the literature search on Scopus, a combination of keywords was used to ensure comprehensive coverage of the topic. The search was structured around two sets of keywords: "water," "energy," or "food," in conjunction with "WEF" or "nexus." This approach allowed for a broad yet focused retrieval of documents related to the WEF nexus, capturing studies that specifically addressed any of these three critical resources in combination with nexus concepts.

A total of 11,775 references covering the time span from 1974 to 2023 were found. By removing references that lack an abstract and eliminating duplicates, 11,460 references remained. Using NLP, references that did not contain any combination of the previously mentioned sets of keywords were also removed. After these screening processes, a total of 9,359 references were selected for review.

9.3.1.2 Process 2: Pre-Processing the Text Data of Literature

The text data from the abstracts of selected literature was pre-processed using NLP techniques to remove punctuation and stop words, convert text to lowercase, and tokenize the content.

9.3.1.3 Process 3: Extraction of Key Phrases From SDG Text

In a parallel process to processes 1 and 2, N-grams—including single words, bi-grams, and tri-grams—were extracted from SDG texts using NLP techniques. The extracted SDG phrases were then manually screened to remove any that were irrelevant.

9.3.1.4 Process 4: Calculation of the Frequency of SDG Key Phrases From Literature Text Data

Using NLP techniques, SDG key phrases were extracted from the pre-processed text data of literature from process 2, and their frequencies were calculated.

9.3.1.5 Process 5: Mapping SDG Key Phrases and Calculation of Co-Occurrence Frequency

Using NLP techniques, SDG key phrases were mapped in pairs if they appeared simultaneously in the same sentence. The co-occurrence frequency of each pair was then calculated.

9.3.1.6 Process 6: Building a Network Model for the WEF–SDG Interlinkage System

Using NLP techniques, a network model was constructed and visualized to depict the interlinkages between WEF and SDGs, including SDG key phrases and their relationships within the context of the WEF nexus.

9.3.2 A Network Model of WEF–SDG Interlinkages

The network model illustrates the key SDGs and their relationships within the WEF–SDG interlinkage system. Key SDGs are displayed in a word cloud map based on the frequencies of SDG key phrases extracted from the literature (refer to Figure 9.2). Prominent SDGs identified in the network include "land" (Target 15.3), "resources" (cross-cutting SDGs 6, 12, 14, and 15), "renewable energy" (Target 7.2), "economic growth" (Target 8.1), "climate change" (Target 13.1), "trade" (Target 17.10), "health" (Goal 3), "agriculture" (Goal 2), and "technologies" (Target 17.7), among others.

The network model comprises 136 SDG targets that are linked with all 17 goals, featuring over 5,043 linkages among them. Figure 9.3 displays a simplified graph that describes the minimum

Figure 9.2 Word cloud highlighting key SDGs in the WEF–SDG interlinkage system.

Note: Each phrase represents either an SDG target (e.g., Target 15.3 for land) or a broader goal (e.g., SDG 6 for water).

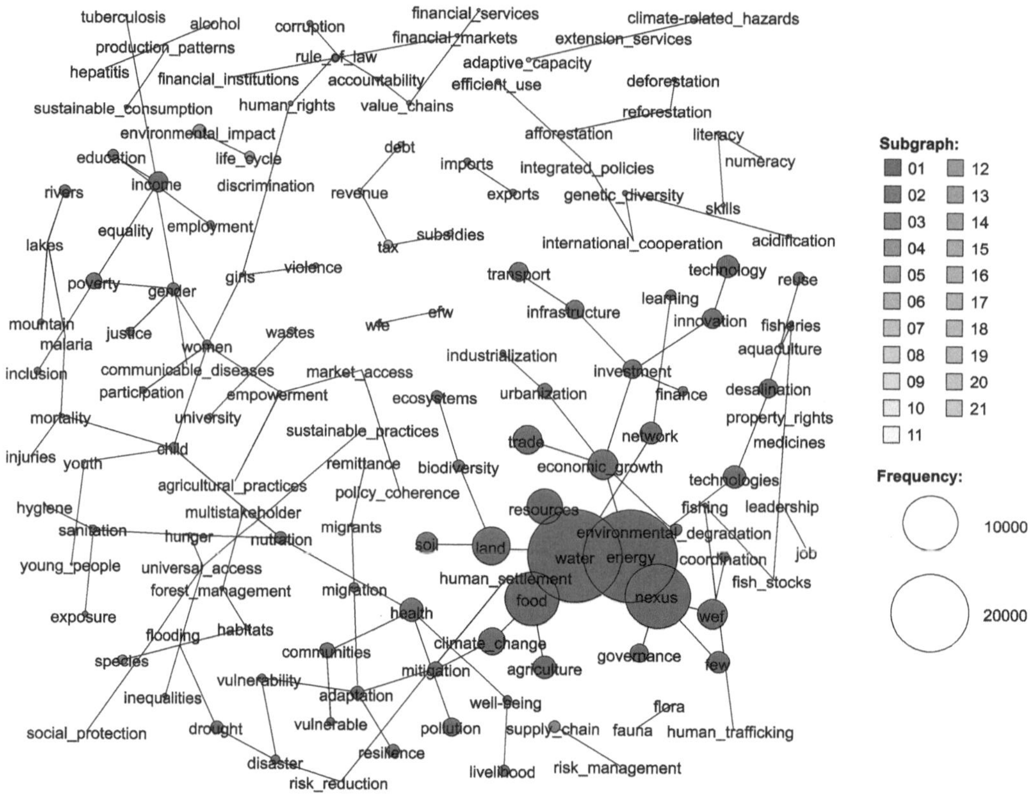

Figure 9.3 Network model of the WEF–SDG interlinkage system.

Note: Due to visualization constraints, a simplified network model featuring a minimum spanning tree is shown. Nodes represent SDG targets (e.g., Target 15.3 for land) or broader goals (e.g., SDG 6 for water), while edges illustrate the connections between nodes.

spanning tree of the top 200 key linkages based on their co-occurrence frequency. The top 50 key linkages are presented in Figure 9.4.

The results of key SDGs and key linkages in the network model provide a knowledge base on the WEF–SDG interlinkages. When planning WEF resource management and using the WEF nexus approach to leverage SDG progress, these key factors and linkages should be taken account together with the local context.

9.4 WEF Nexus at the River Basin Scale: A Case Study in the Ganges River Basin

Ganges, the most populous river basin in the world, is home to 600 million people of South Asia (World Bank, 2015), connecting three countries of the region, namely, India, Bangladesh, and Nepal. The Ganges River basin stands as one of the pivotal hubs of economic advancement in the region, contributing USD 700 billion to India's GDP (World Bank, 2015). Over 40% of the population of the region is either directly or indirectly dependent on water from this river for drinking, agriculture, and energy generation purposes (Anand et al., 2018). For instance, agriculture alone

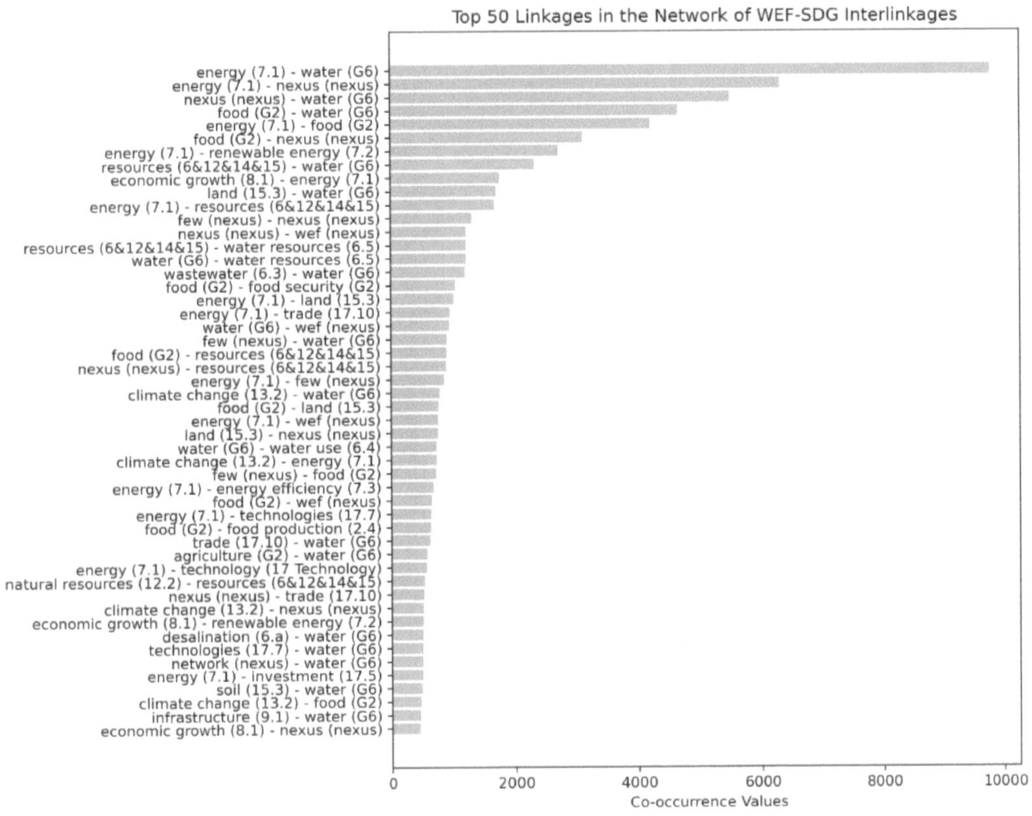

Figure 9.4 Top 50 linkages in the network of WEF–SDG interlinkages.

Note: Codes in parentheses represent specific SDG targets (e.g., Target 15.3 for land management) or broader goals (e.g., SDG 6 for water sustainability).

accounts for approximately 90% of the total water usage in the area (Sinha, 2014). The Ganges River is a vital water source for numerous thermal power plants, boasting a capacity exceeding 50 gigawatts (Sinha, 2014). Consequently, any fluctuations in water availability in the Ganges will profoundly affect the region's electricity security.

With a 40% share of the total electricity generation capacity in the area, the river basin plays a crucial role in power production (Sinha, 2014). However, per capita energy consumption in riparian countries remains comparatively low, with figures of 609 kilowatt-hours in Bangladesh (MOF, 2023), 1,327 kilowatt-hours in India (CEA, 2023), and 305 kilowatt-hours in Nepal (WECS, 2023), significantly below the global average of 3,130 kilowatt-hours (World Bank Database, on "Electric Power Consumption (kWh per capita)," available at https://data.worldbank.org/indicator/EG.USE.ELEC.KH.PC). This indicates considerable potential for growth in the electricity sector in riparian countries for years to come. Nevertheless, coal and gas-based thermal power dominates the electricity fuel mix in the region, comprising around 73% of the total electricity generation capacity (NPP, 2023; MPEMR, 2023). Thermal power plants (TPPs), relying on conventional cooling technology, demand substantial amounts of water for cooling purposes. The impact of cooling technology selection on water demand for power generation is vital for the optimization of water

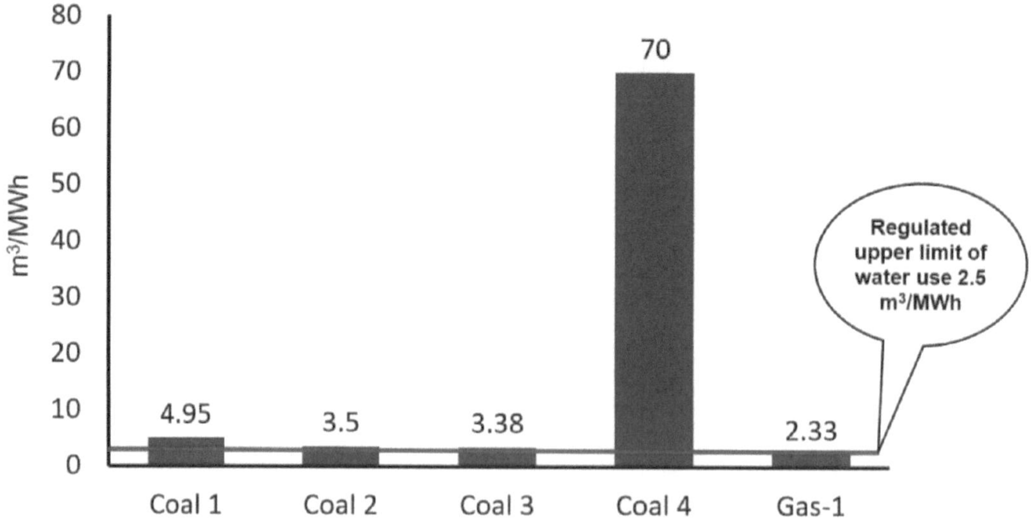

Figure 9.5 Water use intensity in the existing power plants.
Source: Mitra et al. (2021).

use in thermal power generation (Mitra et al., 2021). Power plants employing open-loop cooling systems exhibit a water use intensity of about 70 m^3/MWh (Coal 4), whereas those with closed-loop cooling systems range from 3.4 m^3/MWh to 5.0 m^3/MWh (Coal 1–3) (Figure 9.5). With ample reserves of indigenous coal and gas in India and Bangladesh, future power generation is expected to heavily lean on thermal sources. However, the sustainability of thermal power generation faces significant challenges due to climate-induced water resource variability (Gosain et al., 2011). For instance, drought events between 2013 and 2016 necessitated the shutdown of 14 major power plants due to cooling water scarcity, resulting in potential revenue losses of at least USD 1.4 billion (Luo et al., 2018). This highlights the precariousness of water availability for thermal power generation and underscores the operational risks faced by power plants in this river basin.

The future water risk situation in many sub-basins of the Ganges is alarming. For instance, the water surplus in the Damodar sub-basin will decrease by 15% in 2050, and the water deficit in the Yamuna is projected to surge by 86% by 2050 (Mitra et al., 2021). In contrast, in some basins, such as in Chambal and Damodar, water risks will diminish, where surplus water is anticipated in the future. As illustrated in Figure 9.6, among the 23 districts in the Chambal sub-basin, climate change is projected to have moderate to high positive impacts, reducing water risks in 19 districts by 2050. Conversely, power plants in the Yamuna and Gandak sub-basins are poised to face escalating water risks in the future. By 2050, the number of districts experiencing moderate negative impacts is projected to increase to 19. Notably, all existing and planned future power plants in the Yamuna and Gandak sub-basins are situated in districts where water risks are expected to rise. Particularly in the Gandak sub-basin, several planned thermal power installations are anticipated to encounter severe water risks during operation. Thermal power plants, whether existing or planned, situated in regions characterized by high water risks may encounter significant water shortages, thereby jeopardizing the stability of their operations. Districts within the Chambal and Damodar sub-basins experiencing moderate to high positive impacts on water risks from climate change could be deemed suitable locations for future electricity generation projects. This implies that the

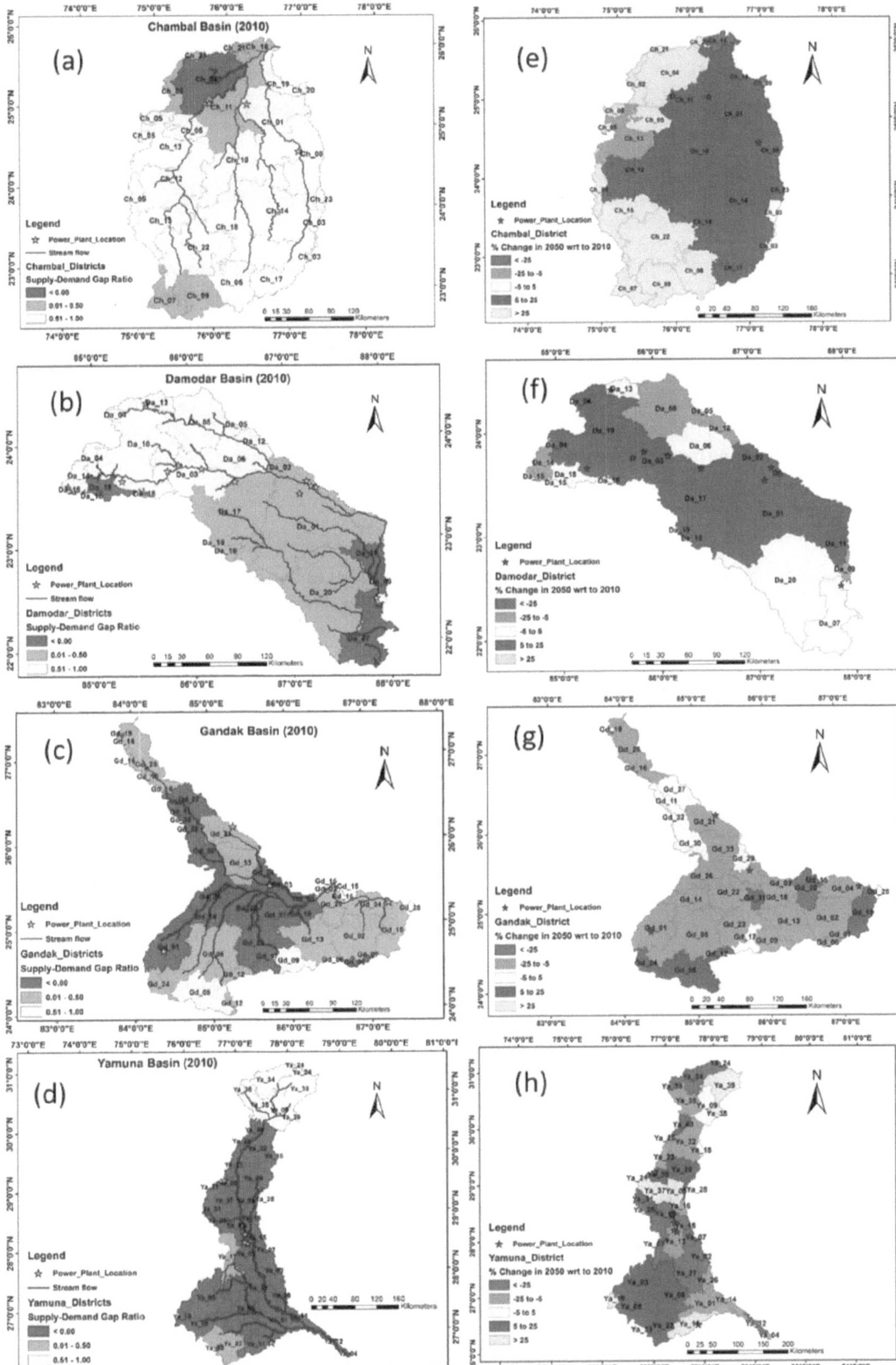

Figure 9.6 Water risks in 2010 (a, b, c, d) and the percent changes in the supply–demand ratio in 2050 in four sub-basins compared with 2010 (e, f, g, h).

Source: Mitra et al. (2021).

spatial distribution of the water–energy nexus risks should be considered as one of the key criteria for fuel and site selection of future power plants.

9.5 Case Studies on the Synergies of the WEF Nexus and the SDGs in Asia

As emphasized, a WEF nexus–based approach is imperative for the simultaneous achievement of diverse SDGs, as it helps capture synergies across different sectors of resource management, while also addressing the underlying trade-offs. Yet the grassroots-level implementation of nexus approach remains to be challenged by impediments, like discrete governance and lack of resources. Correspondingly, there is also a limited progress in the timely realization of the SDGs, especially in developing countries of Asia. In order to introduce resource-efficient pathways and cost-effective innovations, the following sub-sections present six case examples from Asia (summarized in Table 9.1) which reflect how the synergistical implementation of WEF nexus–based approaches can contribute toward the diverse SDGs. In reference to identified performance parameters, Table 9.1 presents a summarized description of the anticipated co-benefits through different nexus approaches and the corresponding SDGs.

9.5.1 Installation of Solar Power Systems at Agricultural Farms in Ulaanbaatar, Mongolia

9.5.1.1 Key Challenge

Ulaanbaatar, the capital city of Mongolia, is home to over half of the country's population. Between 2000 and 2017, an average of 19,000 persons/year migrated to the capital city (IOM, 2018). Despite the government of Mongolia's ban on migration into the capital city, the number of unregistered migrants may continuously increase, which is mostly driven by unequal economic and social development. Resultantly, the demands for basic services and infrastructure development are on the rise. In particular, the energy sector in Mongolia relies on coal-based power plants, which leads to large amounts of CO_2 emissions.

9.5.1.2 Nexus-Based Solution

To reduce its dependency on high-CO_2-emitting coal-based power generation, Mongolia has set up a target of generating 20% of total energy from renewable energy sources (GCF, 2019). As a nexus-based solution, a wholesale distributor, Everyday International LLC, initiated a solar power generation project which was registered by the government through bilateral cooperation under the Joint Crediting Mechanism as a mitigation project. The project aimed to improve power supply in the city and contribute in CO_2 mitigation by installing solar power systems at agricultural farms in the suburbs of Ulaanbaatar. These power plants can generate 3,352 MWh of electricity, which translates as a significant reduction in reliance on coal-fired thermal power, and thus avoidance of CO_2 emissions. Moreover, the solar power system also helps improve food supply by capturing synergies between agricultural and solar power generation technology.

9.5.1.3 Synergies With SDGs

The symbiotic use of agricultural lands to simultaneously yield food and solar energy not only helps in climate change mitigation but also harnesses clean renewable energy, together with improved protection of crops from extreme weather events. In that manner, this project not only

Table 9.1 SDG co-benefits of WEF-based solutions in selected cases of Asia

Performance parameters	*Anticipated co-benefits*	*Relevant SDGs*
1. Installation of solar power systems at agricultural farms in Ulaanbaatar, Mongolia		
CO_2 mitigation	2,682 tons CO_2/year	SDG 2
Renewable energy generation	3,352 MWh/year (estimated)	SDG 7
Agriculture production	Projected to improve	SDG 11
		SDG 13
2. Optimizing water, energy, and land resources in Ulaanbaatar, Mongolia		
Energy efficiency improvement	Heat energy consumption reduced by 50%	SDG 6
Wastewater management	Projected to improve	SDG 7
Awareness on nexus solutions		SDG 8
Partnerships between public and private actors		SDG 11
Value of land plots and economic growth		SDG 12
		SDG 13
		SDG 17
3. Capturing energy and reuse of treated domestic wastewater in Da Nang, Vietnam		
CO_2 mitigation	Captured methane from wastewater	SDG 2
Renewable energy generation	Energy generation from captured methane	SDG 6
Use of by-products	Use of treated wastewater for irrigation and processed sludge for urban agriculture	SDG 7 SDG 11
Awareness on urban nexus solutions	Projected to improve	SDG 12
Cross-sectoral interactions		SDG 13
Cost performance		
4. Introduction of high-efficiency water pumps in Da Nang City, Vietnam		
Increased efficiency of pumps	Intake pump: from 50% to 83%; distribution pump: from 64% to 90%	SDG 6 SDG 7
Improvement of supply capacity	Intake pumps: from 18 to 20 million m^3/year; distribution pumps: from 84 to 101 million m^3/year	SDG 11 SDG 12
Electricity saving	Electricity consumption for water production reduced by 3,400 MWh/year	SDG 13
CO_2 mitigation amount	Contributes to annual reduction in emissions of approximately 1,145 tCO_2/year	
Reduction of operation cost	Total electricity costs of operating pumps in 2018 reduced by about 5.77 billion VND/year; estimated payback period only three years	
5. Energy-saving technologies for wastewater treatment in Palembang City, Indonesia		
GHG mitigation	Estimated CO_2 mitigation of 745 tons/year through energy savings; CH_4 emissions reduced through stimulated aerobic process (GEC, 2015)	SDG 6 SDG 7 SDG 11
Energy efficiency improvement	Estimated reduced energy consumption of 931 MWh	SDG 13
Effluent quality	Meets effluent standard of Indonesia	
6. Reuse of municipal wastewater in power plants to mitigate water crisis in Delhi, India		
Amount of wastewater treated and reused	20 MLD of wastewater	SDG 6
Amount of treated wastewater supplied for power generation	Intake of 20 MLD treated wastewater	SDG 7 SDG 11
Cross-institutional collaboration	Annual operation and maintenance cost of sewage treatment plant borne by the power company	SDG 13

contributes toward multiple SDGs (SDG 2, SDG 7, SDG 11, and SDG 13) but also helps progress toward the NDCs of Mongolia.

9.5.2 Optimizing Water, Energy, and Land Resources in Ulaanbaatar, Mongolia

9.5.2.1 Key Challenge

More than 60% of Ulaanbaatar's 1.4 million residents live in disorganized ger-based settlements (*ger*: a portable round tent covered with skins or felt) on the city's periphery. As urban migration continues to rise, these settlements are expanding in a haphazard manner, exacerbating major socio-environmental problems, such as air pollution, as well as soil and groundwater contamination. Even though the residents of Ulaanbaatar's ger districts generally have legal titles to their plots, a majority desire to live in detached homes with basic infrastructure connections.

9.5.2.2 Nexus-Based Solution

Working with limited capacity and budgetary shortages, the city of Ulaanbaatar launched a community-driven land readjustment effort of about 180 participating households on 12 hectares of land to create an "eco village" in Khoroo No. 19 in the Bayanzurkh District, next to the "Tsaiz" wood market. Through this redevelopment, the city incorporated energy-efficient, ecological development principles and decentralized heating and wastewater management, with the objective of improving land use, raising the land value, and strengthening energy and waste management infrastructure. The redevelopment efforts are projected to reduce heat energy consumption and CO_2 emissions by 50%.

9.5.2.3 Synergies With SDGs

The creation of an eco village generates multiple co-benefits toward wastewater management, energy efficiency, land management, sectoral partnerships for economic growth, etc. It also contributes toward the simultaneous realization of multiple SDGs (SDG 6, SDG 7, SDG 8, SDG 11, SDG 12, SDG 13, and SDG 17), together with enhancing sustainable urban and regional development.

9.5.3 Capturing Energy and Reuse of Treated Domestic Wastewater in Da Nang, Vietnam

9.5.3.1 Key Challenge

Da Nang is one of the fastest-growing cities in Vietnam, with a current population of more than 1 million. The city has experienced stable and continued economic growth, mainly driven by rapid increase in the tourism and service sectors. In 2015, it had over 4.5 million visitors, and these numbers are expected to continue rising in the coming years. Da Nang's continued growth and expansion, however, has increased pressures on its infrastructure, including water and energy services, and wastewater treatment. As a coastal city, Da Nang also faces disaster risks associated with sea level rise, floods, and typhoons. In particular, access to adequate sanitation services has become a major challenge, with most households using septic tanks to dispose of wastewater. The wastewater collection system is inadequate, which increases environmental risks, including groundwater pollution and methane emissions.

9.5.3.2 Nexus-Based Solution

In collaboration with the Department of Planning and Investment, GIZ implemented a nexus project to improve wastewater management in the city. An innovative, flood-proof vacuum sewer collection system was introduced that can collect both black- and gray water, as well as capture biogas to produce energy. By-products such as processed sludge could then be used as fertilizer, and the treated gray water could be used for urban agriculture irrigation. The Da Nang People's Committee (DPC) established the Nexus Task Force to ensure institutional coordination, and a technical design was elaborated to install the vacuum sewer system and connect 110 households at an estimated cost of 416,500 euros. Subsequently, DPC expressed interest in the concept and agreed to implement the vacuum sewer system throughout the My An to My Khe area, covering 12,000 households, with support of the World Bank (GIZ, 2017).

9.5.3.3 Synergies With SDGs

The enhanced wastewater management system contributes to national and local climate actions while also enhancing the resilience of Da Nang toward flooding risks. It helps capture methane, generates renewable energy, and also contributes to improving adaptation capacity by supplying treated wastewater for irrigation and processed sludge for urban agriculture. Overall, it contributes to at least six SDGs (SDG 2, SDG 6, SDG 7, SDG 11, SDG 12, and SDG 13).

9.5.4 Introduction of High-Efficiency Water Pumps in Da Nang City, Vietnam

9.5.4.1 Key Challenge

Over the recent decades, the rapid growth of Da Nang for transportation, services, and tourism has increased pressure on the city's aging water supply infrastructure. In general, water demand is rapidly increasing from both local residents and tourists. It is anticipated that by 2030, the city's total water demand on the present water utility will have doubled. Simultaneously, the associated demand of energy for water supply will also increase; it is estimated to rise from 18 to 21 gigawatt-hours over 2015 to 2017 (Ninh, 2019).

9.5.4.2 Nexus-Based Solution

Acknowledging the water–energy interlinkages as well as the potential barriers in enhancing resource use, the Da Nang Water Supply Company (DAWACO) took an initiative to replace low-efficiency conventional pumps with higher-efficiency pumps at the Cau Do water treatment plant. Three existing conventional raw water pumps (50% efficiency) and six existing clean water pumps (64% efficiency) were replaced with higher-energy efficient pumps (83% for intake pumps, and 90.6% for clean water distribution pumps), which have been customized to the specific conditions and requirements of the recipient plants.

9.5.4.3 Synergies With SDGs

The introduction of high-efficiency pumps not only improved the water supply capacity of the pumps but also contributed toward energy saving, carbon mitigation, and cost-effectiveness. In that manner, it has a synergistic contribution toward multiple SDGs (SDG 6, SDG 7, SDG 11, SDG 12, and SDG 13).

9.5.5 Energy-Saving Technologies for Wastewater Treatment in Palembang City, Indonesia

9.5.5.1 Key Challenge

Indonesia has been among the world's largest emitters of GHGs (Climate Watch, 2024). After the Paris Agreement, the country strengthened its commitments to reduce global warming. It set an unconditional target to reduce CO_2 emissions by 32% by 2030 and, with international support, to 43% (Government of Indonesia, 2022). In its NDC, Indonesia identified energy use improvement as a priority area, and as cities are home to over half of the country's population and most economic activity is concentrated in urban areas, this revealed many energy-saving opportunities. Under Government Regulation No. 70/2009, private and industrial sectors are allotted major responsibilities toward energy conservation, and industry is one of the targeted sectors for enhancing energy efficiency through technological transformation. Besides, the government has implemented a set of incentives on energy conservation, including tax exemptions for energy-saving appliances, import tax reductions for energy-saving products, and low interest financing for energy conservation.

9.5.5.2 Nexus-Based Solution

To improve the performance of wastewater treatment facilities of a rubber company, an innovative project was initiated by PT. Aneka Bumi Pratama, located in Palembang City. As a result of the project, the company is now able to meet Indonesia's effluent–wastewater quality standards. Aerator technology is to be installed to improve the wastewater treatment facilities of the rubber company, which will accelerate the mixing of air from the blower to reduce methane emissions. Since there is no pressure loss at the blower, this will likely reduce the energy consumption of wastewater treatment facilities. Also, the aerator requires minimal maintenance to operate.

9.5.5.3 Synergies With SDGs

By increasing energy efficiency and avoiding methane emissions, the aerator technology directly contributes to national and local climate actions, while also helping mitigate water pollution, which can contribute to enhancing adaptation capacity in water management of Palembang City. This solution contributes simultaneously to SDG 6, SDG 7, SDG 11, and SDG 13.

9.5.6 Reuse of Municipal Wastewater in Power Plants to Mitigate Water Crisis in Delhi, India

9.5.6.1 Key Challenge

India's capital city, Delhi, is situated in the Yamuna River basin. The anticipated rapid population growth, and economic development, in Delhi is likely to intensify the water demand in this already water-stressed river basin, along with increased competition for the limited water resources. Concurrently, the demand for energy is also increasing. In Delhi, the power supply capacity has not kept pace with population growth and economic development. Further, the management of wastewater is also a big challenge for Delhi, as the sewage treatment capacity has also not kept pace with the rise in wastewater generation. Over the last four decades, while the treatment capacity increased sevenfold, the volume of wastewater increased twelvefold.

9.5.6.2 Nexus-Based Solution

While the limited water availability is an issue, reuse of wastewater can help mitigate the water crisis as well as reduce the pollution of natural water bodies. If Delhi's wastewater could be properly collected, treated, and reused in power plants, the freshwater demand of the power sector could be cut by up to 75%. Indraprastha Power Generation Co. Ltd. (IPGCL), the state-owned electricity provider of Delhi, is moving toward cleaner and water-smart power generation by using treated municipal wastewater instead of fresh water in the operation of its Pragati Combined Cycle Gas Turbine (CCGT) power plant. This represents a win–win solution for water crisis mitigation and wastewater management. A salient feature of the power plant is that the raw water requirement for operation of the power plant is met by a Delhi municipal sewage water treatment plant. As per an agreement between IPGCL and the Delhi Municipal Sewage Water Treatment Authority, in exchange for 20 million liters per day (MLD) of treated wastewater from the sewage treatment plant without charge, IPGCL covers the operation and maintenance costs of the sewage treatment plant.

9.5.6.3 Synergies With SDGs

The cross-institutional collaboration for the reuse of municipal wastewater in Delhi is a win–win solution which contributes toward several different SDGs, including SDG 6 (Clean Water and Sanitation), SDG 7 (Affordable and Clean Energy), SDG 11 (Sustainable Cities and Communities), and SDG 13 (Climate Action).

9.6 Conclusion

This chapter underscores the transformative potential of the WEF nexus as a pivotal framework for addressing the interconnected challenges of sustainable resource management. As we passed the halfway point toward the 2030 Agenda, it became evident that traditional approaches to resource management are insufficient in meeting the escalating demands and complexities posed by global sustainability objectives. The WEF nexus not only facilitates a deeper understanding of the synergistic and trade-off dynamics among water, energy, and food systems but also serves as a crucial lever for propelling the integrated advancement of the SDGs.

Through a systematic exploration of the WEF nexus, including a focused examination of the Ganges River basin and illustrative case studies from the research works of the Institute for Global Environmental Strategies (IGES), this chapter illuminates the tangible benefits and strategic importance of nexus thinking. These examples vividly demonstrate how integrated management can enhance resource efficiency and bolster other development imperatives, thereby contributing to a sustainable and resilient future.

As we look forward, the adoption of the WEF nexus approach should be seen not just as a strategic necessity but as an opportunity to reframe and deepen efforts toward the SDGs. It challenges us to move beyond siloed thinking and embrace a more interconnected and holistic perspective on resource management. By fostering cross-sectoral collaborations and enhancing policy coherence, the WEF nexus approach aims to optimize resource use, mitigate conflicts, and unlock synergies that can significantly propel sustainable development, particularly in rapidly developing regions.

Ultimately, this chapter aims to serve as a foundational guide for policymakers, researchers, and practitioners, providing them with the insights and methodologies to implement the WEF nexus framework effectively. In doing so, it seeks to catalyze actions that are not only informed by a deep understanding of the complex interdependencies but are also aligned with the overarching goal of achieving a balanced and sustainable future for all.

References

Albrecht TR, Crootof A, Scott CA (2018). The Water-Energy-Food Nexus: A Systematic Review of Methods for Nexus Assessment. *Environ. Res. Lett.* 13(4), 043002. https://doi.org/10.1088/1748-9326/aaa9c6

Anand, J, Gosain, AK, Khosa, R, Srinivasan, R (2018). Regional Scale Hydrologic Modeling for Prediction of Water Balance, Analysis of Trends in Streamflow and Variations in Streamflow: The Case Study of the Ganga River Basin. *J. Hydrol. Reg. Stud.* 36, 32–53.

Carmona-Moreno C, Crestaz E, Cimmarrusti Y, Farinosi F, et al. (Eds.). (2021). *Implementing the Water–Energy–Food–Ecosystems Nexus and Achieving the Sustainable Development Goals.* UNESCO, European Union and IWA Publishing.

CEA. (2023). *Growth of Electricity Sector of India.* www.indiaenvironmentportal.org.in/files/file/Growth%20of%20Electricity%20Sector%20in%20India.pdf

Climate Watch. (2024). www.climatewatchdata.org/ghg-emissions

de Andrade Guerra JBSO, Berchin II, Garcia J, et al. (2021). A Literature-Based Study on the Water–Energy–Food Nexus for Sustainable Development. *Stoch. Environ. Res. Risk Assess.* 35, 95–116.

Ferroukhi R, Nagpal D, Lopez-Pena A, Hodges T, Mohtar RH, Daher B, Mohtar S, Keulertz M (2015). Renewable Energy in the Water, Energy & Food Nexus. *IRENA—International Renewable Energy Agency.*

GCF. (2019). *Country Programme, Mongolia.* Green Climate Fund (GCF) Documentation. www.greenclimate.fund/documents/20182/1688867/Mongolia_Country_Programme.pdf/da4b9c33-75ce-3ed7-69bf-7cdadd84697e

GEC. (2015). *Energy Saving in Industrial Wastewater Treatment for Rubber Industry.* Global Environment Centre Foundation (GEC). https://gec.jp/jcm/projects/p_archive/15ps_ina_01/

Ghodsvali M, Krishnamurthy S, de Vries B (2019). Review of Transdisciplinary Approaches to Food-Water-Energy Nexus: A Guide Towards Sustainable Development. *Environ. Sci. Policy.* 101, 266–278. https://doi.org/10.1016/j.envsci.2019.09.003

GIZ. (2017). *Case Study: Innovative Waste Water Management*, Danang, Vietnam.

GIZ. (2020). *Global Water, Energy and Food Nexus Principles*, The Nexus Regional Dialogues Programme.

Gosain, AK, Rao S, Arora A (2011). Climate Change Impact Assessment of Water Resources of India. *Curr. Sci.* 101, 356–371.

Government of Indonesia. (2022). *Enhanced Nationally Determined Contribution.* https://unfccc.int/sites/default/files/NDC/2022-09/ENDC%20Indonesia.pdf

Hoff. (2011). *Understanding the Nexus. Background Paper for the Bonn2011 Conference: The Water, Energy and Food Security Nexus.* Stockholm Environment Institute: Stockholm, Sweden.

IOM. (2018). *Mongolia: Internal Migration Study, International Organization for Migration.* https://reliefweb.int/sites/reliefweb.int/files/resources/mongolia_internal_migration_study.pdf

Luo, T, Krishnan, D, Sen, S (2018). Parched Power: Water Demands, Risks, and Opportunities for India's Power Sector. *Water Resources Institute.* www.wri.org/publication/parched-power

Mitra, BK, Sharma, D, Zhou, X, Dasgupta, R. (2021). Assessment of the Impacts of Spatial Water Resource Variability on Energy Planning in the Ganges River Basin under Climate Change Scenarios. *Sustainability.* 13, 7273. https://doi.org/10.3390/su13137273

MOF (Ministry of Finance). (2023). *Bangladesh Economic Review 2023.* https://mof.portal.gov.bd/sites/default/files/files/mof.portal.gov.bd/page/f2d8fabb_29c1_423a_9d37_cdb500260002/Chapter-10%20%28English%29%202023.pdf

Mohtar RH, Daher B (2012). Water, Energy, and Food: The Ultimate Nexus. In *Encyclopedia of Agricultural, Food, and Biological Engineering*, 2nd ed. https://doi.org/10.1081/E-EAFE2-120048376

MPEMR. (2023). *Integrated Energy and Power Master Plan (IEPMP) 2023.* https://powerdivision.portal.gov.bd/sites/default/files/files/powerdivision.portal.gov.bd/page/4f81bf4d_1180_4c53_b27c_8fa0eb11e2c1/IEPMP%202023.pdf

Ninh. (2019). *The Joint Crediting Mechanism (JCM) within Framework of JCM Model Project in Da Nang & Recommendations of JCM Application for an Trach Pumping Station.* Proceedings of the Workshop on Promoting Private Sector Participation Towards Low-carbon Project Development in Vietnam, Held in Hanoi on 11st January 2019.

NPP. (2023). *Location Wise Regional Summary of All India Installed Capacity (IN MW) of Powe Station as on 31/10/2023*. https://npp.gov.in/public-reports/cea/monthly/installcap/2023/OCT/capacity1-2023-10.pdf

Sinha, D (2014). *Reviving the Ganga, at the Cost of Its Ecology*! www.indiatogether.org/articles/ganga-river-waterway-reviving-and-impact-on-ecology-environment/print

Stephan RM, Mohtar RH, Daher B, Irujo AE, Hillers A, Ganter JC, et al. (2018). Water–Energy–Food Nexus: A Platform for Implementing the Sustainable Development Goals. *Water Int.* 43(3), 472–479. https://doi.org/10.1080/02508060.2018.1446581

Sukhwani V, Shaw R (2022). An Integrated Governance Approach Towards a Water-Energy-Food Nexus and Climate Change. In Shaw, R (Ed.), *Handbook on Climate Change and Disasters*. Edward Elgar, 562–573. https://doi.org/10.4337/9781800371613.00053

Sukhwani V, Shaw R, Deshkar S, Mitra BK, Yan W (2020). Role of Smart Cities in Optimizing Water-Energy-Food Nexus: Opportunities in Nagpur, India. *Smart Cities* 3(4), 1266–1292. https://doi.org/10.3390/smartcities3040062

Sukhwani V, Shaw R, Mitra BK, Yan W (2019). Optimizing Food-Energy-Water (FEW) Nexus to Foster Collective Resilience in Urban-Rural Systems. *Prog. Disaster Sci.* 1, 100005. https://doi.org/10.1016/j.pdisas.2019.100005

UNDESA. (2023). *The Sustainable Development Report*. Special Edition. https://desapublications.un.org/publications/sustainable-development-goals-report-2023-special-edition.

WECS. (2023). *Energy Synopsis Report 2023*. https://wecs.gov.np/source/Energy%20Synopsis%20Report%2C%202023.pdf

World Bank. (2015). *The National Ganga River Basin Project*. World Bank: Washington, DC. www.worldbank.org/en/news/feature/2015/03/23/india-the-national-ganga-river-basin-project

Zhang C, Chen X, Li Y, Ding W, Fu G (2018). Water-Energy-Food Nexus: Concepts, Questions and Methodologies. *J. Clean. Prod.* 195, 625e639. https://doi.org/10.1016/j.jclepro.2018.05.194

Zhou, X, Jain, K, Moinuddin, M, McSharry, P (2022). *Using Natural Language Processing for Automating the Identification of Climate Action Interlinkages Within the Sustainable Development Goals*. Proceedings of the 2022 Fall Symposium on the Role of AI in Responding to Climate Challenges, Association for the Advancement of Artificial Intelligence (AAAI), 17–19 November 2022, Arlington, VA.

Zhou, X, Moinuddin, M (2023). Chapter 7 An Integrated Approach to the Sustainable Development Goals from an Interlinkage Perspective: Methodology, Decision Support Tool and Applications. In Bali Swain, R, Min, Y (Eds.), *Interlinkages Between the Sustainable Development Goals*. Cheltenham: Edward Elgar Publishing. https://doi.org/10.4337/9781803924946.00012.

Zhou, X, Moinuddin, M, Renaud, F, Barrett, B, Xu, J, Liang, Q, Zhao, J, Xia, X, Bosher, L, Huang, S, Hoey, T (2022). Development of an SDG Interlinkages Analysis Model at the River Basin Scale: A Case Study in the Luanhe River Basin, China. *Sustain. Sci.* 17, 1405–1433. https://doi.org/10.1007/s11625-021-01065-z.

10

CONSERVING WATER RESOURCES TO ATTAIN THE SUSTAINABLE DEVELOPMENT GOALS IN BANGLADESH

A Legal Analysis

Zelina Sultana and Nasrin Akter

10.1 Introduction

Water, one of the world's most significant natural resources, is an integral aspect of all living beings on the planet. However, it is continually polluted and has become invisible in many parts of the world due to pollution (Hasan et al., 2019). Bangladesh is a land of rivers, but it also has other water sources, like lakes, ponds, canals, *haor* (under Section 2(35) of the Bangladesh Water Act [BWA] 2013, *haor* means any saucer-shaped large shallow depression created naturally in between two separate rivers), *baor* (in Section 2(22) of the BWA 2013, *baor* means any kind of oxbow-shaped lake where watercourse has become stagnant in the course of time), and *beels* (Section 2(35) of the BWA defines *beel* as any natural low-land or round area which gets inundated by rainwater or river water and which remains inundated over the year or partially or entirely remains dry for a certain period of a year), which are part of wetlands. These various sources of water are continuously polluted as a result of anthropogenic activities (Adejumoke et al., 2018) and have become incompatible for both plants and human existence (Hasan et al., 2019).

Additionally, cities and towns have been the focal point of anthropogenic activities that have relied on water sources other than rivers since the dawn of civilization, and their unplanned development activities, unsustainable use of water, and polluters who have gone unpunished have been responsible for pollution and water scarcity in recent decades (Hasan et al., 2019; Ahmed, 2018). However, overpopulation, industrialization (Adejumoke et al., 2018), urbanization (Parvin et al., 2020), usage of agrochemical substances in agricultural land (FAO, 2017), oil leaks from different water vessels (Babu, 2020), and domestic and human waste (Uddin and Jeong, 2021) all are the concerning factors for water pollution in Bangladesh (Faroque and South, 2021). Water management and pollution prevention are a theoretical existence and will be unable to tackle the present challenges unless a synergistic strategy is adopted by all relevant water-using sectors.

Climate change, according to prior scholarships, has negative impacts on freshwater sources as well as the health of the inhabitants in the coastal parts of Bangladesh (Abedin et al., 2019). The Intergovernmental Panel on Climate Change (IPCC) also confirmed that, because of climate change implications, increased tidal waves and storm surges would cause excessive salinization of

DOI: 10.4324/9781003285472-11

groundwater, agricultural fields, and many rivers (Khan et al., 2011). A total of 720 kilometers of coastlines and 100 kilometers of domestic ponds, agricultural land, and groundwater have already been encroached by salinity due to changing rainfall patterns, sudden storm surges, sea levels rising, reduced river flow, and upstream withdrawal of water in Bangladesh (Khan et al., 2011). Human health and well-being, on the other hand, are adversely affected by the consumption of saline-tainted water (Khan et al., 2011). It also poses a threat to soil and agricultural goods that are not resistant to saline water, as well as shrimp farming and inland fisheries culture in coastal areas, among other things (Huq et al., 2015). Most importantly, it threatens the hydrological cycle, which delineates the constant water circulation between land, ocean, and the atmosphere. Various transfer processes of water in different scales are referred to as water balance and are essential in maintaining an equilibrium climate in a region (Stagl et al., 2014). As a result of recent accelerated alterations in climate and land use, the world's water cycles are experiencing increased quantities of geographical and temporal fluctuation, which have caused a plethora of water-related challenges that pose a threat to the supply of clean drinking water for mankind (Yang et al., 2021). Over 4 billion people are currently impacted by water scarcity, and the consumption of water resources has already reached alarming levels in many regions of the globe, including Bangladesh (Yang et al., 2021). The increasing need for fresh water by a growing population leads to an alarming increase in the price of fresh water around the world, which is also true in Bangladesh (Hasan et al., 2019).

Unsustainable water usages are degrading the entire ecosystem, resulting in low water flow in rivers, wetland shrinkage, and a deepening of the groundwater level in Bangladesh (Yang et al., 2021). Sustainable water resource management depends on maintaining the quality of water and the proper use or distribution of water by decentralization (Hasan, 2021). Without limiting the resource, *sustainable use* means continuing for future generation (Hasan, 2021). The sustainability of a country's development is also dependent on the preservation of its water resources' quality, given the importance of water resources to the country's agriculture and to the general well-being of its citizens (Flint, 2004). Thus, a proper water management system is needed in Bangladesh in order to ensure the long-term sustainability of the country's water resources as well as in attaining the SDGs.

The Sustainable Development Goals (SDGs) are a comprehensive, ambitious set of goals determined by the United Nations which states have to contend by 2030 (Alcamo, 2019). Alcamo (2019) also mentioned that the targets to maintain the quality of water and its management are interlinked with the other SDG targets, and *vice versa*. Sometimes, achieving some targets of the SDGs, like Zero Hunger (goal 2) and Industry, Innovation, and Infrastructure (goal 9), might hinder water quality and increased pollution (Alcamo, 2019). What is required is a concentrated effort to connect goal 6, which is mostly about water, with the other SDGs. For the Sustainable Development Goals on water to be achieved, collaboration between diverse institutions is essential. This is also true for attaining water quality, preventing pollution, and protecting life below water as well as life on land.

Bangladesh, until recently, has passed many laws, plans, and policies for the protection and management of water resources (Table 10.3) (Babu, 2020). These laws and policies, however, failed to appropriately address pollution consequences and pollution control in Bangladesh, and they were insufficient in compliance with the water quality, water management, and climate change adaptation requirements. Several pieces of legislation dealing with water regulations are out of sync with the criteria for achieving the SDGs while maintaining the quality of both surface and groundwater (Babu, 2020). The failure to implement such existing legislations and policies, on the other hand, opened the door to pollution and unsustainable water consumption from the very beginning. For achieving SDGs such as ensuring clean water, pollution prevention, and the linked targets, Bangladesh is lacking adequate mechanisms to collaborate with key institutions, organizations, and stakeholders.

The main objective of this chapter is to analyze Bangladesh's legal measures for water management and protection from pollution, with the ambit to achieving goal 6 of the SDGs and targets such as 6.1, 6.3, 6.4, and 6.6. It also draws attention to the fact that Bangladesh should take a step forward in achieving SDG 6 (Clean Water and Sanitation) by coordinating with the actions taken by other SDGs that are relevant. The water regulatory landscape should cover all aspects of ensuring clean water, including the protection of water and proper management of water resources, saving from pollution and unsustainable use in one side and maintaining synergies with water-using sectors in the other. The sustainable use and effective management of water would bring beneficial consequences for a number of government activities carried out to achieve other SDGs, as well as the other way around, if they were performed in a synergistic manner. In Bangladesh, development goals such as industrialization, urbanization, infrastructure, and other projects have not been adequately coordinated with water management to ensure clean water and healthy life under it.

To attain the key objective of the chapter, we have taken a policy analysis method based on primary and secondary sources. To identify legal gaps in Bangladesh's implementation of the Sustainable Development Goals (SDGs) of ensuring clean water and preventing pollution, we examined water resource management–related laws, plans, and policies, as well as contemporary long-term development policies. A vast body of literature on the Sustainable Development Goals and water resource management, both domestically and internationally, has been reviewed as well. We analyzed the existing literature carefully and used the relevant data to attain our objective. We find that the government of Bangladesh (GoB) primarily concentrates on the management of water resources through passing legislations rather than emphasizing on the monitoring and accountability of multiple operating entities and organizations, as well as their coordination. A major impediment to the attainment of SDG 6 as well as other relevant SDGs has remained to be inadequate coordination among varied entities that are interconnected on water concerns. As a result, the chapter highly suggests that proper water resource management be implemented in conjunction with connected targets of SDG 6, which is critical to achieving Agenda 2030 as a whole, in addition to a number of other relevant SDGs and global targets, such as food security and biodiversity targets.

The existing scholarships lack a specific focus on the importance of synergistic water management policies, plans, and legislations, as well as the gaps in these policies, plans, and legislations, in order to accomplish the SDGs. However, some of the previous published papers which have already focused on water pollution prevention by legal interventions in the Bangladesh perspective should be credited here. Most of the cases—the examination of water pollution, its causes, and the consequences in Bangladesh, as well as a discussion of the institutional framework in a broad fashion—have been conducted literally without considering the necessity of synergy for the attainment of the SDGs through legislative interventions (Babu, 2020). The impacts of water pollution on human health (Uddin and Jeong, 2021), ecology (Faroque and South, 2021), and agro-economy (Parvin et al., 2020) have also been argued broadly in many published scientific documents (Faroque and South, 2021; Parvin et al., 2020), but focus on maintaining coordination or interlinking of all targets of SDG 6 and other SDGs linked to water concerns is overlooked from the Bangladesh perspective. While budding scientific literatures have proved water pollution (Adejumoke et al., 2018) and contamination (Abedin et al., 2019), little scholarships focused on extant legal regime of Bangladesh for coordinating the obligations of water management with protection and the legislative gaps on these for attaining the SDGs synergistically.

While conducting this research, we observed that there is a lack of cooperation and institutional coordination among the SAARC countries when it comes to achieving the Sustainable Development Goals (SDGs) related to water. However, additional research is required to examine the feasibility of ameliorative legal measures to ensure the achievement of the SDGs, notably goal 6, from a regional viewpoint. This chapter is organized into five sections: "Introduction," "Present Context of Water Resources in Bangladesh," "Legal Regime for the Protection and Management of Water Resources," "Collaborative Approach for Water Resource Management," and "Concluding Remarks." The following section of this article presents a brief overview of water resources of Bangladesh in the present situation.

10.2 Present Context of Water Resources in Bangladesh

In addition to being the largest deltaic plain at the confluence of the Ganges, the Brahmaputra (Jamuna), and the Meghna (GBM) Rivers, Bangladesh is also fortunate in having wide-ranging water resources, such as ponds, lakes, natural depressions (beels, haors, and baors), estuaries, canals, as well as rivers encompassing an area of 4.56 million hectares (DoF, 2005). Moreover, Bangladesh is a privileged country with marine waters of roughly 166,000 square kilometers, of which 141,000 square kilometers are within the exclusive economic zone (Seraj, 2015). Both fresh watercourses and marine areas are teeming with biodiversity and natural resources (non-living), which have received limited attention in terms of legislative protection in Bangladesh. Water resource is undoubtedly a crucial resource for Bangladesh to achieve the SDGs as a whole and the growing food demand (Reliefweb, 2011), but the existing strategies to protect and manage this resource are poorly implemented (Babu, 2020), and the linkage with other SDGs is hardly understood (Benson et al., 2020). Due to unmanaged water resources, it failed to prevent increasing pollution, illegal sand quarrying, and illegal encroachment of water bodies, which reduce the area of water bodies (Babu, 2020). Table 10.1, which follows, presents an overview of the loss of water bodies in Bangladesh over the years.

The pollution of watercourses is beyond control country-wide and weakens water quality and the water ecosystem, which trigger increased water scarcity (Reliefweb, 2011) and ecologically critical areas in Bangladesh. As per Section 5 of the country's main environmental legislation, namely, the Bangladesh Environment Conservation Act 1995, the GoB can declare any area as an ecologically critical area (ECA), which is an environmentally critical situation or is threatened. Furthermore, the government of Bangladesh (GoB) has the authority to specify the activities not permitted to be undertaken or carried on in an ECA. Until now, the GoB has declared 13 areas as ECAs, of which

Table 10.1 Loss of water bodies in Bangladesh within 27 years (1989–2016)

Name of wetlands/water bodies	*Area (in ha) in 1989*	*Area (in km^2) in 2016*
Rivers and streams	480,000	7,497
Haors, baors, beels, and jheels	120,000–290,000	1,197
Seasonal inundable floodplains	5,770,000	54,866
Lake and marsh	120,000–290,000	688
Fishponds and tanks	150,000–180,000	1,469
Large water storage	90,000	—
Total	7,100,000	65,717

Source: *A Directory of Asian Wetlands*, 1989 (https://portals.iucn.org/library/node/5933).

Table 10.2 Ecologically critical areas, particularly water bodies, in Bangladesh, 2020

Ecologically critical areas	*Location*	*Ecosystem*	*Year*	*Area (in ha)*
Hakaluki Haor	Sylhet and Moulvibazar	Freshwater wetland	1999	40,446
Tanguar Haor	Sunamganj	Freshwater wetland	1999	—
Marjat Baor	Oxbow Lake	Jhenaidah and Jashore	1999	326
Gulshan-Baridhara Lake	Dhaka	Urban wetland	2001	101
Buriganga River	Dhaka	River	2009	1,336
Turag River	Dhaka	River	2009	1,184
Sitalakhya River	Dhaka	River	2009	3,771
Balu River and Tongi Khal	Dhaka	River and canal	2009	1,315
Jafong-Dawki	Sylhet	River	2015	1,493
Total ecologically critical areas of Bangladesh (all fresh watercourses)				49,972 ha

Source: Department of Environment, 2019 (www.doe.gov.bd/site/page/9481fd1b-7ca6-4087-890a-886cd226df0e/-).

9 areas are different water bodies and rivers (Table 10.2) (Department of Environment), but so far has initiated inadequate steps to control pollution and manage in an eco-friendly way (DoE, 2019; Babu, 2020). Despite pollution, climate change and the storage of water by upper stream states are all contributing factors in lessening water flow and drying rivers in the summer. Rivers and water bodies now overflow in the monsoon season, dry up in the summer, and turn as misery for the people of Bangladesh. Thus, it adversely affects the country's agro-economy (Babu, 2020) and constrains the achievement of the SDGs as a whole. Alternatively, for the growing demand of food, the use of pesticides and chemicals to increase food security threatens water quality and pollutes it.

From the last decades, a new tendency for the illegal capturing of water bodies and adjacent areas was a pertinent concern in urban areas of Bangladesh. For rapid and unplanned urbanization, industrialization, and infrastructural development, land- and pond-grabbing are rife. When the District Fisheries Department of the Chittagong port city conducted a survey in 1991, the number of water bodies found there was 19,250 (Hussain, 2018). However, a survey conducted by the Chittagong Development Authority (CDA) in 2006–2007 found that the number of water bodies had dropped dramatically to 4,523, a significant decrease from the previous survey (Hussain, 2018). The same situation exists in Dhaka and in the Rajshahi City corporation area. According to the Rajshahi Development Authority, the city had 4,238 canals, ponds, and wetlands in 1961, which were reduced at present to only 214 (Dhaka Tribune, 2014).

It is notable that Dhaka City has lost the maximum number of water bodies, which include ponds, wetlands, lakes, and rivers, due to the growing residential and other infrastructural demands of a huge population. Even though water bodies are crucial in maintaining the environmental balance of a city, only five ponds exist now in Dhaka, and 58 canals have already been occupied illegally by different entities, NGOs, and politicians (Editorial, 2017). Unfortunately, no legislative measures that are now in effect have imposed any kind of stringent punishment for grabbing water bodies in urban areas (Yousuf, 2017), and water bodies–related case filings are only six according to the Chittagong Department of Environment (Hussain, 2018) and in some places are absent. The current situation in Bangladesh is the outcome of poor water resource management and the non-implementation of plans and policies (Babu, 2020). In cases of infrastructural development of urban areas, no coordination exists among the different responsive authorities to recycle water for reuse or not to pollute; rather, they are responsible for declining water resources.

Above all, constant pollutions and chemical contamination create serious barriers in achieving SDG 6, particularly Target 6.1 (Akter et al., 2016), and constrains in improving water quality (Target 6.3), reducing water scarcity (Target 6.4), and improving pollution condition (Target 6.5) (Roy, 2021). Bangladesh failed to ensure an integrated water resource management from policy to execution level, including through transboundary cooperation, and continued to fail in protecting and restoring water-related ecosystems, such as mountains, rivers, forests, aquifers, wetlands, as well as lakes, for rapid and unplanned industrialization and urbanization (Roy, 2021). These failures ultimately chocked the attainment of SDG commitment under goal 6, which ultimately constrains attaining other SDGs as well which are indirectly connected, such as goals 2, 3, 11, 12, 13, 14, 15, and 16, and many of their targets.

Presently, the other part of SDG 6, particularly sanitation for all, is on track in Bangladesh, and the proportion of access to sanitation service increased 46.9% in 2015 (Khatun and Saadat, 2021). Khatun and Saadat (2021) contend that the gap between urban and rural area access has reduced also. However, Bangladesh could not align SDG 6 and its other targets with its water resource–related plans, policies, and legislations and also lacks monitoring and evaluation of water resource and, ultimately, financing for further development (Khatun and Saadat, 2021). Some Asian countries experienced considerable progress in achieving the SDGs by coordinating the existing sectoral development policies and plans with the SDGs' water targets, supporting their forthcoming agendas for development in alignment with the SDGs, and making related alterations in extant legal regime (Khan et al., 2020). The same approach could be extremely efficient in achieving the Sustainable Development Goals relating to water resources in Bangladesh.

The following section of the chapter will shed light on Bangladesh's existing legal frameworks, strategies, and policies implemented so far to support the achievement of the Sustainable Development Goals by 2030.

10.3 Legal Regime for the Protection and Management of Water Resources

Bangladesh has promulgated numerous laws, policies, and plans to improve environmental elements, including water, as per the mandate of its highest law, the Constitution of Bangladesh. Article 18A of the Constitution of Bangladesh declares that "the State shall endeavour to protect and improve the environment and to preserve and safeguard the natural resources, biodiversity, wetlands, forests and wildlife for the present and future citizens." The development of institutions and entities tasked with the responsibility of water management could be traced back to the 1950s (Chan et al., 2016). The East Pakistan Water and Power Development Board Authority was established as well as empowered to manage water issues in East Pakistan (present Bangladesh). The Board was abolished in 1971, following Bangladesh's independence from Pakistan, and replaced by the Bangladesh Water Development Board in 1972. It was designated as the primary authority for managing and controlling water resources for irrigation, flood, and drainage projects (Chan et al., 2016). However, various laws and policies have been passed since then to safeguard water resources and to improve water management, but many significant and substantial water issues remain unaddressed in the existing law.

10.3.1 Legislative Interventions for Water Resources Management and Protection

The Bangladesh Water Act (BWA) 2013 is the most recent addition that the government of Bangladesh has made in reference to an endeavor for the management of water resources. There are also previously enacted legislations and policies that overlap with and are connected to the existing key

Table 10.3 Year-wise list of legislations and policies related to water management

Year	*Existing Legislations and Policies related to Water Management*
1974	The Territorial Waters and Maritime Zones Act
1992	Water Resource Planning Act
1995	Bangladesh Environment Conservation (ECA) Act
1996	Water Supply and Sewerage Authority Act
1997	The Environment Conservation Rules (ECR)
1998	National Policy for Safe Water Supply and Sanitation
1998	Upazila Parishad Act
1999	The National Water Policy
2000	District Council (*Zila Parishad*) Act
2000	Municipal Areas' Playground, Open Space, Park and Natural Water Reservoir Conservation Act
2001	National Water Management Plan
2004	National Policy for Arsenic Mitigation and Its Implementation Procedure
2005	Coastal Zone Policy
2005	Pro-Poor Strategy for Water and Sanitation Sector
2005	National Sector Development Programme for Water Supply and Sanitation
2006	Coastal Development Strategy
2009	Local Government (Union Parishad) Act
2009	Local Government (Pourashava) Act
2009	Local Government (City Corporation) Act
2009	National Vetting Guideline for Water Supply and Sanitation Sub-Sector in Bangladesh
2011	National Water Management Plan: Development Strategy
2011	Sector Development Plan (2011–2025), Water Supply and Sanitation Sector
2011	Integrated Small-Scale Irrigation Policy
2013	The Bangladesh Water Act
2013	Brick Manufacturing and Brick Kilns Establishment (Control) Act
2013	Food Safety Act
2013	National River Protection Commission (NRPC) Act
2014	National Shrimp Policy
2018	The *Bangladesh Pani Bidhimala*
2018	The Groundwater Management for Agricultural Purposes Act

water legislation. Table 10.3 contains a collection of relevant legislations and policies on water resource protection and management in Bangladesh. The Bangladesh Water Act contains provisions for the development, abstraction, management, use, distribution, conservation, and protection of water resources in an integrated manner. Specifically, the National Water Resources Council (NWRC) was established under the Bangladesh Water Act, with the prime minister serving as its chair and an executive committee charged with putting the Council's recommendations into action. The legislation in its Section 13 provides that no one can extract, use, distribute, develop, protect, or conserve water resources without the executive committee's authorization, nor can they build structures that obstruct natural river and stream flow.

BWA 2013 prioritized various water usages in water-stressed areas in the following ways: drinking water, domestic consumption, irrigation purposes, fish culture, biodiversity, wildlife requirements, stream flow, industry, salinity control, power production, recreation, and finally, miscellaneous. Despite the fact that drinking water and home consumption were mentioned as the highest priorities in the act, there are still various concerns that must be addressed in order to achieve this, such as pollution. However, in the event of protecting and conserving water resources from pollution, it has specified that the Bangladesh Environment Conservation Act 1995 should be followed. Furthermore, the Act made no mention of the legal impediments that exist in the environmental court under the Environment Court Act 2010. Neither does it provide any measures for protection against unlawful and unsustainable use of water bodies, surrounding land encroachment, and pollution in water bodies, nor does it provide any specific means for combating these threats (Chan et al., 2016).

Under BWA 2013, large-scale water extraction by individuals as well as organizations for more than domestic usage is required to get permits or licenses. However, some key issues were left unanswered, including a lack of coordination with the Bangladesh Environment Conservation Act of 1995 and the absence of any procedures for punishment in case of industrial spills. Most crucially, the unrestricted power placed in the executive committee, which is not subject to any accountability obligations, has the potential to be abused, in violation of the Act's objectives. In accordance with the Environment Court Act, the people of Bangladesh, who are the actual proprietors of water resources, are not permitted to file complaints or lawsuits against anyone, even government entities (Arifuzzaman et al., 2019).

The Bangladesh Environment Conservation Act 1995 (BECA) was passed to conserve the environment, to improve the standards of the environment, and also to mitigate and control environmental pollution, including that of water. The main limitation of this Act is in how it gives a wide power to the director general (DG), from giving directions to handling complaints or applications from the public for remedies against polluters. The Act failed to control pollution of all elements of the environment, like air, water, and soil, and to ensure the accountability and transparency of the Department of Environment (DoE) and the DG of DoE. According to the Environmental Court Act 2010, in the absence of a written report from an inspector of the DoE or any other person authorized, the environmental court cannot take cognizance of any claim and has no ability to receive any compensation claim. This Act only focused on industrial pollution, leaving water quality, safety, and pollution from other sources unaddressed (Arifuzzaman et al., 2019). Despite the fact that climate change and its potential repercussions have had substantial consequences in the field of water resource management, there has been no attention paid to these issues. It made no specific provision for institutional coordination between the DoE and the Water Resources Planning Organization (WARPO) or any other water agencies (Arifuzzaman et al., 2019).

In 2000, two important Acts have also been passed for the protection of water bodies in Bangladesh. The Bangladesh Water Development Board Act (BWDBA) and the Municipal Areas' Playground, Open Space, Park, and Natural Water Reservoir Conservation Act 2000 were enacted but failed to attain their own objectives stated in their preambles. However, the lack of broadcasting, the poor implementation, the lack of strict monitoring, and the inadequate penalty are limitations in the implementation of the two Acts.

The forerunners of BWA 2013 made significant contributions to a variety of positive endeavors, while also leaving numerous challenges unanswered in the field of water resource management. Examples include the establishment of the Water Resources Planning Institution pursuant to the Water Resources Planning Act of 1992. Under Section 7 of the Act, the Institution's responsibilities include planning for environmentally balanced water resources, setting national standards for water resource management, identifying scientific exploitation and conservation methods, and collaborating with other water resource bodies. According to Section 3 of the Water Supply and Sewerage Authority Act of 1996, the Authority (WASA) responsible for developing water supply and sanitation systems as well as delivering water supply, sewage, and stormwater management services is not concerned with industrial effluents. The importance of preserving water quality has been overlooked in this legislation. Section 8 of the Territorial Water and Maritime Zones Act 1974 empowered the GoB to take whatever actions it considers necessary to prevent and regulate marine pollution in order to protect the environment. The legislation, on the other hand, does not specify what actions should be taken or what the polluter's obligation or penalty should be. The Groundwater Management Act 2018, a recently enacted water-related law, failed to address a number of issues, including safe and sustainable abstraction of groundwater, protection of groundwater, integrated usage of surface and groundwater, lack of a central monitoring system, groundwater quality monitoring, pollution, and water safety concerns (Arifuzzaman et al., 2019).

Control of water pollution is addressed in the amendment of the Chittagong Port Authority Ordinance 1976 and the *Mongla* Port Authority Ordinance 1976 in 1995. These Acts addressed pollution in the port area, but the punishment is not sufficient to the level of pollution in the port area or for the wrongdoers. It is also silent about the nature of pollution of and punishment for marine vessels. The Bangladesh Petroleum Act 1974 provides a provision explaining the duties of persons engaged in petroleum operation, which controls the flow and prevents waste or escape in that area of water, but the penalty for polluters is inadequate. However, the other existing laws regulating mineral resources have not mentioned about the protection of water bodies from pollution as well.

Aside from the implementation of the penalties under the aforementioned legislations, the Environment Court Act 2010 was enacted, repealing the Environment Court Act 2000, which established environmental courts. Currently, the number of environmental courts is only three, and one environmental appellate court in Dhaka. However, the rate of filing of cases is lowest in the environment court of Bangladesh for the stated excessive powers of DG (Sajal and Akhtaruzzaman, 2016). Another problem is, the environment court has an indirect jurisdiction on water pollution according to BWA 2013 and the Environment Court Act 2013. To date, no official gazette notification has been published by the government of Bangladesh incorporating the Water Act 2013 or any other legislation pertaining to the protection of water bodies. Thus, the environment court has no direct jurisdiction to try water resource like other natural resource–related offences (Sajal and Akhtaruzzaman, 2016).

Regarding transboundary water bodies, only one treaty (Ganges Treaty) exists between Bangladesh and India relating to Farakka Barrage, but there remain tons of limitations in the treaty and its implementation. The Convention on the Non-Navigational Uses of International Watercourses 1997 provided the definition of *pollution* and the measures for controlling international watercourse pollution, but Bangladesh has not ratified the convention. However, the Ganges Treaty is not amended with the provision of the Convention under Articles 21, 22, and 23. This tends to increase water scarcity due to the withdrawing of fresh water of transboundary rivers by the upstream states (Karim and Uddin, 2019).

10.3.2 Water Resource Management Policies and Plans

The National Water Policy (NWP) 1999 focused on water management, which addressed some of the significant water management challenges, such as concentrating on water accessibility, users' responsibility, erosion and siltation, water quality deterioration, decentralization, and privatization (Hasan, 2021). Despite NWP, various other policies related to water management, such as the National Shrimp Policy 2014, National Water Management Plan 2001, and National Policy for Safe Water Supply and Sanitation 1998, all mentioned decentralization as a necessary component of effective water management, but this is not the case in practice. Control of pollution for water resources is not mentioned at any point in the objectives. Again, this policy was not updated with the passing of the Water Act 2013 and after taking the SDG commitments. The *Bangladesh Pani Bidhimala* 2018 (if translated, the Bangladesh Water Rules 2018), on the other hand, is a comprehensive instrument for the protection of water resources, but it is a guideline to introduce a new water policy with some provisions for the protection of water from pollution. Section 41 of the Water Act 2013 states about the collection of data relating to the control of water pollution, but a pollution control mechanism is missing there. Finally, legislative interventions constantly fail to protect and manage water resources from pollution and are unsuccessful in punishing the culprits. While several of the legislations, as well as policies, mentioned water pollution issues, some stayed mute; however, one thing that stands out is the absence of policy coherence for water management and in the fight against pollution, which increases the difficulty of combating water pollution.

Bangladesh has adopted a number of policies to attain a long-term development vision which takes into account environmental and climate change crises as well as the challenges they present. The Bangladesh Delta Plan (BDP) 2100 is a long-term vision to make the Bangladesh Delta safe by the end of the 21st century. It has also set its objectives and goals in a broader aspect, making the environment and climate change its priority. The first and second objectives are directly connected, such as ensuring safety from floods and climate change–related disasters and enhancing water security and the efficiency of water usage (Delta Plan 2100, 2020). The Plan identified its priority areas, such as flood protection, river erosion control, river management (including river training and navigability), urban and rural water supply and waste management, and urban flood control and drainage. It allocates a low portion of the budget, only 20% of the total BDP 2100 FY 2031, for improving water and sanitation services in small towns and rural areas (Delta Plan 2100, 2020). However, a total of USD 20,936 million is needed in the water sector, according to the government's Sector Development Plan 2011–2025 (Local Government Division, 2011). Thus, insufficient financing makes the target, including clean water achievement, hard and impossible. Following a review of past projects undertaken by the Ministry of Water Resources (MoWR), it has been noticed that most of those projects have concentrated solely on the construction of high-value infrastructure projects, with little emphasis placed on the implementation of holistic water resource management strategies (United Nations, 2018). It is necessary to shift the focus away from water development and toward comprehensive water management.

One of the key themes of Bangladesh's Perspective Plan for the years 2021–2041 (PP2041) is the integration of environmental and climate change factors into the country's development plan. Bangladesh sought to implement a green growth strategy under the PP2041 framework. Under PP2041, the specific strategies, policies, and institutional reforms include (1) integrating environmental costs into the macroeconomic framework, (2) implementing the Delta Plan to build resilience and reduce vulnerability to climate change, (3) reducing air and water pollution, (4) removing fuel subsidies, (5) adopting a green tax on fossil fuel consumption, (6) taxing emission from industrial units, (7) preventing surface water pollution, and (8) performing a geospatial data analysis for evidence-based decision-making. Despite the fact that the environment is often considered a public good, the public sector cannot finance it on its own. Innovative solutions must be found to ensure a good division of financing options between public and private sectors. PP2041 is focused on various financing options, such as private financing options, public financing policies, tapping the Green Climate Fund (GCF), and mobilizing resources from other global funds. While the reforms that PP2041 seeks to pursue involve water concerns, including the reduction of water pollution, existing legislations and water policies for water protection and management are not being revised to reflect the reformation that is expected to be brought about by PP2041. Bangladesh, on the other hand, is on the verge of adopting a strategic plan (2022–2026Z) for the country that is still in the process of being revised. The country's strategic plan is aligned with the SDGs and the government's eighth five-year plan (July 2020–June 2025) but includes no mention of water issues, pollution, or clean water and water management (World Food Programme 2022–2026, 2021).

10.3.3 Evaluation of Coordination Between Legislations and Policies With SDGs

Water is one of the Sustainable Development Goals' central focuses, as it is vital to sustainable growth and socio-economic development (Moumen et al., 2019). Putting an end to hunger, creating food security, boosting nutrition, conserving the ecosystem, and society's overall health and well-being are all dependent on water, either directly or indirectly. In addition, water is a critical component of risk management strategies for disease outbreaks, migration, and natural disasters. Therefore, a comprehensive and consistent water resource protection and management plan is required, one that takes into consideration not only major water impacts but also cross-sectoral implications. In particular, this involves comprehensive legislative measures for water and the intersections among relevant water-using sectors, in accordance with a realistic arrangement for water quality and supply, as well as strong water governance, in order to make sensible use of the world's finite freshwater resources (WWF, 2015). Clean Water and Sanitation is the 6th of the 17 Sustainable Development Goals, with the aim to ensure everyone has access to clean water and sanitation as well as sustainable water management. Water is also intertwined with the other SDGs, which means it can support cross-sectoral or institutional actions needed to advance the SDGs as a whole (IWA, 2019). The initiatives that have been taken by the government of Bangladesh so far to attain goals 2, 3, 6, 11, 12, and 15 are outlined in Table 10.4. In order to achieve SDGs related to water, an effective water governance that facilitates execution in a consistent manner must be in place. Collaborations among multiple government entities responsible for accomplishing other SDGs that intersect with water challenges could aid in realizing the full potential of water protection, access, and sustainable management (Moumen et al., 2019). Sharing, developing, and adapting innovative solutions necessitate such collaboration, as does avoiding contradictory measures and overriding and overlapping worries. SDG 6 targets offer a guide for cross-sectoral collaborations regarding water access and management (SDG 6).

According to the United Nations Sustainable Development Goal 6 Synthesis Report on Water and Sanitation (United Nations, 2020), there is no one-size-fits-all strategy for adopting an integrated water resource management; therefore, each country has developed its own course based on societal, political, economic, and environmental factors. It took 42 years for Bangladesh to enact a water legislation. A number of authorities, both nationally and locally, have been given responsibility for the management and preservation of water resources, in addition to the aforementioned laws and regulations (as listed in Table 10.5). Thus, there is no shortage of legislative and organizational interventions in Bangladesh, whether it is to safeguard water resources or for their sustainable management. Specifications on water quality, pollution, and drinking water management are lacking in those policies (Water Policy and Public Rights, 2020), as is guidance on how an implementing authority as well as institutions should operate to implement the policies in conjunction with contemporary development, such as the SDGs.

In addition to the Ministry of Water Resources, the local governments and the Ministry of Agriculture (MoA), Ministry of Textiles and Jute (MoT&J), Ministry of Industry (MoI), Ministry

Table 10.4 Sustainable Development Goals and Bangladesh Progress Report 2020

Goals	*Government-adopted efforts*	*Challenges for Bangladesh*
2	• National Social Security Strategy (NSSS) in 2015 • Bangladesh Delta Plan 2100 (BDP 2100) in 2018	• Implementation of the National Food Policy Plan of Action (2008–2015) and the Delta Plan 2100 a big challenge for Bangladesh
3	• 7th Five-Year Plan (2016–2020) • 4th Health, Population, and Nutrition Sector Programme (HPNSP, 2016–2020) • Health Care Waste Management Plan (HCWMP) • The National Strategic Plan for TB Control (2018–2022)	• No coordination between the two related ministries of MOHFW and MOLGRDC • Non-implementation of the mentioned strategies and plans • Failed to remove severe disparities between urban and rural areas in accessing health services
6	• An action plan • A national policy on water, water supply and sewerage, environment protection rules • The Bangladesh Water Act 2013	• Non-implementations of laws and policies • Problem in mobilizing domestic and international finance
11	• The government's *Ashrayan* 2 project for the poor • Increased fund for communication • Implementation of a pilot project, "Programmatic CDM," etc. • Projects to develop environment-based waste management	• Lots of efforts for achieving SDG 11, but challenges remain on affordable housing, sustainable urban transports, air quality improvement, urban resilience, resource constrain, synchronization of policies, strategies, and master plans
12	• Submission of the annual national reports under the Basel Convention and a national implementation plan • Compliance with reporting requirements for the production and consumption of ozone-depleting substances under the Montreal Protocol	• Challenge of documenting the country's macro-policies, regulatory, voluntary, or economic instruments that support the shift toward sustainable consumption and production (SCP) • Lack of adequate monitoring framework • Availability of, and access to, financial resources to support actions
15	• Declaration of ecologically critical area • Declared Ratargul (only freshwater swamp in the haor basin) as a special biosphere reserve in 2015	• Could not effectively implement appropriate management strategies for its protected and critical areas, with full participation of the local stakeholders • Pressure of population on environment, expansion of human settlements and agriculture, shifting cultivation, habitat degradation, over-exploitation of natural resources

Source: Prepared by authors after studying the Sustainable Development Goals: Bangladesh Progress Report 2020.

of Rural Development and Cooperatives (MoLGRD&C), as well as the Ministry of Disaster Management and Relief (MoDM&R), etc. are the primary government water users in the country (Rahman, 2021). Despite the fact that numerous pieces of legislation have been established to address the protection and management of water resources, the question of their proper execution, institutional incapacities, and coordination continues to linger (Rahman, 2021). When the majority

Table 10.5 Activities of different authorities for water protection

Under the authority of. . .	*Departments or authorities*	*Areas for continuing activities*
Ministry of Environment, Forest, and Climate Change	Department of Environment	Management of ecologically critical areas
Ministry of Agriculture	Department of Agricultural Extension Bangladesh Agricultural Development Corporation Barind Multipurpose Development Authority	Agro-biodiversity management Control of chemical fertilizer and pesticides Control of serious degradation of water or eutrophication
Ministry of Fisheries and Livestock	Fisheries Department Marine Fisheries Academy	Freshwater ecosystem Marine fisheries in protected areas Control of eutrophication
Ministry of Water Resources	Bangladesh Haor and Wetland Development Board Water Resource Planning Organization (WARPO) BD Water Development Board (BWDB) Institute of Water Modeling (IWM) Center for Environment and Geographic Information Services	Anti-salinity measure Estuary and coastal management Development of haor and wetlands
Ministry of Land	Land Development Board District administration	Land management in ecologically critical areas Estuary and coastal management Ecotourism Government ponds, shrimp hachari ponds management
Ministry of Industry		Industrial waste management
Ministry of Civil Aviation and Tourism		Ecotourism
Ministry of Disaster and Rehabilitation	Cyclone preparedness Disaster management	Risk reduction, preparedness, and response Flood forecasting, warming center
Ministry of Local Government, Rural Development, and Cooperatives	Local government engineering department Department of Public Health Engineering Water and sewerage authorities	Waste management Risk reduction, preparedness, and response Clean water campaign in rural areas
Joint River Commission	National River Protection Commission	Control of pollution of joint rivers
Ministry of Textile and Jute	Department of Jute Silk Development Board	Processing activities and control of pollution from this

of legislations dealing with water resource protection and management are examined in light of the Sustainable Development Goals, it is discovered that the country has mostly rebranded what it was already doing, rather than introducing truly transformational initiatives. Following the SDG pledge in 2015, no comprehensive policy guideline was developed to execute the existing laws and policies in order to fulfill the targets of SDG 6 and to coordinate with other relevant SDGs implementing government agencies' efforts to achieve it. It would have been the most significant step forward to achieve SDG 6 if the recognition of water's interconnectedness with other water-using sectors, as well as the development of a comprehensive policy guideline to promote policy coherence, had been taken. Collaboration across sectors and with inclusive institutions would have made water a matter for all.

The country is frequently confronted with water accessibility challenges that develop as a result of water degradation, pollution, scarcity, and contamination, yet such access is also not equally distributed (Water Policy and Public Rights, 2020). The pollution of water bodies continues to occur despite the enactment of several regulations and laws, and inhabitants are pushed to use polluted water for their everyday necessities (Water Policy and Public Rights, 2020). Bangladesh joined the rest of the world in 2015 in pledging commitment to the Sustainable Development Goals, which include attaining universal and equitable access to clean and affordable drinking water for all by 2030. The SDG framework allows articulating better the greater contributions of water and the water sector, as well as how the strategies developed by the sector could be applied in other sectors that use water and water resources.

There is a dearth of coordination among water-related regulations across the country (Rahman, 2021). In light of the climate crisis, coordination with the Disaster Management Act 2012 and other relevant policies that were supposed to be absorbed by a water legislation is missing. A number of legislative initiatives act as roadblocks in the protection of water reservoirs. For example, the land acquisition policy of Bangladesh does not take into consideration the community's fresh or drinking water rights, because the state is permitted to acquire any sort of land, including land with water reserves on which the community is reliant. Local fresh drinking water sources are being impacted, especially with the rise of the energy sector and construction initiatives.

The integration of appropriate policies implemented to advance other SDGs is required to ensure water management and, ultimately, SDG attainment. As a result, there is a perceived need for the government to revisit all these regulations and develop comprehensive water policy guidelines aligned with SDG targets so that all key water-using stakeholders, including government agencies, non-governmental organizations, and businesses, are informed about the extent of their involvement and actions.

10.4 Collaborative Approach for Water Resource Management

The water resources of Bangladesh could be well protected in implementing the proper management and monitoring of water resources. However, the following might be helpful for Bangladesh to achieve SDGs linked to water resources.

10.4.1 Implementing Existing Legal Regime

The government has enacted comprehensive legislative measures for water resource management and provided limited protection of water resources from pollution. Bangladesh could strictly implement such existing legislative frameworks for ensuring clean water and monitor the activities of management bodies. However, Bangladesh needs comprehensive legislative measures for protecting surface and groundwater from pollution in light of climate change crisis and in achieving the SDGs. The implementation of the provisions for an environmental impact assessment (EIA) report before passing any development project adjacent to water bodies would be effective in protecting water bodies from development-related hazards. The filing of environmental suits should be more litigant-friendly, ensuring access to environmental justice easily by removing environmental lawsuit barriers.

10.4.2 Ensuring a Proactive Approach of Water Management Bodies

The BWA 2013 is surely a comprehensive legislation for the management of water resources. Under this Act, the government can declare any area, or any part with such water resource, as water-stressed area for a specified time after getting a recommendation from the executive committee. Precaution and prevention are better for the protection of the environment. It is better to become proactive in saving water bodies rather than to declare water-stressed area after pollution. Authorities should monitor the quality of water resources and ensure the punishment of polluters, to save them, rather than to declare them as ECA. Thus, the water management body that is a council, an executive committee, should show a proactive approach to protect water resources and for better management.

10.4.3 Establishing Cooperation Among Different Institutions and SDGs

Bangladesh has the tradition of establishing institutions but no culture to maintain cooperation among the institutions. As water bodies are everywhere in Bangladesh, and as they are in threat, it is necessary to establish institutions from the sub-district to central capital and their coordination as well. Similarly, GoB has adopted several long-term development plans and projects that have been taken for the attainment of the SDGs, but there is no coordination among those plans and policies with water-related institutions and policies, and vice versa.

10.4.4 Establishing Cooperation Among Different Institutions and SDGs

Bangladesh has 57 international watercourses, but only 1 treaty for the management of one river; until now, no treaty is for other 56 rivers. This also causes water scarcity, transboundary pollution, and unsustainable use of water. The absence of proper management of shared rivers has chocked the way to attain the SDGs in time. Thus, Bangladesh should take the initiative to implement an international water law in adopting a bilateral treaty to regulate international watercourses and their protection from pollution. For this, Bangladesh should ratify the UN Convention on the Non-Navigational Watercourse 1997.

10.5 Concluding Remarks

Water resources should be protected synergistically with its management system and the other interlinked SDGs to protect lives below water and on land. Bangladesh may benefit from better understanding the significance of ensuring clean water and its impacts on other SDGs. However, despite several legal measures adopted to conserve water resources, the GoB has been unable to significantly improve institutional synergy and harmony at all levels, which is also the requirement for attaining SDG 16.6. The legislations and different executing authorities established under those legislations have no coordination in doing their respective duties, which triggers extreme pollutions, water scarcity, unsustainable use of water, critical ecology of water bodies, and thus, SDGs 16a and 16b went far from Bangladesh to achieve within 2030. The mandatory legislations and water policies are not updated to meet the requirement of the SDGs and climate change impact on water bodies. Integrating possible climate change impacts into forthcoming water management planning is also essential to achieve the sustainable use of water and the proper management of water resources, as well as to achieve the SDGs. Thus, for all these limitations, Bangladesh is far from attaining the SDGs, particularly goal 6, and also other related SDGs. Existing legal regime and policy assessment leads us to suggest a completely different strategy for the protection of water from pollution and the management of water resources of Bangladesh. Taking a collaborative approach, Bangladesh could enhance its management system with pollution control through monitoring, creating an accountable and transparent authority, financing, and ensuring the sustainable use of water. Legislative measures to attain the targets of SDG 6 must be aligned with the other SDGs and with forthcoming development-related policies. Only then will Bangladesh step closer to achieving the SDGs and other global goals.

References

Abedin, M. A., Collins, A. E., Habiba, U. & Shaw, R. (2019). Climate Change, Water Scarcity, and Health Adaptation in Southwestern Coastal Bangladesh. *International Journal of Disaster Risk Science*, 10, 28–42. Retrieve from: https://link.springer.com/article/10.1007/s13753-018-0211-8#citeas

Adejumoke, A. I., Babatunde, O. A., Abimbola, P. O., Tabitha, A. A. A., Adewumi, O. D. & Toyin, A. O. (2018). Water Pollution: Effects, Prevention and Climatic Impact. *Water Challenges of an Urbanizing World*, 33, 33–47.

Ahmed, M. A. (2018). Water Pollution: Solutions That Actually Work. *The Daily Star* (Bangladesh, 6 January). Retrieve from: www.thedailystar.net/opinion/environment/water-pollution-solutions-actually-work-1515562

Akonda, A. W. (1989). *Bangladesh, A Directory of Asian Wetlands*. Retrieve from: https://portals.iucn.org/library/node/593321

Akter, T., Jhohura, F. T., Akter, F., Chowdhury, R. T., Mistry, S. K., Dey, D., Milan Kanti Barua, M. K., Islam M. A. & Rahman, M. (2016). Water Quality Index for Measuring Drinking Water Quality in Rural Bangladesh: A Cross-sectional Study. *Journal of Health, Population and Nutrition*, 35 (1), 1–12.

Alcamo, J. (2019). Water Quality and Its Inter Linkages with the Sustainable Development Goals. *Current Opinion in Environmental Sustainability*, 36, 126–140.

Arifuzzaman, M., Hannan, M. A., Rahman, M. R. & Rahman, M. A. (2019). Laws Regulating Water Pollution in Bangladesh. *Journal of Sociology and Anthropology*, 3 (1), 15–24. Retrieve from: http://pubs.sciepub.com/jsa/3/1/3/index.html

Babu, K. E. K. (2020). Water Pollution in Bangladesh, Its Causes and Impacts: An Analysis Based on Existing Regulatory and Institutional Framework. *EM International*, 39 (4), 825–829. Retrieve from: www.researchgate.net/publication/354091039

Benson, D., Gain, A. K. & Giupponi, C. (2020). Moving Beyond Water Centricity? Conceptualizing Integrated Water Resources Management for Implementing Sustainable Development Goals. *Sustainability Science*, 15, 671–681.

Chan, N. W., Roy, R. & Chaffin, B. C. (2016). Water Governance in Bangladesh: An Evaluation of Institutional and Political Context. *Water*, 8 (9), 403. Retrieve from: www.mdpi.com/2073-4441/8/9/403#cite

Dhaka Tribune. (2014). *Pond Filling Plagues Rajshahi City* (1 September). Retrieve from: www.dhakatribune.com/uncategorized/2014/09/01/pond-filling-plagues-rajshahi-city-2

DoE. (2019). *Department of Environment, ECA, Bangladesh Paribeshgato Songkataponno Area*. Retrieve from: www.doe.gov.bd/site/page/9481

DoF. (2005). *Fishery Statistical Yearbook of Bangladesh 2003–2004*. Fisheries Resources Survey System, Department of Fisheries, Ministry of Fisheries and Livestock, 46, Retrieve from: http://fisheries.portal.gov.bd

Editorial. (2017). Stop Canal Encroachment. *The Daily Independent* (11 December). Retrieve from: www.theindependentbd.com/printversion/details/127477

FAO. (2017). *Water Pollution from Agriculture: A Global Review*. Retrieve from: www.fao.org

Faroque, S. & South, N. (2021). Water Pollution and Environmental Injustices in Bangladesh. *International Journal for Crime, Justice and Social Democracy*, 11 (1). Retrieve from: www.crimejusticejournal.com/article/view/2006

Flint, R. W. (2004). The Sustainable Development of Water Resources. *Water Resources Update*, 127, 48–59.

Hasan, M. K., Shahriar, A. & Jim, U. K. (2019). Water Pollution in Bangladesh and Its Impact on Public Health. *PMCID*. Retrieve from: www.ncbi.nlm.nih.gov/pmc/articles/PMC6684462/

Hasan, R. S. (2021). Water Policy and Public Rights' Roundtables. *The Daily Star* (Bangladesh 20 December). Retrieve from: www.thedailystar.net/round-tables/news/water-policy-and-public-rights-2001409

Huq, N., Hugé, J., Boon, E. & Gain, A. K. (2015). Climate Change Impacts in Agricultural Communities in Rural Areas of Coastal Bangladesh: A Tale of Many Stories. *Sustainability*, 7 (7), 8437–8460. Retrieve from: www.mdpi.com/2071-1050/7/7/8437

Hussain, A. (2018). Chittagong Ponds, Tanks, and Water Bodies in Death Throes. *Dhaka Tribune* (4 March). Retrieve from: www.dhakatribune.com/bangladesh/environment/2018/03/03/chittgong-ponds-tanks-water-bodies-death-throes

IWA. (2019). International Water Association. *Water and the SDGs*. Retrieve from: www.thesourcemagazine.org/water-and-the-sdgs/

Karim, M. S. & Uddin, M. M. (2019). Swatch-of-No-Ground Marine Protected Area for Sharks, Dolphins, Porpoises and Whales: Legal and Institutional Challenges. *Marine Pollution Bulletin*, 139, 275–281. Retrieve from: www.academia.edu/50869853/

Khan, A. E., Ireson, A., Kovats, S., Mojumder, K. S., Khusru, A., Rahman, A. & Vineis, P. (2011). Drinking Water Salinity and Maternal Health in Coastal Bangladesh: Implications of Climate Change. *Environmental Health Perspectives*, 119 (9), 1328–1332.

Khan, M. R., Khan, T. I. & Sadique, M. Z. (2020). SDG Implementation Progress: What Does the Asian Experience Reveal? *Centre for Policy Dialogue (CPD)*. Retrieve from: http://southernvoice.org/wp-content/uploads/2020/07/SDG-Implementation-Progress-Asia-Rhaman-Khan-and-Sadique-2020.pdf

Khatun, F. & Saadat, S. Y. (2021). *Implementation of the Sustainable Development Goals (SDGs) in Bangladesh: Domestic Challenges and Regional Considerations*. UN ESCAP, South and South-West Asia Development Papers, p. 14. Retrieve from: www.unescap.org

Light Castle Partners. (2018). *Bangladesh Water Sector Network Study Final Report* (Light Castle Partners). Retrieve from: www.netherlandswaterpartnership.com/sites/nwp_corp/files/201901/bangladesh_water_sector_network_studyreport.pdf

Local Government Division. (2011). *Sector Development Plan (FY 2011-25) Water Supply and Sanitation Sector in Bangladesh*. Ministry of Local-Government, Rural Development and Cooperatives Government of the People's Republic of Bangladesh. Retrieve from: https://psb.gov.bd/policies/sdpwsssbe.pdf

Moumen, Z., Idrissi, N. E. A. E., Tvaronavičienė, M. & Lahrach, A. (2019). Water Security and Sustainable Development. *Insights into Regional Development, Entrepreneurship and Sustainability Center*, 1 (4), 301–317.

Parvin, F., Haque, M. M. & Tareq, M. S. (2020). Recent Status of Water Quality in Bangladesh: A Systematic Review, Meta-analysis and Health Risk Assessment. *Environmental Challenges, 6*. Retrieve from: www.sciencedirect.com/science/article/pii/S2667010021003905

Rahman, M. M. (2021). Assessing the Progress and Pitfalls of the Ministry of Environment. *Forest, and Climate Change in Achieving SDGs in Bangladesh*, 29 (2). Retrieve from: http://bjpa.bpatc.org.bd/index.php/bjpa/catalog/category/BJPA29-2.

Reliefweb. (2011). *Ecosystems for Water and Food Security*. Retrieve from: https://reliefweb.int/report/world/ecosystems-water-and-food-security

Roy, D. R. (2021). Pollution of Surface Water Hinders Achievement of SDG 6' Editorial. *Daily Sun* (21 September). Retrieve from: www.daily-sun.com/printversion/details/577823

Sajal, I. A. & Akhtaruzzaman, M. (2016). Common People's Access to the Environment Courts of Bangladesh: An Appraisal. *Journal of Judicial Administration Training Institution*. Retrieve from: www.researchgate.net/publication/317035859

Seraj, S. (2015). Marine Resources in Our Maritime Boundary. *The Daily Star* (8 March). Retrieve from: www.thedailystar.net/marine-resources-in-our-maritime-boundary-34593

Stagl, J., Mayr, E., Koch, H., Hattermann, F. F. & Huang, S. (2014). Effects of Climate Change on the Hydrological Cycle in Central and Eastern Europe. In: Rannow, S., Neubert, M. (eds) *Managing Protected Areas in Central and Eastern Europe Under Climate Change, Advances in Global Change Research*. Springer, Dordrecht, 58. Retrieve from: https://link.springer.com/chapter/10.1007/978-94-007-7960-0_3

Uddin, M. J. & Jeong, Y. K. (2021). Urban River Pollution in Bangladesh During Last 40 Years: Potential Public Health and Ecological Risk, Present Policy, and Future Prospects Toward Smart Water Management. *Heliyon*, 7 (2). Retrieve from: www.sciencedirect.com/science/article/pii/S2405844021002127

United Nations. (2018). *Sustainable Development Goal 6 Synthesis Report on Water and Sanitation*. Retrieve from: https://sustainabledevelopment.un.org/content/documents/19901SDG6_SR2018_web_3.pdf

Water Policy and Public Rights. (2020). *The Daily Stars* (Dhaka, 26 November). Retrieve from: www.thedailystar.net/round-tables/news/water-policy-and-public-rights-2001409

World Food Programme, Bangladesh Country Strategic Plan (2022–2026) (2021). Retrieve from: https://executiveboard.wfp.org/document_download/WFP-0000132226

WWF. (2015). Water Governance in Bangladesh Challenges and Opportunities Around Policy. *Institutional Function and Implementation for a Sustainable Water Future*. Retrieve from: https://d2ouvy59p0dg6k.cloudfront.net/downloads/wwf_hm___water_governance_in_bangladesh_published.pdf

Yang, D., Yang, Y. & Xia, J. (2021). Hydrological Cycle and Water Resources in a Changing World: A Review. *Geography and Sustainability*, 2 (2), 115–122. Retrieve from: www.sciencedirect.com/science/article/pii/S2666683921000213

Yousuf, M. (2017). No One Here Goes to Jail for Filling up Water Bodies. *The Daily Star* (3 November). Retrieve from: www.thedailystar.net/backpage/no-one-here-goes-jail-filling-water-bodies-1485769

11
ACHIEVING SUSTAINABLE MARKETS

Ranjula Bali Swain and Örjan Sjöberg

11.1 Introduction

Are markets up to the task of promoting sustainable development? It is an issue of paramount significance in our pursuit of Agenda 2030 and the attainment of the Sustainable Development Goals (SDGs) and the provisions of the Paris Agreement. Against the backdrop of four aspects of markets, this chapter discusses the challenges posited by climate change and sustainable development. Considering their limitations, should markets as a main vehicle for sustainable development be discarded, or can their shortcomings be overcome? Can markets become sustainable in the sense of fostering sustainability outcomes?

Extensively or completely substituting markets, a backbone of modern societies, would be a most disruptive change, especially since our experiences with other allocation mechanisms are not entirely encouraging. To the extent that markets are not simply seen as devices for optimization, they use optimization mechanisms that rely on discrete decisions by a large number of individuals to solve allocation problems within the society. Given environmental concerns where the quality of life and material consumption are not a zero-sum game, we must produce better (not necessarily less), consume differently (not necessarily less), not limit growth but grow in a sustainable manner in the interest of improving the quality of life (with improving environment). This is the message of Agenda 2030 and its SDG 8 (Decent Work and Economic Growth).

However, unless existing trade-offs across the SDGs are successfully dealt with (Bali Swain and Min 2023), and unless, as Lerpold and Sjöberg (2023) point out, complete decoupling is achieved, the goals of the Paris Agreement and Agenda 2030 will not be reached (see also, for example, Ward et al. 2016). The trick of the matter, then, is to retain the benefits that markets provide while leveraging their allocative services and the ability to structure incentives toward sustainability. If this is achieved, markets can be regarded as sustainable, core characteristics of which would be the ability to support a long-term perspective (and not just being viable and long-lasting as such; Aras 2015) and to encourage choices that help achieve environmental and social justice objectives.

The problem is that current approaches, primarily in the form of rectifying market failures, important as these efforts are, do not quite seem to reach that goal (e.g., Sachs et al. 2019; Moyer and Hedden 2020; Lafortune et al. 2022; Leal Filho et al. 2023; Bin-Nashwan et al. 2024). Whether

DOI: 10.4324/9781003285472-12

this is because the potential of existing tools (e.g., Pigouvian taxes, subsidies, cap-and-trade mechanisms, etc.) is not properly applied, is resisted, or does not mix well with parallel measures of a more direct regulatory type is beyond the scope of our efforts here. Instead, we are interested in the process whereby sustainable outcomes might result from market action.

The objective of this short intervention, therefore, is to explore the sustainability potential of an alternative approach to markets, one that does not primarily see it as a tool for allocating resources efficiently but instead explicitly looks at its ability to deliver sustainability outcomes. Although it would be relevant and entirely feasible to approach markets conceptualized as value-creating structures (Vargo and Lusch 2004) and, hence, stress, for instance, the role of business models, for now we opt to start (albeit not necessarily stay) within the bounds of economics as opposed to business studies. Focusing on choice and sustainability preferences (à la Graciela Chichilnisky) as components of exchange across markets, it shifts the focus to the trade or transaction itself. Will this help us advance the role of markets in the quest for sustainability or at least shore up some of the major shortcomings of markets? Indeed, will it help us bridge the gap between weak and strong sustainability? That binary, which revolves around the substitutability of natural and other forms of capital, in effect, defines the relative standing of the economy versus the natural world that sustains humankind.

11.2 (Not Necessarily) Sustainable Markets

In classical economics, "the invisible hand" metaphor was used by Adam Smith to describe the unintended social benefits resulting from individual actions. The phrase is employed by Smith with respect to income distribution (1759) and production (1776). Whilst the exact phrase is used very sparingly in Smith's writings, it captures his notion that efforts to pursue interest may result in superior outcomes than direct actions that are intended to benefit society. The idea of trade and market exchange automatically channeling self-interest toward socially desirable ends is a central justification for the laissez-faire economic philosophy which lies behind neoclassical economics.

Not just self-interested behavior but also competition is central to the idea of a market that results in efficient resource allocation; prices that reflect relative scarcities, as established by supply and demand, provide the incentives needed to move forward. This, however, presupposes that markets accommodate a number of agents sufficiently large to ensure that no single market actor has the possibility of unilaterally influencing the price. A failure to operate in this manner will result in less-efficient resource allocation, with a commensurate erosion of welfare gains. Markets may also create costs to third parties not involved in the production, exchange, and consumption of the goods or services a market is engaged with. Such negative externalities are a core phenomenon in the creation of environmental problems; if costs are not internalized, an oversupply resulting in more extensive pollution or resource depletion than is societally optimal will result. Conversely, incentives issuing from market exchange might be such that useful goods and services are undersupplied; any positive externalities as might be created, for instance, in the form of ecosystems services, are difficult to capture, with the result that supply will fall short of what is societally desirable. The discussion on externalities, their sources, and how to remedy them is extensive and need not detain us here (but see, for example, Arrow 1970; Wheaton 1972; Chipman and Tian 2012; Rezai et al. 2012). Importantly, though, market failures being a possible outcome, firms and other economic agents will act responsibly if they are made to cover the costs imposed on others. This could be done through taxes or legislation; conversely, some form of inducement is likely needed to make available positive externalities that else might not materialize.

A second aspect that is likely to have an impact on the operations of markets (including with respect to sustainability outcomes) focuses on their geographical range. Markets can operate locally, within the national boundaries, cross-boundary (exports and imports), or a combination of all. Although trade or exchange is conducted on a particular good or service, from the perspective of sustainability, this particular dimension of the markets, within which trade (perhaps quite literally) takes place, can have important ramifications. Trade may shift the burden of negative externalities from one country to the next just as off-shoring of production may imply the off-shoring of emissions, too, yet remain driven by consumption at the original location of production. Furthermore, also as they are locally generated, some externalities have global reach, greenhouse gas emissions being but the most obvious example. Therefore, as Ayres et al. (2001: 167) propose, "regional and national sustainability should be consistent with global sustainability." This does not happen automatically.

Third, yet another dimension of sustainability is time. Intertemporal or intergenerational decisions are the hallmark of most sustainability decisions, and the choice is between the present and future generations or time periods. These intertemporal/intergenerational decisions impose a construct where the decision must be between consumption of a particular good (resource) in the present vis-à-vis consumption in the future. In a setting of finite means, it can be seen as a choice between investment or consumption, the latter potentially eroding the former by reducing the resources available for future generations.

Fourth, technological developments and the introduction of new products and services, especially clean energy and green technology, are critical to solving the environmental problems humankind faces. Investment in research and development, including innovations that benefit the environment, is a high-risk activity and it may prove difficult for the innovator to capture the value created. This is one reason, however, that innovation is not only an issue of technology in a narrow sense. Developments such as new business models might also prove decisive. Indeed, beyond new ways of capturing value that would else accrue to third parties (thereby undermining the initial incentives to innovate), such advances may prove their worth in other ways, for instance, in the form of designs allowing circular economy solutions.

11.3 Toward Sustainable Markets

This leads us on to the possibility of *sustainable markets*, defined as those markets that support outcomes that advance or at least are in line with prevalent notions of sustainability, be it of the environmental, social, or economic variety. Precisely because this has become the preferred way of interpreting sustainable markets, however, there is a certain irony in that early use of the term itself included the ability of the market to retain its integrity as a market. That is, a market, and in particular one that supplies above-normal profits, resistant to competition. This contrasts with the contestable markets that, in principle, can be challenged and therefore induce a measure of constraint lest new entrants might be attracted to it or set their mind to providing workable substitutes. Originating in the work of William J. Baumol and colleagues (Baumol et al. 1977, 1982, 1983; Baumol 1982), it was subject to intense debate in the early 1980s (Schwartz and Reynolds 1983; Weitzman 1983; Shepherd 1984), subsequently being integrated into the thought and vocabulary of neoclassical economics.

Likely as they are to accept the idea of contestability, other economists have turned their back on the associated use of the term sustainable markets, instead opting for the usage present in other disciplines (e.g., Gordon et al. 2011; Aras 2015; Mattsson 2016; Ottosson et al. 2020). A prominent example is Graciela Chichilnisky's research on sustainable markets. Her contribution rests with shifting the attention from the market as such and the generic implications of supply and demand to the transaction across that market. If sustainable markets are markets that deal with the market failures caused by environmental externalities to produce sustainable outcomes instead, how might they be achieved? Can sustainable markets so defined help resolve the four potentially problematic aspects of markets as identified earlier? To what extent is Chichilnisky's contribution able to provide advice on how to proceed? Setting out to address these questions, in the following section, we take a closer look at Chichilnisky's technically advanced (and axiomatic rather than empirical) investigation into the issue of whether markets can indeed further the cause of sustainability. We try to do so in plain language, although this might come at the expense of the finer yet enlightening points raised by her analysis.

11.4 Sustainable Markets à la Chichilnisky

Setting out from the idea that "neither the 'present' nor the 'future' should be favored over the other" (Chichilnisky 1997b: 468; see also, for example, Chichilnisky 1996), hers is a formulation that considers the basic needs of the present without sacrificing the needs of future generations, or vice versa. As such, it is essentially consonant with the definition of *sustainability* made famous by the Brundtland Report (WCED 1987: 41) and in line with Hartwick's (1977) version of what has since become known as weak sustainability. Against that background, Chichilnisky (2012) was the first to define *sustainable markets* as markets where traders have sustainable preferences with possibility for limited arbitrage. She argues that markets with sustainable preferences overcome impatience and thereby embody the needs of present and future generations for sustainable development.

To start with the most obvious of the four issues, then, Chichilnisky (1997a, 1997b, 2012) suggests a resolution to the conflict between short-term market objectives with sustainable development by allowing for prices that at once reflect both the value of instantaneous consumption and the value of the long-run future, thus resulting in a market where the invisible hand delivers sustainable and efficient solutions. She derives the conditions under which this could happen and notes that Koopmans's (1960) impatience axiom—the idea that a dollar tomorrow is worth less than a dollar today—introduces an economic bias against the future (Chichilnisky 1996; Chichilnisky and Heal 1998; Chichilnisky 2012). This has implications for economic *growth*, a body of theory that Chichilnisky (1997b) actively engages with, where a core issue is how to balance investments and consumption within globally finite resources and subject to concurrent budget constraints. More consumption leads to less room for investment, thereby reducing the production capacity tomorrow, in turn implying lower incomes and levels of consumption than would otherwise have been the case. Koopmans's theory also leads to an understanding of markets where the limits and the value of the Earth's resources may be ignored (Chichilnisky 2012). Were it not for the impatience axiom for choice over time, there would be no differentiation between the present and the future—thus resulting in outcomes that are sustainable.

Chichilnisky (2012: 294) then asks the central question: should a society committed to sustainability disregard markets, drop them, or should one try to overcome any market-based bias against the future (i.e., in the form of discounting)? She arrives at the latter. Disregarding markets will not solve the overarching problem as posed; doing away with them "would be a major change [as] markets are a widespread form of organization" (Chichilnisky 2012: 294) and, therefore, not really an option. She further expounds that sustainable markets differ from Arrow–Debreu markets, which do require adherence to the theory of Koopmans, as the traders have sustainable preferences and no bounds on short sales (i.e., a situation where traders speculate on a decline in prices) but with limited possibilities for arbitrage (in the sense developed in Chichilnisky 1994). Sustainable preferences are sensitive to the basic needs of the present without sacrificing the needs of future generations. These sustainable preferences ensure the existence of a sustainable market equilibrium where the invisible hand delivers sustainable as well as efficient solutions. Thus, traders take bounded positions with respect to each other. In sustainable markets, the prices thus have dual roles, where they reflect both the value of instantaneous consumption and the value of the long-run future (Chichilnisky 1996).

While this might seem at quite some remove from real-world markets, it is a coherent framework. It indicates that by shifting the focus away from the market to the trade and traders, it would, at least in principle, be possible to solve the problem of bias in favor of the present. Although it is not self-evident what an attempt to relax the assumptions that "traders have infinite horizons [. . .] and sustainable preferences over time" (Chichilnisky 2012: 297) would yield in terms of changed model results or in terms of realism (or, for that matter, tractability), her line of reasoning opens for two fundamental questions. First, why and how does exchange take place? Second, how do markets for new products develop?

The latter question can be approached from the angle of market design (Roth 2007) or the literature on market shaping (e.g., Callon 1998; Kjellberg and Helgesson 2006; Nenonen et al. 2019). For our present purposes, suffice it to note that markets typically follow the priorities of the traders, and this can be expected also as the assumption of limited arbitrage that has a restrictive impact on "the utility that can be achieved by the traders when trading with each other" (Chichilnisky 2012: 300). On the positive side of the ledger, if businesses are to move toward sustainable development, as indeed they must if the SDGs and climate change mitigation goals are to be met, trade or exchange of goods and services must internalize in their exchange the better part of all externality costs irrespective of source and not just that of a presentism bias. Chichilnisky's set-up allows for that.

Consider new products, the production of which takes place in anticipation of opportunities of trade, not only as a consequence of trade itself (e.g., following prices on existing goods rising on the heels of increased sales). In short, there must be demand for the product or service offered to the market, or at least an actionable expectation that demand will materialize. This is not quite the same as identifying the conditions under which new products can be developed with a view to replacing existing ones, thereby potentially undermining trading in unsustainable goods and services that create negative environmental impacts. Incentives and risks must be balanced, as would investment costs against projected revenues. Is it the case that, by focusing on trade (exchange) instead of markets *per se*, we will be able to address this key aspect of meeting the environmental challenges? Yes, in principle, at the aggregate level. As already noted, externalities impose costs on a non-participating third party due to action of others trading in goods and services. If traders do, indeed, have stable sustainable preferences, as by definition they have in Chichilnisky's framework, this should be taken care of.

However, trying to work out the real-world equivalent, it does beg the question of how conflicting sustainability demands are internalized or managed, trade-offs being a problem often raised in discussions on sustainability, including in the case of the SDGs (e.g., Le Blanc 2015; Spaiser et al. 2016; Pradhan et al. 2017; Barbier and Burgess 2019; Moyer and Bohl 2019; Bali Swain and Ranganathan 2021). Trade-offs may imply uncomfortable compromises or making equally painful priorities. This is not taken care of in the Chichilnisky framework, where sustainable preferences are treated as if they were unitary, synergistic, and unproblematic to aggregate.

Similarly, where do these preferences come from? They are not formed endogenously in the framework (and there is no discussion of any intrinsic value that traders might derive from adopting them). Hence, they appear to originate somewhere else. For instance, consistent demand from customers could be such a source. Moreover, by understanding the motivations, choices, and preferences of the economic agents in trade, policymakers could try to influence the behavior or implement policies to remove the discrepancy between the private and the social costs at the source of the trade creating the externality. However, the question then arises: how can that be attained? Nudging is one possibility, but it is neither foolproof nor a particularly democratic approach (it having been identified, including by Thaler and Sunstein [2004: 4–6], as libertarian paternalism). After all, it is not only market failure that is a well-established phenomenon in this world but so also is policy failure. This remains so irrespective of the interventions that are targeting the formation of motivations and preferences or that try to influence agents at the point of choice. Hence, there is no guarantee that intervention in favor of sustainable preferences will work. On the other hand, the literature on policy failure makes the point that "failure is rarely unequivocal and absolute [. . .] even policies that have become known as classic policy failures also produced small and modest successes" (McConnell 2015: 231). Perhaps unmitigated failure of both the policy and the market variety sits at the extreme end of the spectrum, with success of resolving it in full at the other. There are a multitude of possible outcomes in between that show partial success in rectifying the failures we are faced with in discussing the sources of and solutions to non-sustainable outcomes. Can we establish the conditions under which partial solutions are sufficiently good? Would it be worthwhile to go for something less than all-out success? These are questions that require further research and contemplation.

We have also noted that markets can be local, national, or international. However, trade or exchange is conducted on a particular good or service. This allows us to determine the impact of trade, depending on the scale of operation of the trade, without getting embroiled in identifying markets in various contexts, interesting as that might be for other reasons. But on our reading, Chichilnisky stays clear of this shoal. On the other hand, it is not equally clear that her framework allows for markets at different spatial ranges or at different geographical scales to be linked and thereby act as communicating vessels. We know, for instance, from the literature on externalities that the applicability of public policy to public goods may be local, and Ostrom (2012) argues that multiple smaller-scale polycentric efforts to mitigate climate change should be encouraged. Incidentally, this is especially relevant for the SDGs in general, not just number 13 (Climate Action), and also suggests that choices issuing from similar preferences might need to be differentiated. However, this is not the same as adding up to a sufficiently good outcome at higher levels of aggregation or at a global scale.

This leaves us with the question of technological development. Given the nature of existing technology, as seen from the vantage point of sustainability, much of it invested with undesirable properties, new products and services, especially clean energy and green technology, are critical to solving the environmental problems. While the ultimate and proximate causes of the arrival of new business models and markets are beyond the scope of this contribution, we note that novel ideas about value creation and ditto capture can actively shape both markets and the transactions that take place there. The fact that either may involve the development of, for instance, green technologies should not be neglected. As such, it should at least in principle be consonant with Chichilnisky's focus on sustainable preferences and how they might form.

Finally, turning to the role of preferences in shaping transactions, and possibly also business models and markets, Sen (1997) admonishes us to take the issue of choice seriously, as does Chichilnisky (1996). In effect, choice may take place against the background of preferences. Not all choices derive from an ability or willingness to act in accordance with the preferences held, though. As Sen notes, non-volitional choice implies a severe constraint on individual decision-making and may therefore come in conflict with pre-established preferences, including that for desirable sustainability outcomes. There are also other things that might interfere with the direct influence of preferences on choices made, such as inertia, routine, and a lack of reflection, a list to which we may perhaps also add fear as does (Chichilnisky 2009). Preferences, for their part, may reflect, for instance, objectives, values, and norms, in themselves necessary referents for any assessment of the consistency of choice over settings and time (Sen 1993).

The upshot of this, short of adding to the list of restrictive assumptions, is that there is no one-to-one correspondence between preferences and choice as expressed in anticipated or observed behavior. This should be kept in mind, lest we lapse into thinking that trades reflect rational calculation taking place against a backdrop of crystal clear incentives provided by scarcity pricing alone. On the other hand, it implies that there are not just possibilities to influence preferences but also a potential for making market activities of economic agents align more effectively with desired or necessary sustainability outcomes precisely because the preferences for sustainability might already be in place. In short, we may note that the tools applied with a view to affecting the structure of incentives might not be limited to those that are conventionally employed to rectify market failures (taxes, subsidies, mandatory cap-and-trade mechanisms, etc.).

11.5 Concluding Remarks

To the extent that markets, in our quest to achieve sustainable outcomes, can help resolve the very problems—those of negative externalities—that they may cause, we may speak of sustainable markets. At a time when we are faced with a "market failure on the greatest scale the world has ever seen" (Stern 2007: 27), and as global carbon budgets to meet the Paris Agreement are fast running out (Friedlingstein et al. 2023), we must welcome any advance that can address the consequences of the environment suffering because of its unavoidable public good nature (Siebert 2008). One approach that merits attention is that of sustainable preferences suggested by Chichilnisky (2012). Such preferences ensure the existence of a sustainable market equilibrium where sustainable development could be achieved. It is not just an issue for the climate, though. Such markets can contribute to the meeting of several of the SDGs and their various targets. As noted by Bali Swain and Min (2023), besides SDG 13 (Climate Action), this also includes SDGs 8 (Decent Work and Economic Growth), 9 (Industry, Innovation, and Infrastructure), and 12 (Responsible Consumption and Production), all of which have been shown to have strong links to each other and to many of the other SDGs.

As long as these interlinkages are positive and synergistic, that is, that they reinforce each other, we should be satisfied with the outcome, and our discussion could stop here. However, this does not always appear to be the case. As already noted, research on SDG interlinkages also implies the existence of trade-offs, in turn suggesting a need to balance various desiderata against each other. Still more challenging is the realization that the SDGs are based on weak sustainability reasoning (e.g., Haberl et al. 2020; Bonnedahl et al. 2022). Put differently, the ability of meeting the SDGs hinges on economic growth taking place.

This is problematic. For while it is easy to see that economic growth will make it much easier to meet the demands of most of the social SDGs, it runs counter to the imperative of preserving the integrity of the biosphere. Weak sustainability requires that the prescription of Hartwick (1977) be followed, namely, that rents reaped from natural capital be invested in other forms of capital, allowing the total capital stock to remain at least constant. Yet because of the discounting of future value (e.g., to consumers of the next generation), this will not be enough to ensure intergenerational equity. Add population growth and a sizeable "backlog" in the form of glaring present-day differences in consumption levels across various population groups and it becomes evident that economic growth is needed. SDG 8 recognizes as much, but it is thus likely to come at the expense of the environment and finite natural capital. This therefore not only provides ammunition for advocates of strong sustainability (Daly 1992: 250–252) but also is similarly reflected in the Resolution of the United Nations that launched Agenda 2030. Here, Target 8.4 admonishes signatories to enhance "global resource efficiency in consumption and production and endeavour to decouple economic growth from environmental degradation" (UN 2015: 19).

Haberl et al. (2020) maintain that absolute decoupling is necessary, not just decoupling, implying that greenhouse gas emissions and other adverse environmental impacts increase at a slower pace than does production. If successful, output growth will be accompanied by declining negative impacts.

However, to the extent that the Earth system is running up against absolute planetary boundaries—think CO_2 forcings in the atmosphere, where absolute (albeit approximate) concentrations define impacts in the form of particular projected levels of global warming—it is not quite sufficient. Indeed, as Ward et al. (2016) argue, this is a partial solution, allowing any gains made to dissipate (e.g., through rebound effects). Instead, complete decoupling will be needed: no further resource use or impact can be tolerated. Judging by Agenda 2030, the world community has arrived at an agreement that has not quite internalized this insight. Instead, as Lerpold and Sjöberg (2023: 30) put it, "economic growth is thus conceptually understood as a *sine qua non* for the realisation of the SDGs, [yet] it conveniently overlooks the observation that infinite growth sits uneasily with finite planetary resources."

Chichilnisky (2012) offers a path out of this. She does so by showing that there are ways of circumventing the problems caused by discounting the future. As she does so, she shifts the attention from markets as such to the preferences of traders in those markets: sustainable preferences allow for sustainable outcomes. Although preferences do not lead directly to outcomes—rather, preferences translate into choices, not all of which are straightforward, in turn impacting trades—it opens for discussion on not just the nature of those preferences but also how they are formed and translate into action.

Some further concluding observations on the usefulness of Chichilnisky's approach are warranted. One is the issue of how the lessons of an axiomatic exposition such as hers, logically and elegantly coherent as it is, map onto real-world action. Also, allowing her framework the benefit of the doubt, the presence of trade-offs still seems a problem. Sustainable preferences à la Chichilnisky are well placed to avoid a bias against the future, but can they also accommodate multiple sustainability goals? If sustainable preferences are an amalgam of a variety of worthwhile objectives, would they not be open to the risks that follow from compromises, such as opting for optimization, at the expense of the precautionary principle? This in particular as science cannot be expected to identify the precise thresholds for various outcomes; hence, there is still a need for an extra margin to ensure the safe operating space that the precautionary principle requires (Steffen et al. 2015). And if strict priorities are imposed, say, by setting climate change mitigation above all else, what about other dimensions of sustainability? After all, not just the SDGs include a large set of different goals that supposedly are "integrated and indivisible, global in nature and universally applicable" (UN 2015: 13). Other influential ways of reasoning about our use of finite resources also acknowledge that there are multiple goals worth pursuing. Thus, the current incarnation of the planetary boundaries' framework (Gupta et al. 2023; Rockström et al. 2023), as is also true of some applications of that line of thinking (such as the doughnut economy; Raworth 2017), includes social justice as an important component.

And at a time when because of the urgency of the matter (next to) immediate impact is of the essence—carbon budgets, in line with the objectives of the Paris Agreement, are, after all, running out—we may well ask about the ability of her suggested solution, even if implemented here and now, to meet any such deadline. But then, because of a general inability to incorporate calendar time, this is a more general problem of modelling efforts in the neoclassical economics tradition, not just one that can be noted relative to Chichilnisky's intellectually rewarding approach to the problem at hand.

References

Aras, Güler (2015): "Causality and interaction: Sustainable markets and sustainable business", in Güler Aras (ed.): *Sustainable Markets for Sustainable Business: A Global Perspective for Business and Financial Markets*. Gower, Farnham: 15–40.

Arrow, Kenneth J. (1970): "Political and economic evaluation of social effects and externalities", in Julius Margolis (ed.): *The Analysis of Public Output*. National Bureau of Economic Research, New York: 1–30.

Ayres, Robert, Jeroen van den Bergh and John Gowdy (2001): "Strong versus weak sustainability: Economics, natural sciences, and consilience", *Environmental Ethics* 23 (2): 155–168.

Bali Swain, Ranjula and Yongyi Min (eds.) (2023): *Interlinkages between the Sustainable Development Goals*. Elgar, Cheltenham.

Bali Swain, Ranjula and Shyam Ranganathan (2021): "Modeling interlinkages between sustainable development goals using network analysis", *World Development* 138: art. 105136.

Barbier, Edward B. and Joanne C. Burgess (2019): "Sustainable development goal indicators: Analyzing trade-offs and complementarities", *World Development* 122: 295–305.

Baumol, William J. (1982): "Contestable markets: An uprising in the theory of industry structure", *American Economic Review* 72 (1): 1–15.

Baumol, William J., Elizabeth E. Bailey and Robert D. Willig (1977): "Weak invisible hand theorems on the sustainability of prices in a multiproduct national monopoly", *American Economic Review* 67 (3): 360–365.

Baumol, William J., John C. Panzar and Robert D. Willig (1982): *Contestable Markets and the Theory of Industry Structure*. Harcourt Brace Jovanovich, San Diego, CA.

Baumol, William J., John C. Panzar and Robert D. Willig (1983): "Contestable markets: An uprising in the theory of industry structure: Reply", *American Economic Review* 73 (3): 491–496.

Bin-Nashwan, Saeed Awadh, M. Kabir Hassan and Aishath Muneeza (2024): "Russia–Ukraine conflict: 2030 Agenda for SDGs hangs in the balance", *International Journal of Ethics and System* 41 (1): 3–16.

Bonnedahl, Karl Johan, Pasi Heikkurinen and Jouni Paavola (2022): "Strongly sustainable development goals: Overcoming distances constraining responsible action", *Environmental Science and Policy* 129: 150–158.

Callon, Michel (1998): "An essay on framing and overflowing: Economic externalities revisited by sociology", *Sociological Review* 41 (1S): 244–269.

Chichilnisky, Graciela (1994): "Social diversity, arbitrage and gains from trade: A unified perspective on resource allocation", *American Economic Review* 84 (2): 427–434.

Chichilnisky, Graciela (1996): "An axiomatic approach to sustainable development", *Social Choice and Welfare* 13 (2): 231–257.

Chichilnisky, Graciela (1997a): "A unified perspective on resource allocation: Limited arbitrage is necessary and sufficient for the existence of a competitive equilibrium, the core and social choice", in Kenneth J. Arrow, Amartya Sen and Kotaro Suzumura (eds.): *Social Choice Re-examined*, Vol. 1. Palgrave Macmillan, Basingstoke: 121–174.

Chichilnisky, Graciela (1997b): "What is sustainable development", *Land Economics* 73 (4): 467–491.

Chichilnisky, Graciela (2009): "The topology of fear", *Journal of Mathematical Economics* 45 (12): 807–816.

Chichilnisky, Graciela (2012): "Sustainable markets with short sales", *Economic Theory* 49 (2): 293–307.

Chichilnisky, Graciela and Geoffrey M. Heal (1998): "A unified treatment of finite and infinite economies: Limited arbitrage is necessary and sufficient for the existence of equilibrium and the core", *Economic Theory* 12 (1): 163–176.

Chipman, John S. and Guoqiang Tian (2012): "Detrimental externalities, pollution rights, and the 'Coase theorem'", *Economic Theory* 49 (2): 309–332.

Daly, Herman E. (1992): *Steady-State Economics*. 2nd edition, Earthscan, London.

Friedlingstein, Pierre, Michael O'Sullivan, Matthew W. Jones, et al. (2023): "Global carbon budget 2022", *Earth System Science Data* 15 (12): 5301–5369.

Gordon, Ross, Marylyn Carrigan and Gerard Hastings (2011): "A framework for sustainable marketing", *Marketing Theory* 11 (2): 143–163.

Gupta, Joyeeta, Diana Liverman, Klaudia Prodani, et al. (2023): "Earth system justice needed to identify and live within Earth system boundaries", *Nature Sustainability* 6 (6): 630–638.

Haberl, Helmut, Dominik Wiedenhofer, Doris Virág, et al. (2020): "A systematic review of the evidence on decoupling of GDP, resource use and GHG emissions, part II: Synthesizing the insights", *Environmental Research Letters* 15 (6): art. 065003.

Hartwick, John M. (1977): "Intergenerational equity and the investing of rents from exhaustible resources", *American Economic Review* 67 (5): 972–974.

Kjellberg, Hans and Claes-Fredrik Helgesson (2006): "Multiple versions of markets: Multiplicity and performativity in market practice", *Industrial Marketing Management* 35 (7): 839–855.

Koopmans, Tjalling C. (1960): "Stationary ordinal utility and impatience", *Econometrica* 28 (2): 287–309.

Lafortune, Guillaume, Grayson Fuller, Leslie Bermont Diaz, Adolf Kloke-Lesch, Phoebe Koundouri and Angelo Riccaboni (2022): *Europe Sustainable Development Report 2022. Achieving the SDGs: Europe's Compass in a Multipolar World*, Sustainable Development Solutions Network, Paris.

Leal Filho, Walter, Laís Viera Trevisan, Izabela Simon Rampasso, et al. (2023): "When the alarm bells ring: Why the UN sustainable development goals may not be achieved by 2030", *Journal of Cleaner Production* 407: art. 137108.

Le Blanc, David (2015): "Towards integration at last? The Sustainable Development Goals as a network of targets", *Sustainable Development* 23 (3): 176–187.

Lerpold, Lin and Örjan Sjöberg (2023): "Decoupling and redistribution in realising the Sustainable Development Goals", in Ranjula Bali Swain and Yongyi Min (eds.): *Interlinkages between the Sustainable Development Goals*, Elgar, Cheltenham: 16–36.

Mattsson, Lars-Gunnar (2016): "Bridging gaps between policies for sustainable markets and market practices", *IMP Journal* 10 (2): 339–356.

McConnell, Allan (2015): "What is policy failure? A primer to help navigate the maze", *Public Policy and Administration* 30 (3–4): 221–242.

Moyer, Jonathan D. and David K. Bohl (2019): "Alternative pathways to human development: Assessing trade-offs and synergies in achieving the Sustainable Development Goals", *Futures* 105: 199–210.

Moyer, Jonathan D. and Steve Hedden (2020): "Are we on the right path to achieve the sustainable development goals?", *World Development* 127: art. 104749.

Nenonen, Suvi, Kaj Storbacka and Charlotta Windahl (2019): "Capabilities for market-shaping: Triggering and facilitating increased value creation", *Journal of the Academy of Marketing Science* 47 (4): 617–639.

Ostrom, Elinor (2012): "Nested externalities and polycentric institutions: Must we wait for global solutions to climate change before taking actions at other scales?", *Economic Theory* 49 (2): 353–369.

Ottosson, Mikael, Thomas Magnusson and Hans Andersson (2020): "Shaping sustainable markets—a conceptual framework illustrated by the case of biogas in Sweden", *Environmental Innovation and Societal Transitions* 36: 303–320.

Pradhan, Prajal, Luís Costa, Diego Rybski, Wolfgang Lucht and Jürgen P. Kropp (2017): "A systematic study of sustainable development goal (SDG) interactions", *Earth's Future* 5 (11): 1169–1179.

Raworth, Kate (2017): *Doughnut Economics: Seven Ways to Think Like a 21st Century Economist*. Penguin, London.

Rezai, Armon, Duncan K. Foley and Lance Taylor (2012): "Global warming and economic externalities", *Economic Theory* 49 (2): 329–351.

Rockström, Johan, Joyeeta Gupta, Dahe Qin, et al. (2023): "Safe and just Earth system boundaries", *Nature* 619 (7968): 102–111.

Roth, Alvin E. (2007): "The art of designing markets", *Harvard Business Review* 85 (10): 118–126.

Sachs, Jeffrey, Guido Schmidt-Traub, Christian Kroll, Guillaume Lafortune and Grayson Fuller (2019): *Sustainable Development Report 2019: Transformations to Achieve the Sustainable Development Goals*. Bertelsmann Stiftung and Sustainable Development Solutions Network, New York.

Schwartz, Marius and Robert J. Reynolds (1983): "Contestable markets: An uprising in the theory of industry structure: Comment", *American Economic Review* 73 (3): 488–490.

Sen, Amartya (1993): "Internal consistency of choice", *Econometrica* 61 (3): 495–521.

Sen, Amartya (1997): "Maximization and the act of choice", *Econometrica* 65 (4): 745–779.

Shepherd, William G. (1984): "'Contestability' vs. competition", *American Economic Review* 74 (4): 572–587.

Siebert, Horst (2008): *Economics of the Environment: Theory and Policy*, 7th ed., Springer-Verlag, Berlin.

Smith, Adam ([1759] 2002): *The Theory of Moral Sentiments*. Ed. Knud Haakonssen. Cambridge University Press, Cambridge.

Smith, Adam ([1776] 1977): *An Inquiry into the Nature and Causes of the Wealth of Nations*. University of Chicago Press, Chicago, IL.

Spaiser, Viktoria, Shyam Ranganathan, Ranjula Bali Swain and David J.T. Sumpter (2016): "The sustainable development oxymoron: Quantifying and modelling the incompatibility of sustainable development goals", *International Journal of Sustainable Development and World Ecology* 24 (6): 457–470.

Steffen, Will, Katherine Richardson, Johan Rockström, et al. (2015): "Planetary boundaries: Guiding human development on a changing plant", *Science* 347 (6223): art. 1259855.

Stern, Nichols (2007): *The Economics of Climate Change: The Stern Review*. Cambridge University Press, Cambridge.

Thaler, Richard H. and Cass R. Sunstein (2008): *Nudge: Improving Decisions about Health, Wealth, and Happiness*. Yale University Press, New Haven, CT.

UN (2015): "Resolution Adopted by the General Assembly on 25 September 2015", A/RES/70/1, General Assembly, United Nations, New York, 21 October.

Vargo, Stephen L. and Robert F. Lusch (2004): "Evolving to a new dominant logic for marketing", *Journal of Marketing* 68 (1): 1–17.

Ward, James D., Paul C. Sutton, Adrian D. Werner, Robert Costanza, Steve H. Mohr and Craig T. Simmons (2016): "Is decoupling GDP growth from environmental impact possible?", *PLoS ONE* 11 (10): art. e0164733.

WCED (1987): *Our Common Future: The World Commission on Environment and Development*. Oxford University Press, Oxford.

Weitzman, Martin L. (1983): "Contestable markets: An uprising in the theory of industry structure: Comment", *American Economic Review* 73 (3): 486–487.

Wheaton, William C. (1972): "On the possibility of a market for externalities", *Journal of Political Economy* 80 (5): 1039–1044.

12

HOW BORN-SUSTAINABLE AND SUSTAINABILITY-DRIVEN COMPANIES CONTRIBUTE TO THE SDGS

Svenja Damberg, Morgane Fritz, and Ulla Saari

12.1 Introduction

The vision, mission, and values of companies are key to developing their strategies toward achieving sustainability. Values are particularly necessary to determine the "hows" of doing so (Mirvis et al., 2010). The principal values and drivers of born-sustainable companies and sustainability-driven entrepreneurial efforts are built into the core of the companies' business models. These elements of a sustainable business model are crucial to achieve some of the critical United Nations Sustainable Development Goals (SDGs; United Nations, 2015) in the business context. In this chapter, we explain that emphasizing sustainability values in born-sustainable and sustainability-driven companies from the very start can help create a more positive impact on society and the environment, as this leads to the development of sustainable business models. Previously, the 3Ps (people, planet, and profit) or triple bottom line (TBL) approach developed by John Elkington offered some general guidance to companies on how to consider economic, social, and environmental aspects instead of focusing too much on the financial success of a company. However, this approach is clearly insufficient and has been misused by firms to assess how sustainable they are, leading to a recall by John Elkington himself, because

> the TBL wasn't designed to be just an accounting tool. It was supposed to provoke deeper thinking about capitalism and its future. . . . Its goal was "system change—pushing toward the transformation of capitalism. It was never supposed to be just an accounting system."
>
> (Kraaijenbrink, 2019)

Furthermore, the corporate social responsibility (CSR) activities of many companies have not been adequately incorporated into their respective business models, and stakeholders are requesting that businesses generate a more positive impact. In fact, we argue that businesses focusing mainly on profits are unable to provide safe and sustainable solutions for society and the environment.

The theoretical background of this research is based on the sustainable business literature. We applied a case study methodology to assess companies from the clothes and apparel industry, cosmetics industry, and food sector (i.e., companies offering consumer products in large volumes with a strong social and environmental impact through their business activities). For this purpose, we

DOI: 10.4324/9781003285472-13

used secondary data sources, such as public information on the case companies and their websites and previous research on the companies.

We analyzed how born-sustainable and sustainability-driven companies present themselves, including how they have built their businesses, by focusing on their core sustainability values. Some case companies were established before 2012, meaning, before the SDGs were formed and published during the UN Sustainable Development Summit in 2015. Therefore, we also assessed the extent to which these companies uphold the SDGs, focusing on the goals that are relevant to the sector in which the company operates. Based on our analyses, we address the topic of "building blocks for the post-SDG 2030 Agenda" that could help in transforming the business context and address the key elements in the sustainable business models that future born-sustainable and sustainability-driven companies and start-ups need to consider from the very beginning of their early entrepreneurship and innovation journeys. Moreover, these components could offer some guidance for incumbent companies that need to update and develop their business models to become more sustainable in alignment with the SDGs.

Based on our findings, we present a conceptual model of how born-sustainable and sustainability-driven companies base their business models on a sustainable foundation, which promotes the key SDGs that are relevant for that specific industrial sector. This will help develop and build more sustainable business ecosystems that focus on sustainable quality and social practices and promote more sustainable consumption behavior that does not support overconsumption. Born-sustainable and sustainability-driven companies with sustainable future-oriented business models and practices that are linked to relevant key SDGs could help ensure overall well-being in the society and have a more positive impact on the natural environment.

12.2 Theoretical Background

12.2.1 Sustainable Value Creation

In the past decade, research has focused on the importance of sustainable business models with regard to ensuring a long-term competitive advantage (e.g., Laszlo & Zhexembayeva, 2017), as "[t]hose that create value for society and the environment without trade-offs have an opportunity to create even more value for their customers and shareholders than they otherwise would" (Laszlo & Zhexembayeva, 2017, p. 2). Therefore, sustainable value creation is one of the most important aspects of business modelling. In a recent publication by Jonker and Faber (2021), the authors dedicated an entire chapter to value creation. According to the authors,

> a business model provides a logic for value creation . . . that is translated into a proposition. . . . Transactions, in turn, are the operational acts that demonstrate an appreciation of the value created between parties while also realising it.
>
> (Jonker and Faber 2021, p. 152)

Recently, the sustainable business literature has focused on the dynamic capabilities of so-called born-sustainable companies (Knoppen & Knight, 2022). Also referred to as born-green companies (Hall et al., 2010), born-sustainable companies are seen as those companies that decide to adopt sustainable business models from the beginning—that is, from their founding (Dicuonzo et al., 2020). Often, these companies are developed as small and medium-sized enterprises (SMEs) from individual entrepreneurs or from academic or other spin-offs (Demirel et al., 2019, p. 761).

Value creation is a particularly important aspect for these SMEs for several reasons. First, the company's identity is built on the values of its entrepreneurs (O'Neil & Ucbasaran, 2016). Second, the communicated values are important signals for investors (Schaltegger & Hörisch, 2017). Third, sustainable value creation is important for other stakeholders as well, such as employees in the sense of employer branding, and customers to signal and promote sustainable consumption behavior (Evans et al., 2017).

With respect to a sustainability transition, companies need to take up the responsibility of creating long-term sustainable value for stakeholders and society (e.g., regarding fulfilling the SDGs). Vitols and Kluge (2011) define elements of "the sustainable company" (Vitols & Kluge, 2011, p. 25) to include stakeholder value, sustainability goals and strategies, long-term responsible investors, renumeration oriented toward sustainability, disclosure of social and environmental performance, and stakeholder voice.

12.2.2 Core Values of Sustainability-Driven Companies

Prior research has identified certain core values of born-sustainable companies, looking at the question of what makes a business sustainable. One way to identify whether a business is sustainable is to see whether it aims to "mak[e] the world a better place or if their objectives are primarily rooted in profit maximization" (Demirel et al., 2019, p. 760). Entrepreneurs' own values were also found to be an important element (O'Neil & Ucbasaran, 2016). In this sense, the ethical orientation of the entrepreneur is the basis of a truly sustainable company (Dicuonzo et al., 2020). Integrating sustainability into a firm's core values is essential for coping with sustainability risk management. In some cases, it is even more important than stakeholder pressure (Wijethilake & Lama, 2019). The sustainable core values promoted by managers in a firm also allow employees to take the time to reflect on and contribute to sustainable development (Jollands et al., 2015). Specific sustainable values such as honesty and humility among individuals have been found to be key to a sustainable business (Marcus & Roy, 2019). However, even though the will to do good might exist in a firm, managers often do not know how to do that or where to start (Mirvis et al., 2010).

In this context, Bocken et al. (2014) reviewed the literature to build a conceptual model for sustainable business model archetypes consisting of value proposition, value creation and delivery, as well as value capture (Bocken et al., 2014, p. 43). We further argue that sustainable value creation is important in relation to SDG fulfilment for the theoretical and practical application of value propositions. Prior research on shared value (SV; Porter & Kramer, 2011) has emphasized that companies' value propositions emerge via activities that increase SV (Osorio-Vega, 2019). A variety of methodologies for characterizing sustainable business approaches exist and can be divided into legitimacy-oriented and profit-oriented approaches (Schaltegger & Hörisch, 2017). Communicating and interacting with stakeholders in society is one core dimension of corporate sustainability (Székely & Knirsch, 2005).

In our study, we looked at the element of sustainable value creation with respect to SDG fulfilment. The SDGs, which were introduced by the UN in 2015, are global calls to action to protect the planet, end poverty, and ensure that all people enjoy peace and prosperity by 2030.[1] The official 17 SDGs related to all aspects of our lives are displayed in Chapter 1 (Figure 1.1).

In a highly cited article published in *Nature*, Nilsson et al. (2016) stress the importance of mapping the interactions between the defined SDGs, as these are interlinked and depend on each other. Therefore, it is important to consider how the SDGs are connected to each other to ensure sustainability in all SDG areas. Based on the overview of the literature, we argue that a conceptual model of sustainable business model development aligned with the SDGs is needed. In the case of researchers and business development practitioners, the conceptual model would enable the categorization of companies according to their business models. In the case of entrepreneurs and managers, the conceptual model would enable the analysis of the suitability of their business model with the long-term SDGs to which they are—or need to be—committed. In our analysis section, we assess how the core values of born-sustainable and sustainability-driven B Corporations (B Corps) relate to their mission and vision and are linked to specific SDGs. Based on the results, we present a conceptual model on the way born-sustainable companies' sustainable business models are aligned with the SDGs. The model offers a theoretical contribution to connecting the higher-level missions of UN SDGs with the sustainable business model literature. As for managerial implications for companies, the conceptual model offers insights into how to incorporate the SDGs in practice in the sustainable business development and implementation phases.

12.3 Methodology and Case Descriptions

12.3.1 Data Collection

To collect our data, we used the B Corp platform (www.BCorporation.net/en-us) to find information on born-sustainable and sustainability-driven companies' visions, missions, and values, as well as additional information. We enhanced the data by checking and triangulating with the information retrieved from our case companies' websites that have sustainable values as core values. A *vision* "is the 'What': the picture of the future [the organization] seek[s] to create," a *mission* "is the 'Why': the organization's answer to why [it] exist[s] (purpose)," and *values* "are the 'How': [how the organization] act[s] to achieve [its] vision" (Mirvis et al., 2010, p. 317).

The B Corp certification is awarded by the US non-profit organization B Lab, which was founded in 2006 in Wayne, Pennsylvania, and nowadays operates globally across various sectors. The *B* in "B Corp" stands for "beneficial" in the sense of a non-profit. Thereby, B Corp–certified companies distinguish themselves from purely profit-oriented companies by voluntarily meeting certain standards in terms of transparency, responsibility, and sustainability. Therefore, the corporate objective of B Corps is not solely defined by the generation of financial added value but is linked to the goal of creating added value for society and solving social and ecological problems. The B Corp certificate provides a label for for-profit companies that, at the same time, are building a more sustainable economy (Del Baldo, 2019).

We analyzed a total of 30 cases (i.e., 10 cases each from the clothing and apparel, cosmetics, and food sector, respectively). We chose these sectors because they refer to companies producing necessary consumer goods (food and clothing) or because they refer to companies that otherwise have a significant impact on biodiversity (cosmetics). For instance, the cosmetics sector strongly affects biodiversity on the planet for its "over-exploitation of natural resources and unsustainable farming practices," "toxic chemicals used in these products and pesticides used in growing raw materials," and packaging.[2] We chose the cases based on the oldest B Corp–certified companies in these sectors to ensure that we obtained sufficient information for our analysis and that we incorporated in the analysis companies that have sought to signal their true sustainability core values with the B Corp certification from the very early phases of their business. We analyzed the cases from the B Corp search portal[3] by extracting the companies' descriptions using the following keywords: "clothes," "cosmetics," and "food." This platform allowed us to look only at relevant companies to analyze their core values, and we conducted analysis regarding their SDG fulfilment. We chose the B Corp platform because the B Corp certificate stands for "transforming the global economy to benefit all people, communities, and the planet."[4] We further used company websites and other secondary data to back up our analysis. The list of cases can be found in Table 12.1.

Table 12.1 B Corp analysis results

B Corp companies analyzed	*Core values (either stated directly or taken from vision/mission statement)*	*Core values reflected in mission/ vision?*	*Link to SDGs*
Clothes and apparel			
TS Designs	Quality product, ES impact, short distance	Not clear	SDGs 8, 11, 12
Ditto Hangers	Service, integrity, personal responsibility	Yes	SDG 12
Warby Parker	Social responsibility	Yes	SDG 17
United By Blue	Sustainable production, ethical manufacturing	Yes	SDG 12, 14, 15
Patagonia Works	Minimalism, functionality (repairable, recyclable, durable)	Not clear	SDG 12, 14, 15
Apollis Holding LLC	Local, empowering, partnering, social responsibility	Yes	SDG 8, 17
Bixbee	Equity, education, empowerment, awareness raising	Yes	SDG 4
Oliberte Limited	Respect, quality, fair trade, partnership	Not clear	SDG 3, 8, 12, 17
Oaklandish	Culture and history preservation, partnership	Yes	SDG 8, 11, 17
Karun	Reusability, community, sustainable design, quality, respect	Yes	SDG 11, 12, 13, 17
Cosmetics			
Alima Cosmetics	Safety, well-being of Earth and its inhabitants	Yes	SDG 3, 12
Baubo	Feminism, sustainability, being lovable and eco-friendly	Yes	SDG 3, 5, 12
Beautycounter	Transparency, accountability, advocacy	Not clear	SDG 3
Cheekbone Beauty Cosm.	Innovation, representation	Not clear	SDG 4, 10, 13
Eluma	Individual responsibility, authenticity, transparency, trust, inclusion, fairness, kindness, community, innovation	Not clear	SDG 11
Ethique	Health, sustainability, natural ingredients	Not clear	SDG 12, 13
Kaffe Bueno	Sustainability, health, technology	Yes	SDG 3, 9, 12
The Lekker Company	Health, honesty, ease, cruelty-free	Not clear	SDG 12
The Body Shop	Empowerment, quality, nature, community, sustainability	Yes	SDG 5, 13, 17
Weleda	Holistic nature, responsibility, trust, efficiency, openness, enthusiasm	Not clear	SDG 12
Food			
King Arthur Baking Company	Communities, people and the planet, responsible sourcing, "never bleached" guarantee, close cooperation, sustainability	Yes	SDG 8, 11, 12
Terrafertil US LLC	Human health and lifestyles, high quality and nutritious food, ecologically sound, trustworthy, health partner	Yes	SDG 3, 8

(*Continued*)

Table 12.1 (Continued)

B Corp companies analyzed	*Core values (either stated directly or taken from vision/mission statement)*	*Core values reflected in mission/ vision?*	*Link to SDGs*
One Village Coffee LLC	Co-ownership with farmers, partnership with NGOs	Yes	SDG 2, 17
Numi	Fairtrade, carbon footprint, climate action, plant-based food, clean drinking water, healing power of tea, communities	Yes	SDG 3, 6, 11, 12, 13
Greyston Bakery	Vegan options, healthy products, inclusive employment, workforce programs for job seekers	Yes	SDG 3, 5, 8, 10
Alter Eco	Going organic, clean and green, regeneration of ecosystems, empowering farmers, improving farmer livelihood, mitigating and reversing climate change, fairer trade, low/no waste, carbon-neutral	Yes	SDG 3, 8, 12, 13, 15
Revolution Foods	Communities, healthy food accessible to all, for locals by locals, clean, high-quality ingredient	Yes	SDG 3, 11, 12, 17
Happy Family Organics	Children's health accessible to all, reducing climate impact, sustainable packaging, regenerative agriculture, support families with unique needs, inclusivity	Yes	SDG 3, 12, 13, 15
Larry's Coffee	Fair trade, cooperative roasters/farmers, being organic, solar energy, recycling, rainwater harvesting, propagating pollinators	Yes	SDG 6, 7, 8, 12, 17
The Guayaki Yerba Mate Co	Community, market-driven regeneration business model (personal, social, ecological, and cultural), inclusive employer, sustainable sourcing	Yes	SDG 10, 11, 12

Note: We analyzed the data from the B Corp website (www.BCorporation.net/en-us/) and additional secondary data sources, including the B Corps' websites.

12.3.2 Data Analysis and Findings

To derive the categories for our conceptual model, we applied qualitative content analysis (QCA). *QCA* defines itself within a "framework as an approach of empirical, methodological controlled analysis of texts within their context of communication, following content analytical rules and step by step models, without rash quantification" (Mayring, 2000, p. 2). Therefore, the QCA methodology is applicable in situations in which the goal is a deductive category application, as in the case of our study. We used the software QCAmap (Fenzl & Mayring, 2017; Mayring, 2014) and MAXQDA to analyze our secondary data. This approach helped us structure and then further analyze the data. Using QCA, we defined a variety of themes and categories from the literature and secondary data we collected. Regarding the QCA methodology, researchers are advised to first analyze the data individually and then bring it together for comparison purposes. One of our research team members looked at the clothing industry, another at the cosmetics industry, and one more at the food industry, respectively. We then compared our results and looked for overlapping categories, with a focus on born-sustainable companies' core values and their relation to their mission and vision. Moreover, we analyzed these with regard to targeting the SDGs they tackle. Our findings are summarized in Table 12.1, where we list our resulting main categories and highlight the vision/mission of the analyzed companies and the SDGs to which they contribute.

In the left column of Table 12.1, we see the 30 B Corps we found to be born-sustainable or sustainability-driven companies under study across the three sectors (i.e., clothes and apparel, cosmetics, and food). Looking at the companies' backgrounds and characteristics, we see that most of the companies in the clothes and apparel as well as the food sectors were from the United States and Canada, which is understandable, since the certification is American and we focused on the oldest companies. However, in the cosmetics sector, half of the B Corps were from Europe (i.e., France, Denmark, the Netherlands, the United Kingdom, Switzerland, and New Zealand). The fact that the cosmetics B Corps were spread across the globe might be due to the leadership and knowledge of several European companies in the sector (e.g., L'Oréal) and the necessary organization of the global supply chain, with raw materials originating from different countries and regions. The oldest cosmetics B Corp under study became certified in 2008 (Alima), while some of the other B Corps only recently received their certification. The clothes and apparel B Corps received their B Corp status on average a longer time ago (i.e., between 2008 and 2013). For the food B Corps, this time frame was even earlier (i.e., between 2005 and 2011).

When comparing the qualitative data from the B Corp database and the company websites, we see that the B Corps under study share certain core values. For the clothing sector, the values of "social responsibility," "quality," "partnership," and "respect" were named particularly often. These values are related to SDG 8 (Decent Work and Economic Growth), SDG 12 (Responsible Consumption and Production), and SDG 17 (Partnerships for the Goals). Looking at the cosmetics sector, the core values named most often were "transparency," "trust," "health," and "community." SDG 3 (Good Health and Well-Being) and SDG 12 (Responsible Consumption and Production) seemed to be especially important to this sector. For the food sector, the values of "community/partnerships," "health," and "sustainability" were named particularly often. These are especially linked to SDG 3 and SDG 12.

To some extent, the core values were further reflected in the B Corps' visions and missions. In the case of Karun (clothes and apparel sector), for instance, the B Corp's values are "reusability," "community," "sustainable design," "quality," and "respect," and the mission statement is:

> Karun Shades seeks to literally change the way businesses operate throughout the world. They propose a new way of seeing the world in which respect for profitability is not more heavily weighted than that for people and the environment. Karun Shades demonstrates that you can create the best quality products in the world, making fashionable products based on the concept of social and environmental responsibility.

The stated purpose is "[t]o inspire the world into reflecting about the question: What would the world look like if we understood that we are all nature?" In the case of Patagonia (clothes and apparel sector), the B Corp's core values are "minimalism" and "functionality (repairable, recyclable, durable)," and Patagonia's mission/vision is this: "They seek not only to do less harm, but more good. Use business to protect nature." However, for the cosmetics sector, the link between the stated core values and the mission/vision statements found on the websites was not always clear.

There seemed to be a set of common values that B Corps shared or at least communicated. The two values that were most often named were "(social) responsibility" and "partnership/community." This outcome from the analysis further underlines the fact that born-sustainable and sustainability-driven B Corps focus on social responsibility and collaboration, which are relevant in certain fields of management literature, such as supply chain management, as the social dimension lacks consideration (Carter et al., 2019). From Table 12.1, we further see that "empowering" is important to B Corps operating in the cosmetics and clothes and apparel sectors. However, some values seem to be more sector-specific. For example, "health" plays an important role in the cosmetics and food sectors, but not in the clothes and apparel sector.

This section presented our analysis, with a focus on the core values that make these companies sustainable from their founding. The next section focuses on building a conceptual model from our initial analysis to establish clear linkages to the relevant core SDGs for each specific sector.

12.3.3 Development of a Conceptual Model

From our initial analysis, we developed a conceptual model that links the core values (propositions) of born-sustainable and sustainability-driven companies to the core SDGs that they tackle. As a result, from our initial analysis presented in Table 12.1, we can see that SDG 12 (Responsible Consumption and Production) is the overlapping global goal that is relevant to all three sectors under research.

In Figure 12.1, we therefore illustrated SDG 12 on top of the main identified SDGs in the three sectors under study. This overall result links well with the transformation map provided by the

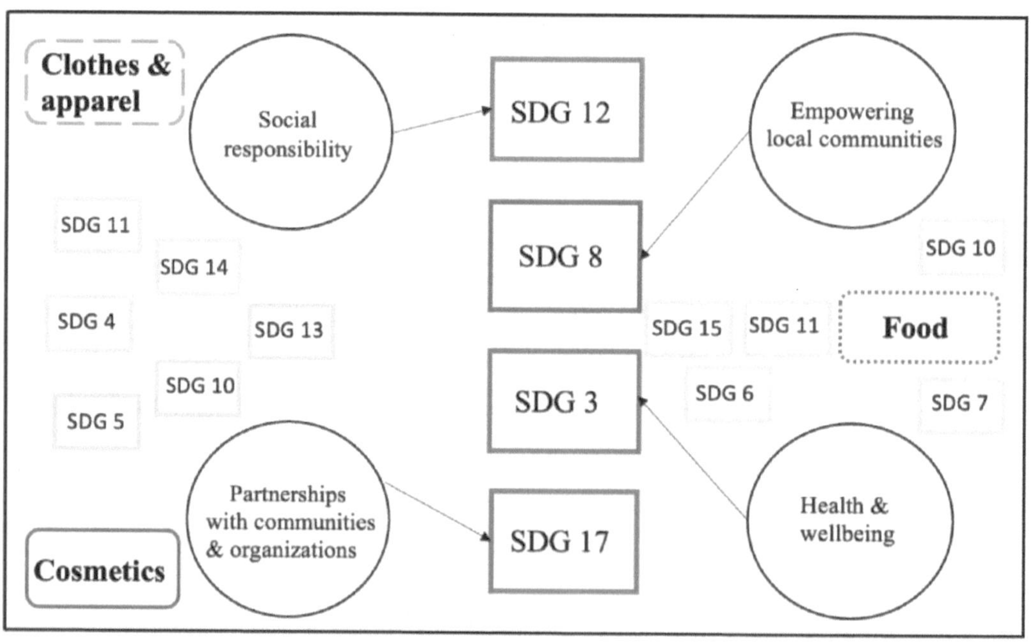

Figure 12.1 Conceptual model of the core values of B Corps and their relationships with the SDGs (clothes and apparel, cosmetics, and food sectors).

Note: Circles = core values of B Corps (clothes & apparel, cosmetics, and food sector); Large boxes = overlapping SDGs across all three sectors; Small boxes = sector-specific SDGs).

World Economic Forum[5] that also positions SDG 12 as a central development goal according to the UN's description of SDG 12 linking businesses and consumers (United Nations, 2023), also regarding taking up responsibility.

When looking at the individual sectors under study, we found that for the clothes and apparel and food industries, SDG 8 (Decent Work and Economic Growth) was also important, whereas for the cosmetics sector, SDG 3 (Good Health and Well-Being) seemed particularly relevant. Moreover, SDG 17 (Partnerships for the Goals) was also related to some of the core values communicated by the case companies in all three sectors. Figure 12.1 graphically presents these linkages for all three sectors.

This graphical illustration shows the four core values that are linked to the four key SDGs that B Corps in the clothes and apparel, cosmetics, and food sectors address. None of the analyzed B Corps had initially named the SDGs on their websites; however, we found clear links, interlinkages, and overlaps for the B Corps under study. Based on this analysis, we argue that B Corps as sustainable businesses could better communicate how they specifically address the core SDGs to inspire other firms to transition toward a more sustainable society and the respect of the Earth.

12.4 Discussion

We started out by speculating whether born-sustainable and sustainability-driven companies share certain core values that make them "truly" sustainable. We could see that they do to some extent, and that these core values can be related to the SDGs.

From our analysis, we built and presented a conceptual model of how born-sustainable and sustainability-driven companies build and link the main drivers of their business models to create a sustainable foundation that promotes the core SDGs relevant to the specific sector. However, not all SDGs were addressed by our 30 cases, which is likely due to the sectors we analyzed. While the literature argues that all SDGs need to be fulfilled, it makes sense that SDG 12 was the most important in the case of the three sectors under study, as we looked at companies focusing on the production of consumer goods. SDG 12 is also related to the overall CSR concept and linked sustainability measures. However, the studied B Corps could also address and tackle more and other SDGs. For example, a food B Corps company might engage in initiatives that tackle SDG 2 (Zero Hunger), relatively water-intense clothing production companies might address SDG 6 (Clean Water and Sanitation), and cosmetics B Corps might focus more on SDG 10 (Reduced Inequalities) when it comes to producing cosmetics for all ethnicities and income groups, or SDG 15 (Life on Land), given its significant impact on biodiversity.

The results and conceptual model show also that sustainability-driven companies focus only on some of the core SDGs in particular. The addressed values furthermore seem to be sector-specific (see Figure 12.1). However, sustainable business development should not be about "ticking boxes" regarding what SDGs should be presented in the sustainability communication of companies, in order not to fall into the same normative approach that led John Elkington to withdraw the TBL concept. Instead, the sustainable value creation and SDG-related activities need to be incorporated into the strategy of companies and organizations. From this perspective, we suggest to policymakers to consider adding a new goal to the list of UN SDGs that would focus more on "doing good with business" and "having a positive impact on society and on Earth." This could be formulated as an SDG on "giving back to the natural environment and/or the local community." Furthermore, the recent decisions taken by entrepreneurs, like Patagonia's founder, Yvon Chouinard,[6] to donate its dividends to organizations that have a true impact in protecting nature and combatting climate change could be used as a source of inspiration to move away from harming capitalism. Indeed, even companies with sustainability in their core values are often pushed to satisfy their shareholders on the short term, whatever the cost might be for people and nature. As Yvon Chouinard states for Patagonia, "instead of 'going public,' [one] could say [they're] 'going purpose.'" Thus, integrating SDGs in sustainable business models is going into the right direction, but it should also be aligned with the firm's/owner's purpose. In this sense, another additional SDG might be about "doing business with positive purpose for people and the planet." In other sectors (i.e., the mining sector), the SDGs are used in a wider sense, looking at all stakeholders around the business case (i.e., local communities) as a way to develop "means, actions and ends to accelerate a sustainable transition" (Lara-Rodríguez and Fritz, 2023) that can inspire not only companies but also policymakers.

Our study contributes to the sustainable business literature by identifying the core value sets of born-sustainable and sustainability-driven companies and by linking them to the UN SDGs. Our analysis and conceptual model can be used and further developed by researchers and used by entrepreneurs striving to develop more sustainable business models, which are clearly needed for a sustainability transition in the corporate world. Furthermore, by analyzing the core values and related SDGs of B Corp–certified companies, other non-certified companies could identify areas for improvement according to their sectors.

12.5 Conclusion and Future Research

The main contribution of our study is an analysis of the key values of born-sustainable and sustainability-driven B Corp companies and their linkages to the SDGs. Based on our findings, we present a conceptual model of how born-sustainable and sustainability-driven companies initially build their business models on a sustainable foundation that also promotes core values and associated SDGs. In the future, this understanding and knowledge will help researchers and companies build more sustainable business ecosystems (i.e., ecosystems that focus on sustainable quality and social practices, as well as promoting sustainable consumption behavior that does not support overconsumption among consumers). Thus, sustainable companies with future-oriented sustainable business models and practices could help ensure overall societal well-being and have a more positive impact on the natural environment.

Next to this, current research debates are pointing toward the need for focusing also on the inner dimension of sustainability, which is linked to personal and cultural values, that is, change from inside out (Horlings, 2015). Companies need to focus on inner development goals among the people in organizations as part of ethics, responsibility, and sustainability competencies (Laasch et al., 2023) that are needed for facing the environmental and social challenges of our time (Ehgartner et al., 2017). We see this competence development of managers and employees as part of a meaningful and necessary toolset for born-sustainable and sustainability-driven companies in the overall sustainability transition. Future research might investigate how this mindset change influences sustainable development both on a micro and a macro level.

Our study has some limitations. The findings are limited by the data obtained (i.e., secondary data from the B Corp platform and the websites of companies). In the future, the analysis and findings should be reinforced using qualitative and quantitative surveys, such as with born-sustainable and sustainability-driven companies and others, to validate and compare the results, versus founders' and customers' perceptions of how the core values and relevant SDGs are fulfilled. However, our initial conceptual model presented in this chapter can be used as a basis for analyzing how to create (more) sustainable companies and for understanding the way core values in sustainable business can be linked to the UN SDGs by not just only being communicated but also getting incorporated into the company strategy.

Notes

1 www.undp.org/sustainable-development-goals (last accessed March 12, 2023).
2 www.cbd.int/article/cosmetic-shift (accessed March 13, 2023).
3 www.B Corporation.net/en-us/find-a-b-corp/ (accessed March 13, 2023).
4 www.B Corporation.net/en-us (accessed March 13, 2023).
5 https://intelligence.weforum.org/topics/a1G0X0000057N18UAE (accessed December 7, 2023).
6 See Yvon Chouniard's declaration here: https://eu.patagonia.com/gb/en/ownership/ (accessed May 17, 2023).

References

Bocken, N. M., Short, S. W., Rana, P., & Evans, S. (2014). A literature and practice review to develop sustainable business model archetypes. *Journal of Cleaner Production*, 65, 42–56.

Carter, C., Hatton, M., Wu, C., & Chen, X. (2019). Sustainable supply chain management: Continuing evolution and future directions. *International Journal of Physical Distribution & Logistics Management*, 50(1), 122–146.

Del Baldo, M. (2019). Acting as a benefit corporation and a B Corp to responsibly pursue private and public benefits. The case of Paradisi Srl (Italy). *International Journal of Corporate Social Responsibility*, 4(1), 1–18.

Demirel, P., Li, Q. C., Rentocchini, F., & Tamvada, J. P. (2019). Born to be green: New insights into the economics and management of green entrepreneurship. *Small Business Economics*, 52(4), 759–771.

Dicuonzo, G., Galeone, G., Ranaldo, S., & Turco, M. (2020). The key drivers of born-sustainable businesses: Evidence from the Italian fashion industry. *Sustainability*, 12(24), 10237.

Ehgartner, U., Gould, P., & Hudson, M. (2017). On the obsolescence of human beings in sustainable development. *Global Discourse*, 7(1), 66–83.

Evans, S., Fernando, L., & Yang, M. (2017). Sustainable value creation—from concept towards implementation. In *Sustainable Manufacturing: Challenges, Solutions and Implementation Perspectives* (pp. 203–220).

Fenzl, T., & Mayring, P. (2017). QCAmap: eine interaktive Webapplikation für Qualitative Inhaltsanalyse. *Zeitschrift für Soziologie der Erziehung und Sozialisation ZSE*, 37, 333–340.

Hall, J. K., Daneke, G. A., & Lenox, M. J. (2010). Sustainable development and entrepreneurship: Oast contributions and future directions. *Journal of Business Venturing*, 25(5), 439–448.

Horlings, L. G. (2015). The inner dimension of sustainability: Personal and cultural values. *Current Opinion in Environmental Sustainability*, 14, 163–169.

Jollands, S., Akroyd, C., & Sawabe, N. (2015). Core values as a management control in the construction of "sustainable development". *Qualitative Research in Accounting & Management*, 12(2), 127–152.

Jonker, J., & Faber, N. (2021). *Organizing for Sustainability: A Guide to Developing New Business Models* (p. 242). Springer Nature.

Knoppen, D., & Knight, L. (2022, in press). *Pursuing Sustainability Advantage: The Dynamic Capabilities of Born Sustainable Firms*. Business Strategy and the Environment.

Kraaijenbrink, J. (2019). *What the 3Ps of the Triple Bottom Line Really Mean*. Retrieved from: www.forbes.com/sites/jeroenkraaijenbrink/2019/12/10/what-the-3ps-of-the-triple-bottom-line-really-mean/ (accessed 7th December 2023).

Laasch, O., Moosmayer, D. C., & Antonacopoulou, E. P. (2023). The interdisciplinary responsible management competence framework: An integrative review of ethics, responsibility, and sustainability competences. *Journal of Business Ethics*, 187(4), 733–757.

Lara-Rodríguez, J. S., & Fritz, M. M. (2023). How does eliminating mercury from artisanal and small-scale gold mining lead to achieving sustainable development goals? *Natural Resources Forum*, 47(2), 214–228.

Laszlo, C., & Zhexembayeva, N. (2017). *Embedded Sustainability: The Next Big Competitive Advantage*. Routledge.

Marcus, J., & Roy, J. (2019). In search of sustainable behaviour: The role of core values and personality traits. *Journal of Business Ethics*, 158(1), 63–79.
Mayring, P. (2000). Qualitative content analysis [28 paragraphs]. *Forum Qualitative Sozialforschung*, 1(2), Art. 20. Retrieved from: http://nbnresolving.de/urn:nbn:de:0114-fqs0002204.
Mayring, P. (2014). *Qualitative Content Analysis. Theoretical Foundation, Basic Procedures and Software Solution*. Retrieved from: http://nbn-resolving.de/urn:nbn:de:0168-ssoar-395173 (accessed 7th December 2023).
Mirvis, P., Googins, B., & Kinnicutt, S. (2010). Vision, mission, values. *Organizational Dynamics*, 39(4), 316.
Nilsson, M., Griggs, D., & Visbeck, M. (2016). Policy: Map the interactions between Sustainable Development Goals. *Nature*, 534(7607), 320–322.
O'Neil, I., & Ucbasaran, D. (2016). Balancing "what matters to me" with "what matters to them": Exploring the legitimation process of environmental entrepreneurs. *Journal of Business Venturing*, 31(2), 133–152.
Osorio-Vega, P. (2019). The ethics of entrepreneurial shared value. *Journal of Business Ethics*, 157(4), 981–995.
Porter, M. E., & Kramer, M. R. (2011). Creating shared value: Redefining capitalism and the role of the corporation in society. *Harvard Business Review*, 89(1/2), 62–77.
Schaltegger, S., & Hörisch, J. (2017). In search of the dominant rationale in sustainability management: Legitimacy-or profit-seeking? *Journal of Business Ethics*, 145(2), 259–276.
Székely, F., & Knirsch, M. (2005). Responsible leadership and corporate social responsibility: Metrics for sustainable performance. *European Management Journal*, 23(6), 628–647.
United Nations (2015). *The Sustainable Development Agenda*. Retrieved from: www.un.org/sustainable development/development-agenda/ (accessed 7th December 2023).
United Nations. (2023). *Goal 12: Ensure Sustainable Consumption and Production Patterns*. Retrieved from: www.un.org/sustainabledevelopment/sustainable-consumption-production/ (accessed 7th December 2023).
Vitols, S., & Kluge, N. (2011). *The Sustainable Company: A New Approach to Corporate Governance* (Vol. 1). ETUI.
Wijethilake, C., & Lama, T. (2019). Sustainability core values and sustainability risk management: Moderating effects of top management commitment and stakeholder pressure. *Business Strategy and the Environment*, 28(1), 143–154.

13

THE TRANSFORMATION OF THE GLOBAL FINANCIAL SYSTEM TO SUSTAINABILITY

Sebastian Thomas and Angela Bruckner

13.1 Introduction

Human activities are now the most significant driver of environmental change on the planet.[1] Most rivers and major deltas have been totally restructured, ocean chemistry has been severely altered and its biodiversity depleted, coral reefs are close to disappearing as functional ecosystems, over half of the land surface is devoted to livestock and crop agriculture, and species are disappearing at an unprecedented rate in the sixth mass extinction event in the geological record.[2–5] Global warming and climate change, driven by anthropogenic greenhouse gas emissions, are disrupting ecosystems in new ways, altering species behaviors and interactions, driving migrations, remaking habitats, producing disease, and causing cascading effects that are impossible to measure and predict.[6, 7] These challenges represent an extraordinary threat to human societies and, in fact, to global civilization.[8, 9] The planetary climate system is now entering a state that human societies have never previously experienced, and there is no way to know how or whether we will be able to adapt.[2, 10, 11]

It is now well understood that the global economic system—and the societal structures that it serves—must transition to a net-zero emissions industrial base, build capabilities to move to "true zero" and beyond, reverse biodiversity decline and restore ecological values and capital, and support just and equitable social outcomes through these transformation processes.

While governments and intergovernmental organizations develop policy frameworks to address the challenges of climate change and ecological collapse that threaten social stability and economic continuity, the private sector is seeking to play a part by understanding climate change, nature dependency, and wider social and cultural sustainability issues through its fundamental principles of risk and return. There are global movements to integrate climate and nature considerations into financial thinking, and these have implications for how organizations invest and do business. The integration of environmental, social, and governance (ESG) risks into corporate strategies has been primarily driven by consideration of the financial risks to organizations and made operational through platforms such as the Sustainability Accounting Standards Board (SASB), the Taskforce on Climate-Related Financial Disclosure (TCFD), and the Taskforce on Nature-Related Financial Disclosure (TNFD). These platforms provide disclosure frameworks for companies to identify and publicly report the risks and opportunities to their businesses—and, by implication, the potential

DOI: 10.4324/9781003285472-14

for instability and disruption in national and global economies—that arise from sustainability issues, climate change impacts, and nature dependency in supply chains.

Innovation has been recognized as a critical driver of business performance for more than a century.[12] As public awareness of environmental and social challenges has grown, innovations in sustainable business practices have become increasingly important to firms, and this trend is continuing.[13] There has been a shift in the understanding of corporate responsibility from "creating profit for shareholders" to a contemporary focus on the creation of environmental, social, and financial value.[14–16] A key question is whether sustainability innovations, including ESG strategies, have a demonstrable impact on corporate performance measured in financial and non-financial ways. If addressing ESG risks means achieving positive commercial outcomes, including financial returns, then the private sector will act.

This chapter presents the business case for sustainability by demonstrating the outcomes of ESG integration into corporate strategies and behaviors. This supports the existing paradigm of transformation rather than revolution in our commercial and economic systems. Rather than building alternative economic models, the global community is seeking to keep much of the current economic architecture but adapt this to incorporate additional (and previously unaccounted and externalized) values and metrics. We then describe some of the tools of the transition, the taxonomies and disclosure frameworks that are guiding organizational behaviors within these sustainability constraints. The chapter then argues that while voluntary actions and even the emerging regimes of mandatory disclosures are necessary components of a move toward a sustainable economic system, they are not sufficient to achieve it. More is required, and in more jurisdictions.

We conclude by describing the macroeconomic scenarios currently being explored by key actors in the financial system, assessing their utility and robustness and considering their implication for the future of the global economy and the Sustainable Development Goals.

13.2 Evolution, Not Revolution—The Business Case for Sustainability Transition

Sustainability has long been understood by the business community in terms of activities that balance three key considerations: economic outcomes, social issues, and environmental impact.[17] While this triple bottom line approach has been a useful tool, it is also conceptually inaccurate and frequently impractical. The relationship between human financial and economic systems and the natural world is not reciprocal and equal but dependent and interacting.[18–20] Triple bottom line thinking does not acknowledge that an economy does not exist without a society and no society can exist without a supporting environment.[21] Further, when economic, social, and environmental issues are assessed as equal considerations in decision-making, it is often the case that one or more will be compromised to achieve corporate priorities, which have traditionally been measured in financial terms.[22] This is now changing, as environmental, social, and governance issues are increasingly included as part of market regulation and directors' legal responsibilities.

The ecosystem services provided by nature are critical to human existence. Without healthy, functional supporting and regulating services (such as soil production and nutrient cycling), agricultural productivity collapses and supply chains dissolve. Without robust provisioning services (such as provision of fresh water and plant growth), people face food and water shortages and lack the resources necessary to sustain communities. Without nature, societies fail. Financial and economic systems are functions of healthy human societies and, therefore, fundamentally reliant on the goods and services provided by nature. Healthy ecosystems are critical to economic growth but facing unprecedented and existential threats in a climate-changing world.

The value of global ecosystem services in 1995 was estimated to average US$46 trillion per year, measured in 2007 US dollars.[23] An updated estimate for 2011 found a total value of $125 trillion per year including updated unit values and changes to biome areas and $145 trillion per year assuming only unit values changed. This implies that the loss of ecosystem services due to land use change from 1997 to 2011 was $4.3–20.2 trillion per year.[24] Environmental degradation and destruction of natural capital are costing astronomical amounts.

While it is difficult and arguably absurd to estimate the financial value of natural assets, such valuations can usefully inform discussions around decision-making and policy.[24] For example, Australia's Great Barrier Reef (GBR) has been estimated to have an economic, social, and iconic asset value of AUD$56 billion.[25] The GBR supports at least 64,000 jobs and directly contributes more than $6 billion to the Australian economy every year, and these numbers do not necessarily include the extensive ecosystem services provided by the GBR, including coastal protection, water purification, and habitat for commercially important fisheries.[26]

The number of public companies reporting ESG information grew from fewer than 20 in the early 1990s to 8,500 by 2014.[27] By the end of 2021, almost 4,000 institutional investors managing over US$120 trillion in assets had signed the UN Principles for Responsible Investment (UNPRI).[28] Research on the impacts of ESG strategies and cultures on investment performance and other business metrics is extensive and demonstrates a robust long-term association between sustainability and investment outcomes.[13, 27, 29–32]

Multiple large-scale, long-term studies have demonstrated that ESG performance is not only material but also correlated with financial predictability and investment outcomes.[33] These include diverse types of analyses. For example, tests of transmission channels (cash flow, idiosyncratic risk, valuation) using MSCI ESG Ratings data and financial variables across 1,600 stocks in the MSCI World Index showed that ESG information was directly linked to valuation and performance, both through systematic risk profile (lower costs of capital and higher valuations) and idiosyncratic risk profile (higher profitability and lower exposures to tail risk).[34] An analysis of a large sample of US companies included in the Russell 3000 from January 2008 to July 2019 showed that not only do ESG rating changes have a consistent impact on equity performance but also that the market seems to reward companies operating in industries with a high level of concentration of ESG materiality.[33] In other words, having fewer but more financially relevant material issues is rewarded by the market. Further, the impact of ESG materiality differs depending on the industry and the type of information. Polluting or emissions-intensive industries are more affected by their material ESG performance, and industry-specific factors should therefore be considered when assessing the impact of ESG performance on a firm's valuation.[35]

However, do businesses that create social and environmental value through their corporate activities demonstrate improved financial, reputational, and competitive advantage?

A 2021 systematic review of research on the relationship between sustainability innovations and competitiveness found that 64 of 100 studies (published 2005–2020) identified a positive relationship between sustainability innovations and firm competitiveness; 29 reported mixed results that included positive, neutral, and/or negative effects; 5 studies were inconclusive; and 2 reported negative effects.[36] The relationship is complex and mediated and moderated by national-, market-, industry-, and firm-level factors, but the overwhelming majority of evidence supports the idea that sustainability improves firm performance.

A 2022 study of 194 mining companies found that sustainable environmental strategies relate positively to firm competitiveness, and that firm competitiveness affects the relationship between sustainable environmental strategy and financial performance.[37] In other words, being environmentally proactive generates competitive advantage in the mining sector, and this translates into improved financial returns. Similarly, as far back as 2004, a study of the European manufacturing sector found that firms with shareholder value–oriented strategies had a more positive relationship between environmental and economic performance.[38] More recently, a 2019 quantitative analysis of the financial materiality of ESG information in Europe demonstrated that social, environmental, and governance strategies are key elements of firm competitiveness.[39]

In China, a 2012 study of 217 firms found that proactive environmental strategies were a key driver of dynamic capabilities—meaning, an organizational culture of strategic innovation—and had a significant positive effect on competitive advantage.[40] A 2020 analysis of 232 firms in Vietnam found that the adoption of a proactive environmental strategy generated competitive advantage through both market differentiation and cost leadership.[41] Interestingly, it is also the case that participation of female executives in corporate decision-making has had positive effects on sustainable competitive advantage, by both inhibiting unethical environmental behavior and stimulating proactive environmental strategies.[42]

One of the more recent large-scale meta-analyses investigating the relationship between ESG and financial performance is a report authored by the NYU Stern Center for Sustainable Business with colleagues at Rockefeller Asset Management.[32] The analysis considered more than 1,000 papers published from 2015 to 2020 to assess the financial performance (separately) of corporations with sustainability initiatives, ESG funds, portfolios, or indices, and studies that considered a specific theme, such as climate change. The study found a positive relationship between ESG and financial performance for 58% of the corporate studies focused on operational metrics, such as return on equity (ROE) or stock price, with 13% showing neutral impact, 21% mixed results (the same study finding a positive, neutral, or negative results), and only 8% showing a negative relationship.[32]

Some key conclusions about the relationship between ESG and financial performance can be drawn. First, it is well recognized that firms can boost key value drivers, including growth, profitability, capital efficiency, and risk exposure, by improving performance on ESG issues and managing for a low-carbon future.[32, 34, 43] Second, improved financial performance due to ESG becomes more marked over longer time horizons. A 2019 study found that communication of long-term strategy resulted in abnormal positive reactions by the stock market, while a cross-sectional study of firms with strong ESG ratings found returns up to 3.8% higher per standard deviation of ESG score in the mid and long term.[27, 44] Third, ESG strategies appear to provide downside protection, especially during periods of economic downturn or social crisis, such as the global financial crisis of 2009–2009 and the COVID-19 pandemic of 2020–2021 (or ongoing).[45–47]

Finally, it is important to note that ESG disclosure alone does not necessarily drive improved performance, but performance-based ESG initiatives are very likely to do so through a range of factors, including innovation, risk minimization, stakeholder and reputational value, and higher profitability in the long term. Companies committed to ESG are finding competitive advantages in product, labor, and capital markets. Portfolios with integrated material ESG metrics have provided average returns to their investors that are superior to those of conventional portfolios while exhibiting lower risk. Companies that pursue and communicate major sustainability initiatives have generally attracted disproportionate numbers of longer-term shareholders.[27]

The growing evidence base for the commercial benefits of corporate ESG behaviors is encouraging organizations to see the transition to a sustainable finance system as a transformation driven by opportunity rather than a revolution caused by necessity—although there is widespread and growing understanding of the urgency and importance of action on the interconnected climate and biodiversity crises. Voluntary actions are central to this process, though as we discuss further later, mandatory requirements are increasing, and necessary. What, then, are the tools being used by the private sector to implement the transition?

13.3 The Tools of Transition

For the private sector to transform its strategic assumptions and operating models, globally consistent and interoperable frameworks are required. In some cases, transition tools have been developed by industry; in others, public policy has guided action. The focus of this chapter is on the high-level concepts and approaches that are guiding the transition of the global financial system toward sustainability. It is, however, worth mentioning the different examples of industry guidance being implemented at asset class level and the key challenges being faced.

Real assets, such as infrastructure and real estate, are perhaps the most straightforward areas for asset owners and asset managers. Direct and indirect emissions can be most readily accounted, project boundaries are clear, and operational control allows for immediate influence and accountability. In listed equities markets, asset managers can request and require information which is increasingly mandated by regulators. In asset classes, including corporate fixed income, private markets (private capital and credit), and even sovereign bonds, data availability and quality can be highly constrained, meaning, that asset managers have limited capacity to assess sustainability risks. Increasingly, regulators are demanding shifts in corporate reporting and governance. Private sector coalitions have responded and demonstrated leadership within the United Nations climate change policy process by forming alliances, such as the Glasgow Financial Alliance for Net Zero (GFANZ), an umbrella for more sector-specific groups, such as the Net-Zero Asset Owners Alliance (NZAOA), the Net Zero Asset Managers Alliance (NZAM), the Net-Zero Banking Alliance (NZBA), and the Net-Zero Insurance Alliance (NZIA). These groups are industry-led but convened by the United Nations. They develop industry-relevant and industry-specific asset class guidance on decarbonization pathways and transition planning, such as the Partnership for Carbon Accounting Financials (PCAF) Greenhouse Gas Protocol, the Science-Based Targets initiative (SBTi), and the Net Zero Investment Framework (NZIF).

In terms of policy frameworks, the European Union (EU) has led the way with the European Commission's Action Plan on Sustainable Finance. The Plan includes a variety of legislative and regulatory instruments that address disclosures, greenwashing, product labeling, and other business and finance areas where sustainability issues are relevant. The different instruments interact and are regularly updated. There is increasing convergence between public and private sector approaches to transition frameworks. Many other jurisdictions, however, are moving to develop their own regulatory and legislative approaches to transition in the financial sector.

Here, we discuss several key examples of transition approaches, including sustainability reporting and disclosure, sustainable finance taxonomies, and financial instruments designed to support positive sustainability outcomes. We conclude this section with a brief examination of the issue of "materiality." The discussion is not intended to be comprehensive but structured in such a way as to highlight the different types of transition approaches and key examples within those.

13.3.1 Sustainability Reporting and Disclosure

13.3.1.1 Disclosure Frameworks

The Taskforce on Climate-Related Financial Disclosures (TCFD) was convened by the Financial Stability Board (FSB) in 2015 and arose from investors' need to better understand their climate risk exposure and the implications of this for shareholders in particular. This was followed by the formal launch of the Taskforce on Nature-Related Financial Disclosures (TNFD) in 2021 to capture broader nature-related risks across five domains of environmental change: climate change, land fresh water and ocean use change, resource use and replenishment, pollution and pollution removal, and invasive species introduction and removal.

Both the TCFD and TNFD recommendations are structured around four thematic areas of corporate operations: governance, strategy, risk and impact management, and metrics and targets. The TCFD recommendations are intended to apply to both the physical and transition risks arising from climate change. In addition, the TNFD also considers systemic risks arising from the failure of large-scale systems, such as ecosystem collapse or financial market contagion. While the TCFD is focused on assessing climate-derived risks to and opportunities for the business, the TNFD encourages organizations to produce integrated climate–nature disclosures and to develop appropriate risk management processes. A summary of the key TCFD and TNFD recommendations is shown in Table 13.1.

Disclosure regimes such as the TCFD and TNFD, while originally envisioned as voluntary, are increasingly being included in regulation for the disclosure of material financial risks. For example, as of October 2023, the TCFD recommendations have been fully incorporated into International Financial Reporting Standards (IFRS) S1 (General Requirements for Disclosure of Sustainability-Related Financial Information) and S2 (Climate-Related Disclosures), and accordingly, the monitoring of climate risk disclosures has been incorporated into the IFRS Foundation work program. These are discussed further in what follows.

A key element of both the TCFD and TNFD is the use of scenario analysis to understand the risks and opportunities in different possible future scenarios. The scenarios used under the TCFD include, *inter alia*, scientific scenarios developed by the IPCC; market- and sector-based scenarios, such as those developed by the International Energy Agency; and bespoke scenarios developed by organizations such as the Network for Greening the Financial System (discussed later in this chapter).

Table 13.1 Summary of the key TCFD and TNFD recommendations

	Governance	*Strategy*	*Risk (and impact) management*	*Metrics and targets*
TCFD	Disclose an organization's governance structures for climate-related risks and opportunities	Disclose the potential and actual climate-related risks and opportunities on an organization's business activities, strategy, and financial planning, where material	Disclose how an organization identifies, assesses, and manages climate-related risks	Disclose the metrics and targets used to assess and manage climate-related risks and opportunities, where material
TNFD	Disclose an organization's governance structures for nature-related dependencies, impacts, risks, and opportunities	Disclose the potential and actual nature-related risks and opportunities on an organization's business activities, strategy, and financial planning, where material	Disclose how an organization identifies, assesses, and manages nature-related dependencies, impacts, risks, and opportunities	Disclose the metrics and targets used to assess and manage nature-related dependencies, impacts, risks, and opportunities, where material

13.3.1.2 The International Sustainability Standards Board (ISSB)

In the early 1970s, the professional accounting industry associations of numerous developed countries collaborated to develop basic accounting standards and advocate for their global acceptance. Over the next quarter century, this collaboration expanded its membership and established standardized accounting practices, resulting in the creation of a full-time International Accounting Standards Board (IASB), overseen by independent trustees, along with the International Financial Reporting Standards (IFRS) Foundation, in 2000.

The IFRS Foundation is a not-for-profit, public interest organization that develops rigorous, accessible, enforceable, and globally accepted accounting and sustainability disclosure standards. Standards are developed by the International Accounting Standards Board (IASB) and the International Sustainability Standards Board (ISSB).

IFRS S1 and IFRS S2 were issued as exposure drafts in 2022 and finalized in 2023. IFRS S1 requires entities to disclose information about its sustainability risks and opportunities, while IFRS S2 requires the disclosure of climate-related risks. For both standards, entities are required to disclose any in-scope matters that could reasonably be expected to affect its cash flows, access to finance, or cost of capital.

13.3.1.3 European Union Policies

The EU's *Sustainable Finance Disclosures Regulation (SFDR)* empowers the EU Commission to specify how authorities and market participants will meet the requirements of the directive. The SFDR seeks to ensure a level playing field for participants and advisors in financial markets through transparency in relation to ESG risks, the consideration of negative sustainability outcomes in investment processes, and the provision of ESG information for financial products. The SFDR requires asset managers to provide standardized disclosures on how sustainability factors are addressed at both organizational and product level. The SFDR covers sustainability disclosures for financial market participants on websites, in prospectuses, and in periodic reports. The SFDR has been in effect since March 2021, with increasing requirements phased in over time.

The EU's *Corporate Sustainability Reporting Directive (CSRD)* entered into force in January 2023. It updates and strengthens regulations on the social and environmental information that companies are required to report publicly. Large companies, as well as listed smaller firms, are now required to report on ESG topics. Non-EU companies that generate over EUR 150 million on the EU market will also have to report under the CSRD. The CSRD rules are designed to ensure that investors and other stakeholders can assess the impact companies have on people and the environment. Importantly, this allows investors to assess financial risks and opportunities arising from climate change, nature dependencies, and other sustainability issues.

13.3.1.4 The United States Securities and Exchange Commission (SEC)

The landscape of ESG reporting is somewhat different in the United States, where political dynamics play a much greater role and sustainability disclosure regulations are highly contested. The US SEC has developed a rule that would require companies to disclose their climate-related risks, but while finalized in early 2024, the regulation faces substantial legal challenges. Once implemented, the rule will require companies to disclose any risks climate change poses to their business and, in some cases, that medium and larger companies provide information about their greenhouse gas emissions.

The clear trend in the evolution of risk disclosure frameworks has been to ensure that risks and opportunities can be easily identified and articulated in standardized language that facilitates investment decision-making. A summary of key points in the IFRS standards, the EU CSRD, and the US SEC regulation is shown in the Table 13.2.

Beyond risk, sustainable investment requires transparent, standardized definitions of sustainable activities. This brings us to taxonomies.

Table 13.2 Summary of key points in the IFRS standards, the EUCSRD, and the USSEC regulation

	ISSB IFRS S1 and S2	*EU CSRD*	*US SEC Rule*
Jurisdiction	Global (depending on adoption by states)	European Union	United States
Authority	International Financial Reporting Standards (IFRS) Foundation	European Financial Reporting Advisory Group (EFRAG)	United States Securities and Exchange Commission (US SEC)
First reporting year	2024	2024	2025 for climate-related risks; 2026Z for emissions data
Coverage	All environment, social, and governance (ESG) components	All environment, social, and governance (ESG) components	Climate only
Topics	S1 covers general sustainability; S2 covers climate-specific disclosures following TCFD structures	2 general sustainability reporting standards; 10 topic-specific standards (5 environmental, 4 social, 1 governance)	Single climate standard
Relationship to TCFD	Closely aligned to TCFD but provides enhanced depth and detailed guidance	Broadly aligned to TCFD structure and recommendations but provides enhanced depth and guidance	Broadly aligned to TCFD structure but guidance has different approach and fewer requirements
Captured entities	Commercial, for-profit organizations, including state-owned (public sector) enterprises, dependent on national adoption	Organizations subject to the EU CSRD—meaning, nearly all large EU entities and many non-EU entities	Only large listed organizations in the initial phase
Materiality disclosure	Financial materiality only	Both financial and impact materiality (double materiality)	Financial materiality only
Emission scopes required	Scopes 1, 2, and 3	Scopes 1, 2, and 3	Scopes 1 and 2 if considered material
Scenario analysis	Required to assess strategic resilience	Required to assess strategic resilience	Not required unless considered material
Executive compensation	Required disclosure of climate-related executive remuneration	Required disclosure of climate-related executive remuneration	Does not require disclosure of climate-related executive remuneration

13.4 Sustainable Finance Taxonomies

Many countries, regional jurisdictions, and financial services organizations have established, or are in the process of establishing, sustainable finance taxonomies. From early approaches including climate, social, and green bonds standards that promoted transparency and disclosure of environmental or social objectives, many jurisdictions have now developed either voluntary or mandatory approaches to sustainable finance taxonomies. By September 2022, 12 taxonomies were in place, and 15 were in development.[48] The proliferation of taxonomies has demonstrated the importance of considering national characteristics within globally consistent and interoperable science-based contexts. This is necessary to provide transparency in financial markets and transnational capital flows.

The European Union (EU) taxonomy for sustainable activities is central to the EU's sustainable finance framework and arguably the leading example of this approach to transition finance policy. It provides market transparency and directs investments to economic activities necessary for the energy and sustainability transition, supporting the objectives of the European Green Deal. The taxonomy defines criteria for activities that are aligned with a net-zero trajectory by 2050 and broader sustainability goals beyond climate change mitigation.

The Taxonomy Regulation establishes six climate and environmental objectives:

1. Climate change mitigation
2. Climate change adaptation
3. The sustainable use and protection of water and marine resources
4. The transition to a circular economy
5. Pollution prevention and control
6. The protection and restoration of biodiversity and ecosystems

The taxonomy defines environmental sustainability criteria for activities rather than financial products. This means that companies in and outside of the finance sector can share a common definition of *environmental sustainability*. This facilitates investment in climate-friendly projects, creates security for investors, prevents or mitigates greenwashing, and enables companies to implement decarbonization strategies in strategic and commercial pathways.

13.5 Financial Instruments

Capital is increasingly linked to sustainable outcomes through the use of thematic bonds, sukuks (a financial certificate that complies with Islamic religious law, similar to a corporate or sovereign bond), and loans. Although the specific allowable investments are typically defined by the term sheet of each individual instrument, these can generally be categorized as follows:

- *Green instruments* apply to investments that are used to fund environmental goods, which can include renewable energy, green transportation, energy- and thermally efficient buildings, and recycling and waste reduction activities.
- *Blue instruments* apply to ocean-based investments, such as sustainable fisheries and marine protected areas.
- *Social instruments* apply to investments in social goods, such as health, education, and access to sustainable freshwater and sanitation systems. A subset of social bonds is gender bonds, which specifically invest in programs to reduce gender-based inequality.
- *Sustainable instruments* apply to a broader range of both green and social objectives and may also be linked to the Sustainable Development Goals.

In China, investments that are classified as "green" are set out in the Green Bond Endorsed Projects Catalogue. This provides an exhaustive list of products, processes, and services that are eligible for investment by way of a green bond or instrument.

13.6 Materiality

The concept of materiality has been an important tool for firms seeking to assess what is relevant to their business. Materiality has traditionally been interpreted as relating to external environmental factors that might affect the organization's financial performance—and this has expanded to include not only financial but also reputational, legal, and other commercial impacts. As sustainability becomes increasingly important, however, the idea of materiality is evolving to include not only impacts *from* an ESG issue *to* the organization but also the impacts *of* the organization's activities *on* external environments and stakeholders, including biodiversity, communities, and climate.

The EU Taxonomy Regulation has established a "double materiality" concept, which requires organizations to consider how an activity is helping to meet one or more of the taxonomy objectives. Further, firms must demonstrate that their actions "do no significant harm" (DNSH) to the other objectives. This requires a two-directional perspective on impact; any indicator or data point could be material. This double materiality view contributes to the consideration of "principal adverse impacts" (PAIs) described in the EU Sustainable Finance Disclosures Regulation (SFDR), which impacts both investors and their intermediaries. A related concept is the requirement for "biodiversity net gain" (BNG) established in the United Kingdom's Environment Act 2021. All these concepts have major implications for organizations in terms of data requirements, operational procedures, and resourcing.

A *principal adverse impact* (PAI) is any effect of investment decisions or advice that results in a negative sustainability outcome in areas including environmental, social, and governance or stakeholder concerns; modern slavery and human rights; anti-corruption efforts; and anti-bribery matters. For advisors and participants in financial markets, providing required PAI information is among the most challenging requirements of the SFDR.

Biodiversity net gain (BNG) is focused on requiring development activities to include the creation and improvement of natural habitats. BNG requires development to demonstrate a quantifiable positive impact (net gain) on biodiversity compared to what existed previously. Developers, local planning authorities, and land managers seeking to trade in the UK's nature markets need to understand the requirements of BNG. Developers must deliver a BNG of 10%, meaning, any development must result in more or higher-quality natural habitat than there was before the development activity.

13.7 Voluntary Action, Mandatory Disclosure—Necessary, But Not Sufficient

In many jurisdictions, emission reduction policies are driving changes in the operations of firms in the industrial, agricultural, transport, and resource sectors. Examples include the European Union Emissions Trading Scheme, the California *Clean Energy and Pollution Reduction Act* (Senate Bill 350), the China Certified Emission Reduction Scheme, and the Australian government's Safeguard Mechanism, among many others. These policy frameworks mandate mitigation outcomes that require changes in business practices as well as technological and other types of innovation. For the finance sector, however, action on climate change mitigation has occurred largely through voluntary frameworks for the disclosure of financial risks and opportunities. Similar voluntary approaches have been developed for nature-based risks, including the Taskforce for Nature-Related Financial Disclosure referenced previously.

As countries define their national commitments through the United Nations Framework Convention on Climate Change (UNFCCC) process, including establishing—and regularly enhancing—their nationally determined contributions (NDCs), or commitments and road maps to decarbonize their economies, private sector organizations are increasingly required to align with national priorities. Legislation, such as the mandated emission trading schemes in China and the European Union, Australia's Safeguard Mechanism, California's Climate Commitment, and many other policy frameworks globally, is driving private sector initiatives, such as the Carbon Offsetting and Reduction Scheme for International Aviation (CORSIA), the first global market-based measure that establishes a cooperative, industry-centric approach to orderly decarbonization, driving finance toward emission reduction activities that offset aviation sector emissions, where technological improvements, operational improvements, and sustainable aviation fuels are unable to do so.

The policy and regulatory frameworks discussed earlier promote private sector investment in climate mitigation and nature-focused initiatives. Perhaps most importantly, policy ambitions legitimize a diverse range of initiatives and approaches specifically as investment opportunities.[49]

This dynamic interaction between international agreements, national policy frameworks, industry ambitions, and organizational commitments occurs under the umbrella of transition to net-zero and nature-positive economies. Net-zero targets are central to the global effort to mitigate, manage, and reverse the climate crisis. Net-zero targets are generally science-based—although it is likely that assumptions in scenarios used for target setting are conservative at least, if not drastically inadequate[6, 50, 51]—but overcoming their implementation challenges now becomes key.[52]

There are clear challenges to the implementation of decarbonization approaches and other sustainability transition priorities by the private sector, and financial institutions in particular. These include availability of data (for instance, the emissions profiles and other quantifiable sustainability impacts of investments), limits to operational control or influence over portfolio companies' actions (in the case of private equity investments), and the extent of indirect financed emissions in a value chain (such as the emissions from aviation in the Scope 3 inventory of an airport company). All these challenges must be overcome to achieve an effective and efficient transformation of the global financial system to sustainability.

A helpful resource on the progress of countries toward net-zero economies—including nationally determined contributions (NDCs) and legislation enacting these—is the Climate Watch NDC Tracker, available at www.climatewatchdata.org/2020-ndc-tracker.

Beyond regulation and investor demands, companies are increasingly shifting toward more sustainable operations, in part from an increased intrinsic desire to support sustainable initiatives, and in part from increased consumer and public expectations. This has led to an explosion in voluntary sustainability accreditation schemes and alliances, such as the Forest Stewardship Council (https://fsc.org/en), the UN Global Compact (https://unglobalcompact.org/), and many others, including the net-zero alliances noted previously.

However, substantial gaps in corporate actions and priorities remain. The ocean, for example, is considered far less than other sustainability priorities and yet represents a significant sustainability risk as we consider seabed mining and growth opportunities in the blue economy. The question is whether governments and the private sector are being realistic in their perception of and response to sustainability risks. There is still a large gap between what science is saying and what corporates are doing.

In short, industry and corporate initiatives for sustainability transformation are important and necessary, but they are far from sufficient to address the required socio-economic changes that are needed to avoid a possible global societal disruption or collapse.

13.8 Scenarios of the Climate-Changed Future

Scenario analysis is a key tool for exploring the possible implications of climate change. As noted earlier in this chapter, there are several organizations that develop and publish climate-related scenarios. Reference scenarios have been produced by the climate science community and collated by the IPCC since its First Assessment Report.[53] The most recent collection of scenarios developed during the IPCC's Sixth Assessment Report includes five scenarios that capture both climate scenarios and different potential societal responses to climate change.

Another group of scenarios produced by the International Energy Agency considers different configurations of the global energy system. Although these are not strictly climate scenarios, they can be useful, given that energy-related emissions account for the majority of global greenhouse gas emissions.

The Network for Greening the Financial System (NGFS) is a coalition of central banks and financial market supervisory bodies, with the aim of strengthening the response of the global financial system toward meeting the goals of the Paris Agreement. As of April 2024, the NGFS had 138 members and 21 observer organizations from every inhabited continent.

The NGFS has published a range of possible climate-based macroeconomic scenarios and has also made some recommendations toward nature-based scenarios. However, the NGFS scenarios (particularly those that purport to be aligned to a 1.5-degree pathway) have been criticized for their slow reduction of fossil fuel investment; a correspondingly heavy reliance on unproven (and currently uneconomical) mechanisms, such as carbon capture and storage; and overly optimistic predictions of the impacts of extreme global weather events and systemic shocks.[50, 54] This has been partially addressed by the addition of a "low-demand" scenario in 2023, but it remains the case that mainstream scenarios of macroeconomic climate change impacts are at least optimistic, if not extremely naïve and conservative.

Regardless of their source, climate scenarios have several limitations. They largely fail to capture physical and socio-economic climate tipping points (negative or positive) and, as a result, are likely to underestimate the true risks of climate change. They can also incorrectly be treated as forecasts, rather than a set of plausible futures. Finally, scenarios often do not capture non-climate trends that can impact on decarbonization and societal climate responses, such as conflict.

13.9 Conclusion

While many ESG actions remain voluntary, companies that choose to disclose climate change risks achieve a higher valuation post-disclosure, suggesting that investors value transparency with respect to firms' exposure to climate change risks.[55] The significant majority of research evidence from the United States, Europe, China, and globally demonstrates a clear positive link between environmental and social sustainability strategies and firm competitiveness and financial performance. The positive relationship between sustainability innovation, ESG strategies, and corporate competitive advantage and financial performance is well established and recognized by the growth in business sustainability programs in the university sector globally.[13] Firms that ignore the role of social and environmental strategies in corporate performance are not only accepting unnecessary risk but also failing to achieve optimal returns.

The private sector now understands that sustainability is a material risk. Climate change (SDG 13) and environmental degradation and biodiversity (SDGs 14 and 15) are now recognized to directly threaten business activities, supply chains, and commercial outcomes. Similarly, social issues, such as modern slavery, human rights, and well-being (SDGs 2, 3,6, 8, and 10), are understood to affect business performance and legitimacy. Investing in energy transition (SDG 7), innovation (SDG 9), and the circular economy (SDG 12) provides demonstrated returns and contributes to wider sustainable development.

Business is acting, and the necessary frameworks for the transition to sustainability are being implemented. Whether they will be sufficient and in time remains to be seen.

References

1. Steffen, W., et al. The Anthropocene: From Global Change to Planetary Stewardship. *AMBIO* **40**, 739–761 (2011).
2. Steffen, W., et al. Trajectories of the Earth System in the Anthropocene. *PNAS* **115**, 8252–8259 (2018).
3. Figueres, C., et al. Three Years to Safeguard Our Climate. *Nature* **546**, 593–595 (2017).
4. IPCC. *Climate Change 2014: Synthesis Report. Contribution of Working Groups I, II and III to the Fifth Assessment Report of the Intergovernmental Panel on Climate Change*. 151 (2014).
5. Pörtner, H.-O., et al. *The Ocean and Cryosphere in a Changing Climate. IPCC Special Report on the Ocean and Cryosphere in a Changing Climate* (2019).
6. Pörtner, H.-O., et al. *Climate Change 2022: Impacts, Adaptation and Vulnerability. Summary for Policy Makers* (2022). https://report.ipcc.ch/ar6wg2/pdf/IPCC_AR6_WGII_SummaryForPolicymakers.pdf.
7. Lancet, T. Predicting Pandemics. *The Lancet* **388**, 2960 (2016).
8. Butler, C. D. Climate Change, Health and Existential Risks to Civilization: A Comprehensive Review (1989–2013). *International Journal of Environmental Research and Public Health* **15**, 2266 (2018).
9. Kareiva, P. & Carranza, V. Existential Risk Due to Ecosystem Collapse: Nature Strikes Back. *Futures* **102**, 39–50 (2018).
10. Lederer, M. & Kreuter, J. Organising the Unthinkable in Times of Crises: Will Climate Engineering Become the Weapon of Last Resort in the Anthropocene? *Organization* **25**, 472–490 (2018).
11. Coates, J. F. Risks and Threats to Civilization, Humankind, and the Earth. *Futures* **41**, 694–705 (2009).
12. Hospers, G.-J. Joseph Schumpeter and His Legacy in Innovation Studies. *Knowledge, Technology & Policy* **18**, 20–37 (2005).
13. Hoffman, A. J. The Next Phase of Business Sustainability . *Stanford Social Innovation Review* **16**, 34–39 (2018).
14. Hart, S. L. & Milstein, M. B. Creating Sustainable Value. *Academy of Management Executive* **17**, 56 (2003).
15. Hoffman, A. J. Climate Change Strategy: The Business Logic Behind Voluntary Greenhouse Gas Reductions. *California Management Review* **47**, 21–46 (2005).
16. Friedman, M. The Social Responsibility of Business Is to Increase Its Profits. *The New York Times Magazine* (1970).
17. Silvius, G. Sustainability as a New School of Thought in Project Management. *Journal of Cleaner Production* **166**, 1479–1493 (2017).
18. Reid, W. V. *Millennium Ecosystem Assessment* (2005).
19. Díaz, S., et al. The IPBES Conceptual Framework—Connecting Nature and People. *Current Opinion in Environmental Sustainability* **14**, 1–16 (2015).
20. Pascual, U., et al. Valuing Nature's Contributions to People: The IPBES Approach. *Current Opinion in Environmental Sustainability* **26–27**, 7–16 (2017).
21. Milne, M. J. & Gray, R. W (h) Ither Ecology? The Triple Bottom Line, the Global Reporting Initiative, and Corporate Sustainability Reporting. *Journal of Business Ethics* (2013).
22. Halpern, B. S., et al. Achieving the Triple Bottom Line in the Face of Inherent Trade-Offs Among Social Equity, Economic Return, and Conservation. *Proceedings of the National Academy of Sciences* **110**, 6229–6234 (2013).
23. Costanza, R., et al. The Value of the World's Ecosystem Services and Natural Capital. *Nature* **387**, 253–260 (1997).
24. Costanza, R., et al. Changes in the Global Value of Ecosystem Services. *Global Environmental Change* **26**, 152–158 (2014).
25. Deloitte Access Economics. *At What Price? The Economic, Social and Icon Value of the Great Barrier Reef* (2017). https://apo.org.au/node/96031.

26. Stoeckl, N., et al. The Economic Value of Ecosystem Services in the Great Barrier Reef: Our State of Knowledge. *Annals of the New York Academy of Sciences* **1219**, 113–133 (2011).
27. Kotsantonis, S., Pinney, C. & Serafeim, G. ESG Integration in Investment Management: Myths and Realities. *Journal of Applied Corporate Finance* **28**, 10–16 (2016).
28. United Nations Principles for Responsible Investment. About the PRI. (2022).
29. Lashitew, A. A. Corporate uptake of the Sustainable Development Goals: Mere Greenwashing or an Advent of Institutional Change? *Journal of International Business Policy* **4**, 184–200 (2021).
30. Verheyden, T., Eccles, R. G. & Feiner, A. ESG for All? The Impact of ESG Screening on Return, Risk, and Diversification . *Journal of Applied Corporate Finance* **28**, 47–55 (2016).
31. Busch, T. & Schwarzkopf, J. Carbon Management Strategies—A Quest for Corporate Competitiveness. *Progress in Industrial Ecology* **8**, 4–29 (2013).
32. Whelan, T., Atz, U., van Holt, T. & Clark, C. Whelan, T., Atz, U., Van Holt, T. & Clark, C. ESG and Financial Performance. *Uncovering the Relationship by Aggregating Evidence from 1000 Plus Studies Published Between 2015–2020*. 19 (2021).
33. Consolandi, C., Eccles, R. G. & Gabbi, G. How Material Is a Material Issue? Stock Returns and the Financial Relevance and Financial Intensity of ESG Materiality. *Journal of Sustainable Finance & Investment*, 1–24 (2020).
34. Giese, G., Lee, L.-E., Melas, D., Nagy, Z. & Nishikawa, L. Foundations of ESG Investing: How ESG Affects Equity Valuation, Risk, and Performance. *The Journal of Portfolio Management* **45**, 69–83 (2019).
35. van Heijningen, K. The Impact of ESG Factor Materiality on Stock Performance of Firms. (2019).
36. Hermundsdottir, F. & Aspelund, A. Sustainability Innovations and Firm Competitiveness: A Review. *Journal of Cleaner Production* **280**, 124715 (2021).
37. Adomako, S. & Tran, M. D. Sustainable Environmental Strategy, Firm Competitiveness, and Financial Performance: Evidence from the Mining Industry. *Resources Policy* **75**, 102515 (2022).
38. Wagner, M. & Schaltegger, S. The Effect of Corporate Environmental Strategy Choice and Environmental Performance on Competitiveness and Economic Performance: An Empirical Study of EU Manufacturing. *European Management Journal* **22**, 557–572 (2004).
39. Taliento, M., Favino, C. & Netti, A. Impact of Environmental, Social, and Governance Information on Economic Performance: Evidence of a Corporate 'Sustainability Advantage' from Europe. *Sustainability* **11**, 1738 (2019).
40. Li, D. & Liu, J. Dynamic Capabilities, Environmental Dynamism, and Competitive Advantage: Evidence from China. *Journal of Business Research* **67**, 2793–2799 (2014).
41. Do, B. & Nguyen, N. The Links between Proactive Environmental Strategy, Competitive Advantages and Firm Performance: An Empirical Study in Vietnam. *Sustainability* **12**, 4962 (2020).
42. Pan, C., Guo, H., Jiang, Y., Wang, H. & Qi, W. The Double Effects of Female Executives' Participation on Corporate Sustainable Competitive Advantage Through Unethical Environmental Behavior and Proactive Environmental Strategy. *Business Strategy and the Environment* **29**, 2324–2337 (2020).
43. Schramade, W. Integrating ESG into Valuation Models and Investment Decisions: The Value-Driver Adjustment Approach. *Journal of Sustainable Finance & Investment* **6**, 95–111 (2016).
44. Dorfleitner, G., Utz, S. & Wimmer, M. Patience Pays Off—Corporate Social Responsibility and Long-Term Stock Returns. *Journal of Sustainable Finance & Investment* **8**, 132–157 (2018).
45. Wu, J., Lodorfos, G., Dean, A. & Gioulmpaxiotis, G. The Market Performance of Socially Responsible Investment during Periods of the Economic Cycle—Illustrated Using the Case of FTSE. *Managerial and Decision Economics* **38**, 238–251 (2017).
46. Das, N., Chatterje, S., Ruf, B. & Sunder, A. ESG Ratings and the Performance of Socially Responsible Mutual Funds: A Panel Study. *Journal of Finance Issues* **17**, 49–57 (2018).
47. Fernández, M. S., Abu-Alkheil, A. & Khartabiel, G. M. Do German Green Mutual Funds Perform Better Than Their Peers? *Business and Economics Research Journal* **10**, 297–312 (2019).
48. ASFI. *Analysis of International Taxonomies and Considerations for Australia* (2022). www.asfi.org.au/publications/international-scoping-paper.
49. Papari, C.-A., Toxopeus, H., Polzin, F., Bulkeley, H. & Menguzzo, E. V. Can the EU Taxonomy for Sustainable Activities Help Upscale Investments into Urban Nature-Based Solutions? *Environmental Science & Policy* **151**, 103598 (2024).
50. Neal, T. *The Importance of External Weather Effects in Projecting the Economic Impacts of Climate Change*. SSRN Scholarly Paper (2023). https://papers.ssrn.com/abstract=4471379.

51. Popescu, I.-S., Hitaj, C. & Benetto, E. Measuring the Sustainability of Investment Funds: A Critical Review of Methods and Frameworks in Sustainable Finance. *Journal of Cleaner Production* **314**, 128016 (2021).
52. Rogelj, J. Net Zero Targets in Science and Policy. *Environmental Research Letters* **18**, 021003 (2023).
53. Cointe, B. The AR6 Scenario Explorer and the History of IPCC Scenarios Databases: Evolutions and Challenges for Transparency, Pluralism and Policy-Relevance. *NPJ Climate Action* **3**, 1–10 (2024).
54. Demekas, D. & Grippa, P. Walking a Tightrope: Financial Regulation, Climate Change, and the Transition to a Low-Carbon Economy. *Journal of Financial Regulation* **8**, 203–229 (2022).
55. Flammer, C. Corporate Social Responsibility and Shareholder Reaction: The Environmental Awareness of Investors. *AMJ* **56**, 758–781 (2012).

14

FRUGAL INNOVATION AND SUSTAINABLE DEVELOPMENT AT THE BOTTOM OF THE PYRAMID

Bibhuti Ranjan Bhattacharjya and Hiranya K. Nath

14.1 Introduction

By generating cost-effective solutions, frugal innovation addresses the unmet needs of millions with low incomes, the people at the bottom of the (income or wealth) pyramid. This approach efficiently reduces resource utilization, encompassing raw materials, energy, and financial resources, making technology more affordable and accessible while easing ecological strain. Consequently, frugal innovation is closely tied to economic, social, and ecological sustainability. This chapter conducts a thorough review of existing literature, exploring the interplay between frugal innovation and sustainability.

Given the recent incorporation of "frugal innovation" into academic research, the literature primarily focuses on conceptualizing the term and examining its impacts on the economy, society, and environment. Originating in discussions about developing countries, frugal innovation aligns with the macroeconomic goal of sustainable development, as enshrined in the United Nations' 17 Sustainable Development Goals (SDGs). Despite its potential to contribute significantly to SDGs and sustainable development at the bottom of the pyramid, mainstream economics researchers and journals have not fully explored frugal innovation, with existing studies mostly concentrated in the fields of business management, technology, engineering, ecology, and sustainability.

This chapter aims to fill this gap by reviewing the current understanding of frugal innovation and its relationship with sustainability. It includes an analysis of existing studies across various fields and examines five cases of frugal innovation that have not been previously studied. *Frugal innovation* is characterized by substantial cost reduction, a focus on core functionalities, and optimized performance, while *sustainability* is defined as development meeting present needs without compromising future generations' ability to meet their own needs and framed by the triple bottom line (TBL) and the SDGs.

The literature presents overwhelming evidence supporting a positive relationship between frugal innovation and sustainability, with limited cases showing a negative impact. Frugal innovation is found to contribute to multiple SDGs, with SDGs 8 (Decent Work and Economic Growth), 3 (Good Health and Well-Being), 10 (Reduced Inequalities), 12 (Responsible Consumption and Production), and 9 (Industry, Innovation, and Infrastructure) being the most affected. This conclusion is further supported by the cases studied in this chapter.

DOI: 10.4324/9781003285472-15

The subsequent sections are organized to provide a comprehensive exploration of the subject. The next segment defines and discusses two foundational concepts—frugal innovation and sustainability. Following that, the next section focuses on the relationship between frugal innovation and sustainability. The subsequent section examines the association between frugal innovation and specific SDGs, drawing evidence from existing literature case studies and analyzing five cases of frugal innovations from India. The conclusion offers summarizing remarks and discusses potential avenues for future research.

14.2 Basic Concepts

This chapter focuses on examining the interconnection between two fundamental concepts: frugal innovation and sustainability, with the former being a relatively recent addition. Within this section, we engage in a detailed discussion of these concepts, emphasizing areas where our understanding has not yet reached a solid consensus.

14.2.1 Frugal Innovation

The term "frugal innovation" reportedly caught the attention of academic researchers following an article in *The Economist* magazine in 2010 (Wooldridge, 2010) discussing lessons from a frugal innovation in India's health care.[1] Since then, an expanding body of academic literature has employed this term to describe innovations that are distinct from traditional ones. In essence, *frugal innovation* refers to solutions that significantly reduce costs, prioritize core functionalities, and optimize performance.[2] These innovations stem from a desire to meet the unmet demands of individuals with limited means, often referred to as the "bottom of the pyramid" in developing societies.

Further elaboration is necessary to understand the three defining characteristics of frugal innovation mentioned earlier. The substantial cost reduction may arise from various sources, often linked to the availability and affordability of resources. Scarcity or unavailability of necessary productive resources may lead to seeking less-expensive and locally available alternatives. Even if resources are available, the lack of affordability for many consumers may drive innovative solutions using less costly and locally available resources. However, existing literature examines this substantial cost reduction criterion from the customer's perspective, emphasizing affordability, covering both the purchase price and the cost of ownership. While consensus on the specific percentage for substantial cost reduction is lacking, Weyrauch and Herstatt (2020) propose a reduction of one-third or more in the purchase price or ownership cost compared to currently available solutions in the local market or assumed costs of, for example, importing the solution.

A strategy to reduce costs in designing a solution is focusing on one or two core functionalities that address essential needs. Frugal innovations, therefore, adopt a minimalist approach, creating user-friendly and less-complex solutions. However, this minimalism does not imply a compromise in quality; rather, it signifies generating maximum customer benefits using minimal resources. The focus on core functionality(ies) may be driven not only by cost reduction but also by considerations, such as ease of use (Andel, 2013; Wooldridge, 2010), resource conservation (Barclay, 2014; Rao, 2014; Tiwari & Herstatt, 2012), minimizing environmental impact (Basu et al., 2013; Jancke, 2014; Sharma and Iyer, 2012), or enabling a specific lifestyle or consumer behavior (Flatters and Willmott, 2009).

The criterion of optimized performance extends beyond technical characteristics; it involves contextualizing performance concerning its actual purpose and local conditions. The performance level must ideally align with the intended purpose and the specific requirements of the environment where the frugal innovation will be utilized.

While many definitions align with the preceding explanation, there are some notable differences identified in various studies. Some scholars take a product-centric view (Winterhalter et al., 2017), others conceptualize it as a process (Knizkov and Arlinghaus, 2021), and some define it as a business strategy or a combination of unique techniques, goods, and services (Bound and Thornton, 2012; Hossain, 2021; Hossain et al., 2021). Frugal innovations, whether by introducing a novel product architecture (Rao, 2013; Lim and Fujimoto, 2019) or tapping into a new market (Zeschky et al., 2011), can at times even disrupt established mainstream solutions (Rao, 2013). Despite the diverse approaches to frugal innovation, there is a consensus on its ability to offer quality and cost-effective solutions to address the needs of underserved communities in resource-constrained settings.

Following Leliveld and Knorringa (2018), we may categorize frugal innovations into three groups:

1. *Top-down frugal innovations.* Developed by large firms or multinational corporations (e.g., General Electric's health-care products, Tata Nano by Tata Motors, ChotuKool by Godrej), these market-centric innovations aim to expand business by offering simple, affordable, and functional products. They target resource-constrained households, particularly low-income customers.
2. *Social enterprise innovations.* Developed and implemented by social enterprises to address societal challenges faced by marginalized groups (e.g., Grameen Shakti, SELCO, Husk Power), these organizations pursue frugal innovations to create both social and economic value (Ananthram et al., 2022).
3. *Community-led frugal innovations.* Resulting from individual or collaborative efforts, these innovations involve local-level organizations and community members (Gupta, 2012). Motivated by not-for-profit intentions, they stem from a deep understanding of the local community's needs (Bhaduri and Talat, 2020) and address complex societal challenges noted at the bottom of the pyramid (Sharma and Kumar, 2019).[3]

14.2.2 Sustainability

The concept of sustainability has deep historical roots, with its initial mention attributed to Hans-Carl-von-Carlowitz-Gesellschaft in 1713, focusing on restricting forest harvest to the rate of new forest growth (Albert, 2022). Over the years, the definition and application of *sustainability* have evolved across various fields. Notably, sustainable development gained prominence in the past four decades, as defined by the World Commission on Environment and Development (1987) as development meeting present needs without compromising future generations' ability to meet their own needs.

Operationalizing this concept, the triple bottom line (TBL or 3BL) framework emerged around the same period, emphasizing economic, social, and environmental/ecological dimensions as the three pillars of sustainability.[4] Enshrined in the United Nations' 17 Sustainable Development Goals (2015), sustainable development spans diverse themes, like equity, justice, responsibility, regeneration, resilience, long-term maintenance of systems, and response to global challenges.

In the realm of empirical research, various approaches are employed to quantify sustainability. Some researchers utilize specific indicators, while others align their measures with the three pillars of sustainability (Levanen et al., 2015; Pansera and Sarkar, 2016; Rosca et al., 2017; Wohlfart et al., 2016). Additionally, some exclusively concentrate on social indicators (e.g., Khan, 2016). Furthermore, another group of researchers associates their indicators with the Sustainable Development Goals (Hyvarinen et al., 2016; Levanen et al., 2015; Pansera and Sarkar, 2016). This diversity in methodological approaches reflects the multifaceted nature of sustainability assessment in empirical studies.

The defined dimensions of sustainability and the TBL framework underpin the 17 Sustainable Development Goals adopted in 2015 by the United Nations. This agenda serves as a global road map to end poverty, protect the planet, and address inequalities, offering a comprehensive approach to tackle violent conflict, human rights abuses, climate change, and environmental degradation. The SDGs emphasize the integration of economic growth, social well-being, and environmental protection, weaving these dimensions into each goal. The responsibility for achieving these goals extends to all stakeholders, including governments, the private sector, civil society, the United Nations system, and others, emphasizing a collaborative partnership to ensure that no one is left behind in the pursuit of sustainable development.

14.3 Frugal Innovation and Sustainability

The prevailing consensus in the literature leans toward a positive correlation between frugal innovation and sustainability. Several authors propose that frugal innovations inherently contribute to sustainability, establishing a fundamentally positive relationship. However, to gain a more nuanced understanding of this relationship, it is crucial to delve into the connection of frugal innovation with the three distinct dimensions of sustainability. It is noteworthy that a single frugal innovation may intricately relate to more than one of these dimensions, emphasizing the complexity and interconnectedness inherent in the relationship between frugal innovation and sustainability.

14.3.1 Frugal Innovation and Social Sustainability

With a foundational drive to cater to underserved populations, frugal innovation emerges as a force that champions social equity and justice. It actively seeks to empower and include those at the bottom of the population pyramid, thereby fostering inclusivity.[5] The impact of frugal innovation extends to various dimensions, contributing to the cessation of poverty through the creation of employment opportunities, particularly for local communities and women, while also nurturing entrepreneurship.

Beyond economic implications, frugal innovation plays a crucial role in preserving the social fabric of communities. It establishes a resilient framework by fostering endogenous and, consequently, autonomous solutions. Furthermore, it promotes more horizontal and democratic mechanisms for managing technology and delivering products and services. In developed countries, frugal innovation can address challenges posed by rapidly aging societies.

The cost-saving opportunities created by frugal innovation directly benefit the bottom of the pyramid (BOP), reducing expenditures for low-income individuals. For instance, providing affordable electricity is a tangible way frugal innovation positively impacts the financial burden of those at the BOP. By offering basic, life-improving, and essential services and products in sectors such as food, health, information and communication technology, water, and transportation at remarkably low prices without compromising quality, frugal innovations elevate standards of living. This improvement extends to living conditions, health, access to information, global knowledge, knowledge networks, education, education materials, and training for the BOP.

The affordability and accessibility of frugal innovation, whether directly or indirectly, generate positive outcomes that underscore the connection between social sustainability and frugal innovation. In essence, frugal innovation catalyzes positive social change, enhancing the overall well-being and opportunities for those often marginalized and economically disadvantaged.

14.3.2 Frugal Innovation and Economic Sustainability

Economic systems founded on market principles are inherently reliant on innovation for their sustainability. Therefore, a natural and positive relationship emerges between frugal innovation and economic sustainability. This connection is particularly pronounced as market expansion unfolds predominantly in the developing world, where a significant portion of the population faces financial constraints.[6] In this context, frugal innovation becomes imperative for sustaining economic growth and ensuring profitability for firms of various sizes and organizational structures.

Frugal innovation serves as a catalyst in enhancing the competitive advantage of a firm, positioning it strategically to weather future crises. The impact of frugal innovation on market performance is multifaceted, as it offers higher value to customers, augments a company's competitiveness, yields positive cost effects, and boosts sales revenues. Beyond its immediate economic implications, frugal innovation also serves as a potential entry point for the private sector to actively contribute to promoting sustainable development, thereby elevating a company's sustainability performance.

Moreover, frugal innovation is instrumental in driving higher productivity and reducing the cost per unit across various facets of procurement, production, and distribution. By optimizing these operational aspects, frugal innovation not only contributes to economic sustainability but also positions businesses to navigate challenges efficiently in the ever-evolving market landscape. In essence, the positive relationship between frugal innovation and economic sustainability underscores the vital role innovative strategies play in fostering resilience, competitiveness, and long-term viability within dynamic economic frameworks.

14.3.3 Frugal Innovation and Ecological Sustainability

Frugal innovation plays a pivotal role in advancing environmental sustainability through judicious resource management across various dimensions, including raw materials, production resources, energy, fuel, water, and waste.[7] This approach, characterized by prudent resource utilization, positions frugal innovation as a significant contributor to sustainable environmental practices. By addressing environmental constraints and incorporating ecological considerations as inherent attributes, frugal innovation emerges as an exemplary model for developing green products.

Numerous studies (e.g., Molina-Maturano et al., 2020; Pedroso et al., 2023; Shahid et al., 2023) affirm the environmental benefits of frugal innovation, characterizing such innovations as more climate-neutral and emphasizing their role in maintaining the sustainability of the planet's climate. Frugal innovation is seen as a proactive measure to protect the planet's capacity, mitigate the impacts of global crises, preserve the environmental integrity of communities, and align with green marketing objectives. Furthermore, it is instrumental in enhancing green supply chain initiatives, reducing negative external effects, and improving input resources, value activities, and overall sustainability management within companies.

While the focus of frugal entrepreneurs on economic and social sustainability is evident, some studies, such as those by Rosca et al. (2017), contend that the ecological impact of frugal innovation might be perceived as a spillover effect or a side effect. Nevertheless, the positive ecological impact of frugal innovation arises from resource conservation, achieved through economic, frugal, and limited resource use, as well as the utilization of renewable and local resources.

Frugal innovation exemplifies value creation from waste, embraces the reuse of existing components and materials, and prioritizes recycling over sourcing. The modularity and ease of repair associated with frugal innovations extend their effective life, contributing to resource conservation. Additionally, by economizing on energy and material usage, frugal innovations often result in lower emissions and a reduced carbon footprint compared to conventional solutions.

Various authors highlight that frugal innovation has a low impact on the environment, minimizes ecological footprints, and generates fewer emissions. Consequently, frugal innovations emerge as a positive force in advancing ecological sustainability, embodying an environmentally conscious approach to innovation that resonates with contemporary imperatives for resource conservation and sustainable practices.

14.3.4 Negative Relationship Between Frugal Innovation and Sustainability

Several studies assert and provide evidence suggesting a negative or non-existent relationship between frugal innovation and sustainability. Weyrauch and Herstatt (2016) contend that there is no conceptual connection between frugal innovation and sustainability. Levanen et al. (2015) also express concerns about equating frugality and sustainability at the conceptual level. Empirically, several authors demonstrate that frugal innovation does not inherently yield sustainability impacts. While the impact of frugal innovation on economic and social sustainability lacks clear articulation for many, its ecological impact is often perceived as a spillover effect, as noted by Rosca et al. (2017). In numerous instances, environmental sustainability does not emerge as a primary motivator for frugal innovation.

Documented evidence points out that many frugal innovations involve ecologically and socially unsustainable practices, including the extraction of raw materials in environmentally constrained settings and unsustainable, low-cost manufacturing with associated waste generation. In such cases, the exclusive focus on resource availability and affordability may divert attention from sustainability effects. Concerns also arise regarding the consumption of frugal innovation, with the rebound effect playing a role; the increased availability and affordability of frugal innovations may lead to heightened material use, energy consumption, and waste generation, potentially exacerbating environmental damage.

This scenario illustrates a trade-off between a potential decline in ecological sustainability and improved social sustainability, driven by greater inclusiveness, enhanced well-being, better livelihoods, improved health, and poverty reduction. Additionally, even frugal innovations designed with ecological sustainability in mind can contribute to environmental issues during the end-of-life stage, particularly through improper disposal practices (Albert et al., 2017).

At the firm level, frugal innovation poses potential sustainability challenges. For instance, the introduction of cheaper frugal innovations may cannibalize sales of the company's existing high-end offerings, leading to a reduction in overall sales volume. Moreover, if a frugal innovation fails to match the quality of the company's established regular offerings, it can negatively impact the company's brand. Thus, the relationship between frugal innovation and sustainability is complex, with potential ecological, social, and economic trade-offs that require careful consideration.

14.4 Frugal Innovation and SDGs

The preceding discussion highlights the potential of frugal innovations to contribute to various dimensions of sustainability concurrently. Consequently, it becomes apparent why these innovations can play a role in advancing multiple SDGs. In this section, we initially explore the existing literature's perspectives on the connection between frugal innovation and the SDGs. Subsequently, we delve into less-explored cases that have not received considerable attention from academic researchers, analyzing the potential impact these cases might have on specific SDGs.

14.4.1 Findings of the Existing Studies

Several studies employ the Sustainable Development Goals (SDGs) and their corresponding indicators as a framework to assess the sustainability impact of frugal innovations. Albert (2022) provides a comprehensive summary and review of prominent studies, encompassing a total of 70 cases, of frugal innovation. Table 14.1 offers an overview, consolidating the information presented in Table 5 of Albert's study. According to this table, the top five SDGs most influenced by frugal innovations are 8, 3, 10, 12, and 9, in that order.

Thus, as the table shows, most frugal innovations contribute to the SDG of promoting sustained, inclusive, and sustainable economic growth, full and productive employment, and decent work for all (SDG 8). As we have already discussed, by offering cost-effective solutions, frugal innovation makes consumption and production more affordable and therefore inclusive. Like most innovations, frugal innovation increases the productivity of workers, making the job more rewarding and effective. Next in importance is the SDG of ensuring healthy lives and promoting well-being for all at all ages (SDG 3). Many frugal innovations (e.g., portable

Table 14.1 Summary of contributions of 70 cases of frugal innovation to the SDGs

Variables	*Sustainable Development Goals*																
	1	*2*	*3*	*4*	*5*	*6*	*7*	*8*	*9*	*10*	*11*	*12*	*13*	*14*	*15*	*16*	*17*
Number of cases allocated to different SDGs	18	12	48	10	4	11	14	58	28	42	21	29	15	0	6	9	8
Percentages of total allocations (out of 334)	5.4	3.6	14.4	3.0	1.2	3.3	4.2	17.4	8.4	12.6	6.3	8.7	4.8	0	1.8	2.7	2.4
Percentages of total distinct cases (out of 70)	25.7	17.1	68.6	14.3	5.7	15.7	20.0	82.9	40.0	60.0	30.0	41.4	21.4	0	8.6	12.9	14.3

Source: Adapted from Table 5 of Albert (2022).

ECG machine, telemedicine) have improved access to health care, while others have enhanced well-being by reducing environmental pollution. By providing access to low-cost goods, services, and technology, frugal innovations help achieve the SDG of reducing inequality within and among countries (SDG 10). Further, by reducing waste and economizing on the use of resources, most frugal innovations contribute to the SDG of ensuring sustainable consumption and production patterns (SDG 12). Finally, many frugal innovations contribute to the SDG of building resilient infrastructure, promoting inclusive and sustainable industrialization, and fostering innovation (SDG 9) by building cost-effective, eco-friendly infrastructure, developing low-cost machines and equipment, and promoting entrepreneurship. The impact of frugal innovations on other SDGs, as explored in the literature review, is limited. Notably, none of the frugal innovation cases examined in the surveyed studies directly contributes to SDG 14, which focuses on conserving and sustainably using the oceans, seas, and marine resources for sustainable development.

14.4.2 Findings From Studies on Unexplored Cases

Within this subsection, we analyze five cases of frugal innovation originating from India. The pertinent details about cases 1, 2, and 3 are sourced from the National Innovation Foundation—India portal (https://innovation.nif.org.in/). The data for case 4 is derived from a blog article published by TRCFS (Ray and Bhattacharjya, 2021). The case of pedal-operated pottery *Chaak* (case 5) is reported by Bhattacharjya et al. (2023). These innovations share three key characteristics: significantly reduced cost, emphasis on core functionalities, and optimized performance. A concise description of each case is provided, accompanied by a brief discussion on the SDG(s) it influences.

14.4.2.1 Case 1: Chanekee, Extra Weft Insertion Device

Weaving is crucial to the rural economy of the northeastern state of Assam in India, particularly in producing unique fabrics with high demand at home and abroad. However, the traditional method of manually inserting weft threads into jacquard looms is slow and intricate, posing challenges for younger weavers and resulting in wastage of thread. To address these issues, an innovator designed an accessory tool to streamline the weft insertion process in jacquard looms, making it more efficient and automated. Tested in the Sualkuchi weaving cluster, this tool has garnered recognition, with several national and international awards.

Key features of the tool include thus:

1. Reducing fabric design time by one-third compared to traditional methods
2. Automating warp thread selection and lifting for intricate designs
3. Decreasing weaver drudgery and enhancing productivity by 60%
4. Enabling unskilled weavers to create designed fabric in jacquard looms
5. Ensuring uniform fabric quality, boosting product value

The innovation holds the potential to generate job opportunities in rural areas, attract the younger generation to weaving, and curb forced migration for better economic prospects. By improving product quality and reducing drudgery, it aligns with SDG 8 (Decent Work and Economic Growth) and SDG 3 (Good Health and Well-Being) in Assam, India.

14.4.2.2 Case 2: Rural Egg Incubator

In rural India, numerous young individuals engage in poultry farming, but they often face financial constraints and lack the infrastructure to invest in commercially available egg incubators. While these incubators are effective, they arere not cost-effective for small and medium-sized poultry operations. Moreover, the unreliable electricity supply in many rural areas hinders the adoption of technology-dependent egg incubators by smaller firms. Diesel-powered alternatives, while available, are expensive to run and raise environmental concerns. As a result, small and medium-sized poultry enterprises often opt to buy young chicks from larger firms. However, producing chicks in-house could potentially offer better economic returns for these farmers.

To address these challenges, an innovator in Assam has developed a cost-effective egg incubator. This innovator utilized locally available materials, such as plywood and plain sheets, to create a device with two compartments. The upper compartment houses the egg tray and a humidity source, while the lower compartments are equipped with thermal sources. The device includes a speaker for alarms in case of sudden temperature and humidity changes, as well as control switches. Notably, the device can use a kerosene lamp as a heat source, offering a solution for areas with inadequate electricity supply. It incorporates a regulator from an electric fan to adjust light intensity and features a provision for water heating and evaporation, maintaining the required warmth in the chamber. Air circulation is facilitated through four holes from the bottom to the top cabin.

This innovation is priced at INR 42,000 (approx. USD 500 at current exchange rate) for 500 eggs, significantly more affordable than market alternatives priced at INR 96,000 (approx. USD 1,200) for the same capacity. The use of local materials simplifies the device compared to high-tech alternatives, making it a viable solution for numerous poultry farmers in countries like India. The incorporation of kerosene lamps as a heat source exemplifies frugal innovation, addressing the specific needs of users with limited access to electricity. Consequently, this innovation aligns with the SDGs, particularly SDG 8, promoting decent work and economic growth at the bottom of the pyramid.

14.4.2.3 Case 3: Bamboo Windmill for Irrigation

Although manual irrigation is a laborious and time-intensive task, many farmers in Assam, India, opt for manual methods due to the high costs associated with diesel-powered irrigation systems. Seeking to alleviate the challenges of manual irrigation and enhance efficiency, two innovators, who happen to be brothers, devised a frugal irrigation system using a bamboo-based windmill. This system utilizes natural materials like bamboo, wood, and locally available waste, such as old tire strips and tin sheets. The bamboo windmill, costing INR 4,500 (approx. USD 55), stands in stark contrast to commercially available windmills priced at INR 60,000 (approx. USD 725). The simplicity of the innovation allows for local carpenter units to manufacture it in just four days. Acknowledged nationally, the technology was further tested in Gujarat, a western Indian state, specifically among salt farmers who manually labored for hours with water pumps due to the high initial and running costs of diesel-powered pumps.

Recognizing its utility, the Alston Foundation, in collaboration with the National Innovation Foundation, India, enhanced the windmill's design. This innovation exemplifies how frugal solutions can be scaled up globally, addressing regional needs and challenges. The use of local materials, as opposed to sophisticated engineering materials, not only reduces technology costs but also simplifies manufacturing processes. This approach facilitates local production in carpentry units, overcoming barriers to mass diffusion in remote areas and enabling more accessible maintenance services. Ultimately, the bamboo windmill contributes to enhancing productivity for farmers and promotes sustainable agriculture, aligning with SDGs 1, 2, 3, and 8.

14.4.2.4 Case 4: Mobile Digital Bus

In India, providing quality education in remote areas encounters challenges, such as a shortage of skilled human resources and a lack of essential infrastructure and resources, including computers and learning software. The complexity intensifies for science education and skill-based teachings, which necessitate high-quality laboratories. To tackle this, the SRF Foundation, a CSR (corporate social responsibility) initiative of SRF Limited, has transformed a conventional bus into a classroom/laboratory.[8] The bus is equipped with teaching tools, including audiovisuals and interactive e-learning modules, addressing the specific needs of science education in remote villages with limited infrastructure.

A trained pool of volunteers travels on the bus to deliver lectures and assist rural students in laboratory activities. The bus features internet dongles, interactive panels for video conferencing, and digital learning tools. Additionally, it incorporates the Indian government's digital e-governance tools, providing villagers with access to services that would be challenging due to poor internet connections in villages.

Although launched before the COVID-19 pandemic, the program gained significance during the pandemic-induced school closures. An integrated LCD screen in the bus allowed students to attend classes while maintaining social distancing, a crucial requirement during the pandemic. The bus is equipped with solar power capabilities that enable it to serve areas with limited or irregular electricity supply. This innovative approach serves as a commendable effort to bridge the rural–urban gap in science education in India, aligning with the Sustainable Development Goal of delivering quality education (SDG 4).

14.4.2.5 Case 5: Pedal-Operated Pottery Chaak

In India, traditional pottery artisans use a manually operated Pottery wheel, known as *Chaak*, requiring physical effort to rotate. The non-ergonomic posture of *Chaak* leads to health issues, like back pain and musculoskeletal disorders. Artisans resort to painkillers during peak seasons. The low productivity of manual pottery hampers competitiveness in the market. Despite pottery's vital role in rural employment and local culture, artisans abandon traditional pottery due to drudgery and low returns. Electricity-powered *Chaak* variants exist but are impractical in remote areas with insufficient electricity. The Rural Technology Action Group, a government of India initiative, has redesigned *Chaak*. The new mechanism, akin to a manual sewing machine, replaces hand power with leg power to reduce drudgery. The design includes a seating arrangement for proper posture.

The innovation reduces physical strain on artisans, increasing productivity by 1.7 times compared to the traditional Chaak. This pedal-operated Chaak is specifically designed to be manufactured in steel workshops located in villages, making it easier to produce in remote areas and allowing for convenient doorstep maintenance. This innovation supports decent work and economic growth for artisans (SDG 8) and addresses health concerns, promoting well-being (SDG 3).

Table 14.2 presents a summary of the SDG impacts observed in the sample cases of frugal innovation. We have focused on including only those SDGs that are relevant to our cases. Our analysis reveals that the majority of cases (4 out of 5) contribute to SDG 1 and SDG 8, while three out of

Table 14.2 Case studies fit into the Sustainable Development Goals (1, 2, 3, 4, 8)

Goals	*Goal 1*	*Goal 2*	*Goal 3*	*Goal 4*	*Goal 8*
	End poverty in all its forms everywhere	End hunger, achieve food security and improved nutrition, and promote sustainable agriculture	Ensure healthy lives and promote well-being for all at all ages	Ensure inclusive and equitable quality education and promote lifelong learning opportunities for all	Promote sustained, inclusive, and sustainable economic growth, full and productive employment, and decent work for all
Cases					
Case 1: Chanekee, extra weft insertion device	Help increase income of artisans	Not applicable	The innovation to reduce health drudgery of artisans	Not applicable	Productivity enhancement, new job creation for artisans
Case 2: Rural egg incubator	Help increase income of farmers	Help increase farmers' productivity and contribute to sustainable agriculture	Not applicable	Not applicable	Productivity enhancement, new job creation for farmers
Case 3: Bamboo windmill for irrigation	Help increase income of farmers	Help increase farmers' productivity and contribute to sustainable agriculture	Reduce health drudgery and improve farmers' health	Not applicable	Decent work for farmers
Case 4: Mobile digital bus	Not applicable	Not applicable	Not applicable	Facilitate quality education delivery in remote villages	Not applicable
Case 5: Pedal-operated pottery *Chaak*	Help increase income of artisans	Not applicable	The innovation to reduce health drudgery	Not applicable	Productivity enhancement, new job creation for artisans

Source: Authors' compilation.

five cases influence SDG 3. SDG 2 is linked to only two cases, and SDG 4 is impacted by just one case. Despite the small sample size of five cases, these findings underscore the predominance of SDGs 8 and 3 in the effects of frugal innovation. Notably, although SDG 1 was not prominently affected in the literature, it is influenced by most of the cases we have examined.

14.5 Conclusions and Future Research Agenda

This chapter examines the concepts of frugal innovation and sustainability, exploring their interplay in the context of macro developmental goals in the 21st century. It conducts an extensive literature review, revealing overwhelming evidence of a positive correlation between frugal innovation and sustainability, with only a few instances of negative impacts noted. Frugal innovation emerges as a significant contributor to various SDGs, particularly SDGs 8 (Decent Work and Economic Growth), 3 (Good Health and Well-Being), 10 (Reduced Inequalities), 12 (Responsible Consumption and Production), and 9 (Industry, Innovation, and Infrastructure), as evidenced by the cases analyzed in the literature. Additionally, this chapter examines five cases of frugal innovation from India to further support these findings. Our case studies suggest that frugal innovations can also have significant impacts on other SDGs. Therefore, frugal innovations are poised to play pivotal roles in advancing SDGs, especially in developing countries with large populations facing low incomes and limited resources. Overall, frugal innovations also have the potential to generate global benefits by alleviating environmental pressures.

Two critical areas warrant immediate further research. First, there is a need for a formal theoretical examination of frugal innovation and sustainability. This necessitates researchers to establish internally consistent definitions for both concepts. Fortunately, recent literature has already taken steps in that direction. Second, informed by advancements in the first area, there is a need to develop an empirical framework for mapping frugal innovations—often micro-level phenomena to macro-level indicators of sustainable development. While the Inter-Agency and Expert Group on SDG Indicators (IAEG-SDGs) has crafted a global indicator framework for the SDGs, it is vital to create corresponding micro-level indicators to assess the sustainability impacts of frugal innovations. Our forthcoming research endeavors will focus on addressing these two areas.

Notes

1 In a comprehensive review of the historical literature, Jain and Bhaduri (2021) trace the use and appearance of the concept of frugality back several centuries.

2 Table 2 of Hossain (2018) presents some representative definitions of frugal innovation from the literature. At a philosophical level, frugal innovation is interpreted as a mindset or a way of life, as a process, and as an outcome in the form of products or services. (See Soni and Krishnan, 2014; Basu et al., 2013).

3 Hossain (2020) investigates how people at the grassroots level manage to turn their frugal innovations into commercial successes by examining the process of conceptualization, development, and diffusion.

4 This framework was first proposed by Spreckley (1981) and later fully developed by Elkington (1997).

5 Arvind Eye Care is an example of frugal innovation in the health-care sector, where the primary objective is social inclusion. Similarly, Jaipur Foot has been contributing to social justice by allowing people to participate in society through artificial limbs at an affordable cost. Both innovations have created a significant social infrastructure so that needy people can receive health-care services at an affordable price. Another example of frugal innovation that promotes social sustainability is the hygienic sanitary pad promoted by Arunachalam Muruganantham. It supports social sustainability among women in marginalized communities.

6 For example, M-pesa in Africa has created an ecosystem that enables transactions (deposit, withdrawal, and money transfer) using limited infrastructure, resources, and skills. Thus, it promotes sustainable economic growth at the BOP.

7 For example, organizations like Grameen Shakti Biogas and SELCO India have been contributing to ecological sustainability by supplying frugal energy solutions to needy people at the grassroots.
8 SRF Limited is an Indian multi-business chemicals conglomerate engaged in the manufacturing of industrial and specialty intermediates.

References

Albert, M., 2019. Sustainable frugal innovation—The connection between frugal innovation and sustainability. *Journal of Cleaner Production* 237, 117747.

Albert, M., 2022. Assessing the sustainability impacts of frugal innovation–A literature review. *Journal of Cleaner Production* 365, 132754.

Albert, M., Breler, J., Hüsig, S., 2017. Expansive learning through contradictions of sustainability. In: Arevalo, J.A., Mitchell, S.F. (Eds.), *Handbook of Sustainability in Management Education—in Search of a Multidisciplinary, Innovative and Integrated Approach*. Edward Elgar Publishing.

Ananthram, S., Luo, Y., Peng, M.W., 2022. Social entrepreneurship and frugal innovation: A composition-based view. *Journal of Social Entrepreneurship*, published online: June 29, 2022.

Andel, T., 2013. Frugal price: Virtue or vice? *Material Handling & Logistics* 68(11), 4.

Barclay, C., 2014. Using frugal innovations to support cybercrime legislations in small developing states: Introducing the cyber-legislation development and implementation process model (CyberLeg-DPM). *Information Technology for Development* 20(2), 165–195.

Basu, R.R., Banerjee, P.M., Sweeny, E.G., 2013. Frugal innovation: Core competencies to address global sustainability. *Journal of Management for Global Sustainability* 2, 63e82.

Bhaduri, S., Talat, N., 2020. RRI beyond its comfort zone: Initiating a dialogue with frugal innovation by the vulnerable. *Science, Technology and Society* 25(2), 273–290.

Bhattacharjya, B.R., Bhaduri, S., Kakoty, S.K., 2023. Co-creating community-led frugal innovation: An adapted Quadruple Helix? *Technovation* 124, 102752.

Bound, K., Thornton, I., 2012. Our frugal future: Lessons from India's innovation system. *Project Report*. London: Nesta.

Elkington, J., 1997. *Cannibals with Forks: The Triple Bottom line of 21st Century Business*. Oxford: Capstone.

Flatters, P., Willmott, M., 2009. Understanding the post-recession consumer. *Harvard Business Review* 87(7–8), 106–112.

Gupta, A.K., 2012. Innovations for the poor by the poor. *International Journal of Technological Learning, Innovation and Development* 5 (1/2), 28–39.

Hossain, M., 2018. Frugal innovation: A review and research agenda. *Journal of Cleaner Production* 182, 926–936.

Hossain, M., 2020. Frugal innovation: Conception, development, diffusion, and outcome. *Journal of Cleaner Production* 262, 121456.

Hossain, M., 2021. Frugal innovation and sustainable business models. *Technology in Society* 64, 101508.

Hossain, M., Levanen, J., Wierenga, M., 2021. Pursuing frugal innovation for sustainability at the grassroots level. *Management and Organization Review* 172, 374–381.

Hyvarinen, A., Keskinen, M., Varis, O., 2016. Potential and pitfalls of frugal innovation in the water sector: Insights from Tanzania to global value chains. *Sustainability 8(9)*, 888.

Hyvarinen, A.M.J., Keskinen, M., Levänen, J., 2020. Innovation process and uncertainties in resource-constrained environments: A case from the water service sector in East Africa. *Environmental Science and Policy* 114, 242–252.

Jain, S., Bhaduri, S., 2021. The many facets of frugality: Insights from a quasi-comprehensive literature survey. *Journal of Scientometric Research* 10(2), 265–278.

Janicke, M., 2014. Frugale technik. *Okologisches Wirtschaften-Fachzeitschrift* 29(1), 30–36.

Khan, R., 2016. How frugal innovation promotes social sustainability. *Sustainability* 8(10), 1–39.

Knizkov, S., Arlinghaus, J.C., 2021. Frugal processes: An empirical investigation into the operations of resource-constrained firms. *IEEE Transactions on Engineering Management* 68(3), 667–684.

Leliveld, A., Knorringa, P., 2018. Introduction: Frugal innovation and development research. *European Journal of Development Research* 30(1), 1–16.

Levanen, J., Hossain, M., Lyytinen, T., Hyvarinen, A., Numminen, S., Halme, M., 2015. Implications of frugal innovations on sustainable development: Evaluating water and energy innovations. *Sustainability* 8(4), 1–17.

Lim, C., Fujimoto, T., 2019. Frugal innovation and design changes expanding the cost-performance frontier: A Schumpeterian approach. *Research Policy* 48(4), 1016–1029.

Molina-Maturano, J., Bucher, J., Speelman, S., 2020. Understanding and evaluating the sustainability of frugal water innovations in Mexico: An exploratory case study. *Journal of Cleaner Production* 274, 122692.

Pansera, M., Sarkar, S., 2016. Crafting sustainable development solutions e frugal innovations of grassroots entrepreneurs. *Sustainability* 8(51), 1–25.

Pedroso, I.D., Soares, M.A.F., Dutra, A.R.A., Leal Vieira Cubas, A., Osorio de Andrade Guerra, J.B.S., Brem, A., 2023. Frugal innovation development for sustainability: The case of extractivism of the "Butia catarinensis" in Brazil. *Journal of Cleaner Production* 412, 137318.

Rao, B.C., 2013. How disruptive is frugal? *Technology in Society* 35(1), 65–73.

Rao, B.C., 2014. Alleviating poverty in the twenty-first century through frugal innovations. *Challenge* 57(3), 40–59.

Ray, R.S., Bhattacharjya, B.R., 2021. *Educational Initiatives of SRF Foundation: Implications for Frugal Innovation*. Transdisciplinary Research Cluster on Frugality Studies (TRCFS) Blog Post (November 30). Available at: https://trcfrugal.in/blogpost/sample-blog-post-one/

Rosca, E., Arnold, M., Bendul, J.C., 2017. Business models for sustainable innovation—an empirical analysis of frugal products and services. *Journal of Cleaner Production* 162, 133–145.

Shahid, M.S., Hossain, M., Shahid, S., Anwar, T., 2023. Frugal innovation as a source of sustainable entrepreneurship to tackle social and environmental challenges. *Journal of Cleaner Production* 406, 137050.

Sharma, A., Iyer, G.R., 2012. Resource-constrained product development: Implications for green marketing and green supply chains. *Industrial Marketing Management* 41(4), 599–608.

Sharma, G., Kumar, H., 2019. Commercialising innovations from the informal economy: The grassroots innovation ecosystem in India. *South Asian Journal of Business Studies* 8(1), 40–61.

Soni, P.T., Krishnan, R., 2014. Frugal innovation: Aligning theory, practice, and public policy. *Journal of Indian Business Research* 6(1), 29–47.

Spreckley, F., 1981. *Social Audit: A Management Tool for Co-operative Working*. Beechwood College.

Tiwari, R., Herstatt, C., 2012. *India—A Lead Market for Frugal Innovations? Extending the Lead Market Theory to Emerging Economies*. Working Paper. Technologie- und Innovations Management, Technische Universitat Hamburg-Harburg. No. 67.

United Nations General Assembly, 2015. *A/RES/70/1—Transforming Our World: The 2030 Agenda for Sustainable Development*. Available online at: https://sdgs.un.org/2030agenda.

Weyrauch, T., Herstatt, C., 2016. What is frugal innovation? Three defining criteria. *Journal of Frugal Innovation* 2 (1), 1–17.

Weyrauch, T., Herstatt, C., 2020. What is frugal innovation? In McMurray, A.J., de Waal, G.A. (Eds.), *Frugal Innovation: A Global Research Companion*. London and New York: Routledge.

Winterhalter, S., Zeschky, M.B., Neumann, L., Gassmann, O., 2017. Business models for frugal innovation in emerging markets: The case of the medical device and laboratory equipment industry. *Technovation* 66–67, 3–13.

Wohlfart, L., Bünger, M., Lang-Koetz, C., Wagner, F., 2016. Corporate and grassroot frugal innovation: A comparison of top-down and bottom-up strategies. *Technology Innovation Management Review* 6(4), 6–17.

Wooldridge, A., 2010. First break all the rules: The charms of frugal innovation. *The Economist*, April 17, 3–5.

World Commission on Environment and Development, 1987. *Our Common Future: Report of the World Commission on Environment and Development*. Oxford: Oxford University Press.

Zeschky, M., Widenmayer, B., Gassmann, O., 2011. Frugal innovation in emerging markets. *Research Technology Management* 54(4), 38–45.

15

BORDER CARBON ADJUSTMENTS, THE DEVELOPING COUNTRIES, AND THE SDGS

Partha Sen

15.1 Introduction

Climate change is an existential crisis for life on Earth. It has been caused by anthropogenic factors. The scientific evidence on this is fairly unambiguous. The main culprit is the burning of fossil fuels. This releases carbon into the atmosphere. Over time, the stock of carbon builds up there. World emissions have risen unrelentingly since the industrialization process began. And the problem is further exacerbated by the fact that, over time, the quality of carbon sinks (such as forests and oceans) is compromised.

The IPCC Climate Reports that form a background to the UN Sustainable Development Goals (SDGs) on climate change (especially goal 13) state the severity of the problem thus:

> The global temperature has already risen 1.1°C above the pre-industrial level, with glaciers melting and the sea level rising. Impacts of climate change also includes flooding and drought, displacing millions of people, sinking them into poverty and hunger, denying them access to basic services, such as health and education, expanding inequalities, stifling economic growth and even causing conflict.

In this chapter, we focus on action to combat climate change and its impacts (SDG 13). First, we discuss the climate change problem as it stands today; we then go on to evaluate a specific proposal put forward recently by the European Union (EU) to fight the problem—the so-called border carbon adjustment mechanism (BCA, although sometime the first two letters are interchanged; it is also sometimes called border tax adjustment). This proposal, if implemented, may reduce emissions (and achieve other goals) but will cause distress in the developing countries and affect SDGs 1 (on ending poverty)[1] and 2 (on abolishing hunger)[2] quite directly.[3],[4]

It has been recognized in the climate change fora[5] that the developing countries are not responsible for the climate problem (since they had not industrialized until recently, if at all). Indeed, without having contributed to the problem, they are at the receiving end of it. Since it is the stock of carbon (and other greenhouse gases) in the atmosphere that causes climate change, the primary blame for causing the problem lies with the advanced economies. It is their "dirty" industrialization

DOI: 10.4324/9781003285472-16

that resulted in the overwhelming share of these (stock) pollutants. It is true that in recent years, China and, to a lesser extent, India have had high aggregate pollution discharge accompanying their higher growth. On a per capita basis, these are still small (given their large populations) compared to the rich countries.[6] And however large these emissions may be, these are flows (that is, additions to the total stock of carbon) and are small relative to the stock of carbon already present in the atmosphere.

In passing, it should be noted that while carbon dioxide is seen to be the main driver of climate change, methane is also a very potent cause of it. Methane has a much greater effect on climate change than carbon dioxide. But both in terms of quantity emitted and the durability of its presence in the atmosphere, methane emissions can be ignored (or put on the back burner).

It is clear that all the big economies need to reduce their carbon emissions drastically. This is where blame-shifting starts to occur. Free riding (i.e., a party may derive benefits without paying for it) becomes the norm. And this is used as pretext not to do anything significant about the problem. As a matter of fact, world emissions have been rising in recent years when we needed them to fall over time—the decline needed to be commensurate with the target that the world sets itself for global warming. Time is not on the side of these petty arguments.

We then have to reduce global emission of greenhouse gases while recognizing the stock nature of the pollution. This suggests an intertemporal equity issue flagged earlier, viz, there are the not-so-well-off nations who did not contribute to the problem but are faced with the consequences of global warming as they manifest themselves, for example, in the form increased droughts, increased frequency of cyclones, and desertification. The Kyoto Protocol recognized this and asked industrial nations to make more stringent cuts in their fossil fuel use. Alas, it did not have any legal heft. Some of the big polluters just withdrew from it.

The richer countries promised to make financial transfers to developing economies for them to cope (i.e., to do mitigation and adaptation) with the problem. We delve into the nature of the give-and-take on climate change between the rich and the poor countries. We also look at the actual financial flows to developing countries. In doing so, we shall see that the aggregate flows are much lower than promised, and that adaptation is inadequately funded. Actually, the situation is worse than just the magnitude of transfers to the developing countries; for example, the definition of what constitutes climate finance has not been adequately tackled. The developed countries (barring three) have not come good on their promise of providing 0.7% of their gross national income (GNI) as overseas development aid. The promised climate aid of $100 billion per year was over and above this. This figure is inadequate to make a serious dent into the problem. And as we shall see later, even this low figure has not been provided. It is useful to keep this in the background, since this will enable a more informed discussion of the BCA proposals and what these imply for the SDGs.

The justification for implementing a BCA is that, at a general level, stringent environmental regulation (generally) in the developed countries causes footloose polluting industries to move to countries with lower regulation—that is, those polluting industries that can relocate internationally will do so. Even if relocation does not occur, strict climate change policies cause carbon-intensive sectors to contract in countries where carbon is priced higher. Existing firms' output changes, causing "carbon leakages"—some of the reduced carbon emissions are nullified by increased emissions in jurisdictions where regulation is lax. Hence, we find the United States, Japan, and the EU are net importers of carbon-intensive goods, while India and China (among others) are net exporters. The argument for BCA emphasizes the fact that a tariff on imports based on carbon content would restore competitive parity to domestically produced goods in the imposing countries.

15.2 Climate Change Negotiations and Finance

The Kyoto Protocol came into existence in 1997. It directly addressed the issue of the richer countries being responsible for the stock of carbon in the atmosphere—the so-called common but differentiated responsibilities (CBDR).[7] It was succeeded by the Paris Agreement in 2015. This did assign responsibilities to countries based on past carbon footprints, but the emphasis shifted away from treating countries differently. In this, the mechanism for reducing emissions relied on each country promising a target and sticking to it. These were the so-called national determined commitments (NDCs).[8] The advanced industrial countries promised to pay the developing countries for mitigation[9] and to implicitly transfer technology—this is set out in SDG 7: "Access to affordable, reliable, sustainable and modern energy for all" (especially SDGs 7a and 7b).

This financial commitment was built upon previous agreements, such as the UNFCCC and the Copenhagen Accord, whereby developed countries committed to mobilize US$100 billion per year by 2020 to address the needs of developing countries. This commitment was extended to 2025 in the Paris Agreement, while agreeing to a new collective quantified goal, from a floor of USD 100 billion per year.[10]

Two issues have arisen here: The first concerns the definition of *climate finance*, and whether commercial loans are a part of this. The second is that the $100 billion target was supposed to be over and above the 0.7% of GNI that the developed countries had promised as development aid. If a part of the climate finance was going to be obtained by cutting the 0.7% overseas aid, then this has serious ramifications for all the SDGs. In effect, this means all the development projects have to make room for climate finance.[11]

Let us look at these two points in somewhat greater detail.

First, the first point concerns achieving the set targets. The Glasgow Climate Pact at the COP26 meeting noted "with deep regret that the goal of developed country Parties to mobilize jointly USD 100 billion per year by 2020" had "not yet been met."

Beyond the aggregates, there is the disagreement on the operational definition of "climate finance." This is needed to get clarity on the aggregate picture. Can the developed countries count non-concessional loans to developing countries as part of the promised $100 billion for 2020? Analyses conducted by some organizations, such as OXFAM International and the Organisation for Economic Co-operation and Development (OECD), seem to suggest that loans and other non-grant instruments account for 70–80% of the reported climate finance flows to developing countries. About 40% of loans are made on commercial terms.[12] For instance, of an estimated $59.5 billion annual public climate finance reported in 2017– 2018, only $12.5 billion was provided in the form of grants, $22 billion was provided as concessional loans and other non-grant instruments, and $24 billion was provided as non-concessional loans and other instruments. This means that about 40% of the loans are made on commercial terms.

Non-concessional loans should not be counted as part of climate finance obligations (of the $100 billion) of the developed countries under the UNFCCC and the Paris Agreement structures. This distinction was always going to be important. But as the developing countries emerge from the COVID-19 outbreak, with their economic structures in a very bad shape and their populations in need of basic food and health, this is crucial. The Third World debt is unsustainable; to pile climate change financing on top of it is unconscionable. And of course, it would impede the attainment of SDGs 1 and 2, at the very least.

In addition to the aggregate flows, we have to answer two questions: How was this target of reduced emissions, that is, mitigation, going to be achieved? And how is the world going to cope with the effects of climate change, that is, adaptation?

Most climate financial flows target mitigation and not adaptation. This notwithstanding Articles 4.3 and 4.4 of the UNFCCC, which says: "The developed country Parties . . . shall also assist the developing country Parties that are particularly vulnerable to the adverse effects of climate change in meeting costs of adaptation to those adverse effects." This clearly has not been implemented. Technological innovations and capacity building receive very little finance.

The United Nations Environment Programme (UNEP) mentions the adaptation needs of developing countries to be between $140 billion and $300 billion per year by 2030 and increasing to $280 billion to $500 billion per year by 2050. As is obvious, the $100 billion promised by the developed countries (and the target that was yet to be met, and most of it earmarked for mitigation) is grossly inadequate in the face of these numbers.

It is also worth mentioning that what the developing countries have been campaigning for is a fund for loss and damage. While this was conceded, in principle, at the (latest) COP27 meeting, the details are yet to be fleshed out.

The financial allocation for mitigation is also inadequate. Given these, it was estimated that the world was going to hit 2.4°C above pre-industrial level. On the other hand, if a more stringent target of "net-zero" was to be achieved by 2050, then a 1.8°C target could be achieved. If global emissions were cut by half by 2030, then a 1.5°C target was feasible. A coalition of countries most at risk, the Alliance of Small Island States (AOSIS), wants a 20% cut in emissions by 2025. But as pointed out earlier, the trajectory of aggregate emissions in the world is moving up and not coming down.

The second issue concerns the possibility of repackaging existing overseas aid from the developed countries as climate finance. In 2015, the UN General Assembly set an agenda to secure sustainable development across the world by 2030. The agenda requires that 17 Sustainable Development Goals be met, including the eradication of extreme poverty and hunger as well as achieving gender equality and empowerment of all women and girls. As noted by the IPCC:

> A key agreement was that climate financing should be "new and additional" and not at the cost of the SDGs. Resources prioritising climate at the cost of non-climate development finance increases the vulnerability of a population for any given level of climate shocks, and additionality of climate financing is thus essential.
>
> (IPCC 2022)

The report by CARE (2022) looks at the climate finance reported by 23 rich countries for eight years (2011 to 2018). Only 3 of these, viz, Luxembourg, Norway, and Sweden, have regularly crossed the 0.7% of their GNI as ODA and have provided large per capita amounts additional climate finance.

Ten countries, namely, Italy, Canada, Japan, the United States, France, Spain, Belgium, Portugal, Greece, and Ireland, failed to cross the 0.7% threshold in any of the eight years and provided no additional climate finance.[13]

In what follows, we shall argue that the BCA, by putting impediments in developing countries' exports, would harm their growth process and their ability to tackle poverty and hunger (SDGs 1 and 2). But a reduced ODA cuts potentially all SDGs immediately—it works through a reduced budget outlay.

15.3 The Border Carbon Adjustment Mechanism

After this long background, we turn to the discussion of the BCA mechanism. First, we will explain what it is, and what its proponents believe it will achieve. Then we will turn to the imponderables—first, the measurement issues that must be clarified before its implementation and, second, the problems it is likely to encounter in proving compliance with the international trading regime, WTO. After considering the BCA on its own terms (in a partial equilibrium framework, that is, holding everything else constant), we will bring in the (developing) countries affected by it, the questions of equity (atemporal and intertemporal). This will allow us to pose the question: What does the BCA do to alleviate climate change? Then we will suggest that even if—and that is a big *if*—it passes a narrow interpretation of reducing carbon emissions, what harm does it do to the developing countries' ability to deal with poverty and hunger (SDGs 1 and 2)?

The BCA mechanism is a proposal mooted (in July 2021) by the European Union (EU) that will (starting in 2026Z) allow it to levy tariffs on products deemed to be carbon-intensive, such as cement, steel, aluminum, oil refinery, paper, glass, chemicals, and electricity generation.[14]

The proposal is described as follows:

> Designed in compliance with World Trade Organization (WTO) rules and other international obligations of the EU, the BCA Mechanism system will work as follows: EU importers will buy carbon certificates corresponding to the carbon price that would have been paid, had the goods been produced under the EU's carbon pricing rules. Conversely, once a non-EU producer can show that they have already paid a price for the carbon used in the production of the imported goods in a third country, the corresponding cost can be fully deducted for the EU importer.[15]

If this were not implemented, it is asserted that then "carbon leakage" will occur. This is defined by the EU as follows:

> As we raise our own climate ambition and less stringent environmental and climate policies prevail in non-EU countries, there is a strong risk of so-called "carbon leakage"—i.e. companies based in the EU could move carbon-intensive production abroad to take advantage of lax standards, or EU products could be replaced by more carbon-intensive imports.

Thus, it is expected that the BCA mechanism will lead to the equalization of the carbon price in the EU, and those prevailing in its trading partner countries, while ensuring that carbon leakages do not jeopardize the EU's climate program by an unfair competitive advantage gained by those countries with relatively lax climate policies.

Apart from the EU, the United States, Canada, and Japan are also planning similar measures.[16]

It is worth noting, to date, that the BCAs have not been applied at an international level. At the subnational level, in the United States, California implements a BCA mechanism. This makes its power importers liable for the difference in emissions between the state of power generation and those in California.

Given the nature of the products listed for the BCA, the countries whose imports are likely to be most affected by the tariff are Russia, Ukraine, Turkey, India, and China. The UK is outside the EU now but is unlikely to be affected by the BCA, since its regulations are in line with those in the EU. China has a national carbon market, and thus, some part of its costs of goods exported to the EU can be deducted. Among the top 12 exporters to the EU, only 3 have emissions trading systems (ETS) in place.[17]

The EU, then, has provided the intended imposition of the BCA levy with two justifications: a climate target goal and a global competitiveness for its potentially affected industries.

On the first, the EU has promised to cut its carbon emissions by at least 55% by 2030 (compared to 1990 levels). The EU's carbon emissions have already fallen by 24% since 1990. Emissions embodied in imports contribute about 20% of the EU's carbon dioxide emissions. A carbon tax would incentivize other countries to reduce their emissions and further shrink the EU's total carbon usage.

Second, all the EU member states (27 in all) have ETS laws that cap how much emissions individual industrial units are allowed.[18] Those that fail to cap their emissions can buy the unused allowances from those who have made deeper cuts.

As is clear, the BCA proposals allow for no exceptions for developing countries. And there is no provision for an international transfer of tax revenues generated by the proposal (although European Parliament has said some transfers will take place).

Although the EU has an ETS in place, what the implication of this would be in operationalizing the BCA is not clear. As Sapir (2021) reminds us:

> Under the EU ETS, the effective carbon price paid by producers of manufactured products covered by the ETS (such as cement, chemicals or iron and steel) was essentially zero, since they received free allocations of emission permits. It was neither necessary nor possible, to impose a CBAM on similar products imported into the EU, since by definition, a CBAM is an adjustment meant to equalize the price of carbon between domestic products and imports.

The EU has promised to remedy this and phase out free allocation of emission permits.

To summarize, then, three broad reasons for a BCA can be provided (two of them already mentioned earlier): First, it helps reduce emissions (and thereby help the EU and others meet their respective promised targets under the Paris Agreement); second, for the EU and others not to lose competitiveness in the production of carbon-intensive goods; and third, to make targeted countries fall in line and reduce the carbon intensity of production in the targeted sectors. As we discuss in what follows, the last two points probably fall foul of the WTO rules.

In this context, it is worth recalling that William Nordhaus has proposed what is called a "climate club." This involves a critical mass of countries that form a club with a common carbon price for members and a common tariff on **all** imports from non-members. This is stronger than a BCA because it entails a border adjustment according to membership and not the carbon content of the traded products. This is a proposed stick to make the non-members fall in line. Since it has a common tariff for non-members, I prefer to call it Carbon Customs Union.[19]

15.4 Border Carbon Adjustments: The Details

Let us look at the implications of the structure of BCAs. The import taxes are supposed to remove the "unfair" advantage for imports originating from lower-carbon-tax jurisdictions. The domestic consumption in the importing country falls, both due to expensive (carbon tax–inclusive) domestic production and now tariff-inclusive expensive imports (the latter is seen to be reducing global emissions). The sales abroad by domestic producers will fall (due to the carbon tax). To provide competitiveness in foreign markets, there is a proposal for an "export adjustment" (i.e., rebates) proposed to exporters (this is only at a discussion stage, not to be implemented now). Notice that a full border adjustment, that is, both import tariffs and export rebates (when this is implemented), makes the system akin to a value-added tax system[20]—to be paid by the consumers in the domestic economy.

Given this, we need some details before it can be implemented. First, which countries are exempt? These are those that possibly have a national emissions cap, or a sectoral cap, or poor countries (to which the mechanism will not apply—this is subject to further clarifications).

Second, which emissions are included for the levying of the import tax (the system boundary)? Are these direct emissions, or indirect emissions of energy use, or even indirect emissions embodied in inputs like machines? In practice, including the last one may prove difficult. Finally, which products are to be included? Possibly, a narrow coverage with high trade and carbon exposure (something like those mentioned in the list of the EU proposals earlier).

We can identify three channels through which carbon taxation affects the markets. First, we would expect wages and other factor prices to fall and a rise in the prices of goods produced in the domestic economy. The more intensively a sector uses fossil fuels, the larger the rise in the price of the good.[21] Overall, there is a loss of competitiveness for the domestic economy and a corresponding gain for the foreign country.

Second, as the prices of domestic factors fall, we would expect the GDP of the home country to fall and the foreign economy GDP to rise.

Third, at the level of the world economy, there is now an excess supply of fossil fuels. Their prices will fall. And over time, since carbon is more expensive, through endogenous technical progress, carbon intensity of production would fall.[22] And this improved technology could spill over globally.

The BCA mechanism concentrates on the competitiveness channel of leakages. The other channels are general equilibrium effects at the world level (that is, taking into account possible feedback). In this setting, the foreign country gains from free riding—it had not taxed carbon but gains competitiveness in the short run and, over time, gains from lower carbon in the atmosphere and from technical progress.

It is important to note that the BCA literature is partial equilibrium (i.e., ignores feedback effects) in nature; it focuses on the effect of the tax in the domestic economy. Also, the theoretical literature uses static models. One implication of this modelling strategy is that it cannot differentiate between transboundary (flow) pollution at any point in time (e.g., polluting a river that crosses international boundaries) and the fact that climate change requires a model based on stocks of pollutants (that is, climate change involves crossing of boundaries, but the pollution also has a long life).

Let us flesh out the structure, even if it involves some repetition. The analysis starts with the home country taxing emissions, while the foreign country does not. A border adjustment would then seek to help the home economy regain competitiveness in internationally traded goods. An import carbon adjustment will make the home firms' output relatively cheaper (than without the BCA) and expand home output (and thus home emissions). By how much depends on, at least, the elasticity (i.e., price responsiveness) of home demand and the elasticity of substitution between emissions and other inputs (that is, how easy or difficult it is to substitute inputs now that carbon is expensive).

Similarly, if an export rebate is implemented, it would increase the foreign demand for home exports. Again, increased emissions would depend on the elasticity of supply in home firms and the ease with which emissions can be substituted.

The sum of these two effects will give us full border adjustment effects.

15.5 Border Carbon Adjustment and the World Trade Organization

Since the end of the Second World War, the multilateral organization GATT and its later avatar, the World Trade Organization (WTO), have promoted free trade. This was to prevent the world from slipping back into the anarchy that engulfed the world trade regime of the inter-war period (marked by protectionism, competitive devaluations, etc.).

At the outset, the GATT–WTO regime was clear that free trade was its primary concern, and environmental concerns (e.g., a country stealing markets by maintaining lower environmental standards)—climate change was still not an issue—were given short shrift. For instance, sample this quote:

> [It] is not possible under GATT's rules to make one's market dependent on the domestic environmental policies or practices of the exporting country. . . . [This] protects trade relations from degenerating into anarchy through unilateral actions, in pursuit of unilaterally defined objectives . . . however valid they may appear.
>
> (GATT (1992) "International Trade 1990–1991," vol. 1)

In the climate change context, both the Kyoto Protocol and the Paris Agreement applied multilateralism to climate change issues. But unlike the GATT–WTO, the rules proposed by UNFCC were not attractive enough to bring forth voluntary compliance, nor were they binding on the signatories.

The proposed BCA is a unilateral move against the spirit of multilateralism. It would not be unfair to say the Western narratives have moved from putting all the eggs in the free trade basket (with its emphasis on multilateral implementation), and now the eggs have been transferred to the climate change basket (even if that is unilateral).

The WTO, the successor to GATT, has a structure that is very legalistic and leaves only a minimal room for economics-based arguments. As an example, notwithstanding the economic theoretical equivalence between taxes and quotas,[23] in the WTO setup, they are in different silos.

One can easily find many examples where the BCAs and the WTO would be in conflict. WTO does not allow interventions that target production processes rather than products[24]—the very foundation of a BCA. WTO also does not permit export rebates, something that is proposed under a BCA.

Article XX of WTO may allow for exceptions (to free trade) "to protect human, animal and plant lives; protect exhaustible resources." Whether this allows for a BCA-type interpretation is crucial here. Generally, it would be fair to say that the WTO needs a thorough overhaul for it to confront climate change issues.

How will this overhaul be achieved? On climate change, the WTO statutes could be rewritten, but the likelihood of this happening is close to zero. So far, the environmental disputes reaching the WTO have been tackled on a case-by-case basis. The dispute settlement system is moribund due to a lack of quorum—it does not have the minimum number of three judges.

If a systematic overhaul is not possible to introduce climate change issues in the WTO statutes, what are the options left?[25]

One possibility is to have the WTO Committee on Trade and Environment (CTE) move in that direction—so far, though, it has not happened, since climate change is not part of its mandate. The WTO could also seek advice from UNFCCC on climate issues. This is important because of two pressing points, viz, the emergence of BCAs and the presence of fossil fuel subsidies—the latter, although important, is hardly ever discussed.

If the GATT–WTO rules are to be amended, where should the focus be? In particular, we need to look at national treatment (III.2), border tax adjustments (II.2a), Article 3 of Agreement on Subsidies and Countervailing Measures, PPMs (process and product methods) (for like products), and Article XX. In this regard, waivers are not going to be useful, since to be invoked, these have be of a temporary nature and require exceptional circumstances.

There are problems of interpreting the details of a BCA; for example, GATT allows for adjustment of indirect taxes on "like products," but no allowance is made for direct taxes. Also, the carbon content of imports has to be quantifiable. And this cannot be greater than that in comparable domestic production. This leads to the question: "Which domestic technology should be used as the reference point?"

In implementing the BCA, allowances can be made for the least developed countries (that is allowed under the GAT-WTO rules), but these have to be in the form of tariff reductions only (under existing GATT rules).

Commentators have suggested other possibilities: For example, an "authoritative interpretation" in dispute settlement can be used, but it is an interpretation of existing laws and does not override them; similarly, a "peace clause," whereby members will exempt the departure from existing, requires consensus, failing which it can be introduced by a majority.[26]

Even the possibility of amending the harmonized system of product classification is not necessarily straightforward; for example, carbon content of steel can be inferred, but not for aluminum, across production techniques.

To sum up this section, then, one could argue that a BCA would help meet the Paris targets of the imposing countries (e.g., the EU) and could be seen to be reducing carbon emissions while preventing an erosion of competitiveness. But there are problems at the WTO level that are insurmountable, at least in the immediate future.

These problems are with the concepts, measurements, and an absence of numbers on leakages. These are, as explained earlier, incompatible with the WTO (WTO has nothing to say on these at a subnational level, but not at the national level). So what is to be done? Reform the WTO? Unlikely (as argued earlier). Jettison it or bypass the WTO? Apart from being in conflict with SDG 17 (on strengthening global partnership for sustainable development) (especially 17.10, 17.11, and 17.12), this would definitely lead to retaliation and trade wars.

15.6 Border Carbon Emissions, World Emissions, and Developing Countries

We have looked at the merits of the proposed BCAs and the technical and legal issues that confront it. Now we turn to the equity concerns of implementing such a proposal.

So far, the economic analysis has been conducted in a partial equilibrium framework, that is, no feedback or repercussion effects are taken into account. Let us now look at some feedback effects, both in an economic model and from a global political economy perspective.

First, let us look at some cases where the feedback effects could reduce, or even nullify, the intended effects of a BCA.

As a first example under this heading, let us go back to the discussion on the sources of leakages, recalling that if the BCA also includes export rebates (in the first round, that is not proposed, but it is discussed), this would (i.e., with an import tax and an export rebate) make a BCA exactly like a value-added tax (VAT) (albeit with different motives). Lockwood and Whalley (2010) draw our attention to a basic result in the VAT literature, namely, that if we allow exchange rates and prices to adjust, the VAT is neutral in its effect. It causes the exchange rate and price effects which neutralize the first-round partial equilibrium effects. Now, in a more general setup (e.g., with mobile capital), this may have to be modified, but there is no presumption that some of the intended effects would not be diluted.

As a second example, what are the consequences of the reduced fossil fuel demand on the prices of these at the global level? A concern that should be flagged is that, unless thought through (and implemented), a piecemeal application may well lead to a "green paradox," that is, fossil fuel producers will lower prices of their products, and this will stimulate demand in other markets.[27]

That this is not a theoretical curiosum can be gauged from the effects of the Russian attack on the Ukraine. The boycott of Russian oil has not hindered their sales (with the additional quantities being sold in China and India). Some further examples arising from this conflict can be cited: It has also brought US gas suppliers into Europe. Germany is digging for coal, Norway is looking for new oil fields, the UK had allowed fracking, and elsewhere, the Democratic Republic of the Congo and Gabon want to cut down their rainforests to sell the oil that lies below it, now that it is lucrative to do so.

It is also good to remember that the climate change contribution of a BCA depends on the carbon embodied in traded goods (further limited for transparency to big emitters among goods). Thus, it is hoping to target a very small segment of the goods. And if some producers find other markets not covered by the BCA, the aggregate effects could be small over and above the initial round of EU cuts.

Second, let us now turn to some political economy issues. As mentioned earlier, the analytical framework for mitigation is based on a "price of carbon." And carbon emissions are a perfect global externality (since the burning of carbon anywhere in the world affects climate change exactly in the same way). And the global price of carbon seeks to redress this global externality.

Thus, it is all very well to say that we, the world, need to confront the climate change problem by fixing a price for carbon. But how does this work in the absence of a world government or, in the jargon of economics, when there is no "central planner"? At least three problems suggest themselves: (1) How are tax revenues to be shared? (2) How are pollution permits to be allocated (per capita or by GNP)? And (3) recognizing that there are different prices for carbon in different economies, how will trade and climate interaction be handled?[28]

The inequity of BCAs becomes abundantly clear if we superimpose these three points on the incontrovertible fact that the developing countries did not create great climate crisis. And also, that they have very limited financial and technological arrows in the quiver.

A world carbon price is meant to correct a global externality. But in the absence of a world government, as shown in computational general equilibrium models of the world economy, who keeps the tax revenue is important (e.g., if the oil producers are allowed to keep it, the consequences resemble an OPEC-like oil price increase). In the case of the BCA, it is the tariff-imposing country that keeps the revenue. The EU has proposed that it will retire the outstanding debt with the BCA revenue. (The European Parliament does pay lip service to the notion of transferring part—only a part—of the revenue to the poorest developing countries.[29])

An expression that is used frequently is "free riding." Those who view the developing countries through this prism see a proposed BCA as penalizing the free riders. The general understanding of a *free rider* is one who is not contributing its bit although has the means to do so. In this setup, all agents, the EU and those exporting to it, are similar. And a ceteris paribus argument is applicable. But ceteris paribus is not applicable, as evidenced by the universal acceptance of a CBDR. And the patchy record of ODA and climate finance. The epithet of free riding is probably more applicable to the USA, with their record of "sometimes in, mostly out" of global climate action.

From the case for a BCA, then, equity considerations (both atemporal and intertemporal) are missing. The reduction of carbon emissions is the only metric or yardstick. The history of carbon emissions is ignored. Economics is supposed to be optimization subject to constraints. Somehow, the constraints never appear.

There are other constraints to think of. What it does to marginalized communities, or local pollution, is given a short shrift. As examples, think of replacing petrol (high CO_2 content) with diesel (high CO). The building of nuclear power plants (after Fukushima). Hydroelectric facilities (displacing rural communities and playing with fragile eco-systems). Funding for grids to carry power generated in a sustainable way. Why is there no discussion of fuel subsidies? How come the USA, being the biggest emitter of carbon per capita, is a net importer of carbon-intensive products?

Mechanically, meeting the Paris Agreement targets could be at the cost of other SDGs. Where other vulnerable countries are involved, unilateralism becomes even more undesirable.

15.7 Conclusions

The BCA mechanism has been touted as something that will solve the trade and climate issues at one go.[30] One should be careful, in general, about using one instrument to achieve two targets. In this chapter, we have tried to argue that the BCA proposal is not quite a knight in shining armor. There are a lot of difficulties in its implementation. There are the problems of measurement. These are unilateral actions and are not compatible with the WTO regime, at least as it stands today. Of course, the WTO framework is woefully lacking in its ability to tackle environmental issues generally, leave aside climate change. The BCA proposals also fly in the face of global equity considerations.

The SDGs could be affected directly and indirectly. As developed countries try to increase their climate contributions, without meeting the promised 0.7% of GNI as ODA, developing countries would be forced to cut back on expenditure on SDGs. If BCAs prevent developing countries from exporting goods (presumably industrial output), then their incomes will not rise, and they will find it even more difficult to spend on SDGs from their additional incomes.

The CBA proposal is motivated by the desire of a handful of developed countries to meet their Paris Agreement targets without losing competitiveness. This proposal is made without (almost) any thoughts to equity considerations or, for that matter, history.[31] In trying to achieve that, while it may be in the best-case scenario to meet a climate target, it certainly would damage the attainment of the other SDGs.[32]

Notes

1 Especially 1a.
2 Especially 2.1 and 2.2.
3 It would also affect the other SDGs indirectly, for example, SDGs 6, 7, and 10.
4 This is a much-surveyed area of research, possibly because of its immediate policy ramifications. Some of the overview articles are Resources for the Future (2021), Das et al. (2018), Droege and Fischer (2020), Grubb et al. (2022).
5 In the Kyoto Protocol and its successors, the Copenhagen Accord and the Paris Agreement (that is in effect now).
6 "The contribution of low-income developing countries . . . to atmospheric greenhouse gas concentrations is negligible, both in absolute terms and on a per capita basis" (IMF 2017, p. 119).
7 *Common* is about climate change; *differentiated* because of different contributions to the problem.
8 The Paris Agreement sought to keep the average global temperature rise well below 2°C and as close as possible to 1.5°C above pre-industrial levels.
9 "The international community must play a key role in supporting these countries' efforts to cope with climate change—a global threat to which they have contributed little" (IMF 2017, p. 117).
10 In paragraph 53 of Decision 1/CP.21 Adoption of the Paris agreement: in accordance with Article 9, paragraph 3, of the Agreement.
11 CARE estimates $103 billion has been "repackaged."
12 OXFAM Climate Finance Shadow Report 2020 has put it starkly: "[T]he excessive use of loans and the provision of non-concessional finance in the name of climate assistance is an overlooked scandal."

13 Using a "strong" definition of climate finance—see CARE (2022) on this.
14 The tariff will be levied starting in 2026Z, but monitoring and collection of data will start in 2023.
15 See the European Commission's website.
16 And a US and EU carbon-based sectoral arrangement in steel and aluminum is expected by 2024.
17 Nearly 35% of India's exports would be covered by the BCA in the EU.
18 The EU ETS system has 30 European Economic Area countries as member—27 EU plus Norway, Liechtenstein, and Iceland. It covers about 3.2% of the world's emissions.
19 Like a customs union, it will have trade creation (trade between members expands) and trade diversion (trade with non-members contracts) effects.
20 A value-added tax is intended for the domestic market. The foreign producers are taxed for sales in the domestic market, and domestic producers do not pay for their output to be sold in a foreign market.
21 This depends on the elasticity of substitution among inputs.
22 Some substitution between fossil fuels and other inputs could happen without technical change. It all depends on the elasticity of substitution in the short run and over time.
23 *Equivalence* means we can achieve the same economic outcome either by setting a quota on imports or by putting an appropriate tariff.
24 GATT frowned upon specifying production methods for goods. Its opposition to introducing environmental considerations in the trading world was based on this. BCA starts off by specifying the carbon content of goods.
25 See Das et al. (2018) for a detailed discussion on the options. They discuss, among others, using RTAs in the context of BCAs. They also discuss at length the important, but neglected, topic of fuel subsidies—this is not crucial to the arguments made in this chapter.
26 Cosbey et al. (2019), Droege and Fischer (2020), and Sapir (2021) would concur with the problems posed here.
27 This arises when we have fossil fuel markets that segment geographically or intertemporally.
28 To state what is obvious, climate action has repercussions for trade, and trade affects the attainment of climate goals.
29 This intended transfer is not WTO-compatible, as pointed out earlier.
30 See Mehling et al. (2019), for example. They hear only the music without Scylla and Charybdis.
31 In sharp contrast to the CBDR thinking, where history makes an appearance.
32 And meeting the climate target is not guaranteed if we take into account feedback effects.

References

Climate Change and Resilience Centre (CARE) (2022) *That's Not New Money: Assessing How Much Public Climate Finance Has Been 'New and Additional' to Support for Development*, CARE Climate Justice Center, Denmark.

Cosbey, A., Droege, S., Fischer, C. & Munnings, C. (2019) Comparing Policies to Combat Emissions Leakage: Border Carbon Adjustments Versus Rebates. *Review of Environmental Economics and Policy* 13, 3–22.

Das, K., van Asselt, H., Droege, S. & Mehling, M. (2018) Making Trade Work for Climate: Options for Policymakers. *Climate Strategies*, London.

Droege, S. & Fischer, C. (2020) Pricing Carbon at the Border: Key Questions for the EU. *IFO DICE Report* 18 (1), 30–34.

Gene . . . (2019) Towards a Trade Regime that Works for the Paris Agreement. *Economic and Political Weekly* LIV, 25–30.

Grubb, M., Jordan, N.D., Hertwich, E., Neuhoff, K., Das, K., Bandyopadhyay, K.R., van Asselt, H., Sato, M., Wang, R., Pizer, W.A. & Oh, H. (2022) Carbon Leakage, Consumption, and Trade. *Annual Review of Environment and Resources* 47, 753–795.

International Monetary Fund (2017) *Global Economic Outlook*, Washington, DC.

Kortum, S. & Weisbach, D. (2017) The Design of Border Adjustments for Carbon Prices. *National Tax Journal* 70, 421–446.

Lockwood, B. & Whalley, J. (2010) Carbon-Motivated Border Tax Adjustments: Old Wine in Green Bottles? *The World Economy* 33 (6), 810–819.

Mehling, M., van Asselt, H., Das, K. & Droege, S. (2018) Beat Protectionism and Emissions at a Stroke. *Nature* 559 (7714), 321–324.

Nordhaus, N. (2020) The Climate Club: How to Fix a Failing Global Effort. *Foreign Affairs*, May/June.
Nordhaus, W. (2015) Climate Clubs: Overcoming Free-riding in International Climate Policy. *American Economic Review* 125, 1339–1370.
Pacheco Rodríguez, M.N. & Fernando Rosales, L. (2021) Some Key Elements for Developing Countries in Climate Change Negotiations of COP 26: Climate Finance, Article 6 Negotiations and Implications. *Climate Policy Brief*, www.southcentre.int, Geneva.
Resources for the Future (2021) *Border Carbon Adjustments 101*, Washington, DC.
Sapir, A. (2021) *The European Union's Carbon Border Mechanism and the WTO*, Bruegel-Blogs, Bruegel, Brussels.
Weisbach, D., Kortum, S., Wang, M. & Yao, Y. (2022) *Trade, Leakage, and the Design of a Carbon Tax*, NBER Working Paper No. 30244.

16

SD GOAL 15—LIFE ON LAND

Progress to Date, the Role of Business, and a Case Study in Sustainable Development in Human-Modified Landscapes

Robert Wallis and Graeme Coles

16.1 Introduction

Sustainable Development (SD) Goal 15 is concerned with the sustainable management of terrestrial (and freshwater) ecosystems. It includes the restoration of degraded habitats, biodiversity protection, and ensuring government planning integrates ecosystem and biodiversity considerations. The goal has 12 targets to protect and restore life on land and in freshwater ecosystems, and an overarching goal: to protect, restore, and promote sustainable use of terrestrial ecosystems; sustainably manage forests; combat desertification; and halt and reverse land degradation and biodiversity loss.

In this chapter, we examine the progress toward achieving these targets by 2030 and the emerging role that the corporate sector is playing in ensuring sustainability of the natural environment. We then illustrate work toward SD Goal 15 using a case study in Australia—the use of raptors to disperse pest birds that hinder progress toward sustainable agriculture—and link our work to other SD Goals.

16.2 Why Is Nature Conservation So Important?

In its latest global risk report, the World Economic Forum (WEF) ranked biodiversity loss and ecosystem collapse as one of the top five global risks that humanity faces over the next decade. Stakeholders surveyed by the WEF rated this risk the 4th highest (of 32 most significant global risks) in terms of risk preparedness (World Economic Forum 2023).

More than half of the world's economic output, estimated at US$44 trillion, is dependent on nature and its services (World Economic Forum 2020). These beneficial ecosystem services include provisioning services (such as food, water, fuels, medicines), regulating services (purifying air and water), flood and disease control, cultural services (spiritual well-being, recreational needs), and supporting services (such as pollination, primary production, nutrient cycling).

16.3 Progress Toward Achieving SD Goal 15

In its 2021 SD Goals report, the United Nations described progress toward some of the targets for SD Goal 15. It noted that biodiversity was declining faster now than ever before, and the level of extinction threat was increasing. Over 40% of the world's amphibians, for example, are threatened

DOI: 10.4324/9781003285472-17

with extinction. The protection of the world's key biodiversity areas (KBAs) has been seen as essential for future biodiversity conservation. While progress in this target to 2020 has been satisfactory, the report notes that increases in protected areas have since stalled. Overall, more than half of the world's KBAs remain unprotected.

The loss of forests has also accelerated, despite more existing forests now being sustainably managed. The report notes (p. 57):

> The continuing disappearance of forests signals the need for accelerated action to reduce deforestation, restore degraded lands and implement sustainable forest and land management practices. Such actions will also enhance the resilience of ecosystems to climate change, protect biodiversity and support rural livelihoods.
>
> (United Nations 2021)

Alien species invasions also present a problem. More funding is required from governments to counter introduced and damaging species—with only 55% of countries specifically allocating funds for these tasks. Legislation is also required; while high numbers of countries have laws banning the importation of invasive species that affect agriculture and human health, far fewer have laws relating to the damaging effects on the environment and fisheries (55 and 37%, respectively).

A more recent assessment shows that for SD Goal 15, major challenges remain and progress is stagnating. None of the world's regions were on track to achieve the goal and its targets. Only two nations were expected to achieve the goal by 2030 (Lithuania and the Slovak Republic) (Sachs et al. 2023). Australia rated poorly, with major challenges remaining and decreasing progress being made.

16.4 The Corporate Sector and Moves for Nature-Positive Operations

Businesses face two kinds of nature-related risks. They can produce products, use resources, or generate wastes that can deleteriously affect the environment and, in so doing, reduce their reputation and profitability. Businesses can also be affected by a deterioration in the external environment that presents challenges to their own sustainability.

The Taskforce of Nature-Related Financial Disclosures acknowledges the seriousness of the challenges.

> Future cashflows depend on the resilience of the future flow of these ecosystem services from nature. . . . [I]t is increasingly clear that nature risk, beyond climate change, is financial risk. Business as usual is no longer a viable option and nature must no longer be seen as a corporate social responsibility but a strategic risk management one.
>
> (TNFD 2023)

The recommended disclosures cover governance (3 disclosures), strategy (4), risk and impact management (3), and metrics and targets (3).

Boards of corporations are moving to address the responsibilities to nature. There have been moves for members to become "nature literate" just as directors should all be financially literate (Wallis 2023). Some are appointing directors with specialist skills in science and the environment.

Corporations have also aimed to become "nature-positive," in which their effects on nature are beneficial overall. The WWF defines *nature-positive* as "halting and reversing nature loss, measured from 2020 levels, by increasing the health, abundance, diversity, and resilience of species, populations, and ecosystems so that by 2030 nature is visibly and measurably on the path of recovery" (WWF 2023). They also argue that US$10 trillion in benefits result from engaging in nature-positive business models and give the example of 80 seafood companies (accounting for 30% of the world's production) that have committed to sustainable seafood practices. These initiatives are not just for reputational value but are essential for the continued profitability of the companies. The challenge is that the open ocean is considered a "commons," and if some fishing companies engage in sustainable harvests, others could exploit the benefits of increased availability of seafood.

BHP is a large company specializing in mineral extraction and processing that is headquartered in Melbourne, Australia. It has numerous operations around the world, covering an area about half the size of Switzerland. The company has committed to improve the "biodiversity, ecosystem services and natural capital" of areas under its control and will place 30% of its land under some form of environmental protection (Patrick 2022).

Financial institutions are also aiming to be nature-positive. Australian banks have recently provided loans at discounted interest rates for "green" projects. The largest bank in Australia, the Commonwealth Bank of Australia, has provided interest-free loans for customers wanting to install rooftop solar panels.

16.5 A Case Study in Addressing Targets in SD Goal 15: The Use of Raptors to Disperse Pest Flocks of Native Birds in Australia

For some years, we have been using birds of prey (raptors) to disperse pest flocks of birds that threaten agricultural production, damage infrastructure, cause health challenges, and interfere with sporting and cultural activities (Coles et al. 2019). Our research addresses two targets of SD Goal 15:

- 15.1 To conserve and restore terrestrial (and freshwater) ecosystems.
- 15.5 To protect biodiversity and natural habitats.

Traditional methods of managing pest birds involve lethal and non-lethal techniques. Birds may be killed by poisoning, trapping (and subsequent killing), and shooting.

Non-lethal techniques include exclusion netting, providing decoy food sources to lure birds away from target crops, habitat manipulation, trapping and relocating birds, and scaring the pest birds away from the sites that require protection (Baxter and Allen 2006; Bomford and Sinclair 2002; De Vault et al. 2013; Tracey et al. 2007).

All these traditional techniques have their drawbacks. For example, poisoning can be non-discriminatory and kill non-target species; as well, poison can accumulate up the food chain. Birds can quickly acclimatize to scare devices, such as loud sounds, flashing lights, and models of predators.

Trained birds of prey have been used to control pest flocks of birds at landfill rubbish tips, sporting grounds, and airports with different levels of success (Blokpoel 1976; Olsen 1994; Baxter and Allan 2006; Cook et al. 2008). In the UK, Cook et al. (2008) found falcons and hawks were effective at dispersing gulls at landfill sites. Other studies have found falcons to be especially successful in dispersing introduced passerine birds that consume grapes in New Zealand (Kross et al. 2012), and in the USA, falcons were able to significantly reduce contamination and consumption of leafy greens by pest birds (Navarro-Gonzalez and Jay-Russell 2016).

The success of using falcons to control high densities of nuisance birds seems to depend on site-specific conditions, the species concerned, alternative food sources and habitat, and whether some of the pest birds are killed (Cook et al. 2008; Navarro-Gonzalez and Jay-Russell 2016).

16.5.1 Our Case Study: Dispersing Cockatoos From an Almond Orchard

Flocks of two species of native parrot, the long-billed corella (*Cacatua tenuirostris*) and smaller numbers of sulphur-crested cockatoos (*C. galerita*), were causing significant damage to an almond crop in northern Victoria in Australia. These parrots roosted overnight in trees along the adjacent Murray River and flew to feed on the almonds during the day. The highest densities of parrots occurred on the farms in the mornings. The almond farm is one of Australia's largest, covering about 12,000 hectares.

In 2017, trials began using a trained wedge-tailed eagle (*Aquila audax*)—a natural predator of the parrots—to fly from the river along specific transit routes. Three periods of flyovers, each lasting six days, were conducted during the pre-harvest period. Flyovers consisted of the handlers releasing and receiving the eagle over a distance of about 200 meters at a height of 3–5 meters during early morning and late afternoon, when the parrots were most actively flying to the almond trees, and also during late morning, when some parrots would attempt to return to the orchard. The parrots quickly returned to the river habitat after each flyover rather than dispersing further inland to the other farm blocks. As the trial progressed, fewer parrots would attempt to access the orchards. After the 18 days of flyovers, the parrots did not return to the farm blocks until several months later, when harvesting had finished.

The eagle was trained not to attack the target birds and was enticed back to the handler by a lure containing a food reward when the flyover had been completed.

In 2018, we repeated the trial. Again, about 400 parrots attempted to fly to the orchard each day, especially in the early morning and later afternoon. This time, low numbers (<10) of corellas returned during the trial. We thus checked the farms every 30 minutes during the six-day period and flew the eagle an additional time if parrots were seen in the almond trees, instead of flying the eagle once more during the day late in the morning. This was effective, and after the six days, no parrots were foraging in the orchards.

Non-target birds were unaffected during the trials. We believe one reason for the program's success was that the harvest period is relatively short, and the parrots dispersed for sufficient time before habituating to the threat. Furthermore, once the harvest had finished and the raptors were no longer flown over the site, the cockatoos gradually returned to the orchard and foraged on the decaying fruit that lay on the ground, saving the orchardist from having to manually clean up the site.

The orchard is owned by an international company headquartered in Singapore. The organization has a strong commitment to sustainability, as evidenced by its sustainability goal: "[W]e want to leave the world a better place than we found it." The company also notes regarding its sustainability framework: "Our ability to make an impact on the world is mapped against the UN Sustainable Development Goals (SDGs), and the ten Principles of the UN Global Compact. We also map against stakeholder priorities." Clearly using management techniques to improve harvest and manage pests using raptors aligns with their commitment to sustainable development.

Almond orchards have a significant impact on the environment. Almond growing consumes large volumes of water, and the crop is usually grown as a monoculture that typically requires high pesticide use (Anon 2020). Although the orchard we studied was established on land that had already been cleared for stock grazing, the orchard nevertheless exists where originally a complex and rich habitat existed for native species communities.

Companies have to avoid greenwashing—the misrepresentation of the sustainability credentials of a company through its products or services. In Australia, the first prosecution has occurred, and in nine months in 2023, some 35 actions had been taken against companies by the Australian Securities and Investments Commission for greenwashing (Gudkov 2023).

While this case study exemplifies an approach to preventing biodiversity loss and protecting terrestrial habitats, other SD Goals are relevant. SD Goal 2, for example, is to end hunger, achieve food security and improved nutrition, and promote sustainable agriculture. Almonds are calorie- and nutrient-dense, with most of the fats being mono- and polyunsaturated. The case study company prides itself on working to achieve sustainable agriculture.

SD Goal 12 is also pertinent to our case study. This goal is to ensure sustainable consumption and production patterns. Target 12.6 is to "encourage companies, especially large and transnational companies, to adopt sustainable practices and to integrate sustainability information into their reporting cycle"—the company we have used in this case study certainly reports on its progress toward sustainability. Indeed, since 2016, company sustainability reporting worldwide has tripled (United Nations, undated).

The technique of using raptors to disperse pest flocks of native birds has been demonstrated in our case study to be effective and regardful of animal welfare. It exemplifies compassionate conservation in that there is no harm to wildlife in effecting a solution to a conservation problem. However, it is a labor-intensive operation, as birds have to be reared, trained, cared for, and transported. The handlers must be experts in managing the birds. In some instances, such as in attempts to disperse breeding gulls, the technique is ineffective, as the gulls are fiercely protective of defending their nests and young (Coles et al. 22019; Cook et al. 2008).

16.5.2 Lethal Versus Non-Lethal Control

Using raptors to control pest birds is an example of compassionate conservation, in which no harm comes to animals and cruelty is avoided (Wallach et al. 2018). Our experience is that the public shows great interest in and support for this technique. However, compassionate conservation has been criticized by ecologists as misguided and likely to result in poor biodiversity conservation outcomes (Callen et al. 2020). Lethal control of pests is one tool that could be used with others as part of an integrated pest management operation.

If the raptor kills (or hunts and therefore has the potential to kill) some of the target population, does this increase the efficacy of the control method? There have been few comparisons between lethal and non-lethal use of raptors that allow for meaningful comparison. Baxter and Allan (2006) found peregrine falcons significantly reduced the number of gulls at rubbish tips in England. In about one-third of their trials, the falcons were able to hunt and capture their prey. However, their data analysis did not discriminate between the prey reduction for lethal versus non-lethal trials. The authors surmised the falcons were successful "whether hunting, being flown to simulate hunting. Or more importantly, perceived as potentially present" in deterring the gulls from using the tips (Baxter and Allan 2006, p. 1165). However, the experimental design did not allow any conclusion to be drawn on the relative reduction in gull numbers for three conditions: falcons that did not hunt, simulated hunting, and falcons that actually took prey.

In another study of landfill rubbish tips in the UK, Cook et al. (2008) found that when lethal methods were used (falcons preying on gulls or the gulls being shot), the gulls no longer habituated to the raptors (which occurred when the birds did not hunt their prey). They found lethal techniques (shooting, falconry) which reinforce visual and audio cues with the occasional death of target birds have the opposite effect on habituation, and more gulls were deterred as the trial progressed. This is the most striking evidence that lethal takes (and shooting) do indeed have a positive effect on reducing the number of gulls compared with the non-lethal use of raptors.

In a study on grape damage in New Zealand previously mentioned, Kross et al. (2012) found New Zealand falcons translocated to wine-growing regions preyed on birds that fed on and damaged grapes. The researchers noted that not only did direct predation reduce the numbers of the pest flocks but also that the high predation risk changed the birds' behavior to become more "antipredator." Such actions included avoiding the sites, increased vigilance relative to the time spent actually foraging for grapes, and also relocation by the frugivorous birds to less-favorable sites in which to forage that had lower predation risk.

There seems to be other anecdotal suggestions that if raptors can kill their prey, the control of pest flocks becomes easier, but hard evidence is lacking. In some cases, it seems authors intuitively think this should be the case without any experimental evidence. Thus, Erickson et al. (1990, p. 314) state, "The raptor species used preferably should be a natural predator of the pest bird species as the occasional kills it makes will help reinforce the perception of danger," and cite Grubb (1977) in support. However, Grubb (1977) makes no mention of lethal versus non-lethal control using raptors, referring to differential habituation only in terms of distinguishing between aerial stimuli, such as different shapes of raptors.

Inglis (1980, p. 124) also believes shooting pest birds and the sight of dead or dying birds are aversive stimuli that could disperse pest birds from a site. His argument is based on the literature of avoidance conditioning. However, there are certain elements in its practice that are required:

- The warning stimulus (in this case, birds seeing others taken by a raptor) must retain its potency for long periods (a single kill episode will likely be ineffective).
- It must occur spatially and temporally with the pest bird visits to the site.
- A scaring device (such as the playing of alarm calls) could be used simultaneously with the aversion stimulus (e.g., hazing by falcons) so the pest birds associate the device stimulus with killing.
- Novel scarers are preferable to ones used previously.

Inglis's suggestions are based on theory, and his paper does not produce hard evidence that killing some of the pest birds as aversion stimuli increases the efficacy of pest dispersal. Furthermore, in his discussion, the aversion stimulus of dead conspecifics is by way of shooting and not by raptor take.

Bomford and Sinclair (2002) reviewed the research on bird pests in Australia and listed the circumstances when scaring devices could be effective, including their reinforcement with real danger, such as shooting. Raptor takes of target birds could also be considered such a reinforcement stimulus.

Clearly, there needs to be research undertaken that resolves whether lethal takes of pest birds reduce the impact they cause.

16.6 Conclusion

SD Goal 15 concerns the sustainable development of terrestrial and freshwater ecosystems so that future human communities can thrive without damage to the environment. The goal has 12 targets that include no biodiversity loss, improved sustainable forestry, increases in protected areas, and sustainable use of resources.

Progress across the world in achieving this goal and its targets has been patchy and disappointingly slow. The corporate sector is now demonstrating an awareness that its future sustainability depends on a healthy, productive environment, and many businesses are committing to becoming nature-positive.

Our case study illustrates a technique to manage pest native bird flocks that damage agricultural production. The method of using raptors to disperse the flocks is humane, is effective, and works best when the area covered is discrete and the time short.

Acknowledgments

We thank other members of our research team, Dave Brennan, for his advice, and Fred Spiteri and Jason Jordan, for their valuable work with raptors in the field. All research was carried out with approval from the Federation University Animal Ethics Committee (14-016; 19-004) and under a Victorian government Wildlife Research permit (10007595). Full Flight Birds of Prey has a displayer's license (12529570), also provided by the Victorian government.

References

Anon. (2020). What impact do almonds have on the environment? *Envirotech Online*. www.envirotech-online.com/news/business-news/44/breaking-news/what-impact-do-almonds-have-on-the-enviroment/51374 (accessed 29 October 2023).

Baxter, AT & Allan JR (2006). Use of raptors to reduce scavenging bird numbers at landfill sites. *Wildlife Society Bulletin*, 34, 1162–1168.

Blokpoel, E (1976). *Birds Hazards to Aircraft*. Irwin & Co.

Bomford, M & Sinclair, R (2002). Australian research on bird pests: Impact, management and future directions. *EMU*, 102, 29–45.

Callen, A, Hayward, MW, Klop-Toker, K, Allen, BL, Ballard, G, Broekhuis, F, Clarke, RH, Clulow, J, Daltry, JC, Davies-Mostert, HT, Di Blanco, YE, Dixon, V, Fleming, PS, Howell, LG, Kerley, GIH, Legge, SM, Lenga, DJ, Major, T, Montgomory, RA, Moseby, K, Meyer, N, Parker, DM, Periquet, S, Read, J, Scanlon, RJ, Shuttleworth, C, Tamessar, CT, Taylor, WA, Tuft, K, Upton, RMO, Valenzuela, M, Witt, RR & Wuster, W (2020). Envisioning the future with 'compassionate conservation': An ominous projection for native wildlife and biodiversity. *Biological Conservation*, 241, 108365 https://doi.org/10.1016/j.biocon.2019.108365 (accessed 22 March 2022).

Coles, G, Wallis, R & Brennan, D (2019). Using raptors to disperse pest birds in Victoria. *Australian Field Ornithology*, 36, 132–136.

Cook, A, Rushton, S, Allan J & Baxter, A (2008). An evaluation of techniques to control problem bird species on landfill sites. *Environmental Management*, 4, 834–843.

DeVault, TL, Blackwell, BF & Belant, JL (2013). *Wildlife in Airport Environments Preventing Animal-Aircraft Collisions through Science-Based Management. USDA National Wildlife Research Center—Staff Publications*, 1656. https://digitalcommons.unl.edu/icwdm_usdanwrc/165

Erickson, WA, Marsh, RE & Salmon, TP (1990). A review of falconry as a bird-hazing technique. Pp 314–316 in Davis, LR & Marsh, RE (Eds.), *Proceedings of the 14th Vertebrate Pests Conference*. Davis: University of California.

Grubb, TC (1977). Discrimination of aerial predators by American coots in nature. *Animal Behaviour*, 25, 1065–1066.

Gudkov, A (2023). *Greenwashing in the Spotlight—Regulators Ramp Up Enforcement*. www.aicd.com.au/risk-management/framework/climate/greenwashing-in-the-spotlight-regulators-ramp-up-enforcement.html (accessed 29 October 2023).

Inglis, I (1980). Visual bird scarers: An ethological approach. Pp 121–143 in Wright, EN & Inglis, IR (Eds.), *Bird Problems in Agriculture*. Reading: British Crop Protection Council.

Kross, SM, Tylianakis, JM & Nelson, XJ (2012). Effects of introducing threatened falcons into vineyards on abundance of passeriformes and bird damage to grapes. *Conservation Biology*, 26, 142–149.

Navarro-Gonzalez, N & Jay-Russell, MT (2016). Use of falconry to deter nuisance birds in leafy green fields in Northern California. Pp. 209–216 in Timm, RM & Baldwin, RA (Eds.), *Proceedings of the 27th Vertebrate Pest Conference*. Davis: University of California.

Olsen, J (1994). *Some Time with Eagles and Falcons*. Belconnen, ACT: Olsen.

Patrick, A (2022). What is nature-positive and who is profiting from it? *Financial Review*, 22 October. www.afr.com/policy/energy-and-climate/what-is-nature-positive-and-who-is-profiting-from-it-20221024-p5bsex (accessed 26 October 2023).

Sachs, JD, Lafortune, G, Fuller, G & Drumm, E (2033). *Implementing the SDG Stimulus. Sustainable Development Report 2023*. Dublin, Ireland: Dublin University Press.

Soldatini, C, Albores-Barajas, Y, Torricelli, P & Mainardi, D (2007). Testing the efficacy of deterring systems in two gull species. *Applied Animal Behaviour*, 110, 330–340.

Taskforce on Nature-related Financial Disclosure (TNFD) (2023). *Recommendations on the Taskforce on Nature-related Financial Disclosures*. https://tnfd.global/wp-content/uploads/2023/08/Recommendations_of_the_Taskforce_on_Nature-related_Financial_Disclosures_September_2023.pdf?v=1695118661 (accessed 24 October 2023).
Tracey, J, Bomford, M, Hart, Q, Saunders, G & Sinclair, R (2007). *Managing Bird Damage to Fruit and Other Horticultural Crops*. Canberra, ACT: Bureau of Rural Resources.
United Nations (2021). *The Sustainable Development Goals Report 2021*. https://unstats.un.org/sdgs/report/2021/The-Sustainable-Development-Goals-Report-2021.pdf (accessed 24 October 2023).
United Nations (undated). https://sdgs.un.org/goals/goal12 (accessed 29 October 2023).
Wallach, AD, Bekoff, M, Batavia, C, Nelson, MP & Ramp, D (2018). Summoning compassion to address the challenges of conservation. *Conservation Biology*, 32, 1255–1265.
Wallis, R (2023). The need for education about diversity and threats to ecosystem function in the corporate sector. *Asian Journal of Biology Education*, 15, 2–7. https://doi.org/10.57443/ajbe.15.0_2
World Economic Forum (2020). *More Than Half of the World's Economic Output, Estimated at US$ 44 Trillion, Is Dependent on Nature and Its Services*. www.weforum.org/press/2020/01/half-of-world-s-gdp-moderately-or-highly-dependent-on-nature-says-new-report/ (accessed 24 October 2023).
World Economic Forum (2023). *The Global Risks Report 2023*. 18th edition. Insight Report. Geneva, World Economic Forum. www.weforum.org/publications/global-risks-report-2023 (accessed 24 October 2023).
WWF (2023). *Nature-Positive Business*. www.worldwildlife.org/pages/nature-positive-business#:~:text=Nature%2Dpositive%20is%20defined%20as,on%20the%20path%20of%20recovery (accessed 26 October 2023).

17

SUSTAINABLE DEVELOPMENT GOALS AND THEIR EMPHASIS ON MANAGING PLASTIC POLLUTION

Vinish Kathuria and Nishith Jardosh

17.1 Introduction

The increased consumption of plastic and the resulting pollution have become pervasive, affecting marine life, fresh water, and terrestrial ecosystems globally (Borrelle et al., 2020; Rochman, 2020). Once plastic enters the environment, it continuously breaks down into smaller particles, to as small as a size below 5 millimeters, which are termed microplastics. Due to low degradation and unsustainable use and disposal, plastic and microplastics have also become a transboundary threat to natural ecosystems and human health (Napper et al., 2021). As global plastic production continues to rise, efforts to reduce plastic pollution have also increased through national and international commitments. At the national level, several governments have banned single-use plastics, whereas at the international level, the United Nations (UN) is ensuring commitments from nations to reduce plastic leakage into the environment. One such commitment is through the UN Sustainable Development Goals (SDGs).

The UN launched the SDGs at its summit in New York in September 2015 to address the most critical global threats of the present time to achieve a sustainable future (Walker, 2021). The summit identified 17 universal goals through 169 specific targets and 247 indicators to end poverty, protect the environment, and ensure that all people enjoy peace and prosperity.[1] Among these 17 goals, some indirectly promote less use of plastics (for example, goal 12, through sustainable consumption and production), but it is goal 14 which is specifically concerned about plastics, mainly marine plastics. In Target 14.1, it is mentioned that "[b]y 2024, [nations are to] prevent and significantly reduce marine pollution of all kinds, in particular from land-based activities, including marine debris and nutrient pollution." This SDG sub-target will be measured by an "[i]ndex of [c]oastal eutrophication and floating plastic debris density" (Indicator 14.1.1).

Despite this goal and target, a recent report by the UN mentions that over 17 million metric tons (MMT) of plastics clogged the ocean in 2021, which may double or triple by 2040 (UN, 2023). The data shows plastic production has skyrocketed fourfold in the last 40 years, whereas the recycling rate has remained below 10%. Against this backdrop, this chapter looks into the evolution of the plastic pollution problem and how different SDGs are connected to preventing plastic pollution. The remaining chapter is divided into four sections. Section 2 gives the status of plastic consumption and pollution. Section 3 discusses different policies enacted in different countries to

DOI: 10.4324/9781003285472-18

manage plastic pollution. This is followed by an analysis of how various SDGs affect plastic consumption and pollution in Section 4. The chapter concludes with Section 5, which gives the status of different SDGs linked to plastic pollution and also gives the way forward.

17.2 Plastic Consumption and Plastic Pollution—A Global Perspective

17.2.1 Plastic Consumption

In the 1950s, the global annual production of plastic was around 2 million tons (MT), and over the course of this decade, particularly in the United States of America (USA), a lifestyle emphasizing the widespread adoption of disposable items, such as plastic cutlery and cups, gained popularity (Rochman, 2020). The terms "throwaway culture" and "take-use-throw" were coined during this period only. Since then, there has been a sharp increase in plastic production (and hence consumption), increasing to over 489 million tons in 2023 from 2 million tons in the 1950s (a CAGR of 7.9%). Figure 17.1 plots the trend in plastic production since 1950. The figure also shows the trend in micro- and macroplastics.

As can be seen, the increase has been more rapid in recent periods. The data shows that the total production from 2010 to 2023 was more than in the preceding six decades (OECD, 2022). From now until 2060, global plastic production is expected to reach over 1,230 million tons (a CAGR of

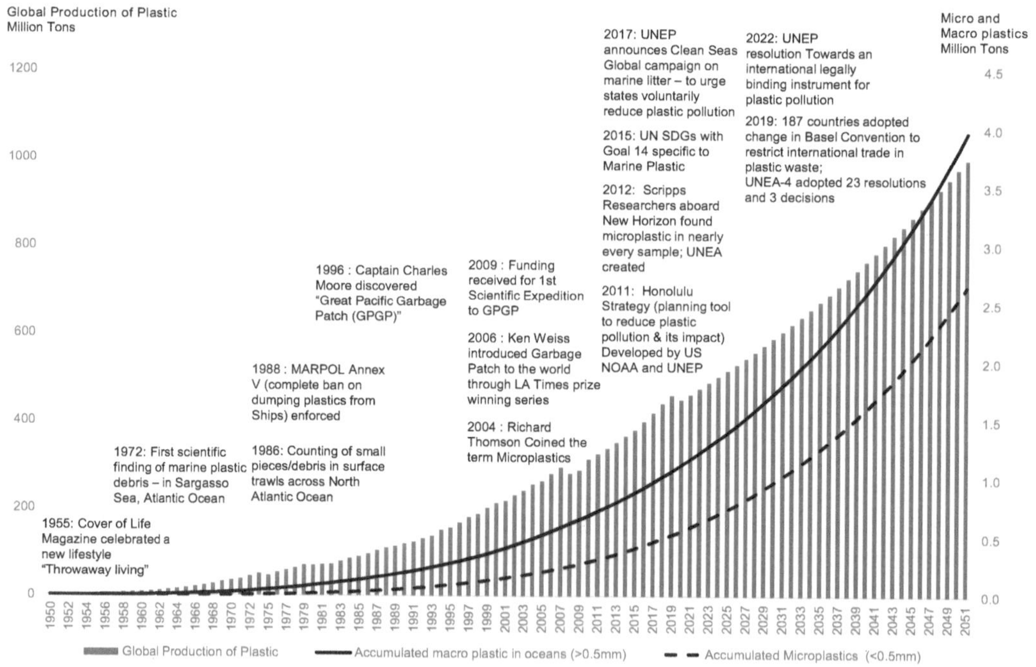

Figure 17.1 Global production of plastics and plastic waste.

Note: MARPOL = International Convention for Prevention of Pollution from Ships); UNEA = United Nations Environment Assembly; NOAA = National Oceanic and Atmospheric Administration; UNEP = United Nations Environment Programme; SDGs = Sustainable Development Goals.

Source: Data from Geyer et al. (2017), Lebreton et al. (2019), and OECD (2022); policy information from multiple sources.

2.5%). Though the growth in plastic production is expected to stabilize, the share of plastic production in developing countries like India and sub-Saharan Africa is expected to rise from 7% to 13% and from 4% to 9%, respectively, while that in developed countries, like the European Union (EU) and the USA, is expected to decline from 18% to 15% and 13%, respectively (Geyer et al., 2017). Since developing countries have limited capability to manage waste, rising consumption in these countries also leads to higher overall waste generation (Kaza et al., 2019).

17.2.2 Plastic Pollution

As plastic consumption and production increased, the problem of plastic pollution simultaneously started emerging. The first scientific evidence of plastic pollution was found in the oceans in 1972 in the Sargasso Sea (Atlantic Ocean), where scientists from the Woods Hole Oceanographic Institution found concentrations of 50 to 12,000 plastic particles (collectively weighing 1.07 kilograms) per square kilometer (Carpenter and Smith, 1972). Since then, in the 1980s, several reports have been compiled by researchers who examined the same area each year, towing nets along the surface twice a day to collect biological samples and ocean debris (Lawrence, 2010). In 1997, Charles Moore discovered the "Great Pacific Garbage Patch" in the middle of the North Pacific Gyre, which led to widespread awareness of the extent of plastic pollution (Rochman, 2020). Since most plastics break down into smaller particles, in 2004, the term *microplastics*, referring to tiny plastic particles of less than 5 mm in size, was coined (Thompson et al., 2004). In terms of weight, macroplastics are significantly more; however, in terms of particle count, both on land and in the sea, microplastics are the dominant source of plastic pollution. In 2009, a similar expedition from the Scripps Institution of Oceanography found microplastics in almost every sample they examined. They emphasized the importance of extending the search for plastic pollution beyond oceans to land. In 2012, Scripps researchers found microplastics nearly in every sample (Goldstein et al., 2012). The finding that most plastic pieces were microplastics illustrated the need to shift from cleanup to preventing plastic pollution from land and other sources. Figure 17.2 gives the historical sketch of the plastic pollution.

Figure 17.3 presents a global material flow of plastics, showing how plastic was consumed, disposed of, and eventually leaked into the oceans in 2019. Of the total of 385 million metric tons

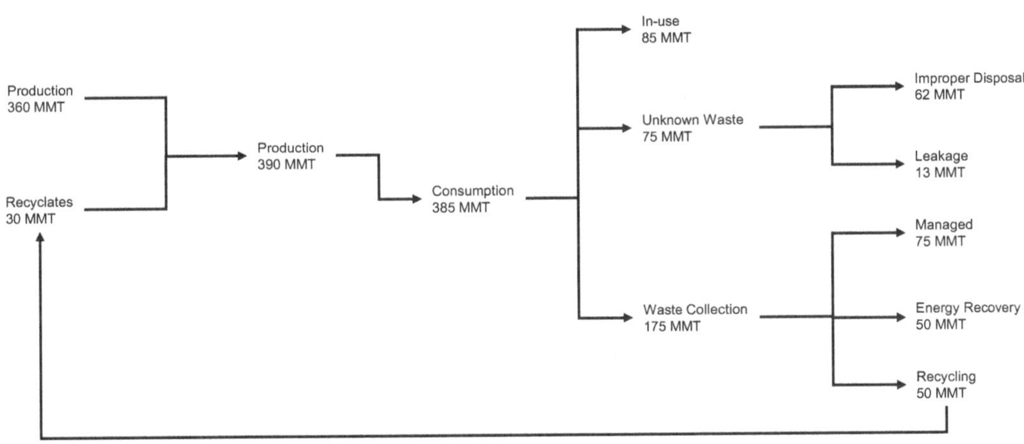

Figure 17.2 Global flow of plastics for 2019.

Note: Data for 44 countries covering 80% of global GDP.

Source: Data from Conversio (2020).

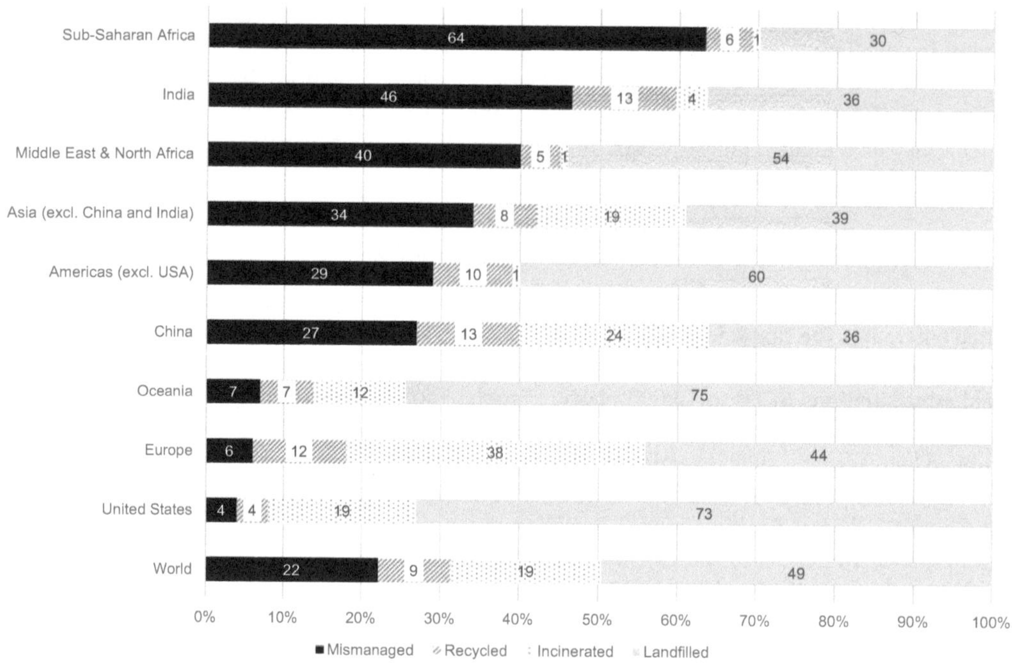

Figure 17.3 Handling of plastic waste.

Source: Data from OECD (2022).

of plastic consumed in 44 major countries covering 80% of global GDP, less than 15% (50 MMT) was recycled, while nearly 20% (75 MMT) was leaked into the environment and mismanaged. The problem of mismanaged plastic waste, as indicated in Figure 17.3, is higher in developing countries, like India, where 40% of the plastic is mismanaged. In contrast, in lower-income countries, like sub-Saharan Africa, over 64% of the plastic waste is mismanaged (OECD, 2022).

17.2.3 Policies to Manage Plastic Pollution

The first scientific evidence of plastic pollution was found in the early 1970s. However, the seriousness, and extent, of the problem was only noticed in the mid-2000s. Until then, plastic pollution was primarily seen as a marine waste problem. After 2010, when evidence of plastic particles was found in all aspects of human life, like drinking water and food chains, there was a quantum leap in research to understand the problem from different angles (Rochman, 2020). As understanding of the problem increased, numerous policy instruments were devised at all levels, from transnational to subnational, to curb plastic pollution. Figure 17.4 gives the trend of policies framed at different levels. As can be seen from the figure, at a transnational level, there have been more commitments and plans, while at regional and national levels, there have been stricter responses in the forms of regulatory instruments to tackle different types of plastics, like bags and single-use plastics.

The data shows that between 2000 and 2023, over 900 policies have been enacted at the transnational, regional, national, and subnational levels (Karasik et al., 2022). Among the 192 countries that are part of the UN, by 2018, 127 had a policy to control plastic bags, 27 had legislation for single-use products, and 63 countries had policies like extended producer responsibility (EPR) and deposit refund schemes in place (UNEP, 2018).

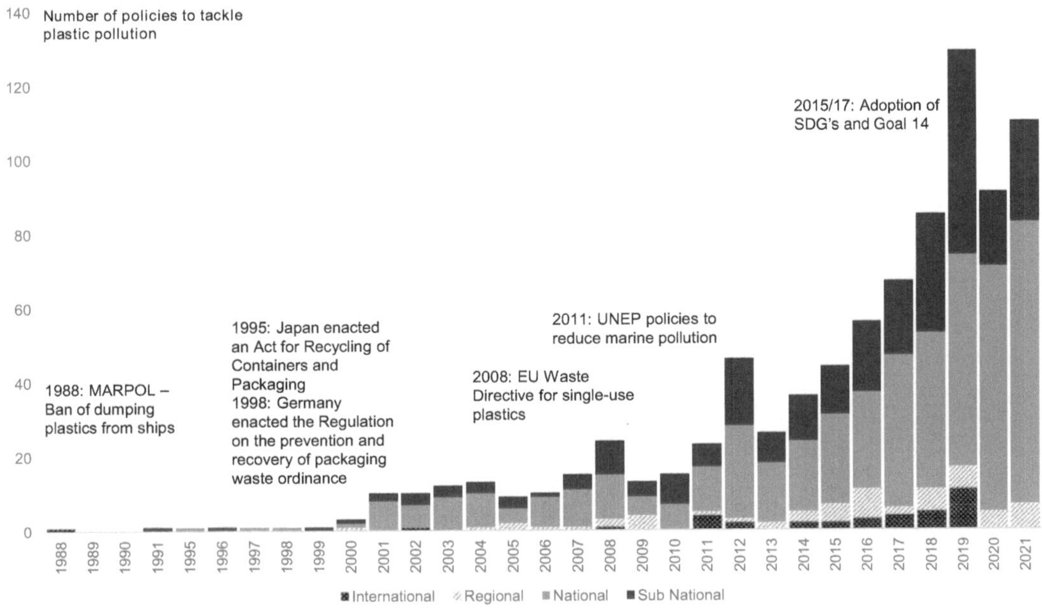

Figure 17.4 Policies to manage plastic waste based on regions.

Source: Data from Karasik et al. (2022).

As expected, most of these policies have been framed in high-income countries, followed by upper-middle-income countries, and then lower-middle-income countries.[2] Figure 17.5 gives the distribution of the enactment of these policies for different income category countries. As can be seen from the figure, the lower-income countries started enacting policies in 2001 only, with the pace picking up in the last few years.

17.2.4 Geographic Classification

Since the plastic pollution problem was initially perceived as a marine problem at a transnational level, in 1988, MARPOL (International Convention for the Prevention of Pollution from Ships) enforced a complete ban on dumping plastic from ships. In 2011, the US National Oceanic and Atmospheric Administration, as part of the UNEP (United Nations Environment Programme), developed a planning tool to reduce marine plastic pollution. Subsequently, numerous resolutions by UNEP and the UN General Assembly (UNGA) through the United Nations Environment Assembly (UNEA) have been adopted to control marine litter (i.e., UNEA Resolution 4/7). Bearing a few policies like the UNEA Resolution 4/10 Addressing "Single-Use Plastic Products Pollution," most transnational policies, including the SDGs adopted in 2015, primarily focus on controlling plastic waste in oceans.

At a regional level, the policies focus on marine plastic pollution and land-based plastic pollution. For example, the East African Community Polythene Materials Control Bill (2015) focuses on plastic bags in eight East African countries, while the EU Directive 2018–852 of the European Parliament and of the Council on Packaging and Packaging Waste (2008) focuses on curbing single-use plastic pollution in the European Union.

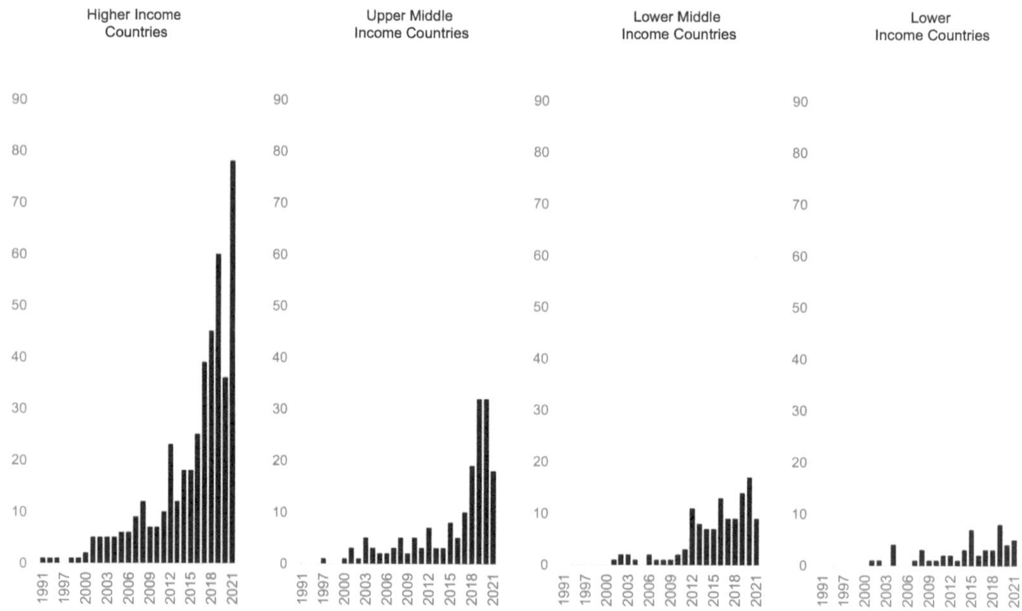

Figure 17.5 Policies to manage plastic waste based on income levels.

Source: Data from Karasik et al. (2022), and income levels from Kaza et al. (2019).

At a national level, in the 1990s, numerous policies were enacted to control packaging waste, of which plastic packaging waste was a significant component. In 1995, Japan enacted an Act for Recycling of Containers and Packaging, and in 1998, Germany passed the Regulation on the Prevention and Recovery of Packaging Waste ordinance. In the early 2000s, several countries, like China, Ireland, and Singapore, enacted laws to control bags and bottle waste.

At the transnational level, the policies are primarily plans/commitments for future action, while at a national level, the predominant policies are regulatory in nature, with bans being a more prevalent strategy.

17.2.5 *Type of Policies to Manage Plastic Waste*

Karasik et al. (2022) classifies the policies to control plastic pollution into three categories—regulatory, economic, and informative. Regulatory policies are further classified into affirmative and prohibitive (e.g., limits and bans); economic policies either provide incentives through subsidies or disincentivize through fees or taxes. Lastly, informative or suasive policies use education and outreach programs to influence behavior. Table 17.1 gives a snapshot of the implementation of these policies in different countries.

Table 17.1 Types of policies to manage plastic waste

Sr No.	Type of policies	Classification	Instruments	Countries	Act
1	Regulatory	Affirmative	EPR, develop new or improve existing processes or product, post-leakage plastic capture	Indonesia, India	Indonesian Act No. 18/2008 on EPR; Guidelines on EPR for Plastic Packaging vide Plastic Waste Management (Amendment) Rules in India (2022)
		Prohibitive	Limit consumption, complete/partial ban plastics	China, USA	China Plastic Limit Order (2008); Florida (USA) Waste (and Plastic) Disposal Limit (2018)
2	Economic	Incentive	Cash for return, subsidies, tax breaks	South Korea	Busan City (S. Korea) Incentive Program for Collecting Marine Debris (2002)
		Disincentive	Fees, taxes, levy, duty	Denmark	Denmark plastic tax on disposable tableware, carrier bags, PVC, and phthalates (2003)
3	Informative (or suasive)		Research, data collection, data reporting, education or outreach programs, labels or placards	Australia, Canada	Melbourne Beach (Australia) Consumer Education Policy (2017); Vancouver (Canada) Education Campaign (2017)

Source: Adapted from Karasik et al. (2022).

17.3 Efforts to Curb Plastic Pollution in an Emerging Economy

India is expected to be one of the world's fastest-growing plastic consumers, with its share of global consumption increasing from 5% in 2023 to over 13% in 2050. Several rules have been framed to tackle the plastic menace in the last few years. For example, the Plastic Waste Management Rules 2016 (updated in 2018 and 2022) (PWM, 2022) focuses on plastic waste generated from packaging. The PWM (2022) has a broader canvass and talks about multiple policy instruments, like plastic bans, EPR, mandatory recycling, mandatory recycled content, and reuse policies. A key aspect of the policy is that all plastic packaging users, which include producers, importers, and brand owners (PIBOs), are required to collect waste equal to their consumption from 2024 onward, recycle 30% of the waste from 2024 onward, use 30% recycled material for rigid packaging from 2025 onward, and reuse 70% of large bottles (above 4.9 liters) from 2025 onward PWM (2022). As of 2022, over 5,400 registered PIBOs have collected over 2.26 million metric tons, which is approximately 23% of overall plastic waste generated (MoEFCC, 2023).

17.4 Sustainable Development Goals (SDG) to Manage Plastic Waste

Despite all these efforts at the national level, plastic pollution is on the rise. As everyone is recognizing this, efforts are being made at the multilateral level to curb plastic pollution. As mentioned, one such effort is the UN SDGs.

In 2015, the member countries of the United Nations adopted the SDGs to be achieved by each country by 2030. It consists of 17 goals with 169 targets, which are measurable against 247 unique indicators. Though initially managing or reducing plastic pollution was not considered as one of the key targets, subsequently, only a few SDGs had a specific mention of plastic pollution (Westerbos, 2019).

In 2017, the UN General Assembly conference supported the implementation of SDG 14, which focuses on conserving and sustainably using the oceans, seas, and marine resources for sustainable development (UNGA, 2017). Though SDG 14 has been regarded as one of the key goals that have a direct impact on plastic pollution (Walker, 2021), lately, there has also been increasing focus on SDG 12, which focuses on sustainable consumption and production patterns to reduce overall demand for plastic (Westerbos, 2019).

SDGs 14 and 12 are, thus, considered key goals in curbing plastic pollution. However, other goals also indirectly impact plastic pollution, as shown in Table 17.2, which tries to fathom the impact of different SDGs on plastic generation and pollution. Walker (2021), in his critical review, has identified at least 12 SDGs that are directly or indirectly impacted by microplastic pollution. Interestingly, the implementation of some SDGs would result in more use of plastic and, hence, plastic waste.

Under SDG 14, subsection 1 aims to accelerate actions to prevent and significantly reduce marine pollution of all kinds, particularly from land-based activities, including marine debris, plastics and microplastics, nutrient pollution, untreated wastewater, solid waste discharges, hazardous substances, pollution from ships, and abandoned, lost, or otherwise discarded fishing gear. It also aims to address, as appropriate, the adverse impacts of other human-related activities on the ocean and marine life. The goal aims to implement long-term and robust strategies to reduce the use of plastics and microplastics, in particular plastic bags and single-use plastics, including by partnering with stakeholders at relevant levels to address their production, marketing, and use.

Under SDG 12, several subsections aim to reduce the consumption of plastics. For example, SDG 12.2 focuses on reducing the consumption of resources, SDG 12.4 emphasizes the reduction of chemical wastes, SDG 12.5 focuses on recycling (and thus reducing waste), and SDG 12.6 encourages companies to follow waste management practices. At the same time, 12.8 aims to develop knowledge and research in the areas of sustainable consumption.

SDGs 1, 5, 8, and 10 indirectly impact curbing plastic pollution as they focus on generating jobs in the recycling/reuse sectors. When these SDGs are promoted, they indirectly help the recycling sector and reduce plastic pollution.

SDG 3 focuses on human health. In terms of plastic pollution, it is indirectly related to SDG 14. Under SDG 14, the target is to reduce marine pollution, which is consumed by aquatic life and plants and becomes a part of the food chain. Thus, when targets under SDG 14 are met, it indirectly meets the target under SDG 3.

SDG 7 promotes energy from renewable sources. Plastic waste is one of the waste streams that can be used as a renewable energy source through waste-to-energy (WtE) plants (Jardosh and Kathuria, 2023). Thus, meeting SDG 7 indirectly helps in reducing plastic waste.

SDGs 4 and 11 focus on increasing awareness of waste collection, segregation, and recycling, thus indirectly helping reduce pollution.

SDGs 9 and 15 promote the development of alternative materials from waste and biobased sources, thus reducing the demand for fossil-based materials.

SDGs 2 and 6 may result in increased plastic consumption, whereas SDGs 16 and 17 may not impact plastic consumption and waste generation.

Table 17.2 SDGs and their impact on controlling plastic pollution

SDG	*Objective*	*Impact on plastic consumption*	*Impact on curbing plastic pollution*	*Details*
1	End poverty in all its forms everywhere	Increases (providing water in plastic bottles and sachets, through plastic industry, trade)	Indirect (+/−)	Plastic industry generates jobs that reduce poverty; jobs (primarily in the informal sector) in the recycling/reuse sector indirectly reduce plastic pollution. Increased plastic trade may affect ecosystem services having negative impact on communities.
2	End hunger, achieve food security, and promote sustainable agriculture	Increases	Indirect (−)	Affordable packaged food increases plastic consumption; no direct impact on reducing plastic pollution but may increase pollution through microplastics in soils, food.
3	Ensure healthy lives and promote well-being for all at all ages	Increases	Indirect (+)	Generates jobs; affordable medicines increase consumption; no direct impact on reducing pollution but may reduce pollution when considered along with SDG 14—less plastic consumed by fish.
4	Ensure inclusive and equitable quality education	No impact	Indirect (+)	Environmental education and conscious use of plastic; selective collection, recycling, etc.
5	Achieve gender equality and empower all women and girls	No impact	Indirect (+)	Produces jobs for women (primarily in the informal sector) in the recycling/reuse sector, thus indirectly reduces plastic pollution.
6	Ensure availability and sustainable management of water and sanitation for all	Increases	Indirect (−)	Plastics are used extensively for irrigation and sanitation, thus increases consumption; presence of microplastics in treated wastewater effluent and drinking water.
7	Ensure access to affordable, reliable, sustainable, and modern energy for all	No impact	Indirect (+)	Promotes the use of waste-to-energy systems that are suitable for certain kinds of plastic waste, but may lead to GHG emissions.
8	Promote sustained, inclusive, and sustainable economic growth	No impact	Indirect (+)	Produces jobs (primarily in the informal sector) in the recycling/reuse sector, thus indirectly reduces plastic pollution.
9	Build resilient infrastructure and promote sustainable industrialization and foster innovation	Increases	Indirect (+)	New materials like lightweight panels increase consumption; use of environmentally friendly and sustainable plastics produced from industrial waste and biomaterials reduces consumption of fossil fuel–based polymers.

(*Continued*)

Table 17.2 (Continued)

SDG	*Objective*	*Impact on plastic consumption*	*Impact on curbing plastic pollution*	*Details*
10	Reduce inequality within and among countries	No impact	Indirect (+/−)	Produces jobs (primarily in the informal sector) in the recycling/reuse sector, thus indirectly reduces plastic pollution; recyclable plastic exports to poor countries may increase plastic pollution to these countries.
11	Make cities inclusive, safe, resilient, and sustainable	Increases	Indirect (−/+)	New materials for construction, sanitation increase consumption; inadequate waste management may increase pollution; focus on educating citizens in the correct management and recycling of waste, thereby reduces pollution.
12	Ensure sustainable consumption and production patterns	Reduces*	Direct (+)	Plastics used for packaging and agriculture increase consumption; focus on developing a circular economy, reducing end-of-life waste, and increasing domestic recycling, thus reducing pollution.
13	Take urgent action to combat climate change and its impacts	Reduces	Indirect (+)	Promotes sustainable use of plastics to reduce GHG emissions; promotes the use of biodiesel (sometimes made from plastic waste), thus reducing related pollution.
14	Protect life below water	Reduces	Direct (+)	Specifically addresses plastics floating in water, with greater emphasis on controlling microplastics that impact human health.
15	Protect life on land	Reduces	Indirect (+)	Supports the development of safer polymers that can be recycled easily and reduces the dependency on land resources.
16	Promote peace, justice, and strong institutions	No impact	No impact	None.
17	Strengthen global partnership for sustainable development	No impact	No impact	None.

Note: (+) means helps reduce pollution, and (−) means increases pollution.

*Advancement in irrigation and longer shelf life for packaged food will increase consumption, and sustainable manufacturing practices will reduce, so the net effect is reducing plastic consumption.

Source: Adapted from Sousa (2021), Walker (2021), and own inputs.

17.5 Discussion and the Way Forward

17.5.1 Overall Impact of the SDGs

The UN established 17 SDGs in 2015 as a global framework for achieving sustainable development by 2030. As the program reaches the halfway mark, abundant research attempts to assess the impact of these goals. Long et al. (2023) estimate that, as a whole, there would be around half the progress toward achieving SDGs, with significant challenges persisting for 28 vulnerable countries, particularly in areas like education, sanitation, and child health. Further, while SDGs help improve coordination among governments, weak governance and regulations hinder the effective deployment of tools necessary for implementation. Thus, it is often quoted that SDGs have facilitated discussions that have helped increase coordination among member countries and have also initiated actions; however, the scale of actions needs to be more substantial to fix the issues highlighted earlier (Long et al. 2023). Biermann et al. (2022) analyze over 3,000 scientific studies published between 2016 and 2021 and classify the impact of the SDGs from these documents as discursive, normative, and institutional changes. *Discursive impact* refers to the SDGs' effect on shaping the overall discussion and debate on sustainability. *Normative impact* is the effect it has had on designing the regulatory framework in line with the goals, while *institutional impact* is the effect SDGs have had in creating separate departments or institutions within countries to achieve their targets. Together, these three are called transformative effects. The study concludes that SDGs have had some political impact on institutions and regulations, primarily in shaping the discussion on sustainability. However, more profound changes, such as concerted action or allocating larger resources toward these goals, remain rare. Overall, the evidence suggests that the SDGs so far have only had limited transformative political impact (Biermann et al. 2022).

Carlsen and Bruggemann (2022) rank 102 countries based on their achievement of the SDG targets and point out that countries focused on SDGs 12, 13, 14, and 17, that is, goals primarily focused on the environment, have higher overall rankings. SDGs 1, 6, and 7, focusing on poverty, clean water, and clean energy, have an overall lower impact on rankings, which also signifies that member countries have given these SDGs less importance. While Medina-Hernández et al. (2023) point out that higher-income countries will fare better at achieving their overall SDGs compared to lower-income countries, Carlsen and Bruggemann (2022) point out that high-income countries are expected to see a relatively high achievement of SDG 5 and a low achievement for SDG 12. Similarly, OECD countries tend to have achievements for SDGs 12 and 13, while SDG 3 receives a relative achievement.

Though there is a consensus that the overall objectives of the SDG might not be achieved, SDGs have benefitted countries at different levels. At a transnational and regional level, SDGs have helped shape the political debate in the direction aimed by the SDGs (Biermann et al., 2022) as well as have helped improve the coordination effort between countries (Medina-Hernández et al., 2023). Countries are encouraged to submit their voluntary national review (VNR) at a national and subnational level. VNRs help benchmark a country's individual sustainability targets with the larger SDGs. By 2022, 187 member countries had submitted at least one VNR assessing a country's overall progress (Long et al., 2023). As VNRs become more transparent and detailed, countries are coaxed into delivering on the targets.

17.5.2 The SDGs' Impact on Plastic Pollution

The story of plastic pollution began decades ago in remote central ocean gyres; the plastic debris is now pervasive across the environment, in oceans, animal habitats, and food chains. The diverse waste of plastics in the environment comes from hundreds of different products with diverse polymer backbones and chemical additives, making plastic pollution complex and different from other chemical pollutions, like pesticides or trace metals (Rochman, 2020). The cumulative total production of plastic (1950 to 2015) is over 8.3 billion tons, of which 6.3 billion tons are waste (Geyer et al., 2017). Of this global waste generated so far, only 9% has been recycled, 12% has been incinerated, and 79% has been diverted to landfills or leached into the natural environment (Geyer et al., 2017). The question now is not whether there is plastic pollution but, instead, to what extent plastic debris affect global change and how much effort we can put into solving the problem of plastic pollution.

A recent study by Borrelle et al. (2020) has estimated that of the 191 million tons of plastic waste generated in 2016, 11% (i.e., nearly 21 MT) has entered the aquatic ecosystem. This may increase to over 53 million tons every year by 2030. This has implications for the success of SDGs 2, 3, 6, 12, and 14 (Walker, 2021).

Despite the acute nature of the problem, curbing plastic pollution was not one of the explicitly mentioned objectives. It was only added in 2017, under one of the seven parts of SDG 14. While Walker (2021) points out that this SDG is insufficient in isolation to control overall plastic pollution as it focuses only on microplastics and contamination in oceans, Haas (2023) points out that targets mentioned in these SDGs are vague and often not easily quantifiable. The OECD (2022) assesses that only a globally coordinated effort that includes international legally binding instruments can curb plastic pollution.

Among the seven targets within this SDG, the timeline for four of the seven targets, namely, 14.2, 14.4, 14.5, and 14.6, focused on the protection of coastal ecosystems and preventing overfishing has already expired, with no country having achieved all four of them (Haas, 2023). Even Indicator 14.1.1b, which addresses plastic debris, does not address microplastics. Even the quantification of floating plastic marine debris (>2.5 cm) is mainly based on visual observations by citizen scientists (Ambrose et al., 2019). This implies that no specific target is mentioned to reduce microplastics and also suggests a lack of baseline data to monitor any reduction. Thus, while overall achievement of the SDGs is expected to be around the halfway point, SDG 14 is regarded as the least-analyzed and least-implemented SDG, with challenges across different income groups and geographic regions (Haas, 2023).

Notes

1 Source: https://wesr.unep.org/article/sustainable-development-goals-0#:~:text=The%20SDG%20 framework %20hs%20a,including%20the%20interactions%20between%20topics, accessed in January 2024.

2 The income classification is the World Bank definition, where high-income countries have per capita income (PCI) of greater than US$12,695, upper-middle have PCI from US$4,096 to US$12,695, lower-middle-income countries have PCI falling between US$1,096 and $4,095, and lower-income countries have PCI below US$1,096 (Kaza et al., 2019).

References

Ambrose, K., Box, C., Boxall, J., Brooks, B., Eriksen, M., Fabres, J., Fylakis, G. & Walker, T. (2019). Spatial trends and drivers of marine debris accumulation on shorelines in South Eleuthera, The Bahamas using citizen science. *Marine Pollution Bulletin*, 142: 145–154.

Biermann, F., Hickmann, T., Sénit, C.A., Beisheim, M., Bernstein, S., Chasek, P., Grob, L., Kim, R., Kotze, L., Nilsson, M., Llanos, A., Okereke, C., Pradhan, P., Raven, R., Sun, Y., Vijge, M., Vuuren, D. & Wicke, B. (2022). Scientific evidence on the political impact of the Sustainable Development Goals. *Nature Sustainability*, 5: 795–800.

Borrelle, S.B., Ringma, J., Law, K.L., Monnahan, C.C., Lebreton, L., McGivern, A., Murphy, E., Jambeck, J., Leonard, G.H., Hilleary, M.A. & others (2020). Predicted growth in plastic waste exceeds efforts to mitigate plastic pollution. *Science*, 369(6510): 1515–1518. https://doi.org/10.1126/science.aba3656.

Carlsen, L. & Bruggemann, R. (2022). The 17 United Nations' sustainable development goals: A status by 2020. *International Journal of Sustainable Development & World Ecology*, 29(3): 219–229.

Carpenter, E.J. & Smith, K.L. (1972). Plastics on the Sargasso sea surface. *Science*, 175(4027): 1,240–1,241. https://doi.org/10.1126/science.175.4027.1240.

Conversio (2020). *Global Plastics Flow 2018*. Prepared by Conversio Market & Strategy GmbH, Mainaschaff, Germany. Report available on www.conversio-gmbh.com/res/News_Media/2020/Summary_Material-Flow-Analysis-_Plastics_Germany_2019_EN.pdf last accessed on 15 Jan 2024.

Geyer, R., Jambeck, J. & Law, K. (2017). Production, use, and fate of all plastics ever made. *Science Advances*, 3(7): e1700782.

Goldstein, M., Rosenberg, M. & Cheng, L. (2012). Increased oceanic microplastic debris enhances oviposition in an endemic pelagic insect. *Biology Letters*, 8: 817–820.

Haas, B. (2023). Achieving SDG 14 in an equitable and just way. *International Environmental Agreements: Politics, Law and Economics*, 23: 199–205.

Jardosh, N. & Kathuria, V. (2023). Social cost--benefit analysis of solid waste management options with application to Mumbai, India. *Waste Management & Research*. https://doi.org/10.1177/0734242X241231401.

Karasik, R., Bering, J., Griffin, M., Diana, Z., Laspada, C. & Schachter, J. (2022). *Annual Trends in Plastic Policy: A Brief (NI PB 22-01)*. Published by Duke University, Durham, North Carolina, USA. Report available on https://nicholasinstitute.duke.edu/sites/default/files/publications/Annual-Trends-in-Plastics-Policy-A-Brief_2.pdf last accessed on 15 Jan 2024.

Kaza, S., Yao, L., Bhada-Tata, P. & Woerden, F. Van. (2019). *What a Waste 2.0*. World Bank Group, Washington, DC. Report available on http://hdl.handle.net/10986/30317 last accessed on 15 Jan 2024.

Law, K., Morét-Ferguson, S., Maximenko, N., Proskurowski, G., Peacock, E., Hafner, J. & Reddy, C. (2010). Plastic accumulation in the North Atlantic subtropical gyre. *Science*, 329(5996): 1185–1188.

Lawrence, D. (2010). *Plastic Particles Permeate the Atlantic*. Article published in Oceanus—Journal of Our Planet by the Woods Hole Oceanographic Institution, MA. Article available on www.whoi.edu/oceanus/feature/plastic-particles-permeate-the-atlantic/last accessed on 15 Jan 2024.

Lebreton, L., Egger, M. & Slat, B. (2019). A global mass budget for positively buoyant macroplastic debris in the ocean. *Scientific Reports*, 9(12922).

Long, G., Censoro, J. & Rietig, K. (2023). The sustainable development goals governing by goals, targets and indicators. *International Environmental Agreements: Politics, Law and Economics*, 23, 149–156.

Medina-Hernández, E.J., Guzmán-Aguilar, D.S., Muñiz-Olite, J.L. & Siado-Castañeda, R. (2023). The current status of the sustainable development goals in the world. *Development Studies Research*, 10(1): 2163677. https://doi.org/10.1080/21665095.2022.2163677.

MoEFCC (2023). *Extended Producer Responsibility (EPR) for the Year 2022–23*. Published by the Ministry of Environment, Forest and Climate Change, Government of India, New Delhi, India. Press Release available on https://pib.gov.in/PressReleasePage.aspx?PRID=1900282 last accessed on 15 Jan 2024.

Napper, I.E., Baroth, A., Barrett, A.C., Bhola, S., Chowdhury, G.W., Davies, B.F.R., Duncan, E.M., Kumar, S., Nelms, S.E., Hasan, N., Nishat, B., Maddalene, T., Thompson, R.C. & Koldewey, H. (2021). The abundance and characteristics of microplastics in surface water in the transboundary Ganges River. *Environmental Pollution*, 274(116348).

OECD (2022). *Global Plastics Outlook*. OECD Publishing, Paris, France. Data available onwww.oecd-ilibrary.org/environment/data/global-plastic-outlook_c0821f81-en last accessed on 15 Jan 2024.

PWM (2022). *Plastic Waste Management Rules 2022*. Published by the Ministry of Environment, Forest and Climate Change, Government of India, New Delhi, India. Rules available at https://environment.delhi.gov.in/ sites/default/files/inline-files/pwm_epr_1.pdf last accessed on 15 Jan 2024.

Rochman, C. (2020). The story of plastic pollution. *Oceanography*, 33(3): 60–70. https://doi.org/10.5670/oceanog.2020.308 last accessed on 15 Jan 2024.

Sousa, F. (2021). The role of plastic concerning the sustainable development goals: The literature point of view. *Cleaner and Responsible Consumption*, 3(100020).

Thompson, R., Olsen, Y., Mitchell, R., Davis, A., Rowland, S., John, A., McGonigle, D. & Russell, A. (2004). Lost at sea: Where is all the plastic? *Science*, 304(5672): 838.

UN (2023). *Goal 14: Conserve and Sustainably Use the Oceans, Seas and Marine Resources Published by United Nations*. Details available on www.un.org/sustainabledevelopment/oceans/#:~:text=Worryingly%2C%20 marine %20pollution%20is%20reaching,harmful%20type%20of%20ocean%20pollution last accessed on 4 Mar 2024.

UNEP (2018). *Legal Limits on Single-Use Plastics and Microplastics: A Global Review of National Laws and Regulations*. Report published by United Nations Environment Program (Nairobi, Kenya) Developed in Partnership with the World Resources Institute, Washington, DC. Report available on https://wedocs.unep.org/bitstream/handle/20.500.11822/27113/plastics_limits.pdf?sequence=1&isAllowed=y last accessed on 15 Jan 2024.

UNEP (2022). *End Plastic Pollution: Towards an International Legally Binding Instrument*. Published by United Nations Environment Program, Dakar, Senegal. Document available on https://wedocs.unep.org/bitstream/handle/20.500.11822/39812/OEWG_PP_1_INF_1_UNEA%20resolution.pdf last accessed on 25 Feb 2024.

UNGA (2017). *Resolution Adopted by the United Nations General Assembly on 6 July 2017, Titled Our Ocean, Our Future: Call for Action (71/312.)*. Resolution available onhttps://ggim.un.org/documents/a_res_71_313.pdf last accessed on 15 Jan 2024.

Walker, T. (2021). (Micro)plastics and the UN Sustainable Development Goals. *Current Opinion in Green and Sustainable Chemistry*, 30(100497).

Westerbos, M. (2019). *How the Sustainable Development Goals Can Help Fight 'Plastic Soup'*. This article is partly based on articles published on plasticsoupfoundation.org and was originally published by Impakter.com. Article available on https://bthechange.com/how-the-sustainable-development-goals-can-help-fight-plastic-soup-3ce746fc1cba last accessed on 15 Jan 2024.

18

SOCIAL INNOVATION AND LOCAL SUSTAINABLE DEVELOPMENT

A Literature Review

Maria João Ruela, João Simão, and Filipe Almeida

18.1 Introduction

Innovation, including social innovation, has been recognized as a key factor in promoting sustainable development (Nijnik et al., 2019). By restructuring social practices, it addresses social and environmental challenges in a responsive manner. There exist worldwide examples of social innovation linked to every Sustainable Development Goal: Poverty elimination in Ghana (Osei & Zhuang, 2020), responsible consumption and production in Australia (Barraket, 2020), the fight against energetic poverty in Europe (Manjon et al., 2022) and climate change impact in Nepal (Khadka et al., 2018), sustainable city contributions in Brazil (Andion et al., 2022), and peacebuilding efforts in Colombia (Castillo & Valderrama, 2017) exemplify this diversity and multiplicity. This fact has justified the European Union's funding of socially innovative projects, as they provide an opportunity for interaction between public policies, the state, and society. However, theoretically, there remain lack of clarity and competition between meanings, which has implications for defining the object of study and framing the concepts of social innovation and sustainable development.

Based on a literature review on the relationship between the two concepts, it is proposed to explore this link at the local level, referring to three challenges identified in the implementation and governance of the SDGs: collective action, trade-offs, and evaluation.

18.2 Social Innovation: Concept and Public Policies

We have selected a definition of *social innovation* that combines components reflecting its pragmatic and systemic nature while considering its different unique aspects, such as objectives, scope, actors, and dynamics. Thus, *social innovation* presupposes a new idea or solution (product, service, model, or process) that simultaneously responds to a social need (more effectively than existing alternatives), leading to the creation of new forms of social relationships or collaboration and/or the optimization of the use of goods or resources. These innovations advance the welfare and ability of communities to enact change in behavior and governance models within social systems in which the innovations occur (Hubert, 2011; Murray et al., 2010; Westley & Antadze, 2010). It

DOI: 10.4324/9781003285472-19

is a definition that considers the duality of process and result and incorporates the territorial aspect of social innovation. This aspect provides an advantageous setting for both formal and informal participatory dynamics that arise from the empirical nature of the practice. It is assumed that the adopted definition of *social innovation* addresses three objectives: change, intersectoral collaboration, and participation. It is further assumed that the intended change is directly linked to the sustainable development concept.

The idea of social innovation has been included in the European Union's strategic documents since the early 21st century (Sabato et al., 2017). Differences existed prior to and subsequent to 2010, which is the year of publication of the BEPA[1] report (Hubert, 2011). The report concluded the necessity of actively promoting the widespread adoption of social innovations as a component of the EU's toolbox to effectively address poverty, generate sustainable wealth and well-being, and promote a learning and participative society. This argument was utilized in the Europe 2020 strategy options.

This social investment agenda has two consequences. Firstly, it alters the method of evaluating public policies by emphasizing outcomes in terms of social impact and fostering collaborations with other social investors. Secondly, it enhances proximity by implementing and experimenting with solutions at a decision-making level that is closer to the issue (Maduro et al., 2018).

The promotion of innovative public policies highlights the connection between social innovation and co-creation.[2] By advocating for self-organized communities as a substitute for conventional public service models, citizens are considered as "initiating co-creators"[3] of novel public services. This is particularly true in key sectors, such as health care, housing, and renewable energy access (Bekkers et al., 2014; Voorberg et al., 2015).

However, citizen participation in public processes remains insufficient to overcome the symbolic dimension. In other words, co-creation is viewed only as a value in itself, with limited potential to effect meaningful change. It is often implemented as part of political legitimization strategies for issues perceived as having democratic deficits, such as sustainability. The opposite side of this coin suggests that co-creation processes alone do not guarantee adequate citizen participation in the public response to their needs. This is why international governance bodies, like the European Union, have introduced policies to promote social innovation. A deeper examination of this issue reveals the need to investigate the concrete outcomes of co-creation and social innovation projects in promoting active citizen participation, which we term the "Voorberg Gap."

When social innovation is used to co-create public initiatives, it establishes a relationship based on networked governance. The development process involves multiple actors from the public, private, and social sectors. Utilizing both top-down and bottom-up strategies, the goal is to democratize the public sector, with social innovation being both the stimulus and the product of this approach. The defense of this conceptualization of social innovation lies with researchers who focus on the connection between social innovation and territorial development. They have verified empirically that successful governance partnerships exist on a territorial scale (Moulaert et al., 2019). The analysis of the territorial context of social innovation is pertinent to the discussion on its relationship with sustainable development.

18.3 Sustainable Development

Sustainable development has undergone a conceptual evolution toward becoming an agenda that governs the balance between economic and social progress and environmental limits that must not be exceeded (Dias et al., 2020; Dryzek, 2005; Redclift, 2005). In this confrontation, outlined since the last quarter of the 20th century, sustainable development can be described as a good idea, an ideal and conciliatory vision for global society, a guiding principle for action (Hugé et al., 2013)—although this good idea, as it has been commonly accepted, contains paradoxes and contradictions (Baker, 2006; Lele, 2013), accommodates different discourses (Hugé et al., 2013) and can be seen as an oxymoron (Robinson, 2004; Redclift, 2005).

The 2030 Agenda draws from the experience of the Millennium Development Goals, a comprehensive and ambitious worldwide blueprint for enhancing human well-being, protecting the environment, and promoting prosperity (UN, 2015). Sachs (2015) defines the SDGs as a milestone in the era of sustainable development. He states that sustainable development should determine the interactions between the economy, environment, politics, and culture, and their influence on social prosperity and environmental sustainability. Moreover, he emphasizes the importance of a holistic approach toward human well-being, including economic progress, social relations, and environmental sustainability.

In the meantime, we may be observing the emergence of a new period, which potentially presents an exceptional opportunity to assess the approaches taken toward achieving sustainable development. An opportunity has emerged from the scientific and social disruptions caused by the pandemic, further amplified by recent global movements like Black Lives Matter and the student-led climate strikes of Fridays for Future or Extinction Rebellion. This potential is enhanced by advancements in technology and communication platforms, allowing for greater global mobilization. It was also an opportunity to affirm the importance of multidisciplinary approaches to sustainable development that are well suited for interpreting the dynamics of the new reality. Implementing sustainability policies in the post-Brundtland era requires articulating the concept of sustainable development with a diverse array of epistemological perspectives, various practices and interpretations of development, and a spatial dimension that spans from local to global. Sustainable development can take the form of a normative principle, a political objective, or a contested argument (Sneddon et al., 2006). The intellectual stance is valuable for assessing the concept of sustainable development in spheres that extend beyond the scientific, academic, institutional, and political realms. This approach reaches citizens who possess the ability to voice their concerns and take action, often using innovative tools (Leipold et al., 2019). Explanations of the political and institutional aspects, as well as inquiries into why citizens opt to support sustainable development as a means to lead a more meaningful life or adhere to their present values, hold equal weight in significance. In this context, social innovation has facilitated the bridging of two agendas, an observation that has been noted by scholars, including Edwards-Schachter and Wallace (2017) and Eichler and Schwarz (2019), as well as by institutions, such as the European Commission (SCU, 2014). The SDG agenda has enabled the convergence of sustainable development and social innovation, allowing for the development of public incentive programs, community action, and private or social initiatives.

Thus, achieving sustainable development necessitates a call to action that mobilizes all stakeholders through a political system that guarantees active citizen participation in decision-making (WCED, 1987; UN, 2015). A collaborative approach is viewed as a vital component of attaining sustainable development. An integrated aspect of sustainable development supports the significance of creating partnerships between different sectors (public, private, and social), along with governance models. In this context, one can assume a correlation with social innovation, as the participation of the general population is crucial for the success of sustainability endeavors, even only symbolically. Baker and Mehmood (2015, p. 328) classify the effort undertaken by A21L as an example of social innovation: "[W]e hold that LA21 is an example of social innovation, one that is particularly interesting from the point of view of promoting collective action for sustainable place making."

The participation of populations also highlights the connection between governance and sustainable development and reinforces the likelihood of similar discourses and possibilities. This has led to a significant change in the ideological and paradigmatic conversation about sustainability governance. Although political actors, whether at the national or local level, retain power, the discourse that prioritizes citizens' active participation or co-creation is gaining prominence. The discourse serves several purposes, such as legitimizing policies, reducing the risk of conflicts, providing additional sources of information or ideas, and ensuring increased expertise of interlocutors in specialized fields, like environmental issues (Kemp & Gibson, 2005).

The idea of participation is reintroduced in this context, and it can be interpreted as a "normative principle" for sustainable development (Baker, 2006). The author incorporates the principle of participation into a set of values that have been linked with the discussion of sustainable development since 1987. These values have become a crucial resource for the definition of *sustainable development*, such as common but differentiated responsibilities, intergenerational equity, intragenerational equity, justice, gender equality, and the principle of participation.

To address the challenge of participation, the European Union has promoted the pursuit of good governance at the institutional level: "openness and participation, accountability, effective coherence, efficiency (proportionality) and greater sensitivity to the immediate context that is promised by subsidiarity" (CEC, 2001). Arguments that are also used as justification for the political subscription to the social innovation agenda.

18.4 Social Innovation and Sustainable Development

While the interrelationship between social innovation and sustainable development grows increasingly evident, only a limited number of scientific articles examine this relationship (Haskell et al., 2021). The analysis recovers the duality between strong and weak sustainability[4] (Ayres et al., 1998; Hugé et al., 2013; Neumayer, 2003), providing a basis for a conceptual differentiation of social innovation for strong and weak versions of sustainable development. In the first instance, social innovation promotes alternative solutions to the ecological crisis, particularly when executed on a localized level, with collaboration as the focal point.

The concept of the "Anthropocene" as a heuristic device for social innovation in theory and practice concerns the link to strong sustainability and entails a fresh perspective for action. Social innovation should not just aim to modify human relations but should principally innovate the relations between humanity and the planet, as well as between social systems and ecosystems. Thus, the analysis scope must consider the temporal urgency dimension when practicing social innovation (Olsson et al., 2017). Efforts to promote a "good Anthropocene" revive the notion of the social system as a domain for social innovation. According to Westley and Antadze (2010), a *social system* may be defined as an organized assembly of human resources, beliefs, and procedures that are united and regulated by interaction or interdependence and ultimately geared toward accomplishing a set of specific functions. Social innovation is seen as a solution to issues resulting from the interactive dynamics of socio-ecological systems, contributing to creating more resilient communities.

Additionally, there is a proposal for a third dimension of social innovation, encompassing a value dimension to incorporate the ethical commitment of sustainable development, which prioritizes the well-being of future generations. The concept of "irreplaceable goods" serves as a conceptual tool for sustainable social innovation. To achieve sustainable development, this concept must be translated into actionable ideas, processes, actions, or results that promote the preservation of irreplaceable goods. The goal is to ensure an adequate level of well-being for future generations.

Examples of such initiatives include platforms for reporting environmental crimes and microcredit instruments (Vasconcellos Oliveira, 2021).

Other scholars (Parra, 2013; Baker & Mehmood, 2015) have highlighted that social innovation is implicitly viewed in the context of social sustainability, which is linked to the material conditions ensuring inter- and intragenerational equity from a macro and normative perspective of sustainable development. Nonetheless, it fails to emphasize governance and participation, leading to limited academic attention to the association between social innovation and sustainable development. Moreover, social innovation's empirical orientation associates it with the micro-level aspect of sustainable development. These authors assert that social innovation ought to be examined in the context of the correlation between collaborative governance and local sustainability, which is one of the top five areas of scholarly attention in examining both concepts (Galego et al., 2021).

Theorizing the interconnection between the two concepts should commence from a micro level, emphasizing the significance of sustainable development's local dimension as a project of a particular community reliant on its context's material conditions. However, do various social innovation strategies exist within diverse social or territorial contexts? Baker and Mehmood (2015) identify two preconditions necessary for the emergence of social innovation in an enabling environment. The first precondition is the presence of networks among civil society organizations, institutions, dedicated public agencies, ecosystems, investment instruments, and social investors with a territorial dimension. The second precondition is the existence of a supportive environment that facilitates social innovation.

At a specific moment in a community's history, there is an opportunity to counteract a certain trend toward globalization and technology that undermines localism. Additionally, the presence of natural and ecological resources in a given location is an essential aspect of its material existence. The neglect of these resources as integral to community life has contributed to the current environmental crisis.

However, variations exist in the material conditions, enabling social innovation versus justifying its adoption and diffusion by communities. Achieving harmony between the goals of social innovation and sustainable development additionally entails evaluating the conditions and incentives that motivate communities to undertake sustainability projects. In this regard, Mota (2020) acknowledges the significance of the profile of local political leaders, who ought to possess certain characteristics or strive for change in this direction. In order to promote a truly inclusive approach, it is essential to listen to minority voices and possess a high level of empathy, while also knowing how to bring together local institutions and social movements in order to build coalitions of interests. It is important to promote a new rationality that goes beyond the market, rescuing a form of rationality that is rooted in the common good. Additionally, it is essential to have the ability to manage the different values that are present in the decision-making process, both cultural and social ones. It is recommended to replace accountability with a form of dialogical justification in order to generate a permanent process of collective learning. Lastly, it is crucial to promote shared leadership that mobilizes the resources and skills of different communities.

In conclusion, it is possible to establish points of contact between social innovation and sustainable development by emphasizing the empirical foundation of both concepts and promoting cross-sector partnerships to ensure community participation.

18.4.1 Social Innovation and the Sustainable Development Goals

Another line of analysis of the relationship between social innovation and sustainable development focuses on the operational lens of the SDGs. Social innovation and sustainable development are closely intertwined, as recognized in European Union documents[5] and in the United Nations' effort to implement the 2030 Agenda. Millard (2018) contends that the increasing conversation among the two spheres holds broader significance, as it fosters a fresh model of innovation and knowledge by acknowledging civil society's significance.

Eichler and Schwarz (2019) propose the SDG matrix as a classification scheme for social innovation since they have found this link in most published case studies. The connection between a social innovation project and specific goals is most apparent concerning goals that address so-called direct needs and in most cases is identified in relation to SDG 1, poverty eradication, and SDG 3, quality health. The advantage of adopting the SDGs as a categorization system for social innovation is their broad scope, including social problems more directly related to the well-being of communities, and also those that address indirect problems, such as SDG 17, partnerships for development. However, this classification presents a view of social innovation that primarily emphasizes outcomes while neglecting the process dimension, which is more comprehensive, systemic, and adaptable to its various contexts. This proposal also enables the perpetuation of the "cherry-picking" phenomenon, wherein certain SDGs are prioritized and chosen over others through empirical means (Hacking, 2019).

To incorporate the process dimension of social innovation, we suggest examining the link between social innovation and sustainable development through three dynamics that coincide with three challenges identified for governing the 2030 Agenda (Bowen et al., 2017) and ultimately achieving the SDGs: the encouragement of collective action, the necessary negotiations among the three pillars of sustainable development and subsequent trade-offs, and the assessment of the SDGs.

This is a proposal based on Meuleman and Niestroy's (2015, p. 12,297) broad definition of *governance*. Its aim is to translate global objectives to the local and national levels. The proposal assumes the principle of "common but differentiated governance" and emphasizes the need to adapt commonly accepted principles to different scales. These scales include multiple levels—global, regional, national, local, and subnational levels—and sectors, such as the various pillars of sustainable development. The proposal implies finding solutions to the three challenges identified.

18.4.2 Collective Action

Putting sustainable development on the ground, as proposed by the United Nations agenda for the decade, leads to the scale and level of application of the SDGs—from the global to the local. The localization of sustainable development began to be considered through the application of Local Agenda 21 (A21L), an action program resulting from the Rio Summit in 1992 (UNCED, 1992) for implementation in specific territories, with the expectation that local authorities (elected officials and local technicians) would put into practice the ideas of participation and collaborative governance and work in partnership with all the actors in the community (citizens, associations, and interest groups).

Promoting collective action and generating diverse decision-making spaces are primary challenges of implementing and governing the SDGs. This challenge entails making politically and collaboratively informed decisions while considering the particularities and potential interconnections of each goal and the nature and prominence of each involved party or actor. The proposal of implementing mechanisms to enhance trust relationships or foster innovation commitments has been suggested as a possible solution to the issue: "[A]mbitious thinking is needed to develop formal institutional structures for policy integration and to identify ways of bridging current gaps" (Bowen et al., 2017, p. 91).

But localization seems to remain the Achilles' heel of the 2030 Agenda, dependent not only on translating global goals into local action but also on the mobilization and commitment of communities and institutions at the local level (Ansell et al., 2022). Thus, the defense of co-creation as a strategy for an effective localization of the SDGs can promote the involvement of local structures in the cause of sustainability at different levels (Figure 18.1).

The success of achieving results in social innovation is dependent upon the development of processes that rely heavily on successful cross-sector partnerships. In the realm of social innovation, the "pure" social innovation as described by Pol and Ville (2009), where market failure is assumed, public stimulus for innovation is increasing. This is particularly notable in the European Union, where attention toward co-creation has been encouraged. Co-creation is a non-linear process that involves multiple actors and stakeholders in ideating, implementing, and assessing products, services, policies, and systems for improved efficiency, effectiveness, and satisfaction of participants. This is defined by the SISCODE program. Co-creation is a generic characteristic of social innovation and a measure of success (Meister Broekema et al., 2021).

Figure 18.1 Cocreation as a strategy of localization of the SDGs.

Source: Ansell et al., 2022.

As a community action strategy, social innovation has adopted and fostered co-creation principles, making it a viable tool for accomplishing the SDGs. Given our prior reference to the "Voorbeg Gap," utilizing co-creation as an approach to localize the SDGs may also evaluate the effects of social innovation initiatives on sustainability at the local level. In this instance, evaluating governance partnerships that have been successful on a territorial scale—as proposed by Moulaert et al. (2019)—and identifying their most effective characteristics in defending sustainability along with their potential for replication—as Haskell et al. (2021)—could enable us to reconcile the outcome aspect of social innovation, as gauged by directly applying the SDG framework to the relevant project, with the process dimension.

18.4.3 Negotiation and Trade-Offs

The sustainable development debate generally centers on the limits of trade-offs between ecological objectives and economic and social growth. It is important to acknowledge the inevitability of negotiations and resulting trade-offs between these objectives. Pragmatically, the solution must involve designing paths toward compatibility (Nilsson et al., 2016; Sachs et al., 2019).

Theorizing on the intersection of social innovation and sustainable development has emphasized the requirement for an ongoing discourse between the macro and micro levels of the two concepts. This facilitates a consequential assimilation between the discourses conveyed and practical implementation.

The role of social innovation in attaining the SDGs, specifically the implementation of co-creation tactics at the local level, necessitates a micro analysis that considers the terms of negotiation and any corresponding compromises.

The adoption of "shared socio-economic pathways (SSPs)" as an example aims to localize the SDGs and project the various levels of SDG compliance into the future of communities. Each SSP includes "a narrative storyline incorporating locally specific ideas from the community" (Szetey et al., 2021). Additionally, the integration of technology with traditional knowledge to create "win–win" solutions (Nijnik et al., 2019) highlights the potential for social innovation and emphasizes the ideas of participation and co-creation. However, the results obtained have not been systematically analyzed and theorized, resulting in two consequences: (1) innovation's limited visibility, which fails to gain scale or replication in implementation contexts, and (2) impeding the macro-evaluation of innovation's contributions to the governance of the 2030 Agenda.

Fostering communication between the micro and macro aspects of these concepts presents a challenge when attempting to establish a new knowledge production paradigm. This challenge requires questioning processes, rediscovering ontology, and paying careful attention to the political implications of knowledge. Additionally, promoting epistemological debates regarding the centrality of practices to understanding knowledge production is vital (Nunes, 2008). This has been the epistemological challenge presented by the analysis of public policies promoting social innovation. The success of these processes depends on appropriate public policies, which are subsequently informed by the results of the same processes. This shapes the contemporary circularity premise, as identified by Maduro et al. (2018) (Figure 18.2).

18.4.4 Evaluation

The epistemological debate previously mentioned emphasizes the importance of focusing on informal civil society networks that produce information related to sustainability. The assessment of the implementation of the 2030 Agenda cannot be divorced from the availability of genuine progress

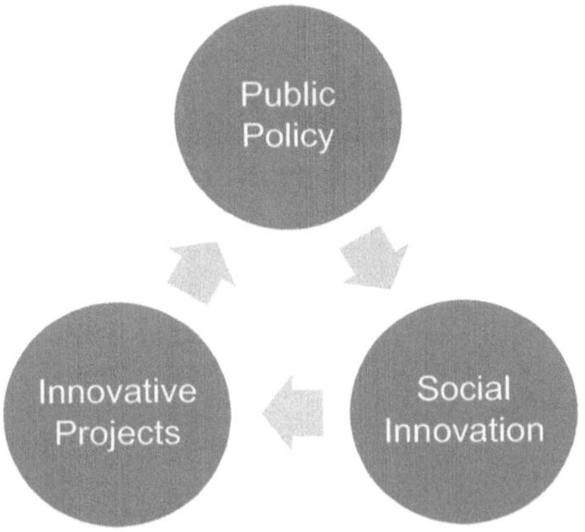

Figure 18.2 Contemporary circularity model.
Source: Data from Maduro et al. (2018) and Moore et al. (2012).

data and the criteria for obtaining it, which may involve engaging non-traditional stakeholders, such as civil society organizations and other research communities (Biermann et al., 2017).

The quality of information regarding the implementation of the SDGs has implications for involving local communities and shaping public opinion (Caiado et al., 2018). At this level, analyzing the increasing influence of social networks on information dissemination and the formation of informal collaborative networks is crucial.

The significance assigned to informality facilitates a connection to the anthropological dynamism of social innovation (Lesemann, 2012; Ilie & During, 2012). It also presents an opportunity for public mobilization on environmental issues and introduces a potential novel research area on the social construction of sustainable development (Leipold et al., 2019).

On the other hand, validation of information is more feasible and prompt in local contexts, within the daily lives of the communities themselves. Local authorities hold a significant role. Apart from their political legitimacy and diverse range of competencies, they hold an advantageous position in terms of their knowledge of the territory and their ability to coordinate and organize the entities working in the field. This position enables them to establish connections with national institutions that devise funding programs (Pinto et al., 2021).

18.5 Conclusions

Social innovation has been recognized as a major force behind sustainable development, facilitating the reconfiguration of behavior and effectively addressing the social and environmental issues faced by communities. This relationship can be evaluated through the two dimensions of social innovation—as a process or an outcome—in direct alignment with the Sustainable Development Goals (SDGs) adopted by the United Nations for sustainable development governance.

The territorial context of social innovation has been a key platform for establishing a connection with the concept of sustainable development. However, the limited concentration of the studied cases poses a constraint for theoretical analysis.

Dialogue between the micro and macro dimensions of the two concepts can uncover social innovation examples that have effective impacts on local sustainable development and can be scaled up, replicated, and evaluated globally.

Co-creation, a generic characteristic of social innovation, exemplifies the connection between the two concepts and serves as a strategy for achieving the SDGs. However, civil society participation in co-creation processes falls short of overcoming a symbolic dimension, resulting in what we refer to as the "Voorberg Gap" in sustainability research. Further research is necessary in this area, as the empirical practice, influenced by public policies, has not fully delivered outcomes.

Notes

1 BEPA's (Bureau of European Policy Advisers) mission was to provide the president and the European Commission with advice on topics relevant to future European policies. From 2014, it was renamed the European Political Strategy Centre.
2 The active participation of citizens in public management processes in which they are both end users is called co-creation.
3 Citizens take the initiative to develop a certain project in which the public responsible is only the next actor (Voorberg et al., 2015).
4 The authors argue that the conclusions of the IPCC (IPCC, 2021, as cited in Haskell et al., 2021) that new technologies, market solutions, and regulations have been insufficient to prevent the transformations imposed by man on the planet suggest the need for profound social change to face the ecological crisis, targeting behaviors and in line with the idea of strong sustainability—environmental qualities cannot be replaced by other types of economic or social qualities but must be sustained. On the contrary, the weak sustainability view admits trade-offs between natural, social, and economic capital.
5 Call for expression of interest for indirect management of the Social Innovation+, Annex B—Explanatory memo: ESF Social Innovation+. July 2021, p. 5. Available at: https://ec.europa.eu/european-social-fund-plus/en/selection-entrusted-entity-indirect-management-initiative-esf-social-innovation.

References

Andion, C., Alperstedt, G. D., Graeff, J. F., & Ronconi, L. (2022). Social innovation ecosystems and sustainability in cities: A study in Florianópolis, Brazil. *Environment, Development and Sustainability*, *24*(1), 1259–1281.

Ansell, C., Sorensen, E., & Torfing, J. (2022). *Translating global goals to local contexts. Co-creation for sustainability* (pp. 41–56). Emerald Publishing Limited.

Ayres, R. U., van den Bergh, J. C. J. M., & Gowdy, J. M. (1998). *Viewpoint: Weak versus strong sustainability*. Tinbergen Institute discussion paper 98/103-3.

Baker, S. (2006). *Sustainable development*. Routledge.

Baker, S., & Mehmood, A. (2015). Social innovation and the governance of sustainable places. *Local Environment*, *20*(3), 321–334.

Barraket, J. (2020). The role of intermediaries in social innovation: The case of social procurement in Australia. *Journal of Social Entrepreneurship*, *11*(2), 194–214.

Bekkers, V. J. J. M., Edelenbos, J., Nederhand, J., Steijn, A. J., Tummers, L. G., & Voorberg, W. H. (2014). *The social innovation perspective in the public sector: Co-creation, self-organization and meta-governance*.

Biermann, F., Kanie, N., & Kim, R. E. (2017). Global governance by goal-setting: The novel approach of the UN Sustainable Development Goals. *Current Opinion in Environmental Sustainability*, *26*, 26–31.

Bowen, K. J., Cradock-Henry, N. A., Koch, F., Patterson, J., Häyhä, T., Vogt, J., & Barbi, F. (2017). Implementing the "Sustainable Development Goals": Towards addressing three key governance challenges—collective action, trade-offs, and accountability. *Current Opinion in Environmental Sustainability*, *26*, 90–96.

Caiado, R. G. G., Leal Filho, W., Quelhas, O. L. G., de Mattos Nascimento, D. L., & Ávila, L. V. (2018). A literature-based review on potentials and constraints in the implementation of the sustainable development goals. *Journal of Cleaner Production*, *198*, 1276–1288.

Castillo, A. L. C., & Valderrama, L. T. R. (2017). Aprendizajes desde iniciativas de innovación social para la construcción de paz en Colombia: narrar la paz desde abajo. *Rumbos*, *16*, 155–178.

Dias, R. C., Seixas, P. C., & Lobner, N. (2020). Sustainable development and ecological modernization: Boundary discourses between "strong" and "weak" approaches. *Journal of Sustainable Development*, *13*(4).

Dryzek, J. S. (2005). *The politics of the earth: Environmental discourses*. Oxford University Press.
Edwards-Schachter, M., & Wallace, M. L. (2017). "Shaken, but not stirred": Sixty years of defining social innovation. *Technological Forecasting and Social Change*, *119*, 64–79.
Eichler, G. M., & Schwarz, E. J. (2019). What sustainable development goals do social innovations address? A systematic review and content analysis of social innovation literature. *Sustainability*, *11*(2), 522.
Galego, D., Moulaert, F., Brans, M., & Santinha, G. (2021). Social innovation & governance: A scoping review. *Innovation: The European Journal of Social Science Research*, 1–26.
Hacking, T. (2019). The SDGs and the sustainability assessment of private-sector projects: Theoretical conceptualisation and comparison with current practice using the case study of the Asian Development Bank. *Impact Assessment and Project Appraisal*, *37*(1), 2–16.
Haskell, L., Bonnedahl, K. J., & Stål, H. I. (2021). Social innovation related to ecological crises: A systematic literature review and a research agenda for strong sustainability. *Journal of Cleaner Production*, *325*, 129316.
Hubert, A. (2011). *Empowering people, driving change: Social innovation in the European Union, the Bureau of European Policy Advisers*. http://ec.europa.eu/bepa/pdf/publications_pdf/social_innovation. pdf.
Hugé, J., Waas, T., Dahdouh-Guebas, F., Koedam, N., & Block, T. (2013). A discourse-analytical perspective on sustainability assessment: Interpreting sustainable development in practice. *Sustainability Science*, *8*(2), 187–198.
Ilie, E. G., & During, R. (2012). *An analysis of social innovation discourses in Europe: Concepts and strategies of social innovation*.
Kemp, R., Parto, S., & Gibson, R. B. (2005). Governance for sustainable development: Moving from theory to practice. *International Journal of Sustainable Development*, *8*(1–2), 12–30.
Khadka, C., Aryal, K. P., Edwards-Jonášová, M., Upadhyaya, A., Dhungana, N., Cudlin, P., & Vacik, H. (2018). Evaluating participatory techniques for adaptation to climate change: Nepal case study. *Forest Policy and Economics*, *97*, 73–82.
Leipold, S., Feindt, P. H., Winkel, G., & Keller, R. (2019). Discourse analysis of environmental policy revisited: Traditions, trends, perspectives. *Journal of Environmental Policy & Planning*, *21*(5), 445–463.
Lele, S. (2013). Rethinking sustainable development. *Current History*, *112*(757), 311–316.
Lesemann, F. (2012). Social innovation, informality and social sciences. *CSI Challenge Social Innovation*, *23*, 1–9.
Maduro, M., Pasi, G., & Misuraca, G. (2018). Social impact investment in the EU. *Financing strategies and outcomes oriented approaches for social policy innovation: Narratives, experiences, and recommendations. JRC Science for Policy Report*.
Manjon, M. J., Merino, A., & Cairns, I. (2022). Business as not usual: A systematic literature review of social entrepreneurship, social innovation, and energy poverty to accelerate the just energy transition. *Energy Research & Social Science*, *90*, 102624.
Meister Broekema, P., Horlings, L. G., & Bulder, E. A. M. (2021). Understanding the value of co-creation for social innovation interpretations of social innovation and co-creation in European policy-related documents between 1995 and 2018. *Innovation: The European Journal of Social Science Research*, 1–18.
Meuleman, L., & Niestroy, I. (2015). Common but differentiated governance: A metagovernance approach to make the SDGs work. *Sustainability*, *7*(9), 12295–12321.
Millard, J. (2018). How social innovation underpins sustainable development. In J. Howaldt, C. Kaletka, A. Schröder & M. Zirngiebl (Eds.), *Atlas of social innovation: New practices for a better future* (pp. 41–43).
Moore, M. L., Westley, F. R., Tjornbo, O., & Holroyd, C. (2012). The loop, the lens, and the lesson: using resilience theory to examine public policy and social innovation. In *Social innovation: Blurring boundaries to reconfigure markets* (pp. 89–113). London: Palgrave Macmillan UK.
Mota, J. C. (2020, fevereiro). Democracia "bottom-linked". *José Carlos Mota—Works*. https://jcmotaworks.wordpress.com.
Moulaert, F., MacCallum, D., Van den Broeck, P., & Garcia, M. (2019). *Bottom-linked governance and socially innovative political transformation*.
Moulaert, F., Martinelli, F., Swyngedouw, E., & Gonzalez, S. (2005). Towards alternative model (s) of local innovation. *Urban Studies*, *42*(11), 1969–1990.
Mulgan, G. (2006, Spring). The process of social innovation. *Innovations*, 145–162.
Murray, R., Caulier-Grice, J., & Mulgan, G. (2010). The open book of social innovation. The young foundation. *The Young Foundation*, NESTA, UK.

Neumayer, E. (2003). *Weak versus strong sustainability: Exploring the limits of two opposing paradigms*. Edward Elgar Publishing.
Nijnik, M., Secco, L., Miller, D., & Melnykovych, M. (2019). Can social innovation make a difference to forest-dependent communities? *Forest Policy and Economics*, *100*, 207–213.
Nilsson, M., Griggs, D., & Visbeck, M. (2016). Policy: Map the interactions between Sustainable Development Goals. *Nature*, *534*, 320–322. https://doi.org/10.1038/534320
Nunes, J. A. (2008). O resgate da epistemologia. *Revista Crítica de Ciências Sociais*, *80*, 45–70.
Olsson, P., Moore, M. L., Westley, F. R., & McCarthy, D. D. (2017). The concept of the Anthropocene as a game-changer: A new context for social innovation and transformations to sustainability. *Ecology and Society*, *22*(2).
Osei, C. D., & Zhuang, J. (2020). Rural poverty alleviation strategies and social capital link: The mediation role of women entrepreneurship and social innovation. *Sage Open*, *10*(2), 2158244020925504.
Parra, C. (2013). *10 social sustainability: A competing concept to social innovation*?
Pinto, H., Nogueira, C., Guerreiro, J. A., & Sampaio, F. (2021). Social innovation and the role of the state: Learning from the Portuguese experience on multi-level interactions. *World*, *2*(1), 62–80.
Pol, E., & Ville, S. (2009). Social innovation: Buzz word or enduring term? *The Journal of Socio-Economics*, *38*(6), 878–885.
Redclift, M. (2005). Sustainable development (1987–2005): An oxymoron comes of age. *Sustainable Development*, *13*(4), 212–227.
Robinson, J. (2004). Squaring the circle? Some thoughts on the idea of sustainable development. *Ecological Economics*, *48*(4), 369–384.
Sabato, S., Vanhercke, B., & Verschraegen, G. (2017). Connecting entrepreneurship with policy experimentation? The EU framework for social innovation. *Innovation: The European Journal of Social Science Research*, *30*(2), 147–167.
Sachs, J. D. (2015). *The age of sustainable development*. Columbia University Press.
Sachs, J. D., Schmidt-Traub, G., Mazzucato, M., Messner, D., Nakicenovic, N., & Rockström, J. (2019). Six transformations to achieve the sustainable development goals. *Nature Sustainability*, *2*(9), 805–814.
SCU—Science Communication Unit, University of the West of England, Bristol. (2014). *Science for environment policy in-depth report: Social innovation and the environment*. Report produced for the European Commission DG Environment, February 2014. Disponível em http://ec.europa.eu/science-environment-policy
Sneddon, C., Howarth, R. B., & Norgaard, R. B. (2006). Sustainable development in a post-Brundtland world. *Ecological Economics*, *57*(2), 253–268.
Szetey, K., Moallemi, E. A., Ashton, E., Butcher, M., Sprunt, B., & Bryan, B. A. (2021). Co-creating local socioeconomic pathways for achieving the sustainable development goals. *Sustainability Science*, *16*(4), 1251–1268.
U.N. (2015). *Transforming our world: The 2030 agenda for sustainable development*. Disponível em https://sustainabledevelopment.un.org
Vasconcellos Oliveira, R. (2021). Social innovation for a just sustainable development: Integrating the well-being of future people. *Sustainability*, *13*(16), 9013.
Voorberg, W. H., Bekkers, V. J., & Tummers, L. G. (2015). A systematic review of co-creation and co-production: Embarking on the social innovation journey. *Public Management Review*, *17*(9), 1333–1357.
WCED—World Commission on Environment and Development. (1987), *Our common future*. Disponível em https://sustainabledevelopment.un.org/content/documents/5987our-common-future.pdf.
Westley, F., & Antadze, N. (2010). Making a difference: Strategies for scaling social innovation for greater impact. *Innovation Journal*, *15*(2).

19

SDGS IN THE AGE OF DISINFORMATION

Marijane Luistro Jonsson

19.1 Introduction

The tragedy of the commons offers a framework for understanding resource depletion and the challenges that most SDGs address, such as climate change, overfishing, deforestation, and crossing over our planetary boundaries. The metaphor of an overgrazed pasture often makes us associate the tragedy with the overharvesting and depletion of natural resources, such as the atmosphere, wildlife, oceans, and forests (Hardin, 1968). At the core of the tragedy lies a social dilemma rooted in choosing between individual short-term self-interest and the long-term common good. What is often neglected is how the traditional concept of the tragedy of the commons can also apply to knowledge and information landscapes that we often take for granted. Similar to polluted oceans and harvested pastures, threats such as the rise of fake news in social media can lead to a tragedy in the form of the distortion of truth and an increase in fragmentation and demise of democratic processes in some societies (e.g., McKay & Tenove, 2021). The subsequent increased ignorance and polarization also pose a threat to societies at large. It should be kept in mind that stakeholders and policymakers behind the different SDGs are not exempted from consuming fake news and resorting to disinformation to pursue their agenda. For instance, the Global Climate Coalition in the 1990s, which represented the oil, coal, auto, utilities, and rail industries, was able to change the narrative on climate change through successful PR strategies (McMullen, 2022). Keeping these points in mind, revisiting the SDG agenda through the lenses of the knowledge commons appears to be relevant in an age of disinformation.

Disinformation, especially during the COVID pandemic and Ukraine invasion, also exposes the urgency and significance of the often ignored last two SDGs—"promoting just, peaceful, and inclusive societies (SDG 16)" and "partnership for the goals (SDG 17)." The consequential waves of conflict-related migration, extrajudicial killings, and deprivation of justice and social services have altogether been triggered by disinformation and have taken a toll on economic development and environmental conservation, as well as threaten the ongoing progress of most SDGs. While both SDG 16 and SDG 17 are considered to dampen the obstacles in fulfilling the other goals, they need to separately address their own challenges and targets as well. Reports (e.g., Global Alliance for Reporting Progress on Peaceful, Just and Inclusive Societies, 2019; UNDP OGC, 2021; Hope, 2020) have depicted SDG 16 as an "enabler goal," showing its interlinkages to other goals, such

DOI: 10.4324/9781003285472-20

as reducing poverty (SDG 1) and inequality (SDG 10). Similarly, SDG 17 is also depicted as a tool or means for achieving sustainable communities and cities, or SDG 11 (MacDonald et al., 2018). While positive interlinkages exist to achieve the other goals, there are also negative or constraining effects. For instance, increasing judicial accountability can lead to elite protection that can entrench inequalities, and representation may be limited as voters in advanced democracies fail to punish incumbents for rising income inequality (UNDP OGC, 2021). As the synergies and constraints of SDG 16 and SDG 17 are made known, it also exposes the familiar trade-offs involved in the tragedy of the commons dilemma—choosing between immediately enabling other goals or taking prudence in dealing with the different threats to peace, justice, inclusiveness, and partnerships. Disinformation is an example of a threat which erodes the foundation of democracy, creates factions, and crumbles the moral foundations of societies. Often advanced by organized groups such as climate change deniers, anti-vaxxers, conspiracy theorists, and other self-interested alliances, it also poses a threat to the other SDGs.

Given the collaborative nature of SDG 16 and SDG 17, let us take a step back and investigate the dynamics on how disinformation can affect realizing these goals. For SDG 16 to work, targets require that "governments, civil society and communities must work together to implement lasting solutions to reduce violence, deliver justice, combat corruption and ensure inclusive participation at all times" (Hope, 2020). Moreover, freedom to express views must be guaranteed, and disputes need to be resolved through functioning justice systems. Working toward these objectives entail cooperation and partnerships among various actors (SDG 17). One can think that both goals are linked and facilitate achieving other goals—the synergy and coordination between partners (SDG 17) for the goals to ensure peace, justice, and inclusivity (SDG 16) are necessary to securely achieve other goals. However, the complexities in the decision interactions to achieve the goals are often hidden and simplified, assumed to automatically lead to synergies, especially with the time pressure to achieve them. Similar to the tragedy of the commons, individuals and organizations involved face a cooperation dilemma of choosing between contributing to the common good toward a just and peaceful world or to prioritize other partnership agendas, and their decisions would often be dependent upon how other actors decide to cooperate. Depending on cooperation decisions, the trade-offs, conflicts, and inconsistencies can become more prominent than the synergies it promises.

Thus, this chapter takes a closer look into the complexities of the decision interactions, particularly the handling of information, that are involved in pursuing the SDGs. It aims to show how synergies, conflicts, and inconsistencies can emerge, revealing that transformation for sustainability is not as straightforward and enduring. By taking the case of how a threat, such as the spread of fake news, affects the realization of the SDGs, it elucidates the interplay of factors affecting the consumption of information favoring one's goals. "Fake news" is not only used as a political weapon by politicians during elections, but as we have witnessed, self-interested groups have also used them as tools to derail the progress of various SDGs.

The following section will discuss the cooperation dilemma involved and present a game theoretical discussion of knowledge commons and information sharing. This will serve as the basis for the various simulation scenarios that will be presented. The simulations aim to show how the interplay of cooperation, the volume of disinformation, and the level of sanctions can pollute knowledge commons and derail the progress of the SDG agenda. Finally, the study concludes with insights and implications regarding the regulation of fake news and the institutionalization of disinformation. After all, failures in securing the targets of SDG 16 and SDG 17 expose not only corrupt states, weak institutions, and futile partnerships but also the vulnerabilities of global security from existential threats to humanity as well.

19.2 Knowledge Commons and the SDGs

Knowledge and information, in their general form, can be considered as a public good, especially with the onset of its digitalization, since information becomes easily available for everyone (non-excludable) and different users can access it simultaneously (non-rivalrous). In this view, knowledge becomes a "knowledge and information common" available and accessible for oneself and everyone. In these common spaces, interactions involve dilemmas that make polluting the commons with fake news possible and consequently require appropriate institutional systems to be in place (e.g., Stiglitz, 1999; Hess & Ostrom, 2007). To exemplify, a complex knowledge environment has emerged in social media, where tweets, shared Facebook links, and news update messages become the main source of information (e.g., Marchi, 2012). While this development has resulted in a faster pace of disseminating information, it also exposes one's knowledge base to include unverifiable sources and mis- and disinformation. The knowledge commons in social media that has been positively envisioned to connect people, raise awareness, foster innovation, and serve as platforms for deliberative democracy has instead resulted in polarized bubbles and fake news pollution that drown real information. Disinformation campaigns have contributed to system-level anti-deliberative elements of epistemic cynicism, affective polarization, and pervasive inauthenticity, which hinder communication using facts and logic, moral respect, and democratic inclusion (McKay & Tenove, 2021). The resulting distortion of facts and knowledge gives way to the tragedy of the knowledge commons, where the quest for truth slowly gets depleted.

Measures to address disinformation through country-specific interventions, in the form of misinformation or fake news legislation, which impose fines and sanctions have been on the rise. Empirical evidence of their effectiveness has, however, been elusive and scarce. The impacts of international instruments, such as the UN Secretary General's Roadmap to Digital Cooperation (2020), appear to be also vague, as it contends with various approaches of different government set-ups. In lieu of empirical evidence to elucidate the intricacies of implementing and deciding on these measures, this study uses a game theoretical approach to present the effectiveness and limitations of fake news legislation and similar interventions. Through a theoretical analysis and by presenting simulations that capture counterfactual situations, it shows how regulatory measures addressing disinformation problems surrounding SDGs are not always the optimal solution to the problem.

Before we immerse into theoretical abstractions, basic concepts need to be clarified. *Fake news*, in its multiplicity of definitions, can be seen as a type of knowledge pollutant as it intentionally distorts the truth and can lead to historical revision. A typical definition is that it is a type of disinformation, characterized by the "deliberate presentation of (typically) false or misleading claims as news, where the claims are misleading by design" (Gelfert, 2018, p. 108). Similarly, *fake news* is commonly alluded to as "news articles that are intentionally and verifiably false, and could mislead readers" (Allcott & Gentzkow, 2017, p. 213), or "fabricated information that mimics news media content in form but not in organizational process or intent" (Lazer et al., 2018, p. 1094). It comes in different forms, depending on its level of facticity and deceptions, and ranges from news satire, news parody, fabrication, manipulation, and advertising to harmful propaganda (Tandoc et al., 2018). In the remainder of this study, we use the expressions *fake news* and *disinformation* interchangeably. *Disinformation* generally refers to "false information that is knowingly shared to cause harm" (Wardle & Derakshan, 2017, p. 5). The analysis will begin by illustrating the information dilemma surrounding stakeholders and policymakers pursuing independent SDGs, then extended to include the probability that the news is fake, and finally show what happens when legislation is introduced.

In examining the dilemma involved when making a simple individual decision between disseminating a new piece of information or not, let us assume a situation with two agents—a policymaker, a bureaucrat, or a stakeholder—pursuing independent SDGs (i.e., *SDG A Agent* and *SDG B Agent*). Each must decide if they will or will not disseminate a piece of news information that they stumbled upon, relevant to their respective SDGs. In the case where the information is not fake but the users do not know it, they are faced with the following predicament: If both decide to wait and verify the information, then both benefit by being assured of knowing the truth; if they both decide to share the information at once, they will get a higher payoff compared to the previous situation since they will be the first to share the information before it becomes old, gaining reputational rewards or conformity benefits. In case *SDG A Agent* decides to share at once and *SDG B Agent* verifies, the latter may lose out as he/she pays a cost in terms of effort and time to verify the news, while *SDG A Agent* will have a first-mover advantage to their reputation to share the news. In this situation, the best response for any SDG agent, without knowing how others will respond, is to forward the information at once. This is because the payoff involved in a non–fake news dilemma will always be greater for any user to share the information at once, regardless of the strategy of the other.

On the other hand, when the information is fake, the best response for any SDG agent, without knowing how others will respond, is to verify the information. This is because the payoff to the agents become different. If the news is fake and *SDG A Agent* shares at once while *SDG B Agent* verifies, *SDG A Agent* will get a bad reputation when it is eventually found that he/she is a bearer of fake news and might create a setback for the SDG. This will entail a lower payoff for damaging one's reputation for spreading fake news. This is opposite to the situation when the news is not fake, where the first one to share always gets a higher payoff. Additionally, the concern for fake news has also brought up recommendations for interventions, such as legislations and regulations that punish the sharer of fake news. This acts as an added cost, resulting to a situation where the temptation to share immediately if the other one verifies is always smaller than the reward if both verify, and the urge to share if both do it is smaller than the reward for verifying it alone. We find a situation that we call the fake news dilemma.

In the real world, we do not know if the information we stumble upon is fake or not—we do not know whether the decision situation we are facing mimics a non-fake or a fake news dilemma. In calculating and taking into account the presence of probabilities that the news is fake, we find that different probability values give rise to different dilemmas. There are certain tipping points when it is best to share (verify) if the chance of the news being fake is small (big). Different strategies matter, depending on the authenticity of the information and how we assess that information. In situations where there is a significant risk of news being fake (e.g., an effective propaganda machine being employed or pursued by lobbyists), people should take extra caution to verify the news. Moreover, as the influx of fake news in social media becomes more common, consumers of information are left more uncertain of whether that information is true or fake. Thus, the need to verify new information is becoming increasingly important as the probability increases and we become more uncertain.

The tragedy occurs if we make inappropriate decisions (e.g., by misinterpreting situations), resulting to detrimental consequences in the long term. That is, if everyone assumes that the information is not fake and instantly shares it to get high short-term reputational gains, we end up in a non-optimal situation if that information turns out to be fake. With the new decision landscape marked with an increasing volume of fake news, the tragedy of the knowledge commons can easily emerge when individuals taking in information are unaware of whether that information is true or false and doing so without proper coordination and institutional support.

Using the abstraction of potential outcomes discussed earlier, we see the complexities involved in a simple decision of forwarding or verifying a new piece of information. With that serving as a basis of our analysis, we can look further into the effects of more intricate yet realistic situations, such as information asymmetries, in the production and supply of fake news. That is, some individuals in a population always share fake news no matter what, giving rise to a new breed of paid cyber troops, trolls, and bots aiming to promote propaganda and manipulate public opinions and even election and legislation outcomes (Bradshaw & Howard, 2017; Subramanian, 2017). In addition, others selectively choose information to feed their own biases or blindly share due to their cognitive limitations. Inherent factors such as our confirmation biases (Illeris, 2003; Wikforss, 2017), politically motivated reasoning (Kahan, 2015; Kahan et al., 2017), conformity, group-thinking norms, choice blindness (Hall et al., 2013), and stark emotions (Cooke, 2018; Weeks, 2015; Vosoughi et al., 2018) blur the decisions of what we assume to be rational individuals. These features affect the utility and benefit people get as they add more incentives for people to share information immediately rather than verifying it first. The level of importance and sensitivity of the information to the SDG would also affect the decisions to share or verify. The effects and interplay of some of these factors are explored in the next section.

19.3 Simulating the Knowledge Commons With Fake News

Given the pair-wise interaction conditions described earlier, how will repeated interactions of different SDG policymakers, stakeholders, and supporters affect the overall circulation of fake news? How do knowledge commons evolve through time, given different probabilities of exposure to fake news? How will the information commons evolve in societies with more sanctions and laws regulating fake news for certain SDGs? In this section, we use an agent-based model with evolutionary assumptions in answering these questions. Previous studies that have used the same approach similarly study the diffusion of information but focus on the effects of parameters such as types of deception (Kopp et al., 2018), threshold of interventions (Brainard et al., 2020), and social networks (Ross et al., 2019). In contrast, this study explores how sharing of information will evolve in different types of environments (i.e., varying fractions of fake news to the overall information flow) when interventions are made, such as implementing sanctions and regulations that would create cost for the sharer, and when fake news is systematically manufactured and institutionalized (i.e., diverse distribution of individual traits in the population). This implies looking into the following parameters: the different probabilities that a random news is fake and the different levels of costs and sanctions as discussed in the previous section, as well as if a certain percentage of the population always shares fake news no matter what (e.g., paid trolls). The simulated population consists of 2,000 agents interacting randomly pair-wise over 50 rounds in deciding to either verify or share the news. For the initial news in the beginning, we assign 50% of the population to share and 50% to verify. There are 1,000 news being decided upon in each round, randomly taken from a pool of news that have an assigned probability that the news is fake. In the succeeding interactions, we make the evolutionary assumption that the SDG agents copy a strategy from a randomly chosen agent, and for each agent, the probability of being copied is proportional to the payoff in the previous round (i.e., there is higher probability to mimic the more successful SDG agents).

19.3.1 Different Fake News Environments

As predicted, Figure 19.1 shows that, in time, the news sharing behavior spreads faster in information commons with less fake news (i.e., low proportion of fake news to the total news or, likewise, low probability that the news is fake), but SDG agents will be more careful in sharing news as the proportion of fake news increases. This is intuitive as SDG agents in environments exposed to more fake news (e.g., proportion of fake news is 90%) would be more cautious to share to protect

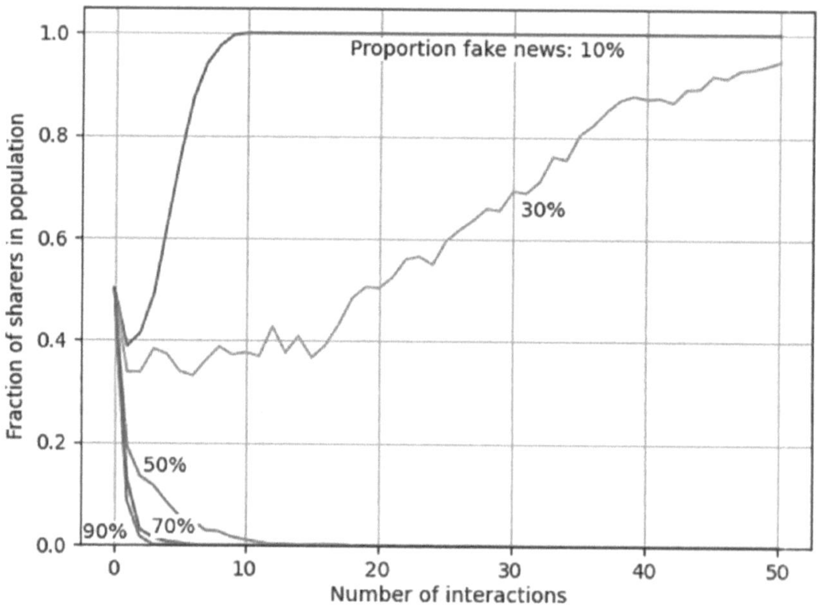

Figure 19.1 Sharing of news in environments with different proportions of fake news.

their reputation. On the other hand, SDG agents in more news-trustworthy environments quickly share knowledge to increase their knowledge status. This finding is consistent to what other studies have shown (e.g., Altay et al., 2019). This is depicted by the lines that have higher proportions of fake news converge toward zero, and those with low ones eventually end up sharing all the time.

19.3.2 Various Levels of Sanctions

The intuition of sharing all the time is tempered, as shown in what follows, when sanctions are implemented. We define *sanctions* as costs to the individuals of sharing news that are fake and only cover effective fake news regulation implementation. Thus, environments with intricate laws but poor and selective implementation would be represented by environments with low sanctions, since the risk of getting caught is low. As can be seen in Figure 19.2, environments that, for instance, have a proportion of 30% fake news will increase verification as the level of sanctions increases.

However, it is important to note that even in the presence of very high sanctions, the sharing strategy will never be eradicated in environments with low proportion of fake news (i.e., sharing is sustained when the proportion of fake news is 10%). This is because people will always be tempted to share the news immediately for reputational and first-mover benefits, especially if the risk is low of getting caught. SDG agents who receive sanctions are never copied by others in that round, explaining why the verification strategy dominates in time in most environments.

19.3.3 Diverse Presence of Trolls

In environments where there are constant sharers, Figure 19.3 shows us that the tendency to verify takes a turn and disappears as the proportion of fake news increases. Constant sharers, which represent fake news trolls and bots, are agents who consistently share no matter what and

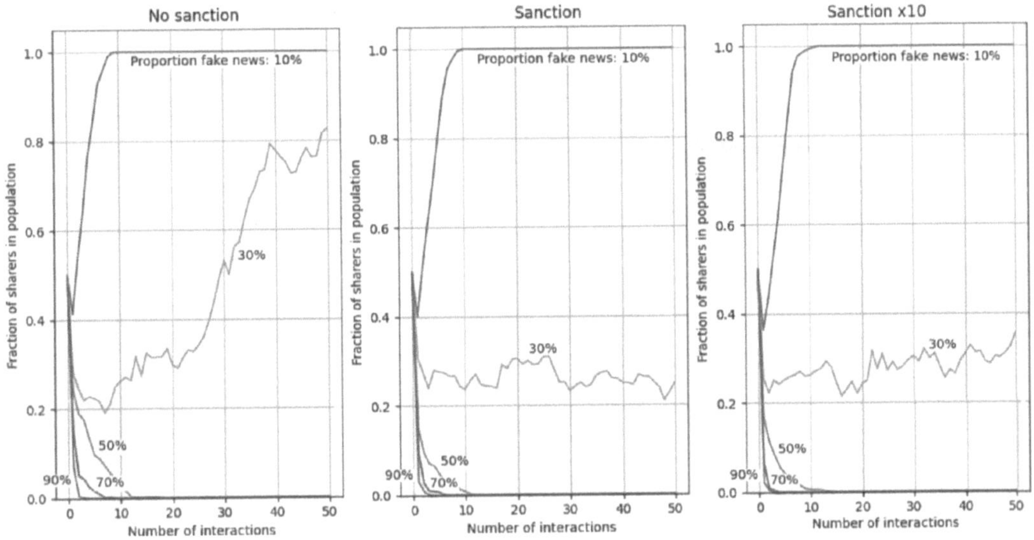

Figure 19.2 Sharing of news in environments with different proportions of fake news and levels of sanctions.

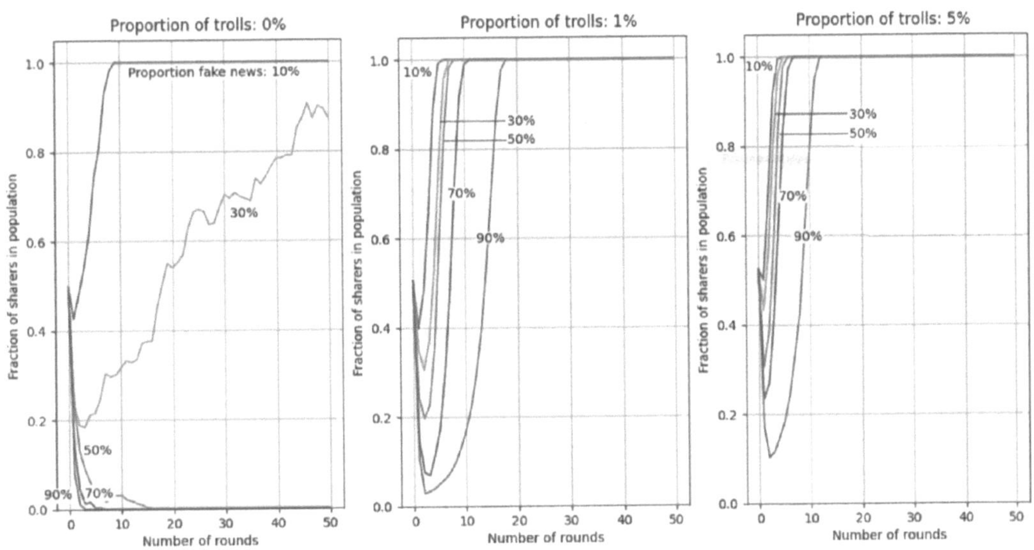

Figure 19.3 Sharing of news in environments with different proportions of fake news and trolls/bots.

do not copy the strategy of others, unlike the rest of the population in our model, who mimics successful strategies in time. Our results further show that even a minuscule presence of 1% fake news trolls will eventually make sharing behavior dominate faster and outcompete the verification strategy. As the presence of trolls increases, there is also faster convergence to sharing the news all the time.

19.3.4 Sanctions and Trolls

Having seen the respective effects of sanctions and trolls in environments with different proportions of fake news, what happens if they are both present? The following figures show how news is shared when there are various levels of sanction in different fake news environments, given 1% (Figure 19.4) and 5% (Figure 19.5) are trolls. We find that the effects of having trolls

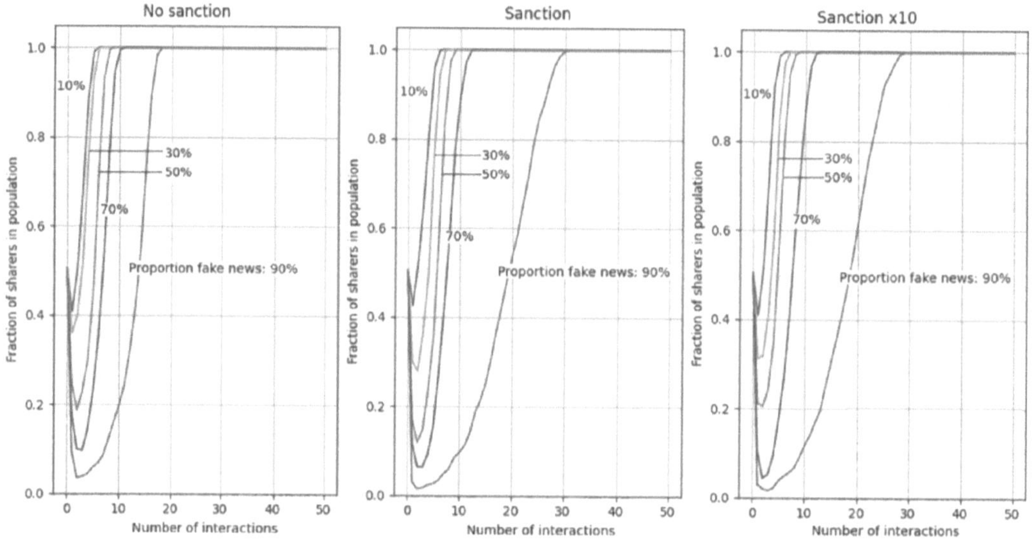

Figure 19.4 Sharing of news in different fake news environments with various levels of sanctions and 1% of the population are trolls.

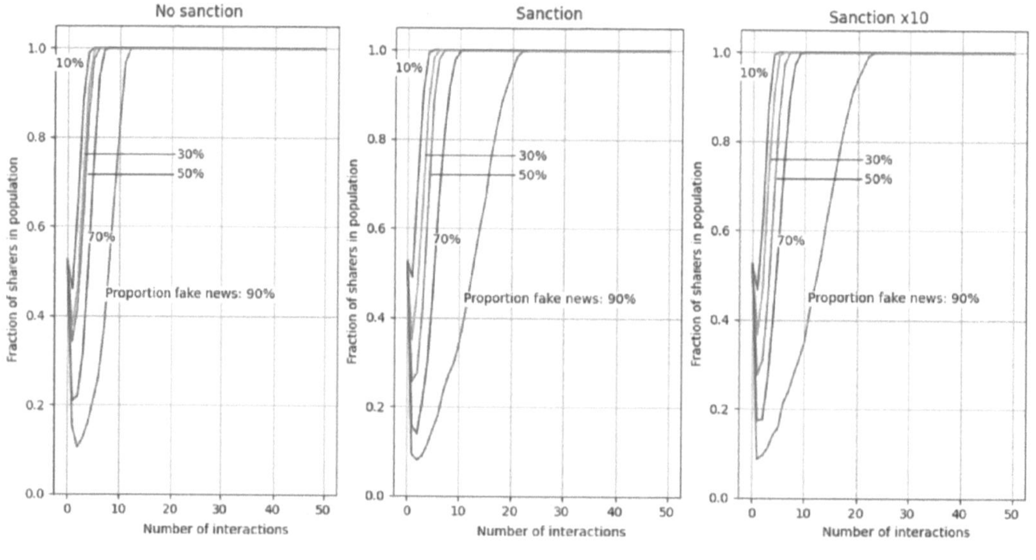

Figure 19.5 Sharing of news in different fake news environments with various levels of sanctions and 5% of the population are trolls.

(i.e., converge to sharing all the time) are stronger than the sanctions (i.e., converge to verification). Moreover, the tendency to share happens faster as the number of trolls increase, as can be seen in how quickly the lines accelerate in Figure 19.5 compared to Figure 19.4.

19.4 Conclusion

The simulations present several insights on how SDG agents possibly share information in this age of disinformation. First, news sharing becomes limited as the information commons becomes increasingly polluted with fake news. Secondly, news sharing is tempered with higher sanctions and costly regulations. However, the sharing strategy will never be eradicated, even in the presence of high sanctions in environments with a low proportion of fake news. Thirdly, when fake news is constantly produced in environments with disinformation machineries, the tendency to verify takes a turn and disappears as the proportion of fake news increases. Finally, when you combine both factors of having sanctions (i.e., verification becomes higher) and trolls (i.e., sharing of fake news becomes higher), the effect of the latter is stronger than the former. The tendency to share happens faster as the number of trolls or propaganda increases.

These results altogether impart how the spread of fake news and information can be daunting, even in the presence of sanctions. In this age, where fake news machineries abound and SDG agents push the urgency of their themes, the spread of fake news will inevitably find its way despite costly sanctions and good intentions to verify. This points to several implications for the attainment of the SDGs in general, and SDG 16 and SDG 17 in particular. First, it is important to control the source and not only sanction the sharing of fake news. To enhance SDG 16's creation of strong institutions, continuous vigilance and reversal of measures geared toward institutionalizing information manipulation, historical revision, and removing ethics and gatekeepers of news are urgently needed. Unfortunately, societies that are vulnerable to fake news are far behind SDG targets and have weak democracies that can easily be silenced by autocratic procedures. More often, they are the forerunners of fake news, as people in power cling to and usurp their powers. Addressing the sources of disinformation can therefore be tricky, especially among vulnerable societies in need of strong democracy to flourish.

Secondly, while much of the discussion and debate is on monitoring and measuring SDGs, this study also exposes the underlying necessity for maintaining an unpolluted knowledge commons resource. Disinformation needs to be initially addressed to produce trustworthy data that can be monitored and measured as well as shared with integrity among partnering institutions. Dealing with disinformation requires awareness among institutions and individuals involved to go beyond their agenda and take a knowledge commons perspective. As the overall SDG agenda aims to provide a "shared blueprint for peace and prosperity for people and the planet, now and into the future," it recognizes that "ending poverty and other deprivations must go hand-in-hand with strategies that improve health and education, reduce inequality, and spur economic growth—all while tackling climate change and working to preserve our oceans and forests" (https://sdgs.un.org/goals). As such, knowledge sharing for providing information and creating solutions between the different SDG agents and institutions should be encouraged yet guarded against fake news that can pollute this knowledge base. Finding this balance of being open yet wary is a challenge but necessary to prevent the creation of polarized groups and uncoordinated measures to achieve the goals. It aids in avoiding an endgame where SDG agents will less likely share information and stagnate the much-needed progress toward the fruition of the SDG agenda.

Finally, propagating the perspective of the tragedy of the knowledge commons and disseminating information on the decision complexities can make decision-makers more reflective in handling information and trade-offs in this age of disinformation. This is necessary for SDG 16 and SDG 17 to truly enable other goals without compromising the long-term work behind establishing peace, justice, and inclusiveness, as well as building enduring partnerships. This study uncovers possible scenarios on how few factors such as fake news machineries and sanctions can affect the knowledge commons. However, in the real world, there are more confounding factors that are involved, making SDG attainment even more elusive. For instance, changing the minds of self-interested groups with strong convictions will not happen in an instant, due to psychological factors, and might require time for standpoints to move and mindsets to evolve. Including the disinformation problem into the SDG discussion, unraveling its intricacies, and getting into its core, as what we have done in this study, can be an important step, though, in stewarding the knowledge commons we share.

References

Allcott, H., & Gentzkow, M. (2017). Social media and fake news in the 2016 election. *Journal of Economic Perspectives*, *31*(2), 211–236.

Altay, S., Hacquin, A. S., & Mercier, H. (2019). Why do so few people share fake news? It hurts their reputation. *New Media & Society*, 1461444820969893.

Bradshaw, S., & Howard, P. (2017). *Troops, trolls and troublemakers: A global inventory of organized social media manipulation*, 2017(12), 1–37. Oxford Internet Institute.

Brainard, J., Hunter, P. R., & Hall, I. R. (2020). An agent-based model about the effects of fake news on a norovirus outbreak. *Revue d'Épidémiologie et de Santé Publique*.

Cooke, N. A. (2018). *Fake news and alternative facts: Information literacy in a post-truth era*. American Library Association.

Gelfert, A. (2018). Fake news: A definition. *Informal Logic*, *38*(1), 84–117.

Global Alliance for Reporting Progress on Peaceful, Just and Inclusive Societies. (2019). *enabling the implementation of the 2030 agenda through SDG 16+: Anchoring peace, justice and inclusion*. United Nations, New York.

Hall, L., Strandberg, T., Pärnamets, P., Lind, A., Tärning, B., & Johansson, P. (2013). How the polls can be both spot on and dead wrong: Using choice blindness to shift political attitudes and voter intentions. *PLoS ONE*, *8*(4), e60554.

Hardin, G. (1968). The tragedy of the commons. *Science*, *162*, 1243–1248.

Hess, C., & Ostrom, E. (Eds.). (2007). *Understanding knowledge as a commons*. Cambridge, MA: MIT Press.

Hope, K. R. (2020). Peace, justice and inclusive institutions: Overcoming challenges to the implementation of Sustainable Development Goal 16. *Global Change, Peace & Security*, *32*(1), 57–77.

Illeris, K. (2003). Towards a contemporary and comprehensive theory of learning. *International Journal of Lifelong Education*, *22*(4), 396–406.

Kahan, D. M. (2015). The politically motivated reasoning paradigm (Part1): What politically motivated reasoning is and how to measure it. In *Emerging trends in the social and behavioral sciences: An interdisciplinary, searchable, and linkable resource* (pp. 1–16).

Kahan, D. M., Landrum, A., Carpenter, K., Helft, L., & Hall Jamieson, K. (2017). Science curiosity and political information processing. *Political Psychology*, *38*, 179–199.

Kopp, C., Korb, K. B., & Mills, B. I. (2018). Information-theoretic models of deception: Modelling cooperation and diffusion in populations exposed to "fake news". *PLoS ONE*, *13*(11), e0207383.

Lazer, D. M., Baum, M. A., Benkler, Y., Berinsky, A. J., Greenhill, K. M., Menczer, F., . . . & Schudson, M. (2018). The science of fake news. *Science*, *359*(6380), 1094–1096.

MacDonald, A., Clarke, A., Huang, L., Roseland, M., & Seitanidi, M. (2018). Multi-stakeholder partnerships (SDG# 17) as a means of achieving sustainable communities and cities (SDG# 11). In *Handbook of sustainability science and research* (pp. 193–209). Springer.

Marchi, R. (2012). With Facebook, blogs, and fake news, teens reject journalistic "objectivity". *Journal of Communication Inquiry*, *36*(3), 246–262.

McKay, S., & Tenove, C. (2021). Disinformation as a threat to deliberative democracy. *Political Research Quarterly*, *74*(3), 703–717.

McMullen, J. (2022, July 23). The audacious PR plot that seeded doubt about climate change. *BBC*. www.bbc.com/news/science-environment-62225696

Ross, R. M., Rand, D. G., & Pennycook, G. (2019). *Beyond "fake news": Analytic thinking and the detection of false and hyperpartisan news headlines*. MIT Creative Commons, Center for Open Science.

Stiglitz, J. E. (1999). Knowledge as a global public good. *Global Public Goods: International Cooperation in the 21st Century*, *308*, 308–325.

Subramanian, S. (2017, February 15). Inside the macedonian fake-news complex. *Wired*. www.wired.com/2017/02/veles-macedonia-fake-news

Tandoc Jr, E. C., Lim, Z. W., & Ling, R. (2018). Defining "fake news" a typology of scholarly definitions. *Digital Journalism*, *6*(2), 137–153.

UNDP Oslo Governance Center. (2021). *SDG 16 interlinkages*. www.sdg16hub.org/system/files/2021-07/Interlinkages%20summary_FINAL.pdf

United Nations General Assembly. (2020). *Speaker's general roadmap for digital cooperation*. https://documents-dds-ny.un.org/doc/UNDOC/GEN/N20/102/51/PDF/N2010251.pdf?OpenElement

Vosoughi, S., Roy, D., & Aral, S. (2018). The spread of true and false news online. *Science*, *359*(6380), 1146–1151.

Wardle, C., & Derakhshan, H. (2017). Information disorder: Toward an interdisciplinary framework for research and policymaking. *Council of Europe Report, DGI, 9*.

Weeks, B. E. (2015). Emotions, partisanship, and misperceptions: How anger and anxiety moderate the effect of partisan bias on susceptibility to political misinformation. *Journal of Communication*, *65*(4), 699–719.

Wikforss, Å. (2017). *Alternativa fakta: om kunskapen och dess fiender*. Fri Tanke.

20

FREE, PRIOR, AND INFORMED CONSENT (FPIC) AND THE RIGHT TO SELF-DETERMINATION

A Case Study of Indigenous Peoples in ASEAN Member States

Muhamad Sayuti Hassan, Rohaida Nordin, Izawati Wook, and Nurul Hidayat Ab Rahman

20.1 Introduction

Indigenous people are one of the vulnerable groups in society and remain marginalized due to the control of the majority (Cobo, E/CN.4 Sub.2/1983/21/Add.8). Despite the fact that the United Nations Declaration on the Rights of Indigenous Peoples (UNDRIP) recognizes the rights of Indigenous peoples, the situation of Indigenous peoples has still been marginalized around the world (Agenda 68, A/Res/61/295). The Indigenous population demands the right to self-determination as a result of oppression and dominance, which has repercussions for the Indigenous population's continued status as a peripheral society (Rehman, 1998). Due to their exposure to oppression and vulnerability, the Office of the High Commissioner for Human Rights, United Nations, has designated the issue of Indigenous people as one of the special themes of human rights that warrant special protection (OHCHR, 2015).

Numerous studies, including those by Cobo (E/CN.4/Sub.2/1982/2/Add.1), Martinez (E/CN.4/Sub.2/1999/20), Anaya (A/HRC/9/9), and other United Nations (UN) rapporteurs, demonstrate that Indigenous people are still vulnerable and marginalized. The right to self-determination is crucial for their survival. It plays a significant role beyond the political and legal spheres and is crucial to ensuring the recognition of their rights (Chavez, E/CN.4/2000/84). The right to self-determination, with a specific focus on free, prior, and informed consent (FPIC), should therefore serve as the basis for this chapter. The notion of the right to self-determination is consistent with the inspiration of Indigenous peoples globally, particularly in the ASEAN region.

The issue of FPIC occurs among the Indigenous peoples in every region worldwide, and it might be a barrier to the attainment of the Sustainable Development Goals (SDGs). For example, Ndlovu and Nwauche (2022) reported that the meaning, scope, and operationalization of FPIC are challenged by states and extractive sectors in the Kenyan legal framework. In a similar vein, Gahman et al. (2020) revealed the dispute over land rights and violation of FPIC that occurred in Southern Belize between Maya communities, the government, and extractives businesses. According to Smith and Mitchell (2020), despite accepting the UNDRIP, the Canadian government retained reservations

DOI: 10.4324/9781003285472-21

over compliance, and endorsing the Declaration does not imply that the Canadian government will amend domestic laws and policies to reflect the UNDRIP's principles, including FPIC. In contrast, the rights of the Sami people made great strides in the Nordic region when the government acknowledged the rights to consultation guaranteed by international law. However, as Ravna (2020) has shown, the proposed consultation statute does not include the Sámi Rights Committee's recommended conditional free, prior, and informed consent conditions.

Therefore, the main objective of this chapter is to critically analyze the law in theory and practice on FPIC as part of the right to self-determination of the Indigenous peoples in ASEAN member states by looking at standards in international law. In relation to this objective, the research sets out to answer the following questions: (1) What is the position of self-determination of Indigenous peoples and FPIC in the international law framework? (2) What is the relationship between FPIC and the SDG? (3) To what extent have the FPIC and the right to self-determination of Indigenous peoples been recognized in ASEAN member states? (4) And how are we to move forward and address the issue of FPIC as a regime of unfulfilled promises? To answer these research questions, the study provides a background of the Indigenous peoples in ASEAN member states and analyzes the concept of FPIC as part of the right to self-determination of Indigenous peoples under international law. As this study focuses on the Indigenous peoples of ASEAN member states, it will evaluate the issues and challenges in applying the principle of FPIC in Malaysia, the Philippines, and Indonesia pursuant to domestic law compared to the international law standards. Finally, this study suggests several legal and institutional reforms to improve the recognition of FPIC as part of the right to self-determination of the Indigenous peoples and will substantially complement the realization of SDG 16 in promoting a just, peaceful, and inclusive society.

20.2 Tracing Indigenous Peoples in the ASEAN

The Association of Southeast Asian Nations (ASEAN) was founded on August 8, 1967, after the ASEAN Declaration (Bangkok Declaration) was signed, consisting of ten member states, including Indonesia, Malaysia, and the Philippines (The ASEAN Secretariat, 2020). As of mid-2020, the ASEAN Statistical Yearbook 2021 discloses that the total population of ASEAN member states is 661,826.8 (The ASEAN Secretariat, 2021). The Indigenous peoples in the Southeast Asia region are referred to as hill tribes, native people, and ethnic minorities (Erni, 2010). As the current study focuses on the Indigenous peoples in Malaysia, the Philippines, and Indonesia, it is argued that the protection and recognition of Indigenous rights differ from one state to another.

In Malaysia, the Indigenous peoples are categorized into two main groups: the Orang Asli in Peninsular Malaysia and the natives of Sabah and Sarawak. However, this chapter focuses on the rights of the Orang Asli. The word *Asli* is an Arabic word that means "original," derived from the root word "Asali" (Carey, 1976). As of December 31, 2023, the total population of Orang Asli in Peninsular Malaysia is 215,215, representing approximately 0.6% of the total population of Malaysia (Jabatan Kemajuan Orang Asli, 2023). Since the early 20th century, the term *Aboriginal peoples* refers to the 18 sub-ethnic groups of Orang Asli, which can then be divided into three major ethnic groups, namely, the Negrito, Proto-Malay, and Senoi, according to their ethnography, respectively (Jabatan Kemajuan Orang Asli, 2011; Williams-Hunt, 1952; Carey, 1970). Orang Asli is a heterogeneous and non-homogeneous tribe because they have uniqueness and distinct socio-cultural and psychosocial-cognitive perspectives (Jabatan Kemajuan Orang Asli, 2011; Musa, 2011). Each Orang Asli sub-ethnic group is unique and distinct, including in language, customs, traditions, practices, and religion (Musa, 2011). One of the famous sub-ethnic groups of the Negrito is the Batek. They can be seen living in harmony with the locals and participating in ecotourism activities in Taman Negara, Malaysia (Khalid et al., 2013)

Unlike the Orang Asli in Malaysia, the Indigenous peoples in the Philippines represent approximately 8.7% of the total Filipino population (The World Bank, 2024). The Indigenous cultural community/Indigenous Peoples (ICC/IP) is a community of approximately 228 classified ethnic groups concerning ethnic linguistics (The World Bank, 2024). The total number of ICC/IP in the 2020 Census of Population and Housing (CPH) is 9.46 million (The World Bank, 2024). ICC/IP can be classified according to certain regions, such as the Igorot tribe, which mainly inhabits the Cordillera region, located in the mountains of Luzon. The Lumad tribe mostly lives in the Mindanao archipelago (Padilla et al., 2013). In contrast, the Mangyan tribe inhabits the central Philippine archipelago (Padilla et al., 2013).

Besides that, Indigenous communities in the Philippines are also referred to as ethnic minorities, tribal Filipinos, or native citizens. However, Indigenous peoples are officially referred to as Indigenous cultural communities (ICC). Referring to Section 2, Chapter I of the Rights of Indigenous Peoples Act 1997 (IPRA), Republic Act 8371, Indigenous peoples are referred to as Indigenous cultural communities and Indigenous peoples (ICC/IP). In addition, ICC/IP has been defined under Section 3(h), Chapter II of IPRA as Indigenous people who have the right to determine their own identity (self-identification), live in groups, and have their own cultural customs and are unique and distinct from majority peoples (MacDonald, 1995).

In Indonesia, Indigenous communities are geographically isolated and known as *komunitas adat terpencil* by the Ministry of Social Affairs (Mamo, 2020). Kadir (2019) adds that "[I]ndigenous people" may be known in different languages, for example, as *pribumi* (native), as opposed to *orang asing* (foreigner). These Indigenous communities consist of 1,128 ethnic groups and are estimated by Aliansi Masyarakat Adat Nusantara (AMAN) to represent approximately 2–2.8% of the total population in Indonesia (Mamo, 2020). Several legislations in Indonesia, such as the Indonesian National Constitution, Act No. 5/1960 on Basic Agrarian Regulation, Act No. 39/1999 on Human Rights, and MPR Decree No. X/2001 on Agrarian Reform, Act No. 27/2007 on Management of Coastal and Small Islands, and Act No. 32/2010 on Environment, recognize Indigenous peoples as *Masyarakat adat* and *Masyarakat hukum adat* (Mamo, 2020). Despite having the uniqueness and diversity of nature and culture that distinguish them from the majority, typically, Indigenous communities of the ASEAN live in rural and remote areas, such as mountainous areas, with limited infrastructure, facilities, and opportunities compared to the majority of people. Although some of the areas inhabited by Indigenous peoples have an abundance of natural resources, their voices receive less attention due to a lack of political participation.

Even though Malaysia, the Philippines, and Indonesia voted in favor of the United Nations Declaration on the Rights of Indigenous Peoples (UNDRIP), the Indigenous peoples in ASEAN member states still face a lack of recognition of their right to customary land issues, are marginalized, and are often dominated by the majority (Human Rights Commission of Malaysia, 2013). The current legal recognition is arguably inadequate to provide comprehensive protection for the Indigenous peoples compared to the standards set out in both UNDRIP and international law. The lacunae in the domestic law weaken the position of the Indigenous peoples, which may affect the legal protection in practice.

20.3 Indigenous Peoples and the SDGs

The 2030 Agenda is an ambitious yet comprehensive framework that complements developed and developing countries since it places people as the focus of the development process. Most

ASEAN countries adopted the Sustainable Development Goals in 2015 and reaffirmed their commitment to support and implement the 2030 Agenda for Sustainable Development. The government has made efforts to integrate 17 SDGs and 169 targets involving the economy, society, and the environment into the national planning framework. It includes localizing SDGs at subnational levels, mobilizing resources and funding through a partnership with stakeholders, and strengthening data readiness to develop a comprehensive mechanism for SDG implementation (Economic Planning Unit, 2017). However, much remains to be done, particularly pertaining to Indigenous peoples and the SDGs. As the core element of the 2030 Agenda sets out clearly to "leave no one behind," all segments of society, including the Indigenous peoples, must be included at every stage, from early public policymaking to implementation and monitoring.

Besides that, as mentioned in the Preamble of the SDG, the new agenda is meant to complete the unaccomplished mission of the last Millennium Development Goals (MDG) 2015, which includes inclusiveness and equality. The previous sustainability plan showed limited achievements across countries due to uneven benefits distribution among people. Based on the MDG 2015 Report, millions of people are left behind, especially the poorest and the disadvantaged, due to reasons such as sex, age, disability, ethnicity, or geographic location (United Nations, 2015). Hence, the SDGs incorporated inclusiveness as the primary key and pledged to empower vulnerable groups, such as Indigenous peoples (Item 23, SDG 2030). These people must be given access to gain adequate knowledge and skills to exploit opportunities and integrate into society. In addition, it will help them realize their rights, capabilities, roles, and functions in the country (Item 25, SDG 2030). Therefore, this study argues that the Indigenous peoples have a significant position in realizing the SDGs and the 2030 Agenda as the subject of international law.

The UN General Assembly has adopted a Resolution on the Agenda 2030 for Sustainable Development highlighting Indigenous people's issues related to goals 2 and 4, Zero Hunger and Quality Education (A/RES/70/1), respectively. However, apart from the two goals, various targets set out in 17 SDGs are highly relevant to Indigenous peoples, with 73 out of the 169 targets in SDGs having significant ties to the UN Declaration on the Rights of Indigenous Peoples (UNDRIP). The Office of the High Commissioner for Human Rights (OHCHR) highlighted five fundamental elements in Agenda 2030 and the SDG framework pertinent to Indigenous peoples: leave no one behind, human rights, equality, participation, and accountability. Furthermore, Indigenous peoples are inextricably linked to the core elements of achieving the SDGs, including fostering peace and inclusive societies in SDG 16.

Nevertheless, the government has inadequately assured its commitment to implement the 2030 Agenda and the Sustainable Development Goals (SDGs) for Indigenous peoples. The lack of recognition and protection of Indigenous people's rights is a source of conflict and will impede the realization of the SDGs and Agenda 2030. Failure to uphold the Indigenous people's rights, particularly their rights to education, land, health, and economics, will hinder achieving SDG 1 to end poverty, SDG 5 on gender equality and peacebuilding, and inclusive societies in SDG 16, especially during the pandemic of COVID-19. The COVID-19 pandemic has worsened existing inequalities within and among nations and regions as it brings unimaginable challenges to the most vulnerable people and the poorest countries. Some of the issues that plague the Indigenous peoples in the ASEAN are injustices and discrimination that are largely influenced by the lack of respect for the Indigenous culture and a lack of representation and participation in the decision-making processes. It is against the principles of international law, particularly the free, prior, and informed consent (FPIC) of the Indigenous peoples.

20.4 Self-determination of Indigenous Peoples in International Legal Framework

Article 1 of both the International Covenant on Civil and Political Rights (ICCPR) and the International Covenant on Economic, Social, and Cultural Rights (ICESCR) recognizes the right to self-determination of all peoples. In the context of Indigenous peoples, Article 3 of UNDRIP recognizes the right to self-determination of Indigenous peoples. By virtue of that right, they freely determine their political status and freely pursue their economic, social, and cultural development. The provision of the right to self-determination covers four main dimensions: the freedom of Indigenous people to determine their (1) political status and the freedom to pursue their (2) economic, (3) social, and (4) cultural developments. Daes (1996) added that these dimensions are closely linked to recognizing the right to natural resources.

The right to self-determination can be valued as independence, integrity, and respect and must be defined according to the principles of justice and equality. According to the National Aboriginal and Islanders representatives, the Legal Services Secretariat of Australia suggested that the right to self-determination is in its original form and genuine based on equality and non-discrimination (Chávez, 1999). Representatives from the Saami Council further described the concept of the right to self-determination as dynamic and not static (Chávez, 1999). This permits the right to self-determination to fit in many situations and be recognized beyond the traditional context, that is, secession.

Furthermore, Daes (2001) outlined the right to self-determination of Indigenous peoples as independence, integrity, and respect for Indigenous peoples. It also allows Indigenous peoples to have the freedom to live according to their customs and beliefs and to be respected by the majority. The implementation of that right can be examined by assessing whether Indigenous peoples have the freedom to choose and determine their way of life (Daes, 2001). It can be implemented through autonomy and self-government by Indigenous peoples, the economic dimension of self-determination, and participation in the decision-making process.

Article 4 of UNDRIP, however, limits the scope of the right to self-determination to include the right to autonomy or self-government only. In other words, it covers matters relating to Indigenous peoples' internal and local affairs, as well as ways and means of financing their autonomous functions. Nordin et al. (2012) argue that Article 4 limits Indigenous peoples' right to self-determination, whereby it applies only to internal aspects, whereas the rights of other people include the right to secession. It reflects an imbalance in the recognition of the right to self-determination of all people, including the right of separation. This approach is contrary to the principle of equality. It discriminates against Indigenous peoples in many situations, even though they may never have dreamed of separation (Daes, 1993).

20.5 Free, Prior, and Informed Consent (FPIC) and Indigenous Peoples

FPIC is recognized and embedded in several international law instruments as part of the self-determination of Indigenous peoples and the decision-making process. Every component in FPIC should be defined for a comprehensive understanding. Firstly, *free* is best understood as free from being influenced by any factor, coercion, or intimidation or from being forced or manipulated by certain parties (Permanent Forum on Indigenous Issues Fourth Session, 2005). The second element of FPIC is *prior*, which requires the consent of Indigenous peoples to be obtained in advance before any project or activity is carried out (Permanent Forum on Indigenous Issues Fourth Session, 2005). Sufficient time should be allocated for the process to obtain permission, and it must involve all segments of society, including women, youth, and children (Permanent Forum on Indigenous Issues Fourth Session, 2005).

Thirdly, Indigenous peoples must be informed of the project or event that may affect them. The third element requires transparency, whereby ASEAN member states oblige to supply detailed information on a project to the Indigenous peoples. To ensure transparency, the states must disclose all sorts of information, such as the nature, size, and scope of the project or activity; its objectives; the period of the project; the locations involved and affected; the assessment of impacts, such as economic, social, cultural, environmental, potential risks, and benefits for equal sharing; the parties involved in executing the project; and the procedures related to the project (Permanent Forum on Indigenous Issues Fourth Session, 2005). Therefore, the information presented to the Indigenous peoples, including women, youth, and children, must be accurate and understandable.

Fourthly, Indigenous peoples' consent may be obtained in two ways, through either consultation or participation (Anaya, 2004). Both methods share the common characteristic that they shall be conducted in good faith and with full participation to form the best mechanism for obtaining consent (Permanent Forum on Indigenous Issues Fourth Session, 2005).

As discussed in many works of literature, such as MacInnes et al. (2017) and Hanna and Vanclay (2013), the recognition of FPIC can be seen in several international instruments. Article 16 of the Conventions Concerning Indigenous and Tribal Peoples in Independent Countries (ILO Convention No. 169) recognizes the application of the principle of FPIC, where the government must seek consent from Indigenous peoples before removing them from their customary lands. This provision requires that the places to which they are moved be equivalent to their places of origin and have a similar quality and legal status (Article 16(4), ILO No. 169). In addition, the government must negotiate with and seek the participation (Article 16, ILO No. 169) of Indigenous peoples regarding development (Article 2, ILO No. 169) in several areas, including the development, natural resources, and soil (Article 15, ILO No. 169), with good faith (Article 16(2) ILO No. 169) (Article 16, 2, 15, ILO No. 169).

Referring to the provisions of FPIC in UNDRIP, Article 10 provides that Indigenous peoples cannot be removed from their ancestral land unless the principle of FPIC is complied with and all compensations have been paid to the affected Indigenous peoples. In the event of Indigenous peoples' land being seized, taken, used, occupied, or damaged without obtaining FPIC, the Indigenous peoples are entitled to claim restitution or compensation that is reasonable and fair (Article 28). If mutual agreement has been reached, compensation may also be paid in the form of land, territories, and natural resources of an equivalent value to what has been taken (Article 32(2)). Compensation in the form of land is important to the Indigenous peoples as it represents their identity. In addition to ancestral lands, the FPIC process is also required to protect the values, customs, and culture of Indigenous peoples. If the government fails to obtain FPIC from Indigenous peoples, they have the right to claim damages because the government has violated their rights of traditions and customs (Article 11(2)).

FPIC is also required if the government plans to introduce a law or any administrative matters that might affect Indigenous peoples. According to Article 19 of UNDRIP, the government must negotiate and cooperate in good faith with the Indigenous peoples. The government must obtain FPIC in good faith before approving any project affecting Indigenous peoples' ancestral lands, territories, and natural resources. It shall be made in accordance with UNDRIP Article 32(2), which requires compliance with the principle of FPIC, particularly concerning the project development, utilization, or exploitation of minerals, water, and other resources. Further, "good faith" in Articles 19 and 32(2) requires the government to be transparent in obtaining FPIC from Indigenous peoples. Article 32(3) of UNDRIP also demands the government to provide an effective mechanism to minimize undesirable impacts, such as pollution, poor economy, and sensitive culture. This is important to preserve the environment as it affects the traditional life of Indigenous peoples, who are reliant upon forests and natural resources. Therefore, governments shall comply with the FPIC process as it forms part of the manifestation of the right to self-determination of Indigenous peoples, including the Indigenous peoples in ASEAN member states.

20.6 The Orang Asli in Peninsular Malaysia

The recognition of the rights of the Orang Asli can be found in various legal instruments. The main legal source is the Federal Constitution, as the highest law of the land (Article 4, Federal Constitution). Article 153 of the Federal Constitution recognizes the special position of Malays and natives of Sabah and Sarawak. However, special protection for the Orang Asli can be found in Article 8(5)(c) of the Federal Constitution, which provides "for the protection, well-being or advancement of Aboriginal peoples of the Malay Peninsula (including the reservation of land) or the reservation to aborigines of a reasonable proportion of suitable positions in the public service." The provisions of Article 8(5)(c) allow discrimination favoring the Orang Asli, emphasizing the development, including reservation of land. The Ninth Schedule, Item 16 of the Federal Constitution, supports this provision, which includes planning for the development; nevertheless, the principle of FPIC is not embedded in this provision.

In addition, the Aboriginal Peoples Act (APA) of 1954 is a significant Act aimed at protecting the rights of Orang Asli. The APA contains 19 sections, including a definition of aboriginal people, the various rights of the Orang Asli, and the administration of Jabatan Kemajuan Orang Asli (JAKOA). The primary purpose of the APA is to provide for the protection, welfare, and progress of the Orang Asli in Peninsular Malaysia. However, the provisions in APA are inadequate to provide full protection for the Orang Asli due to the lack of legal recognition concerning FPIC in this Act.

The APA was initially drafted to prevent the communists from obtaining the help of the Orang Asli and prevent communist propaganda from being spread in the community (Nicholas et al., 2010). Therefore, the government made efforts to prevent any form of dissemination, including control over the appointment of *Batin* (Nicholas et al., 2010). Dentan et al. (1996) were critical of APA as an Act containing "paternalistic" elements. This "paternalistic" approach has negative implications for the Orang Asli to the extent that they are regarded as a wards of the state and not as Indigenous peoples with a right to their ancestral land, in addition to other rights (Idrus, 2010; Nicholas et al., 2010; Nicholas & Baer, 2007). Nordin and Witbrodt (2012) classified the APA as a platform for JAKOA to provide protection to the Orang Asli. The protection of Orang Asli's rights under the APA is insufficient, as Hooker (1996) asserts, particularly in regard to their customary land matters. For example, the APA gives power to the government to revoke, wholly or partly, the aboriginal area of reserved land under Sections 6(3) and 7(3) of the APA. In this regard, the element of FPIC should be observed, particularly when it pertains to land matters.

Furthermore, Malaysia enacted the Access to Biological Resources and Benefit Sharing Act 2017 (ABS, 2017), which came into force on December 18, 2020. Section 23 of ABS 2017 requires the prior informed consent of the communities where traditional knowledge of Indigenous peoples and local communities on the biological resources is used. The benefits gained from its use must be shared with the communities. However, a lack of legal recognition of land would certainly affect the effectiveness of the ABS legislation in achieving its objective of protecting the rights of the Orang Asli.

Besides, the Wildlife Protection Act 1972 (the Wildlife Conservation Act 2010) is also related to the Orang Asli. The Orang Asli was not consulted in the amendment involving the Wildlife

Conservation Act 1972. In 2010, the Act was amended to reduce the number of species that the Orang Asli can hunt (Subramaniam, 2011). The Orang Asli should be consulted prior to any amendment to the Wildlife Conservation Act 1972 because it will directly impact the Orang Asli way of life. The implication of the intended amendment concerns forest products, including hunted animals, that represent a source of livelihood for the Orang Asli. Therefore, Subramaniam (2011) contends that the Orang Asli should be involved in the decision-making and amendment processes.

The government took further steps to introduce the Statement of Policy Regarding the Administration of the Aboriginal Peoples of the Federation of Malaya 1961 (Statement of Policy, 1961), which aimed at strengthening efforts to provide better protection and welfare to the Orang Asli. The objectives of the said policy are the integration and assimilation of the Orang Asli with the mainstream communities in Peninsular Malaysia. This policy also emphasized their equal rights and opportunities to other Malaysians. For that purpose, the government proposed several appropriate standards to protect and develop the lives of the Orang Asli. This policy also highlighted the economic and development progress of the Orang Asli.

While there is no specific provision under Malaysian law recognizing the principle of FPIC, the provisions of Paragraph (d), Policy 1961 indicated the element of FPIC as requiring the consent of the Orang Asli before they are removed away from their ancestral land (Statement of Policy, 1961). However, the provisions of Policy 1961 did not provide clarification on the mechanisms and guidelines for how consent from the Orang Asli could be obtained. In addition, the enforcement of Policy 1961 is ambiguous compared to the APA or other laws. In conclusion, Malaysia's legal position regarding FPIC remains inadequate compared to international law standards. The government should adopt a human rights-based approach to development and implement FPIC law by outlining consent-making processes that ensure effective participation of the Orang Asli as per the recommendation in the National Inquiry into the Land Rights of Indigenous Peoples (Human Rights Commission of Malaysia, 2013).

20.7 FPIC in the Philippines

The primary law that recognizes the rights of ICC/IP in the Philippines is the Rights of Indigenous Peoples Act (IPRA) 1997, which is an Act that can be classified as the most progressive law pertaining to the rights of Indigenous people (Erni et al., 2008). IPRA was drafted to recognize the rights of Indigenous peoples, such as the right to customary land (ancestral domain), the right to self-government and empowerment, and the right to social justice and cultural integrity (Castro, 2008). IPRA is considered the Magna Carta on the rights of Indigenous people, which consists of 84 sections and is intended to recognize, protect, and promote the rights of ICC/IP.

Having said that, the IPRA upholds the rights of all ICCs/IPs to customary land. Section 7 provides the recognition of such rights, including the right to own and develop land and natural resources, the right to inhabit the customary territory, and several others. IPRA can also be considered an Act that complies with international law, particularly Articles 25 and 26 of the UNDRIP, which recognize the right to customary land. The Act also describes the concepts of land ownership, composition, ICC/IP responsibilities toward domains, and customary land.

In line with the spirit of UNDRIP, the law under the IPRA specifically recognizes ICC/IP in the decision-making process through the implementation of the FPIC. Section 3(g) IPRA explains the meaning of the FPIC. The IPRA provides several provisions concerning FPIC as follows:

1. Section 6(c), relating to the right to reside in the territory of the original person
2. Section 32, relating to the rights of the community to intellectual property

3. Section 33(a), on rights to material relating to ICC/IP values, religion, and customs
4. Section 35, relating to access to genetic and biological resources
5. Section 46(a), relating to the management of customary domains or lands
6. Section 58, relating to environmental considerations
7. Section 59, dealing with preconditions of confirmation

Even though the provision of IPRA is comprehensive and meets international standards, the ICC/IP faces several challenges in terms of implementation. Magno (2013) describes that the inadequate systems in place and the implementation setbacks have both been linked to FPIC process flaws. According to the United Nations Special Rapporteur, the developer fails to observe the FPIC as required by law before undertaking development projects, such as logging, dams, and mining projects (Stavenhagen, R, E/CN.4/2003/90/Add.3). Carino (2005) explains that FPIC processes have also been hindered by the failure of industries to consult with local communities prior to construction. Another factor contributing to the setback of the implementation of FPIC is the lack of observation of the Indigenous communities' traditional decision-making processes in line with the FPIC processes (Doyle & Carino, 2013). It may hinder the achievement of SDG 16, particularly target SDG 16.10, to ensure Indigenous people's access to information and protect fundamental freedoms.

20.8 FPIC in Indonesia

Generally, the recognition of Indigenous rights under Indonesian law is arguably inadequate. Referring to the Indonesian National Constitution, all natural resources belong to the government unless granted to an individual for private property. Some existing laws also recognize customary land, such as the Basic Agrarian Law 1930, however, subject to development requirements by the government (Filer et al., 2020).

Unlike ICC/IP in the Philippines, the provision of FPIC in Indonesia is still lagging as there is no comprehensive legislation related to FPIC. Nevertheless, several rules and laws have been introduced to ensure that the government implements FPIC (Elfitra et al., 2019). As a result of NGO insistence, Elfitra et al. (2019) explain that the Indonesian government has included the FPIC element in regulations and laws pertaining to licensing and land acquisition by investors, such as:

1. Law of the Republic of Indonesia No. 2 of 2012;
2. Decree of the Minister of Agrarian Affairs/Head of National Land Agency No. 21 of 1994 Concerning Procedure of Land Acquisition for Companies in the Framework of Investment;
3. Regulation of the Minister of Agrarian Affairs/Head of BPN No. 5 of 2015 about Location Permission;
4. Regulation of the Minister of Environment and Living and Forestry No. 34 of 2017 on Local Wisdom;
5. Government Regulation No. 27 of 2012 on Environmental Permits;
6. Regulation of the State Minister of Environment of the Republic of Indonesia No. 17 of 2012 on Guidelines for Community Involvement in the Environmental Impact Analysis and Environmental Permit Process.

Furthermore, the practice of FPIC in Indonesia was urged by two international bodies, namely, the Forest Steward Council (FSC) and the Roundtable on Sustainable Palm Oil (RSPO), through the development and dissemination of land recognition and dispute management systems (Elfitra et al., 2019). The RSPO endorsed the FPIC Guide for RSPO Members on November 20, 2015, in Kuala Lumpur. These guidelines are essential to ensure that the RSPO is permitted to carry out palm oil activities without any conflict over land. Accordingly, the RSPO establishes clear guidelines by taking into account representation from the local community, respecting the community dispute resolution system, gender equality, and respecting Indigenous customary land rights (FPIC Guide for RSPO Members, 2015).

Even though the RSPO gives guidelines for FPIC, there are still numerous limitations that must be tackled in practice, such as a lack of influence at the grassroots level and also facing a backlog of cases (Filer et al., 2020; Silva-Castañeda, 2015). Furthermore, the RSPO registered 56 complaints from 2010 to 2016, of which 41% were related to violations of FPIC principles in Indonesia (Colchester, 2016). Noting that guidelines have some limitations, the elements of FPIC must be observed when they are put into practice. Therefore, in order for the rights of Indigenous peoples to be effectively recognized, it is essential to eliminate all the impediments that have been encountered about FPIC.

20.9 Concluding Remarks

Observing the experience of Indigenous peoples beyond ASEAN member states in relation to other continents, it is possible to conclude that all Indigenous people face similar obstacles in putting FPIC into practice, which will render FPIC a regime of unfulfilled promises. Having said that, other states, like Norway, are in a better position than others since they have recognized the Sami people's rights to consultation as recognized under international law. In the context of ASEAN member states, the law pertaining to FPIC in IPRA can be said to be more advanced as it recognizes the rights of the ICC/IP of the Philippines. However, implementation challenges indicate that existing laws should be strictly enforced. On the other hand, Malaysia and Indonesia need to take action to provide a more comprehensive legal framework for the government to implement to comply with the FPIC process and safeguard the rights of Indigenous people effectively. Therefore, the relevant law must be amended, in particular, APA 1954. Any such amendment of the domestic law should recognize the right to self-determination and embed the principle of FPIC as enshrined in UNDRIP.

To move forward, states must initiate legal and institutional reform to improve the recognition of FPIC as part of the right to self-determination of Indigenous peoples. These reforms will substantially complement the realization of SDG 16 in promoting a just, peaceful, and inclusive society. The government should address the lack of mechanisms to ensure Indigenous peoples are not left behind by introducing a framework to warrant the inclusive participation of Indigenous peoples in all decision-making processes, especially during the challenging time of the pandemic of COVID-19. UNDRIP and other international law instruments, such as ILO Convention 169, should be the framework to address the target set out in Agenda 2030, including SDG 16 for peacebuilding development. Tauli-Corpuz, UN Special Rapporteur on the Rights of Indigenous Peoples, rightly suggested that the government should introduce special measures and measurements by establishing an appropriate mechanism to protect the rights of Indigenous peoples. Therefore, proactive measures must be taken to achieve the SDGs, enhance the protection of the rights of the Indigenous peoples, and implement FPIC law toward Indigenous peoples globally.

To guarantee the rights of the Indigenous peoples in practice, states and the private sector should observe the laws and regulations that recognize the rights of the Indigenous peoples. It is fundamental to ensure that the protection of the rights of the Indigenous peoples is compatible with the standards of international law and thus assist ASEAN member states in meeting their international law obligations and will substantially complement the realization of SDG 2030 as a whole and goal 16 specifically, in promoting a just, peaceful, and inclusive society. Ultimately, the proposed legal and institutional reform will be of great significance and relevance to accelerate the recognition of Indigenous peoples to participate in the decision-making process at all levels, especially in the unprecedented challenges of the COVID-19 pandemic.

In a nutshell, the recognition of the right to self-determination of Indigenous peoples of ASEAN member states through FPIC is still inadequate compared to international law standards. ASEAN member states must observe the elements of FPIC and consider the participation of Indigenous peoples in decision-making processes at all levels. In the absence of any provision in domestic law that obliges compliance with the elements of FPIC, states have moral obligations to implement and respect the rights of Indigenous peoples as accorded through the spirit of UNDRIP. In line with Target 16.7 of SDG 16, states must ensure the participation of Indigenous peoples is all-encompassing, open, and transparent in the decision-making process at all levels to decide for their future. It will complement the achievement of goal 16 of the SDG in building a just, peaceful, and inclusive society.

Acknowledgments

This publication is being assisted financially by the Research University Grant (GUP-2022-063) and GGPM-2021-043 Universiti Kebangsaan Malaysia.

References

Anaya, S. J. (2004). *Indigenous Peoples in International Law*. Oxford University Press: Oxford.

Anaya, S. J. (2008). *Report of the Special Rapporteur on the Situation of Human Rights and Fundamental Freedoms of Indigenous People, Promotion and Protection of All Human Rights, Civil, Political, Economic, Social and Cultural Rights, Including the Right to Development*, A/HRC/9/9.

A/RES/70/1, *Transforming Our World: The 2030 Agenda for Sustainable Development*, Resolution adopted by the General Assembly on 25 September 2015, Seventieth session.

Carey, I. (1970). The Kensiu Negritos of Baling, Kedah. *Journal of the Malaysian Branch of the Royal Asiatic Society* 43(1). pages 143–160.

Carey, I. (1976). *Orang Asli, the Aboriginal Tribes of Peninsular Malaysia*. Oxford University Press: Kuala Lumpur.

Carino, J. (2005). Indigenous Peoples' Right to Free, Prior, Informed Consent: Reflections on Concepts and Practice. *Arizona Journal of International and Comparative Law* 22(19). pages 19–39.

Castro, N. (2008). Three Years of the Indigenous Peoples Rights Act: Its Impact on Indigenous Communities. *Kasarinlan: Philippine Journal of Third World* 15(2). pages 35–54.

Chávez, L. E. (1999). *Report of the Working Group Established in Accordance with Commission on Human Rights Resolution 1995/32*, Commission on Human Rights, Fifty-Sixth Session, E/CN.4/2000/84.

Cobo, M. J. R. M. (1983). *Study of the Problem of Discrimination against Indigenous Populations, Conclusions, Proposals and Recommendation*. E/CN.4Sub.2/1983/21/Add.8.

Colchester, M. (2016). *Do Commodity Certification Systems Uphold Indigenous Peoples' Rights*? Lessons from the Roundtable on Sustainable Palm Oil and Forest Stewardship Council, Policy Matters 2016: Certification and Biodiversity.

Daes, I. E. (1993). *Report of the Working Group on Indigenous Populations on Its Eleventh Session, Discrimination against Indigenous Peoples*. E/CN.4/Sub.2/1993/29.

Daes, I. E. (1996). The Right of Indigenous Peoples to "Self-Determination" in Contemporary World Order. In Clark, D. (Ed.), *Self-Determination, International Perspectives*. Macmillan Press Ltd.: London. pages 47–57.

Daes, I. E. (2001). *Prevention of Discrimination and Protection of Indigenous Peoples and Minorities, Indigenous Peoples and Their Relationship to Land*. E/CN.4/Sub.2/2001/21.

Daes, I. E. (2002). The Concept of Self-Determination and Autonomy of Indigenous Peoples in the Draft United Nations Declaration on the Rights of Indigenous Peoples. *St. Thomas Law Review* 14(259). pages 259–269.

Dentan, R. K. et al. (1996). *Malaysia and the "Original People", a Case Study of the Impact of Development on Indigenous Peoples*. Allyn and Bacon: Boston.

Doyle, C., Carino, J. (2013). *Making Free Prior & Informed Consent a Reality: Indigenous Peoples and the Extractive Sector, Indigenous Peoples Links, Middlesex University School of Law, and The Ecumenical Council for Corporate Responsibility* (May 2013). www.ecojesuit.com/wp-content/uploads/2014/09/Making-FPIC-a-Reality-Report.pdf (accessed 30.05.2021).

Economic Planning Unit. (2017). *Prime Minister's Department, Sustainable Development Goals Voluntary National Review 2017*, High-Level Political Forum.

Elfitra, et al. (2019). *Free, Prior and Informed Consent (FPIC) as a Conflict Mitigation Instrument: FPIC Applicability for Mitigation of Structural Agrarian Conflicts in Indonesia*, IConShel 2018, September 05–06, Padang, Indonesia. https://eudl.eu/pdf/10.4108/eai.5-9-2018.2281040 (accessed 30.05.2021).

Erni, C. (2010). *ASEAN's Indigenous Peoples, Asia Indigenous Peoples' Pact (AIPP), the International Work Group for Indigenous Affairs (IWGIA), and Asian Forum for Human Rights and Development (FORUM-ASIA)*, Chiang Mai. www.iwgia.org/images/publications/0511_ASEAN_BRIEFING_PAPER_eb.pdf (accessed 30.05.2021).

Erni, C., Padilla, M. T. G., Padilla, S., Nilsson, C. (2008). Country Profile: Philippines. In Erni, C. (Ed.), *The Concept of Indigenous Peoples in Asia, a Resource Book*. International Work Group for Indigenous Affairs (IWGIA), Asia Indigenous Peoples Pact Foundation (AIPPF): Chiang Mai. pages 427–433.

Filer, C., Mahanty, S., Potter, L. (2020). The FPIC Principle Meets Land Struggles in Cambodia, Indonesia and Papua New Guinea. *Land 2020* 9(67). pages 1–21. https://doi.org/10.3390/land9030067.

Free, Prior, and Informed Consent Guide for RSPO Members (2015). *RSPO Human Rights Working Group 2015*, RSPO-GUI-P02-001 V1.0.

Gahman, Levia, et al. (2020). Plunder via Violation of FPIC: Land Grabbing, State Negligence, and Pathways to Peace in Central America and the Caribbean. *Journal of Peacebuilding and Development* 15(3). pages 372–376.

Hanna, P., Vanclay, F. (2013). Human Rights, Indigenous Peoples and the Concept of Free, Prior and Informed Consent. *Impact Assessment and Project Appraisal* 31(2). pages 146–157.

Hooker, M. B. (1996). The Orang Asli and the Laws of Malaysia: With Special Reference to Land. *Akademika* 48. pages 21–50.

Human Rights Commission of Malaysia (2013). *Report of the National Inquiry into the Land Rights of Indigenous Peoples*, Kuala Lumpur.

Idrus, R. (2010). From Wards to Citizens: Indigenous Rights and Citizenship in Malaysia. *Political and Legal Anthropology Review* 33(1). pages 89–108.

Idrus, R. (2011). The Discourse of Protection and the Orang Asli in Malaysia. *Kajian Malaysia* 29(Supp. 1). pages 53–74.

Jabatan Kemajuan Orang Asli (2011). *Pelan Strategik Jabatan Kemajuan Orang Asli*, 2011–2015.

Jabatan Kemajuan Orang Asli (2023). *Total Population of Orang Asli by State*. https://www.jakoa.gov.my/orang-asli/jadual-taburan-etnik-orang-asli-mengikut-negeri/. (accessed 18.11.2024).

Kadir, M. Y. A. (2019). Defining "People" and "Indigenous People" in International Human Rights Law and Its Application in Indonesia. *International Journal on Minority and Group Rights* 26(3). pages 1–36.

Khalid, R. M., Sulaiman, S. S., Bakar, N. A., Rahman, S. A. (2013). Revisiting Legal Framework for Sustainable Tourism in Taman Negara Malaysia. *Research Journal of Applied Sciences* 8(8). pages. 393–397.

Macdonald, C. (1995). Indigenous Peoples of the Philippines: Between Segregation and Integration. In Barnes, R. H., Gray, A., Kingsbury, B. (Eds.), *Indigenous Peoples of Asia*. Association for Asian Studies: MI. pages 345–356.

MacInnes, A., et al. (2017). Free, Prior and Informed Consent: How to Rectify the Devastating Consequences of Harmful Mining for Indigenous People. *Perspectives in Ecology and Conservation* 15. pages 152–160.

Magno, C. (2013). *Free Prior and Informed Consent in the Philippines Regulations and Realities, Oxfam America Briefing Paper*. https://s3.amazonaws.com/oxfam-us/www/static/media/files/fpic-in-the-philippines-september-2013.pdf (accessed 30.05.2021).

Mamo, D., (2020). *The Indigenous World 2020, 34th Edition, The International Work Group for Indigenous Affairs (IWGIA)*. https://iwgia.org/images/yearbook/2020/IWGIA_The_Indigenous_World_2020.pdf (accessed 30.05.2021).

Martínez, M. A. (1999). *Study on Treaties, Agreements and Other Constructive Arrangements between States and Indigenous Populations*. E/CN.4/Sub.2/1999/20.

Musa, M. (2011). The Socioeconomic History of the Orang Kanaq of Johor. *Kajian Malaysia* 29(1). pages 47–74.

National Commission of Indigenous Peoples (NCIP) Indigenous Peoples of the Philippines. www.ncip.gov.ph/index.php/about-us1/vision-mission-mandate (accessed 15.02.2015).

Ndlovu, N., Nwauche, E. S. (2022). Free, Prior and Informed Consent in Kenyan Law and Policy After Endorois and Ogiek. *Journal of African Law* 66(2). pages 201–227. https://doi.org/10.1017/S002185532200002X

Nicholas, C., Baer, A. (2007). Health Care for the Orang Asli: Consequences of Paternalism and Non-Recognition. In Leng, C. H., Barraclough, S. (Eds.), *Health Care in Malaysia, the Dynamics of Provision, Financing, and Access.* Routledge Taylor & Francis Group: London. pages 119–136.

Nicholas, C., Engi, J., Ping, T. Y. (2010). *The Orang Asli and the UNDRIP: From Rhetoric to Recognition.* Center for Orang Asli Concern: Subang Jaya.

Nordin, R., Hassan, M. S., Witbrodt, M. A. (2012). Indigenous Peoples in Asia: Indigenousness and Self-Determination. *The Law Review.* pages 510–529.

Nordin, R., Witbrot, M. A. (2012). Self-Determination of Indigenous Peoples: The Case of Orang Asli. *Asia Pacific Law Review* 20(2). pages 189–210.

Office of the High Commission for Human Rights (OHCHR), Briefing Note on Indigenous Peoples' Rights and the 2030 Agenda, September 2017. www.ohchr.org/Documents/Issues/IPeoples/BriefingPaperIPRights2030Agenda.pdf (accessed 29.05.2021).

Padilla, M. T. G., Villarante, P. B., Trono, W. L. (2013). *Indigenous Peoples in the Philippines.* www.iwgia.org/images/stories/sections/regions/asia/documents/IW2013/Philippines.pdf (accessed 24.03.2014).

Permanent Forum on Indigenous Issues Fourth Session (2005). *Report of the International Workshop on Methodologies Regarding Free, Prior and Informed Consent and Indigenous Peoples*, E/C.19/2005/3, (New York, 17–19 January 2005).

Ravna, Ø. (2020). The Duty to Consult the Sámi in Norwegian Law. *Arctic Review on Law and Politics* 11. pages 233–255. http://doi.org/10.23865/arctic.v11.2582.

Rehman, J. (1998). International Law and Indigenous Peoples: Definitional and Practical Problems. *Journal of Civil Liberties* 3(3). pages 224–239.

Silva-Castañeda, L. (2015). What Kind of Space? Multi-stakeholder Initiatives and the Protection of Land Rights. *The International Journal of Sociology of Agriculture and Food* 22(2). pages 67–83.

Smith, J. A., Mitchell, T. L. (2020). Development of an UNDRIP Compliance Assessment Tool: How a Performance Framework Could Improve State Compliance. *The International Indigenous Policy Journal* 11(2). https://doi.org/10.18584/iipj.2020.11.2.10713

Stavenhagen, R. (2003). *Report of the Special Rapporteur on the Situation of Human Rights and Fundamental Freedoms of Indigenous People, Mr. Rodolfo Stavenhagen, Submitted in Accordance with Commission on Human Rights, Mission to the Philippines.* E/CN.4/2003/90/Add.3.

Statement of Policy Regarding the Adminstration of the Aborigine Peoples of the Federation of Malaya Department of Orang Asli (1961). Kuala Lumpur.

Subramaniam, Y. (2011). Rights Denied: Orang Asli and Rights to Participate in Decision-Making in Peninsular Malaysia. *Waikato Law Review* 19(2). pages 45–65.

The ASEAN Secretariat (2020). *ASEAN Statistical Yearbook 2020, Jakarta.* www.aseanstats.org/wp-content/uploads/2020/12/ASYB_2020.pdf (accessed 30.05.2021).

The ASEAN Secretariat (2021). *ASEAN Statistical Yearbook 2021, Jakarta.* https://asean.org/wp-content/uploads/2021/12/ASYB_2021_All_Final.pdf (accessed 18.11.2024).

The World Bank (2024). *No Data, No Story: Indigenous Peoples in the Philippines, Washington.* https://thedocs.worldbank.org/en/doc/baa43cc91ec55266a538e9023c528bd7-0070062024/original/No-Data-No-Story-Indigenous-Peoples-in-the-Philippines.pdf (accessed 18.11.2024).

United Nations (UN) (2015). *The Millennium Development Goals Report 2015.* www.un.org/millenniumgoals/2015_MDG_Report/pdf/MDG%202015%20rev%20(July%201).pdf (accessed 25.08.2022).

Williams-Hunt, P. D. R. (1952). *An Introduction to the Malayan Aborigine, Malayan Aborigines.* AMS Press: New York.

PART II

Policy

21

TAIWAN'S RESPONSES TO THE EUROPEAN CARBON BORDER ADJUSTMENT MECHANISM AND SDG 13—CLIMATE ACTION

From the Greenhouse Gas Reduction Act of 2015 to the Climate Change Responding Act of 2023

Anton Ming-Zhi Gao

21.1 Introduction

COVID-19 has claimed many lives worldwide. Nevertheless, it has also contributed to the environment's recovery due to an economic slowdown.[1] However, this is not the case in Taiwan. With virtually no lockdown in the country in early 2020, Taiwan's economic growth (3.11%) surpassed that of other Asian countries.[2] Consequently, concerns regarding greenhouse gas emissions surfaced, leading to the deliberation of revising the Greenhouse Gas Reduction Act (GHG Act) of 2015 to improve carbon emission reduction from mid-April 2020. The primary objective was to transition from the current emission-trading scheme embedded in the Climate Change Act of 2015[3] to a carbon fee scheme.[4] Such measures are related to the EU's proposed carbon border adjustment mechanism (CBAM) under the EU Green Deal since March 2020.[5]

After two legislative proposals provided by the Environmental Protection Administration (EPA) in 2020 and 2021, the cabinet adopted its version in April 2022, which was then submitted to Parliament and adopted by Parliament in January 2023.[6] Parallel with this development, the government adopted "Taiwan's Pathway to Net-Zero Emissions in 2050" on March 30, 2022,[7] and announced the detailed action plans for 12 key strategies: (1) wind/solar photovoltaic power; (2) hydrogen energy; (3) innovative energy; (4) power system and energy storage; (5) energy savings and efficiency; (6) carbon capture, utilization, and storage (CCUS); (7) carbon-free electric vehicles; (8) resource recycling with zero waste; (9) carbon sinks; (10) green lifestyle; (11) green finance; and (12) just transitions.[8] These measures are intended to respond to net-zero, EU CBAM, and Sustainable Development Goal 13 (SDG 13: Climate Action), where EU CBAM can be seen as the EU's efforts to improve SDG 13 for non-EU countries.[9]

Despite the fact that the EU CBAM had not yet been finalized,[10] it had nearly reached its final stage after important Negotiators of the Council and the European Parliament arrived at a

DOI: 10.4324/9781003285472-23

provisional and conditional agreement on the CBAM.[11] Moreover, Taiwan Parliament has adopted the Climate Change Responding Act of 2023 (CCRA) in January 2023. Hence, now appears to be the appropriate time to evaluate the effectiveness of Taiwan's new climate legislation in meeting CBAM challenges and net-zero goals. In addition, *Taiwan could be the first non-EU country to respond to EU CBAM prior to its finalization*, thereby serving as a model for other non-EU nations wishing to comply with EU CBAM.

21.2 Evolution of the Recent Amendments of the 2015 Climate Change Act Since 2020

21.2.1 Early 2020

Despite the COVID-19 outbreak in early 2020, the pandemic in Taiwan was effectively contained/controlled in 2020, with few locally confirmed cases[12] in 252 days. Moreover, unusually prosperous economic activities and economic growth (growing 4.94%) outpaced China for the first time in 30 years.[13]

Amidst the COVID-19 crisis, the EU continued to pursue its climate agenda and launched the CBAM climate agenda as part of the EU Green Deal.[14] As this initiative would influence Taiwan's export products, the competent authority for climate change, the Environmental Protection Administration (EPA), proposed the general idea of revising the GHG Act of 2015 in mid-2020. In early April 2020, the EPA minister proposed to amend the 2015 GHG Act to Health, Welfare, and Environment Committee.[15] This triggered the amendment of the Act. A quick response transpired to the EU's initiative, which was only subject to a public consultation process between March 4, 2020, and April 1, 2020. Taiwan soon responded to this initiative. *Taiwan may be the first country to attempt to respond to such a draft EU CBAM.*

21.2.2 Drafts From Late 2020 to 2022

To facilitate the deliberation of climate change legislation and policy, the EPA created an online platform called Climate Talks (www.climatetalks.tw/). Apart from the emission data on Taiwan's general and six sectors (including energy, manufacturing, transportation, residence and business, agriculture, and environment), the website mainly focuses on the public consultation process of the Phase II emission reduction plan for 2025 and the revision of the 2015 GHG Act.

After the EPA minister's announcement in early April 2020, it completed the draft of the revision and held a public consultation conference in early October 2020. Three meetings were held in southern, middle, and northern Taiwan. After obtaining feedback from various participants, specifically from industries and environmental NGOs, the EPA attempted to integrate all information into its proposal to seek the cabinet's approval and submit it to the legislative process in Parliament.

The EPA draft in early October 2020 inspired environmental NGOs and some pro-green members of Parliament (MPs) to propose their revision. As carbon-neutral or net-zero carbon emissions were not included in the EPA draft, the focus of the NGOs' version was on this perspective. For instance, Mr. Hong, the pro-green MP of the ruling party, who was the deputy secretary of the Green Citizen Action Alliance,[16] collaborated with environmental NGOs and proposed a version to highlight the importance of carbon neutrality by 2050. This draft was announced in late December 2020 and submitted for review on March 1, 2021.[17] The 2015 GHG Reduction Act was renamed the "Climate Change Action Act."

Other MPs of the ruling party also proposed an amendment on March 11, 2021 (MP Su's draft).[18] Finally, the MPs of opposition parties proposed their revisions, including the Kuomintang (KMT),[19] Taiwan People's Party,[20] and New Power Party (NPP).[21]

The EPA promulgated the draft amendment of the Act on October 21, 2021, after considering the advice from all sectors of society. The EPA then held in-depth discussions with legislators, experts and scholars, NGOs, and representatives of major industries on December 20–23, 2021, and January 10, 2022, about the content of the draft to forge consensus, and then revised the draft with reference to the advice from all sectors of society, submitting it to the Executive Yuan for approval on February 7, 2022. After being reviewed by interministerial meetings convened by the Executive Yuan, the draft amendment was approved by the 3,799th meeting of the Executive Yuan on April 21, 2022, and submitted to the Legislative Yuan for deliberation. On May 11 and May 12 of the same year, the draft amendment was examined by the Social Welfare and Environmental Hygiene Committee during the 5th session of the 10th Legislative Yuan.[22] In Taiwan's history, a *total of 30 drafts have been deliberated in Parliament*, with perhaps climate and energy bills garnering the most attention. The Climate Change Response Act was passed by the Taiwanese legislature in January 2023 to establish a carbon fee system for large emitters and to set the goal of reaching net-zero emissions by January 2050.[23]

21.2.3 First Country to Nearly Respond to EU CBAM

The response to EU CBAM features can be reflected in the close timeline and the relationship between the key development stages of CCRA and the EU CBAM. As aforementioned, soon after the EU launched a consultation on the road map of the EU Green Deal (carbon border adjustment mechanism) (March 4, 2020–April 1, 2020),[24] the Taiwan EPA began to consider revising the Climate Change Act of 2015. The timing of proposing the first EPA draft in October 2020 was close to the EU's public consultation (July 22, 2020–October 28, 2020). The adoption of CCRA in January 2023 may be relevant to the important Negotiators of the Council and the European Parliament reaching an agreement of a provisional and conditional nature on the CBAM in December 2022. Therefore, *Taiwan may be the first country to strive to follow the evolution of EU CBAM*!

21.3 Overview of the Amendments of the GHG Reduction Act of 2015 and the Climate Change Response Act of 2023

Apart from the renaming of the title from the GHG Reduction Act (GHG Act) to the Climate Change Response Act in the EPA draft, the following four substantive aspects were highlighted in the public hearing meetings in October 2020:[25]

- Reinforcement of administrative regulations
- Perfection of economic incentives
- Reinforcement of institutional design
- Addition of adaptation measures

21.3.1 Reinforcement of Administrative Regulations

Administrative regulations are included in the 2015 GHG Act. For instance, to pave the way for the emission trading scheme (ETS), basic emission situation and data should be collected first. Under Article 15 of the 2015 GHG Act, a large emitter must conduct annual accounting and registration.

However, the key regulatory tools for industries under Chapter 3 of the Emission Reduction Measures in the Act were either relatively lenient (such as incentive emission performance standard under Article 17[26] or the establishment of climate funds under Article 19) or not implemented (e.g., ETS scheme in Articles 18, 20, 21, 22). A lack of sticks and regulations exist to improve the carbon emission behavior of industries. Therefore, the following measures were introduced to improve the situation:

1. The incentive design of the emission performance standards was revised to include mandatory emission performance standards (Article 17, EPA draft).
2. Requirement of the best available technology (BAT) for new or major changes in existing installations (Article 18, EPA draft).
3. GHG impact assessment scheme: Eligible emitters should submit their GHG emission management plan for approval by competent authority before the operation (Article 19, EPA draft).

However, facing opposition from the industries, only point 1, that is, mandatory emission performance standards, is adopted in the final version of Article 23 of the CCRA of 2023. In addition, mandatory emission performance standards would apply not only to industrial installations but also to new buildings and vehicles. Moreover, despite the opposition to the introduction of BAT, new installation developers must offset certain ratios of their increased emissions by either purchasing offsets or paying fines.[27]

21.3.2 Carbon Pricing: Perfection of Economic Incentives

The ETS scheme was chosen as the carbon pricing scheme in the 2015 GHG Act. However, this approach has not yet been implemented in Taiwan, for unknown reasons. This may be related to the focus of the new government's main policy (after 2016) being the renewable energy and nuclear phase-out policy rather than climate change. Moreover, the relatively small size of Taiwan and the concentrated carbon emissions from several big players may hinder the liquidity of the carbon market.

Therefore, in April 2020, the EPA introduced a carbon fee as a possible response to the EU CBAM carbon price. If such a program does not exist in Taiwan, EU exporters must pay for CBAM. Therefore, such a payment can be made in Taiwan, ensuring that the fee and funding for emissions reduction remain domestic.

Hence, the EPA introduced the idea of the carbon management fee in Articles 29, 31, and 32 of the EPA draft. Interestingly, the ETS scheme provisions remain in the EPA draft in Articles 26, 27, and 28, even before the provisions of the carbon management fee. However, there are no plans to implement the ETS within a short time frame if the Act is passed. Although it remains in the drafting stage, every business has inquired about the possibility of a carbon fee following implementation. Referring to Singapore's carbon price, the possible rate for early 2021 was set at 100 NTD (more than 3 USD) when discussing the potential rate.

The use of a carbon fee instead of a carbon tax remained in the EPA draft of 2021 and the cabinet version of 2022, with most of the drafts proposed by legislators. The carbon fee scheme was finally adopted under Article 28 of the CCRA. However, the exact fee rate remains unclear, as Article 28 of the CCRA does not include the exact rate. According to the minister of EPA in Q&A at a recent conference (October 2022), a rate of 300 NTD per ton of CO_2 is highly probable.[28]

21.3.3 Reinforcement of Institutional Design

Climate change matters typically encompass jurisdictions or the competences of different ministries. Thus, only the EPA may not be able to mobilize the actions of different ministries. In the 2015 GHG Act, the cabinet (the Executive Yuan) coordinates the related cross-ministerial climate matters.[29] Although a coordination body was created at the cabinet level immediately after the 2015 GHG Act, the Office of Energy and Emission Reduction, a part of the Executive Yuan,[30] it is not a legal body, and its role and power remain limited.

In this EPA draft, the role of the Climate Change Council at the central government level was provided in Article 8 of the draft, whilst the role of the Council at the local government level was provided in Article 17 of the draft. Moreover, according to Article 9 of the EPA draft, six ministers have been designated to accommodate the needs of actions in six sectors: (1) the energy sector (Ministry of Economic Affairs, or MOEA); (2) the manufacturing sector (MOEA); (3) the transportation sector (Ministry of Transportation); (4) the building sector (Ministry of Interior Affairs); (5) the agricultural sector (Agricultural Council); and (6) the environmental sector (EPA). Moreover, an advisory committee to set short-term emission reduction goals is provided in Article 11. Finally, the role of local governments in climate change–related matters is reinforced in Articles 17 and 18 of the EPA draft.

However, in the final version of the CCRA, the National Council for Sustainable Development (NCSD)[31] conferred the authority to coordinate climate change–related affairs to the cabinet.[32] Moreover, the CCRA of 2023 specified 19 climate actions and their competent authorities in Article 8(2) of the CCRA. For instance, the transition issue should be addressed by the National Development Council.[33] Finally, the local government must establish a climate change response council to implement climate change measures.[34]

21.3.4 Adding Adaptation Measures

Although no special chapter exists on the adaptation of the 2015 GHG Act, its role is mentioned in certain provisions.[35] Such a design was not considered sufficiently aggressive for the adaptation measures. Therefore, the EPA draft contains a special chapter on adaptation (Chapter 4). The governance of adaptation can be improved through these five provisions:

1. Division of the work of preparing the National Adaptation Action Plan at the central government level (Art. 35)
2. Regular implementation and reporting of the adaptation plan (Art. 36)
3. Scientific approach of central technology-competent authority in adaptation matters (Art. 37)
4. Role of different levels of government in climate adaptation matters (Art. 38)
5. Role of local governments in adaptation matters (Art. 39)

In the final version of CCRA of 2023, a special chapter (Chapter 3) comprising four articles (17–20) has been adopted to emphasize adaptation. In addition, the sequence of the chapter on adaptation (Chapter 3) has been moved before the "emission reduction strategy" (Chapter 4) to highlight the priority of adaptation. As this chapter will focus on net-zero and responding to EU CBAM, adaptation measures will not be discussed in detail.

21.3.5 Summary of One MP's Draft

Of the 30 MPs' versions, Hong's draft is the most relevant and interesting for our discussion, as it is presented in close collaboration with environmental NGOs. Generally, as Hong's version of the MP was proposed after the EPA draft, the aforementioned four key aspects were all integrated. A slight difference was observed compared to the EPA draft. For instance, the emission performance standards remain unchanged and voluntary in Hong's version, instead of mandatory, as in the EPA draft. However, notable differences were identified.

The first is the change in legislation title. Similar to EPA's change to the legislation title, Hong intends to rename it to Climate Change Action Act.

Moreover, the EPA was hesitant to introduce net-zero emissions by 2050 in 2020. Thus, in its draft, the original emission target of 2050 in Article 4 of the 2015 GHG Act remains: "Long-term national GHG emission reduction goal shall be to reduce GHG emissions to no more than 50% of 2005 GHG emission by 2050." Only the phrase "will take efforts to reach net zero vision" was added to this target.[36] However, Hong's version did not accept this weak target and stated that the long-term emissions target of 2050 is net-zero.[37] This MP draft appeared to influence the EPA, causing the EPA's 2021 draft to include net-zero emissions, followed by the cabinet version and CCRA of 2023.

Finally, Hong's version introduces the recent popular idea of the climate change litigation clause. Article 40 of MP Hong's version provides the following:

> When a public or private premise violates this Act or related orders, determined pursuant to the authorisation of this Act and the competent authority is negligent in enforcement, victims or public interest groups may notify the competent authority in writing of the details of the negligent enforcement.
>
> For those competent authorities that have failed to carry out enforcement according to the law within 60 days of the receipt of the written notification, the victims or public interest groups may name the competent authority at issue a defendant and directly file a lawsuit with an administrative court based on the negligent behaviour of the competent authority in fulfilling its enforcement duties to seek a ruling ordering the competent authority to conduct enforcement.
>
> When issuing a verdict on the lawsuit in the foregoing paragraph, the administrative court pursuant to its authority may order the defendant competent authority to pay the appropriate lawyer fees, detection and appraisal fees, and other litigation costs to plaintiffs that have made specific contributions to the prevention of marine pollution.
>
> The central competent authority determines the format of the written notification in Paragraph 1.

21.4 Sufficient Response to the EU CBAM

21.4.1 Evolution of the EU CBAM

As the main purpose of Taiwan's revision is related to the EU's proposal of CBAM in March 2020, a brief overview of the development of EU CBAM should be provided in this chapter. The EU announced the Inception Impact Assessment between March 4, 2020, and April 1, 2020.[38] From July 22, 2020, to October 28, 2020, the EU completed its public consultation process to collect feedback from the industries[39] before the Commission's adoption.

21.4.1.1 The 2021 CBAM Proposal

Several countries have criticized EU's plans to introduce a CBAM and questioned the WTO's compatibility with such measures. The European Parliament adopted a resolution in March 2021 supporting the adoption of the EU CBAM if it was compatible with WTO rules.[40] Finally, on July 14, 2021, the Commission adopted a 291-page proposal for a new CBAM,[41] which will impose a carbon price on imports of a targeted selection of products to prevent "carbon leakage" as a result of Europe's aggressive climate action. This will ensure that European emission reductions contribute to the global decline in emissions, as opposed to displacing carbon-intensive production outside of Europe. It also aims to encourage industries outside of the EU and their international partners to follow suit.[42] Overall, this CBAM is initially applied to imports of the following goods:

- Cement
- Iron and steel
- Aluminum
- Fertilizers
- Electricity

EU importers buy carbon certificates corresponding to the carbon price as if the goods were produced under the EU's carbon pricing rules. However, if a non-EU producer can demonstrate that they pay a price for carbon used in the production of imported goods in a third country, the EU importer can deduct the entire cost.[43]

The price of the certificates depends on the weekly average auction price of the EU emissions trading system (ETS) allowances, expressed in €/ton of CO_2 emitted. Importers of the goods must, either individually or through a representative, register with national authorities, where they can also buy CBAM certificates.

21.4.1.2 May and July 2022 Versions

The CBAM proposal was referred to the ENVI committee, which adopted its report on May 17, 2022, with 49 votes in favor, 33 against, and 5 abstentions. The report amends the scope of the proposal, broadening the products and sectors covered (*organic chemicals*, *plastics*, *hydrogen*, and *ammonia*) and including *indirect emissions*. In addition, it reduces the proposed timeline by modifying the transitional phase and the elimination of free allocations. Revenue from the CBAM would accrue to the EU budget, and an equivalent amount would be earmarked for decarbonization efforts. Furthermore, the report supports the establishment of a centralized administration by an EU CBAM authority and identifies additional cases that can constitute circumventions.[44]

On June 22, 2022, Parliament adopted its position on the regulation, with 450 votes for, 115 against, and 55 abstentions. It foresees the phasing in of the CBAM from 2027, with free allowances ending in the EU emissions trading system by 2032. This supports a broadening of the scope to include organic chemicals, plastics, hydrogen, and ammonia, as well as indirect emissions. It also supports the need for a centralized EU CBAM authority and mentions that revenue from CBAM shall accrue to the EU budget and be reflected in an increased support to the least-developed countries' decarbonization efforts.[45]

21.4.1.3 December 2022 Version

On December 13, 2022, the Council of the European Union ("the Council") and the European Parliament came to a provisional agreement on the final text of a CBAM regulation.[46]

The scope of the CBAM includes iron and steel, cement, fertilizers, aluminum, electricity, and hydrogen, as well as "certain precursors as well as . . . downstream products such as screws and bolts and similar articles of iron or steel."

Indirect emissions would be covered "in a well-circumscribed manner." During a European Parliament press conference, it was announced that the methodology used to calculate indirect emissions would be reviewed at the conclusion of the transition period.[47]

Finally, the exemption is applied to imports already paying an explicit carbon price. Where an explicit carbon price has been paid in the country of production, an importer would be exempted from CBAM. Any CO_2 price that is already paid in a third country is deducted. The price of the certificates will then be calculated depending on the weekly average auction price of EU ETS allowances, expressed in euros per ton of CO_2 emitted.[48]

21.4.2 Evaluation of Taiwan's Legislative Response to EU CBAM

21.4.2.1 From Follower to Pioneer of the Carbon Border Tax

Since April 2020, the Taiwan EPA has closely monitored the development of the EU CBAM. As noted in Section 2.3, Taiwan's legislative pace is highly similar to that of the EU CBAM. However, Taiwan began to outpace the EU CBAM in late 2022 and 2023. In mid-December 2022, despite the non-final version of the EU CBAM, Taiwanese Parliament hosted intensive meetings to deliberate the drafts, which led to the adoption of the CCRA in early January 2023. The CCRA reform was triggered by the EU CBAM but did not wait until the EU CBAM's final version.

Furthermore, inspired by the EU CBAM, Taiwan's CCRA even moved further to introduce the Taiwan CBAM (carbon border tax) scheme in Article 31 of the CCRA. So far, the covered sectors and other details remain to be decided by administrative implementation orders; nevertheless, with this development, *Taiwan becomes the first country to have legalized carbon border tax in the world*!

21.4.2.2 Direct Tool: Carbon Fee

The EU used a carbon price scheme to decide whether there was a need to impose a CBAM on imports from third countries. Non-EU producers who have already paid a price for the carbon used in the production of imported goods in a third country may deduct the corresponding cost. This means that exporters in countries that impose a carbon charge under an ETS, such as Korea and Singapore, can deduct the corresponding carbon charge from the amount due under CBAM when importing covered products into the EU. However, this deduction is not available for imports from countries that do not impose a carbon charge, such as the United States, even if those countries impose other measures to reduce carbon output, such as imposing carbon limits.[49]

Despite the unwillingness of the EU CBAM to intervene in the carbon price scheme in non-EU countries, Taiwan has presumed that carbon fees are the best way to deal with such challenges since April 2020. As noted earlier, even though the GHG Act of 2015 already adopted an ETS scheme, as well as a carbon price scheme, the government decided not to implement the former and instead to switch to the latter.

However, the implementation of a carbon fee may be different from the EU CBAM.

21.4.2.2.1 PRODUCT CATEGORIES

Although it remains unclear which industry categories will be covered by the CCRA in 2023, it is highly likely that it will follow the categories of the GHG inventory and cover those large emitters (an estimated 287 companies) with annual emissions of over 25,000 tons.[50] Consequently, certain sectors, primarily the iron, steel, and cement industries, would resemble CBAM sectors of the December 2022 version. However, this may be insufficient to respond to CBAM, as in Taiwan, downstream products, such as screws and bolts and similar articles of iron or steel, may emit less than 25,000 tons of CO_2 and are therefore exempt from the carbon fee scheme. *This also explains why passing CCRA before the final CBAM is problematic!*

Finally, the CCRA scope may be broader than CBAM sectors because CBAM does not include the semiconductor sectors. Perhaps this could be considered as a way to meet net-zero 2050, but it is not appropriate to meet CBAM.

21.4.2.2.2 INDIRECT EMISSION

Although indirect electricity emissions were considered in the EU CBAM draft, they are not included in the 2021 CBAM proposal. *Indirect emission* reappeared in the May and July 2022 versions but has since been removed from the most recent December 2022 version.[50] Interestingly, the Taiwan EPA has included *indirect emissions* in all its drafts since October 2020. The Taiwanese government has maintained its goal of addressing indirect emissions. Again, this may be useful for achieving SDG 13 and net-zero but not respond to the EU CBAM.

21.4.2.2.3 CARBON FEE RATE

Again, no clear indication exists of a carbon fee in the CCRA, but two prices (100 NTD and 300 NTD) are given by the EPA in 2021 and 2022, respectively. If compared with the lowest carbon price of 58.3 euro emissions allowances (EUA) traded on the European Union's emissions trading system (ETS) in 2022,[51] Taiwan's carbon fee is really low.

If such a low carbon fee is not accepted by the EU, Taiwanese exporters in EU CBAM sectors would face the issue of double payment. For instance, steel exporting to the EU may be subject to the Taiwan carbon fee scheme and the EU CBAM. More intriguingly, the EU may receive the majority of payments. Taking the 58.3 euro as an example, the Taiwanese government can only keep 10 euros in its pocket, whilst leaving the rest to the EU.

21.4.2.2.4 MY OPINION: THE SIMPLEST WAY TO MEETING EU CBAM

From the preceding analysis, the Taiwanese government appears to have adopted a narrow definition of the carbon price scheme requested by the EU CBAM. Moreover, the Taiwanese government seemed to have misunderstood the scope of the EU CBAM, including industrial categories, indirect emissions, and the rate of the carbon fee. All these factors lead to difficulties in responding to the EU CBAM.

From the author's opinion, the CBAM-responsive scheme under the carbon fee/tax is straightforward: One government only has to set the rate indexed to the EU ETS market price (the weekly average auction price of EU ETS allowances, expressed in euros per ton of CO_2 emitted) for CBAM products exported to the EU. However, the product scope remained undecided so far and should wait until the final version of the EU CBAM. Taiwan CCRA may make a mistake in setting not only the wrong rate but also flawed industrial categories, and in covering products exported to non-EU countries. Due to such insufficient response, the Taiwanese government may fail to keep the carbon price revenue in its own country but instead give it to the EU, which contradicts the main purpose of the CCRA reform since April 2020.

21.4.2.3 Indirect Tools to Reduce EU CBAM Payments

Another way to deal with the EU CBAM may be the reduction of carbon intensity locally to reduce payments under the EU CBAM. In this regard, the CCRA of 2023 introduced several measures.

Firstly, *mandatory emission performance standards* for industrial installation can help reduce the carbon industry from production, thereby contributing to low emissions and lower CBAM payments. However, it remains unclear whether such standards would be applied retrospectively to all installations before the CCRA of 2023 or only to new installations after the CCRA of 2023. If only the latter were the case, this would minimally affect the reduction of CBAM payments.

Secondly, new installations must offset a specific proportion of new emissions by buying carbon credit or paying buyouts.[52] In terms of effects on the EU CBAM, buying carbon credit may reduce the CBAM payment. However, in terms of the latter, the author is suspicious that the EU CBAM would admit such buyout payment if such buyout payment is set too low compared to the EU ETS carbon price.

Thirdly, the CCRA of 2023 permitted the purchase of international carbon credit to reduce the local carbon fee payments.[53] Theoretically, it would also lead to lower CBAM payment. However, if the EU does not intend to admit some of or all these international carbon credits, there would be no or little CBAM reduction effects for Taiwan exporters.

Fourthly, to avoid the carbon leakage resulting from the carbon fee implementation, CCRA established the Taiwan CBAM for a specific (yet-to-be-determined) type of import (not-subject-to-carbon-price-scheme) products to be subject to Taiwan CBAM scheme.[54] For these products, they must buy local offsets or pay buyout payment in Taiwan. Again, if these products were reprocessed in Taiwan and exported to the EU, part of the carbon cost should be deducted.

Finally, the ETS scheme provided in CCRA can play a role in reducing CBAM payment.[55] However, due to the continuous debate that the ETS is unsuitable for a small country such as Taiwan and the unwillingness to implement ETS in the short term, it may not contribute to reducing CBAM. Nonetheless, I believe that as long as the carbon price under the carbon pricing scheme in individual countries is equivalent to the EU, either the ETS or the carbon tax would work for the EU. However, it seemed that the Taiwan EPA has no intention of implementing the ETS scheme (whose legal basis has been provided in the 2015 GHG Act) but has enthusiastically moved to the carbon fee scheme without any good reason. Moreover, the EPA seemed to ignore the fact that a good mixture of the ETS and the carbon tax scheme would create additional *flexibility* for the industry to meet the EU CBAM challenges. For instance, for an export product, Taiwan's industries can only pay the carbon fee to the EPA in Taiwan. However, in ETS countries, such as China and South Korea, the industry may have other options to buy carbon credit in the ETS market. This option would lead to additional flexibility in the industry to meet such challenges.

21.5 Conclusion

In recent years, the EU has played an ambitious role in climate change policy by expanding its influence on other countries and contributing to the SDG 13 at the global level. For instance, the requirement of foreign flights to the EU to be integrated into the EU ETS scheme has drawn considerable attention. To resolve the long-unresolved carbon leakage problem when the ETS is becoming more rigid in the EU,[56] CBAM seems unavoidable and necessary.

The EU's unilateral attempt would highly affect export-oriented countries, specifically manufacturing-based Asian countries, such as China, Taiwan, South Korea, Japan, India, Vietnam, and Thailand. Thus, each country conducts an impact assessment of its CBAM on its local exports.[57]

Before the final CBAM version, various estimates of the likely final EU CBAM version existed. It appears that most Asian countries exaggerated the damages, leading to an overreaction to the EU CBAM. *This chapter explains how Taiwan reacted or insufficiently reacted to the CBAM since March 2020.*

The main purpose of climate change legislation is not to respond to EU exports' needs. As global village members, industrialized countries, such as Taiwan and South Korea, must also fulfil their emission reduction obligations and net-zero commitments. This can also explain why additional stringent measures were introduced in the EPA and CCRA drafts in Taiwan. However, the flawed design in the current 2023 CCRA may result in an extremely low carbon fee and the leakage of "carbon funding" to the EU, which may not be conducive to boosting local financial support for low carbon investments.

Acknowledgments

This chapter was completed with the funding support of the Ministry of Science and Technology. Project number: 112-2410-H-007 -013 -MY2, 113-NU -E-007 -002 -NU.

Notes

1 See, for example, Priya, S. S., Cuce, E., & Sudhakar, K. (2021). A perspective of COVID-19 impact on global economy, energy, and environment. *International Journal of Sustainable Engineering, 14*(6), 1290–1305.
2 Asia's top-performing economy in 2020 could grow even faster this year, February 22, 2021, www.cnbc.com/2021/02/23/taiwan-asias-top-performing-economy-in-2020-could-grow-faster-in-2021.html
3 Article 20–21 of the Greenhouse Gas Reduction and Management Act CH. Announced Date: 2015-07-01, https://law.moj.gov.tw/ENG/LawClass/LawAll.aspx?pcode=O0020098
4 The reason for not adopting ac arbon tax will be provided in the latter section.
5 EC, Carbon Border Adjustment Mechanism, https://taxation-customs.ec.europa.eu/green-taxation-0/carbon-border-adjustment-mechanism_en
6 *Taipei Times*, Legislature passes climate amendments, January 11, 2023, www.taipeitimes.com/News/taiwan/archives/2023/01/11/2003792395
7 National Development Council, Taiwan's Pathway to Net-Zero Emissions in 2050, March 30, 2022, https://ws.ndc.gov.tw/Download.ashx?u=LzAwMS9hZG1pbmlzdHJhdG9yLzEwL3JlbGZpbGUvMC8xNTA0MC8xM2E3NjIyNi00NzhkLTRkODMtYTI5NS1mNjQyN2ZmNmZlMTUucGRm&n=VGFpd2Fu4oCZcyBQYXRod2F5IHRvIE5ldC1aZXJvIEVtaXNzaW9ucyBpbiAyMDUwLnBkZg%3d%3d&icon=.pdf
8 National Development Council, 12 Key Strategies and Action Plan to Realize "Taiwan's Pathway to Net-Zero Emissions in 2050," January 2023, www.ndc.gov.tw/Content_List.aspx?n=6BA5CC3D71A1BF6F
9 Hancock, L., & Wollersheim, L. (2021). EU carbon diplomacy: Assessing hydrogen security and policy impact in Australia and Germany. *Energies, 14*(23), 8103.
10 The agreement must be confirmed by ambassadors of the EU member states, and by the European Parliament, and adopted by both institutions before it is final. See *CBAM: Statement by the Presidency of the Council of the EU*, December 13, 2022, https://czech-presidency.consilium.europa.eu/en/news/cbam-statement-by-the-presidency-of-the-council-of-the-eu/
11 See EU Climate Action: Provisional Agreement Reached on Carbon Border Adjustment Mechanism (CBAM), December 13, 2022, www.consilium.europa.eu/en/press/press-releases/2022/12/13/eu-climate-action-provisional-agreement-reached-on-carbon-border-adjustment-mechanism-cbam/
12 See, for example, CORONAVIRUS/Taiwan reports first domestic case of COVID-19 in 253 days (update), December 22, 2020, https://focustaiwan.tw/society/202012220013
13 Nikkei Asia, Taiwan GDP Growth Outpaces China for First Time in Three Decades, January 29, 2021, https://asia.nikkei.com/Economy/Taiwan-GDP-growth-outpaces-China-for-first-time-in-three-decades
14 EU,EUGreenDeal(CarbonBorderAdjustmentMechanism):AboutThisInitiative,https://ec.europa.eu/info/law/better-regulation/have-your-say/initiatives/12228-EU-Green-Deal-carbon-border-adjustment-mechanism-_en

15 EPA Minister Chang, Responding to Climate Change in Taiwan: The Legal Reform of Climate Change Act of 2015, April 8, 2020, Health, Welfare and Environment Committee of the Parliament. http://rwd-ghgrule.demo.acubedt.com/admin/resource/files/c924db1e2b815660bf4da45f5d8ff76f.pdf
16 Green Citizens' Action Alliance, INTRO, www.gcaa.org.tw/
17 MP Hong's Draft Bill on Revision of 2015 GHG Act, https://misq.ly.gov.tw/MISQ/IQuery/misq5000QueryBillDetail.action?billNo=1100106070200300
18 MP Su's Draft Bill on Revision of 2015 GHG Act, https://misq.ly.gov.tw/MISQ/IQuery/misq5000QueryBillDetail.action?billNo=1100429070202200
19 KMT Draft Bill on Revision of 2015 GHG Act, https://misq.ly.gov.tw/MISQ/IQuery/misq5000QueryBillDetail.action?billNo=1100503070201100
20 Draft Bill of Taiwan People's Party on Revision of 2015 GHG Act, https://misq.ly.gov.tw/MISQ/IQuery/misq5000QueryBillDetail.action?billNo=1100526070200100
21 NPP's Draft Bill on Revision of 2015 GHG Act, https://misq.ly.gov.tw/MISQ/IQuery/misq5000QueryBillDetail.action?billNo=1100510070202100
22 EPA, Greenhouse Gas Reduction and Management Act, www.english-climatetalks.tw/climate-governance
23 Taiwan Passes Climate Law with Carbon Fee System, 2050 Net-Zero Goal, January 10, 2023, https://focustaiwan.tw/politics/202301100019
24 EUROPA, EU Green Deal (Carbon Border Adjustment Mechanism), https://ec.europa.eu/info/law/better-regulation/have-your-say/initiatives/12228-EU-Green-Deal-carbon-border-adjustment-mechanism-_en
25 EPA, the Key Points on the Draft Bill of Climate Change Response Act of 2020, January 20, 2021, www.scribd.com/document/491412863/20210120%E6%B0%A3%E5%80%99%E6%B3%95%E5%BA%A7%E8%AB%87%E6%9C%83%E7%B0%A1%E5%A0%B1
26 Article 17 of the GHG Act of 2015: "The central competent authority may establish a GHGEPS to incentivize designated Emission Sources to reduce GHG emissions prior to the cap-and-trade scheme. The EPS pursuant to the foregoing paragraph shall be determined and reviewed regularly by the central competent authority in conjunction with the central industry competent authorities, considering facilities, products or other raw materials, output, and consumptions of Emission Sources."
27 Art. 24 of the CCRA of 2023.
28 UDN News, Carbon fee is likely to be 300 NTD/tonne of CO2, June 10, 2022, https://money.udn.com/money/story/12926/6665559
29 Article 8 of the GHG Act of 2015: "The Executive Yuan shall invite relevant central government agencies, non-governmental organizations, experts and scholars to determine and review the division, integration, implementation and compilation of the work of GHG reduction and climate change adaptation."
30 www.ey.gov.tw/oecr/.
31 https://ncsd.ndc.gov.tw/Fore/en/Taiwansdg#about.
32 Art. 8(1) of the CCRA of 2023.
33 Art. 8(2)–18 of the CCRA.
34 Art. 14 of the CCRA.
35 Article 3 of the 2015 GHG Act: "This glossary of terms are to be interpreted with the following meanings: 2. Climate change **adaptation** means responses to actual or expected climate stimuli and their effects, which moderate harm or exploit beneficial opportunities. Adaptation may include 'preventive and/or responsive actions,' 'private and/or public actions', and 'autonomous and/or programmatic' adaptations." Article 8 of the GHG Act: "The Executive Yuan shall invite relevant central government agencies, non-governmental organizations, experts, and scholars to determine and review the division, integration, implementation, and compilation of the work of GHG reduction and climate change **adaptation**. Relevant central government agencies shall promote GHG reduction and climate change adaptation through the following actions."
36 Art.4(1) of the EPA Draft of 2020.
37 Art. 5 of MP Hong's version.
38 EC, EU Green Deal (carbon border adjustment mechanism), https://ec.europa.eu/info/law/better-regulation/have-your-say/initiatives/12228-EU-Green-Deal-carbon-border-adjustment-mechanism-_en.
39 EC, EU Green Deal (carbon border adjustment mechanism): About this Consultation, https://ec.europa.eu/info/law/better-regulation/have-your-say/initiatives/12228-Carbon-Border-Adjustment-Mechanism/public-consultation_en.
40 European Parliament resolution of March 10, 2021, towards a WTO-compatible EU carbon border adjustment mechanism (**2020/2043(INI)**), www.europarl.europa.eu/doceo/document/TA-9-2021-0071_EN.html

41 Proposal for a REGULATION OF THE EUROPEAN PARLIAMENT AND OF THE COUNCIL establishing a carbon border adjustment mechanism (Text with EEA relevance) {SWD(2021) 643 final}—{SWD(2021) 644 final}—{SWD(2021) 647 final} {SEC(2021) 564 final} https://ec.europa.eu/info/sites/default/files/carbon_border_adjustment_mechanism_0.pdf.
42 EC, Carbon Border Adjustment Mechanism, https://ec.europa.eu/taxation_customs/green-taxation-0/carbon-border-adjustment-mechanism_en.
43 EC, Carbon Border Adjustment Mechanism: Questions and Answers, https://ec.europa.eu/commission/presscorner/detail/en/qanda_21_3661.
44 European Parliament, AT A GLANCE Plenary—June 11, 2022, EU carbon border adjustment mechanism, www.europarl.europa.eu/RegData/etudes/ATAG/2022/729462/EPRS_ATA(2022)729462_EN.pdf.
45 Carbon border adjustment mechanism as part of the European green deal In 'A European Green Deal', www.europarl.europa.eu/legislative-train/theme-a-european-green-deal/file-carbon-border-adjustment-mechanism.
46 David E. Bond, Guillermo Giralda Fustes, William De Catelle, Matt Solomonm European Parliament and Council Reach Agreement on Carbon Border Adjustment Mechanism and EU Emissions Trading System, December 27, 2022, www.whitecase.com/insight-alert/european-parliament-and-council-reach-agreement-carbon-border-adjustment-mechanism-0.
47 The CBAM–the European Union's Next Climate Action Step, December 21, 2022, www.velaw.com/insights/the-carbon-border-adjustment-mechanism-cbam-the-european-unions-next-climate-action-step/.
48 Kerstine AppunnJulian Wettengel, FACTSHEET: The EU's Carbon Border Tax—panacea or recipe for trade war? February 2, 2023, www.cleanenergywire.org/factsheets/emission-reduction-panacea-or-recipe-trade-war-eus-carbon-border-tax-debate.
49 EU Reaches Provisional Agreement on Carbon Import Charge, January 26, 2023, www.akingump.com/en/news-insights/eu-reaches-provisional-agreement-on-carbon-import-charge.html.
50 "'**Product Carbon Footprint Methodology:** The CBAM declarations must contain the total quantity of covered goods imported during the calendar year expressed in megawatt hours for electricity and in metric tons for other goods, multiplied by the embedded emissions. **The embedded emissions include direct and indirect (electricity) emissions released during the production of a goods.** They would also include the emissions of input products within the system boundaries that the Commission would define."' Thomas Reilly, The EU's Green Deal Industrial Plan for the Net-Zero Age, February 3, 2023, www.globalpolicywatch.com/2021/06/twelve-things-to-know-about-the-upcoming-eu-carbon-border-adjustment-mechanism/
51 European Union Emissions Trading System (EU-ETS) carbon pricing in 2022 (in euros ***per metric ton)***, www.statista.com/statistics/1322214/carbon-prices-european-union-emission-trading-scheme/.
52 Art. 24 of the CCRA of 2023.
53 Art. 28 of the CCRA of 2023.
54 Art. 31 of the CCRA.
55 Art. 34 and 35 of the CCRA.
56 EP, Carbon leakage: prevent firms from avoiding emissions rules, June 23, 2022, www.europarl.europa.eu/news/en/headlines/society/20210303STO99110/carbon-leakage-prevent-firms-from-avoiding-emissions-rules.
57 For example, Japan's challenge to respond to EU CBAM, see: www.eu-japan.eu/sites/default/files/imce/ercst_2021.6.24.pdf.

22
UNDERSTANDING LEGAL FRAMEWORKS FOR CULTURAL HERITAGE
A Comparative Analysis of SDG 11 in Practice

Kenadi E. Mitchell

22.1 Legal Tools for SDG 11 and Heritage Protection

The Sustainable Development Goals (SDGs) address a broad spectrum of objectives aimed at promoting environmental sustainability, social inclusion, and economic growth. SDG 11 is dedicated to making "cities and human settlements inclusive, safe, resilient, and sustainable." Central to this goal is Target 11.4, which mandates nations to "strengthen efforts to protect and safeguard the world's cultural and natural heritage," specifically emphasizing the financial resources allocated to these efforts. This directive underscores the critical role that adequate funding plays in the preservation and conservation of cultural and natural heritage, contributing to sustainable urban development.

Despite the ambitious scope of the SDGs, integrating environmental and developmental objectives presents substantial challenges. One significant gap is the narrow focus on financial commitments within Target 11.4, which often overlooks broader considerations of how these funds are implemented and managed. Aside from the financial aspect, there is a need for more explicit inclusion of heritage protection across all relevant SDGs, especially in terms of legal and administrative frameworks (Petti, 2020).

At the local and national levels, the implementation of these global targets underscores the importance of robust legal frameworks. Such frameworks are essential for effectively translating the financial commitments of Target 11.4 into actionable strategies that not only advance urban sustainability but also prioritize the preservation of cultural heritage. Research indicates that while some targets related to economic growth and infrastructure see adequate funding and achievement, others, particularly those addressing climate change and education, lag behind due to insufficient integration of cultural considerations (Halkos and Petti).

Furthermore, the effectiveness of SDG 11.4 is heavily reliant on how well financial resources are allocated and utilized. The current methodologies within the UNESCO Framework for Cultural Statistics have limitations in measuring and valuing cultural heritage, impacting the ability to monitor progress and implement this target effectively. Studies by Wang (2018) and Petti (2020) suggest that enhancing these measurement methods could significantly improve the comparability of data across different contexts, thereby facilitating the development of more effective solutions for the implementation of SDG 11.4.

DOI: 10.4324/9781003285472-24

By strengthening the legal frameworks that support SDG 11, particularly concerning the financial aspects of cultural heritage protection, we can ensure a more effective and integrated approach to sustainable urban development. This effort requires not only refining existing legal instruments to better manage and oversee the allocated funds but also broadening their scope to ensure comprehensive coverage across the SDGs. This approach will align global financial commitments with local realities and needs, enhancing the protective measures for cultural and natural heritage within the framework of sustainable development.

22.2 Legal Frameworks: Essential for Effective SDG Enforcement

This chapter underscores the necessity of integrating robust legal frameworks to effectively implement and enforce Sustainable Development Goal 11.4, which specifically calls for significant financial commitments to protect and safeguard the world's cultural and natural heritage. The target points to a broader challenge within the SDG framework: the need for substantial financial resources, which must be managed and overseen through comprehensive legal strategies to ensure their effective use.

Klopp (2018) highlights several systemic issues, such as the lack of standardized and accessible data, weak data collection capabilities at local levels, and inconsistent application of the SDGs across various contexts. These challenges complicate the enforcement of SDG 11.4, as they hinder the accurate assessment of how financial resources are allocated and utilized, emphasizing the need for stronger legal oversight.

Moreover, the SDGs, while aimed at transformative change, often face the risk of being co-opted by existing political and economic agendas. Siegel (2020) and Bexell (2021) critique the implementation of these goals, noting that nations may engage in "cherry-picking" by selectively prioritizing those SDGs that align with their established policies rather than adopting a more integrated approach. This selective prioritization can undermine the goals' broader objectives, particularly the nuanced demands of SDG 11.4, which require dedicated financial and legal commitments to cultural heritage protection.

To address these issues, there is a clear need to develop and enhance existing legal frameworks that can effectively manage the specific commitments required under SDG 11.4. Such frameworks should ensure that funds are not only allocated but also spent transparently and effectively to achieve the intended outcomes, along with meaningful consideration and deference given to historical context. Strengthening these legal mechanisms will support the accountability and effectiveness of SDG implementation, ensuring that the ambitious goals of sustainable and inclusive urban development are met across all contexts.

22.2.1 Interactions and Synergies Among SDG Targets

Understanding the interdependencies among SDG targets is crucial for enhancing the legal frameworks necessary to protect cultural heritage, as emphasized by Pradhan et al. (2017). This analysis shows that while synergies between different SDGs can enhance overall outcomes, trade-offs may need careful management through legal reforms, especially to support the financial commitments outlined in SDG 11.4. This interconnected approach ensures that efforts to protect cultural heritage are not isolated but are integrated with broader sustainability goals.

As cities continue to grow, the challenge of localizing SDGs to suit specific urban environments becomes more pronounced. Aust and du Plessis (2018) highlights how urban governance can leverage SDG 11 to promote broader sustainability objectives, such as poverty reduction and climate action, through strategic planning and robust legal enforcement. This strategy underscores the necessity of embedding strong legal mechanisms within urban planning processes to ensure that financial resources dedicated to cultural heritage under SDG 11.4 are effectively utilized and managed.

Furthermore, adapting SDGs to local contexts requires a nuanced approach that goes beyond mere compliance and reporting. It involves tailoring these goals to the specific cultural, social, political, economic, and environmental needs of each community (Lucci, 2015; Jiménez-Aceituno et al., 2019). Adopting a place-based approach, as advocated by Sterling et al. (2020), enhances the relevance and effectiveness of SDG implementation by ensuring that legal frameworks are not only supportive but also reflective of local realities and capable of managing the designated funds for cultural heritage protection effectively.

To truly harness the potential of the SDGs for sustainable urban development, it is crucial that legal instruments are not only reinforced but also intricately designed to reflect the specific socio-economic and environmental contexts they are meant to serve. This means that legal frameworks must be dynamic and responsive, capable of evolving with changing global and local conditions, to ensure the ongoing relevance and effectiveness of SDG implementations.

Therefore, strengthening legal frameworks, particularly those supporting financial commitments under SDG 11.4, is not merely beneficial—it is a critical requirement for the sustainable transformation of our global society. By embedding these legal tools across all SDGs, we enhance the overall capacity of these frameworks to manage the diverse challenges of sustainable development, ensuring that the goals are not only met but are also sustained over time.

22.3 Theoretical Foundations of SDG Implementation Discussion

This section elaborates other theories by scholars that similarly contend that SDG 11 cannot be fully achieved without a more integrated approach involving the amendment of domestic legal frameworks, particularly as they relate to cultural heritage and the broader context of sustainable urban development. It sets the stage for how this chapter will be examining the interaction between legal frameworks and community histories, emphasizing the importance of understanding both to effectively preserve cultural heritage, even in communities within "industrialized" nations like the United States.

22.3.1 Defining Sustainable Development and Its Legal Implications

The concept of sustainable development, as defined in the widely cited 1987 Brundtland Report, is "development that meets the needs of the present without compromising the ability of future generations to meet their own needs" (WCED, 1987; Halkos and Gkampoura, 2021). This definition emphasizes a balance between current and future needs, providing a foundation for integrating cultural heritage preservation into modern development practices through legal reforms.

22.3.2 The Place-Based Approach to Sustainability

Furthering this foundation, Sterling (2020) advocates for a place-based approach to sustainability, which emphasizes planning and decision-making processes that are deeply rooted within the cultural and social frameworks of communities. This method highlights the importance of localization in decision-making, recognizing that effective SDG implementation requires legal frameworks that are not only adaptive but also deeply integrated into the local socio-economic and cultural landscapes.

22.3.3 Political Dimensions of SDG Implementation

Bexell (2021) explores the influence of political structures on SDG implementation, suggesting that the quality of democracy and governance significantly affects how these goals are pursued and achieved. Key aspects such as legitimacy, responsibility, and accountability are crucial for gaining political ownership and ensuring effective policy outcomes. These elements are vital for the legal frameworks needed to enforce SDG mandates, particularly those related to cultural and environmental sustainability.

22.3.4 Integrating the Water–Energy–Food (WEF) Nexus

Olawuyi (2020) introduces the water–energy–food (WEF) nexus as a comprehensive framework for implementing SDGs, addressing the interdependencies between these critical sectors. This approach advocates for an integrated strategy that promotes efficient resource management, which is essential for supporting sustainable urban and rural development, underscoring the need for legal frameworks that facilitate such integration.

22.3.5 Challenges and Legal Necessities in Achieving the SDGs

Lastly, Soergel et al. (2021) critiques the reliance on climate policies and economic development alone for achieving the SDGs, suggesting that a broader package of sustainable development measures is necessary. This perspective underscores the critical need for legal reforms to support and operationalize these measures, ensuring they are effectively implemented and capable of achieving desired outcomes.

By drawing on these theoretical foundations, the chapter argues for the essential role of adaptable, enforceable, and culturally aware legal frameworks in achieving SDG 11 and, by extension, other SDGs. This integrated approach ensures that legal strategies not only support but also enhance the efforts to preserve cultural heritage within the sustainable development agenda.

22.3.6 The Role of Legal Frameworks in Cultural Heritage Protection

As this chapter progresses into specific case studies, it becomes evident that understanding and integrating the historical and cultural narratives of communities are vital for the effective preservation of cultural heritage. Legal frameworks do not merely support these goals; they are essential to ensuring that development is genuinely sustainable, respectful of cultural identities, and inclusive of all community segments. This holistic approach, supported by the discussed theories, provides a forceful base for advancing global sustainability and cultural preservation simultaneously through legal framework.

Despite the array of theories and approaches to implementing SDGs, the necessity of malleable legal frameworks remains clear. Legal structures are indispensable for enforcing SDG targets, particularly SDG 11.4, with its focus on the protection of cultural heritage. Without legal enforcement, policies and initiatives risk remaining theoretical, without tangible benefits to the communities they aim to serve.

22.4 International Legal Instruments and the Enforcement of the SDGs

International legal instruments play a pivotal role in the enforcement of the Sustainable Development Goals (SDGs), offering a framework for accountability and cooperative action among nations. These instruments are particularly crucial in harmonizing efforts across borders to protect cultural and natural heritage, as outlined in SDG 11.4.

22.4.1 Role of International Frameworks in Supporting National Efforts

The High-Level Political Forum (HLPF) and various international environmental agreements serve as key mechanisms under which SDG commitments, including those related to cultural heritage, are reviewed and enforced. These international platforms facilitate the sharing of best practices, challenges, and progress among member states, ensuring that efforts to protect cultural heritage are aligned with global sustainability objectives.

The establishment of a monitoring, accountability, and review framework within the context of the SDGs is crucial for ensuring that nations adhere to their commitments. This framework, while fostering global accountability, also respects the sovereignty of nation-states, balancing international oversight with national autonomy (Karlsson-Vinkhuyzen et al. 2018).

These international agreements and forums provide a legal basis for actions taken under SDG mandates, promoting a coherent approach that supports national legal systems. For instance, the integration of international standards into national legal frameworks can help ensure that financial commitments toward cultural heritage protection are met and effectively utilized.

By providing a structured yet flexible framework for international cooperation, these legal instruments help translate global SDG commitments into actionable strategies at the national and local levels. They enhance the capacity of countries to implement SDG 11 effectively, particularly in contexts where cultural heritage intersects with urban development and environmental sustainability.

22.5 Comparative Analysis: SDG Implementation in Northern and Southern Contexts

This section explores the application of Sustainable Development Goals (SDGs) within the contrasting socio-economic and political landscapes of Northern and Southern communities. By examining how these global goals are adapted and implemented across diverse global regions, we can better understand the dynamics that influence effectiveness and inclusiveness.

22.5.1 Theoretical Frameworks in North vs. South Dynamics

The implementation of the SDGs often reflects the existing political, social, and economic structures of a country. Siegel (2020) emphasizes that the effectiveness of the SDGs in promoting inclusiveness is profoundly influenced by domestic politics, pre-existing institutions, power relations, and the capacity and experience of civil society and subnational governments. This assertion underscores the variability in SDG success rates between Northern and Southern nations, which often possess differing levels of resources and institutional frameworks.

Biermann et al. (2017) suggests that the success of SDG governance hinges on the formalization of commitments, on the establishment of clear benchmarks, and on measurable pledges. However, the mere presence of higher socio-economic status does not guarantee alignment with SDG guidelines or their effective implementation.

For instance, Koch (2018) observes that, in Germany, adjustments to the original SDG targets and indicators, particularly those related to SDG 11, were necessary to address local sustainability challenges.

In Finland, SDG mainstreaming has predominantly been limited to foreign affairs, with a significant division between development policy and sustainability issues (Bexell, 2021). This indicates a compartmentalized approach where sustainability is sometimes siloed rather than integrated across all government functions.

Vilas (2021) analyzed the integration of the SDGs into EU cultural policies. Findings suggest that while the EU does not explicitly focus on SDGs in cultural policies, there is implicit inclusion of goals related to cultural heritage, diversity, and social inclusion. This reflects a more holistic but less direct approach to SDG integration in policy frameworks.

The varying application and success of the SDGs in their Northern and Southern implementation highlight the importance of adapting strategies that align with local conditions and capacities. While Northern countries may possess the resources to implement complex analytical methods but may not fully integrate SDGs into urban planning, Southern countries may struggle to balance all SDG initiatives and choose to prioritize SDGs that align more closely with immediate national needs and existing policies.

Understanding these dynamics is crucial for designing implementation strategies, capacity building efforts, and policy advice that are sensitive to the distinct needs and capabilities of different global regions. This comparative analysis next uses case studies to illustrate the nuances involved in international cooperation, accountability, and SDG implementation that is both equitable and effective.

22.5.2 Environmental and Cultural Conflicts in Africatown, Alabama

Africatown, located in Mobile, Alabama, represents a moving example of how historically marginalized "Southern" communities within developed "Northern" countries struggle to confront environmental challenges that threaten their cultural heritage. Founded by the last known group of enslaved Africans illegally brought to the United States aboard the *Clotilda*, Africatown is not only a neighborhood but also a testament to the survival and resistance of enslaved persons in the United States (Kemp-Rotan, n.d.). Today, Africatown faces significant pressures from surrounding industrial activities that repeatedly jeopardize its community health and cultural integrity.

22.5.3 Historical Significance and Environmental Challenges

Originally established by 32 West Africans, Africatown grew into a culturally rich community. However, its proximity to various emitters (due to the zoning of Africatown as an "industrial" area despite thriving residential neighborhood qualities) has led to severe environmental degradation, with pollutants affecting air and water quality impacting residents' health and the preservation of the community's historical sites (Moura, 2021; Anderson, 2021).

22.5.4 Legal Frameworks and Their Application: The Role of NEPA

The National Environmental Policy Act (NEPA) is a fundamental legal framework in the United States designed to protect environments like that of Africatown by requiring federal agencies to assess the environmental impacts of their actions (NEPA, 1969). NEPA aims to ensure that all communities, particularly vulnerable ones, have a voice in the development processes that affect them. However, the implementation of NEPA in Africatown has faced challenges. Despite NEPA's requirements, the community has struggled with effective engagement in environmental decision-making processes, often due to the complex, resource-intensive nature of NEPA proceedings and a lack of adequate local advocacy resources (Sharpe, 2021).

The dynamics observed in Africatown are indicative of broader patterns where economic development priorities overshadow the needs for environmental protection and cultural heritage preservation. This case study exposes the gaps in the protection mechanisms intended to safeguard such communities. The outcomes observed in Africatown raise questions about the efficacy of legal tools like NEPA in real-world applications, particularly in settings where economic interests are deeply entrenched and potentially opposed to community well-being (Koch, 2018; Bexell, 2021).

These socio-legal dynamics underscore a systemic issue in the application of environmental laws in the United States—while laws like NEPA are designed to protect communities, their effectiveness often depends on the ability of those communities to influence the legal and regulatory processes. This engagement is frequently hindered by disparities in resources, legal knowledge, and political power, which can marginalize vulnerable communities like Africatown.

The case of Africatown not only highlights the environmental conflicts faced by marginalized communities in more developed nations but also underscores the need for significant reforms in how environmental laws are implemented and enforced. Strengthening NEPA and similar legislation to ensure a more robust community participation and impact assessment can lead to more equitable outcomes and better protection for communities' cultural and environmental heritage. This case advocates for a systemic overhaul to make environmental and cultural heritage protection more inclusive and effective, ensuring that laws like NEPA fulfill their promise to all communities, irrespective of their socio-economic status or historical background. The following case studies will illustrate examples of effective policy creation and amendment as mechanisms to improve SDG implementation.

22.6 Heritage Protection in Mexico and Ecuador

This next case study explores the dynamics of environmental conflicts impacting cultural heritage in Southern countries, focusing on local and national efforts in Mexico and Ecuador to safeguard their rich archaeological and cultural landscapes. Through these examples, we examine how local and national legal frameworks support or hinder heritage protection, and draw parallels and contrasts with the challenges faced in Africatown, Alabama.

22.6.1 Sustainable Policy Development in Mexico

In the southeastern region of Mexico, specifically in the states of Yucatán, Campeche, and Quintana Roo—areas rich in Mayan archaeological sites—a study by Majio (2021) highlights the development of sustainable public policies tailored to manage and preserve cultural and archaeological heritage. The proposed policies are intricately designed to align with the Sustainable Development Goals (SDGs), emphasizing comprehensive and regional planning, land use, cultural tourism, and resource management. These policies encompass a variety of strategic areas, including disaster planning, operations, marketing, interpretation, budgetary issues, and financing, aiming for a holistic approach to cultural tourism that is sustainable and beneficial to local communities. The success of these policies hinges on strong regional governance, transparency, accountability, and effective corporate governance.

22.6.2 Indigenous Governance and "Buen Vivir" in Ecuador

In Tungurahua, Ecuador, Indigenous communities have pioneered a unique approach to governance and sustainable development through the philosophy of "buen vivir" (good life), which integrates the well-being of nature, community, and individuals into a unified governance model (Guardiola and García-Quero, 2014; Kauffman and Martin, 2014). This model prioritizes collective rights, economic solidarity, and sustainable management of resources over traditional development models that often prioritize economic growth at the expense of environmental and cultural integrity. The local governance system in Tungurahua showcases how Indigenous norms can reshape and challenge dominant international models of sustainability, providing a potent alternative that is deeply rooted in cultural and environmental stewardship.

22.6.3 Analysis and Insights

Both Mexico and Ecuador illustrate the potential successes of leveraging the amendment of legal tools to protect heritage. In Mexico, the integration of the SDGs into local policy frameworks demonstrates a proactive approach to aligning national development goals with global sustainability objectives and is tailored to the unique cultural context of Southern Mexico.

In Ecuador, the application of "buen vivir" principles highlights the effectiveness of incorporating local and Indigenous knowledge systems in constructing effective legal governance. This approach mirrors the community-driven aspects that are undervalued in Africatown's available legal protections for environmental justice, emphasizing the universal need for community engagement and culturally informed policymaking in future laws concerning sustainable development.

While these examples from Southern countries show promising approaches to integrating sustainable and cultural preservation goals, they also face challenges similar to those encountered in Africatown, such as navigating complex legal frameworks and ensuring community involvement in decision-making processes. These cases underscore the importance of adopting flexible, culturally sensitive strategies that can adapt to diverse legal and environmental conditions while ensuring the protection of invaluable cultural heritage.

22.7 Environmental Policy Conflicts: The Case of Waste Management in Varanasi

The last case study explores a significant environmental challenge involving waste management in Varanasi, India, a city revered for its cultural and religious significance. The conflict arises due to inadequate and inefficient solid waste management practices, which not only pose environmental hazards but also threaten the cultural heritage of this historic city.

In Varanasi, one of India's oldest cities, the conflict between environmental management and the protection of cultural heritage highlights a significant oversight in SDG 11.4. This case study examines how financial commitments aimed at urban development and environmental sustainability sometimes undermine efforts to protect and preserve cultural heritage, illustrating a critical gap in SDG 11.4.

22.7.1 Environmental and Cultural Challenges in Varanasi

Varanasi's waste management crisis poses a severe threat not only to the environment but also to its rich cultural and historical sites. The city generates approximately 650 metric tons of solid waste daily, with much of it ending up in poorly managed dumpsites. These sites are often near areas of historical significance, causing pollution that can physically damage these heritage sites and detract from their cultural value (Kumar, 2019; Rai, 2018).

22.7.2 Financial Commitments Overlooking Cultural Heritage

The focus on financial commitments within SDG 11.4 often prioritizes infrastructural and environmental solutions that do not always consider the preservation of cultural heritage. For instance, funding may be directed toward creating more efficient waste disposal mechanisms, like incinerators or landfill expansions, which, while solving one problem, might exacerbate the degradation of nearby cultural sites. This approach reflects a broader issue where the economic aspects of sustainability are given precedence, with insufficient weight given to cultural dimensions.

22.7.3 Legal and Policy Frameworks: Bridging the Gap

To address this disconnect, there is a pressing need for integrated legal and policy frameworks that align financial commitments with cultural heritage protection. These frameworks should mandate comprehensive impact assessments that consider both environmental and cultural impacts before the implementation of any development project. Additionally, policies must enforce stringent regulations to protect cultural sites from unintended consequences of urban development projects.

Varanasi's challenges call for collaborative approaches that involve local stakeholders, cultural experts, and urban planners in the decision-making process. Such collaboration can ensure that solutions to environmental challenges are developed with a conscious effort to protect and enhance the city's cultural heritage. International guidelines and funding should support these integrated solutions, encouraging a balance between economic development and cultural preservation.

22.7.4 Comparative Analysis of Legal Strategies in Cultural Heritage Conflict

The case studies reveal the varied applications and effectiveness of legal frameworks in protecting cultural heritage amid environmental conflicts. In Africatown, Alabama, the use of NEPA (National Environmental Policy Act) showcased the challenges of enforcing protections in a historical African American community. In contrast, the sustainable public policy implementation in Mexico's archaeological zones highlighted how aligned legal strategies can preserve cultural heritage while promoting development. Next, the governance adaptations in Ecuador illustrated the integration of Indigenous practices within legal frameworks to protect cultural heritage against environmental degradation. Finally, Varanasi shows us the difficulty in balancing financial interests of the state and cultural protection.

22.7.5 Lessons Learned on Legal Framework Effectiveness

From these analyses, several lessons emerge about the effectiveness of legal frameworks in cultural heritage protection: First, successful legal frameworks are frameworks that adapt to local cultural, historical, and environmental contexts. For instance, in Mexico, incorporating the SDGs into local policy development proved effective in protecting archaeological sites, emphasizing the need for legal frameworks to be culturally and contextually relevant.

Secondly, it is imperative to leverage the interplay of the SDGs as suggested by Pradhan (2017); identifying and prioritizing positive correlations among SDGs can enhance the effectiveness of legal strategies. For example, buen vivir–aligned goals related to both Indigenous knowledge and economic growth and collectively supported broader legal reforms aimed at protecting cultural heritage. The case studies underscore the necessity for a holistic approach that considers legal, educational, and policy-based tools to address the multifaceted challenges of protecting cultural heritage in environmentally sensitive situations.

Reflecting on the chapter's thesis, it is evident that while law is a crucial tool in safeguarding cultural heritage, it requires continual reform to remain effective in the evolving landscapes of

SDG-compliant nations. Legal reforms need to address the dynamic interactions among SDGs, ensuring that trade-offs do not undermine the preservation of cultural heritage. Furthermore, the integration of legal measures with educational programs can enhance public understanding and support for heritage protection, creating a more sustainable framework for achieving SDG targets.

The insights from these case studies not only inform specific legal strategies but also contribute to a broader understanding of how countries can effectively integrate SDG targets into their legal and policy frameworks. By acknowledging the interdependencies among SDGs and the critical role of law and education, nations can better plan and implement strategies that achieve sustainable development while protecting and valuing their cultural heritage. This approach ensures that the pursuit of environmental and economic objectives does not come at the expense of cultural identity and heritage, promoting a balanced and inclusive path toward global sustainability.

22.8 Strengthening Legal Frameworks for Heritage Protection

This section outlines recommendations for improving legal frameworks to safeguard cultural heritage amidst conflicts. It considers perspectives from both Northern and Southern regions, advocating for a balanced approach that respects international law principles and addresses the unique challenges posed by modern sustainability goals.

22.8.1 Critique of Current Legal Approaches

Lempert (2017) critically assessed the Sustainable Development Goals (SDGs), arguing that they often reassert outdated colonial approaches that can undermine the principles of international law, such as cultural diversity and sovereignty. This perspective suggests that the current SDG framework may not sufficiently protect cultural heritage in the face of environmental conflicts, calling for a significant re-evaluation of how these goals are implemented globally.

22.8.2 The Potential for International Legal Innovation

The discussion concludes with an emphasis on the potential for international cooperation and legal innovation to address the unique challenges identified in the case studies. By fostering dialogue between diverse cultural voices and integrating these perspectives into legal and policy frameworks, the international community can enhance its approach to protecting cultural heritage in an increasingly interconnected and environmentally sensitive world. This cooperative approach not only enriches legal strategies but also ensures that they are robust enough to handle the complexities of modern environmental conflicts involving cultural heritage.

22.9 Conclusions

This chapter has explored a range of case studies that illuminate the complexities of protecting cultural heritage within the context of environmental conflicts and the implementation of the Sustainable Development Goals (SDGs). Each case study, from Africatown in the United States to the archaeological zones in Mexico and Indigenous governance systems in Ecuador, has underscored the pivotal role of legal frameworks in addressing these challenges. Through these examples, we have examined how varying legal strategies can either support or hinder efforts to safeguard cultural and historical sites in diverse environmental and socio-economic settings.

22.9.1 Embedding Legal Tools Across All SDGs

The chapter reaffirms the need to embed robust legal tools across all SDGs to ensure comprehensive protection of cultural heritage. Legal frameworks must be adaptable and sensitive to the nuances of local contexts while being capable of addressing the broader implications of global environmental and cultural challenges. This approach aligns with the holistic and multidimensional perspective on development advocated by the SDGs, as highlighted by Pradhan (2017), emphasizing the interconnectedness of all goals and the importance of leveraging synergies to enhance sustainability efforts globally.

The role of legal frameworks in bridging the gap between Northern and Southern countries is critical. As Morita et al. (2020) suggests, achieving the SDGs requires concerted efforts from all countries, each recognizing their achievements and challenges and striving to improve their governance systems. Legal frameworks that are inclusive and equitable can facilitate this process by ensuring that both Northern and Southern countries can protect their cultural heritage against the backdrop of global environmental challenges.

To truly achieve the goals of SDG 11, we need to look beyond mere monetary benchmarks of a country's investment. It is crucial to include indicators that assess the legal frameworks in place, which are necessary to activate domestic cultural protections as stipulated by governing laws. This approach ensures that solutions comprehensively address the objectives of sustainable urban development, by incorporating legal mechanisms that safeguard cultural heritage.

References

Abastante, F., Lami, I. M., & Gaballo, M. (2021). Pursuing the SDG11 Targets: The Role of the Sustainability Protocols. *Sustainability*, *13*(7), Article 7. https://doi.org/10.3390/su13073858

Aust, H. P., & Plessis, A. du. (2018). Good Urban Governance as a Global Aspiration: On the Potential and Limits of SDG 11. In *Sustainable Development Goals* (pp. 201–221). Edward Elgar Publishing. www.elgaronline.com/edcollchap/edcoll/9781786438751/9781786438751.00016.xml

Bexell, M., & Jönsson, K. (2021). *The Politics of the Sustainable Development Goals: Legitimacy, Responsibility, and Accountability* (1st ed.). Routledge. https://doi.org/10.4324/9781003043614

Biermann, F., Kanie, N., & Kim, R. E. (2017). Global Governance by Goal-Setting: The Novel Approach of the UN Sustainable Development Goals. *Current Opinion in Environmental Sustainability*, *26–27*, 26–31. https://doi.org/10.1016/j.cosust.2017.01.010

Ferran-Vila, S., Miotto, G., & Rom-Rodríguez, J. (2022). The SDGs in the EU Cultural Policies: An Institutional Communication Perspective. *Communication & Society*, *35*(4), 117–131.

Fiore, D., Butto, A., & Vargas Filgueira, V. (2021). Yagan Heritage in Tierra del Fuego (Argentina): The Politics of Balance. *Heritage*, *4*(4), Article 4. https://doi.org/10.3390/heritage4040208

Gilpin, L. (2021). 'Infrastructure Apartheid': Africatown's Fight Against Toxins and a New Toll Bridge. *Southerly* (September 16, 2021).

Guardiola, J., & García-Quero, F. (2014). Buen Vivir (living well) in Ecuador: Community and Environmental Satisfaction Without Household Material Prosperity? *Ecological Economics*, *107*, 177–184.

Halkos, G., & Gkampoura, E.-C. (2021). Where Do We Stand on the 17 Sustainable Development Goals? An Overview on Progress. *Economic Analysis and Policy*, *70*, 94–122. https://doi.org/10.1016/j.eap.2021.02.001

Jiménez-Aceituno, A., Peterson, G. D., Norström, A. V., Wong, G. Y., & Downing, A. S. (2019). Local Lens for SDG Implementation: Lessons from Bottom-up Approaches in Africa. *Sustainability Science*, *15*(3), 729–743.

Karlsson-Vinkhuyzen, S., Dahl, A. L., & Persson, Å. (2018). The Emerging Accountability Regimes for the Sustainable Development Goals and Policy Integration: Friend or Foe? *Environment and Planning C: Politics and Space*, *36*(8), 1371–1390. https://doi.org/10.1177/2399654418779995

Kauffman, C. M., & Martin, P. L. (2014). Scaling up Buen Vivir: Globalizing Local Environmental Governance from Ecuador. *Global Environmental Politics*, *14*(1), 40–58. https://doi.org/10.1162/GLEP_a_00639

Kemp-Rotan, R. (n.d.). *If We Can Save the Ship, We Can Save the Town (American Roundtable report on Africatown, Alabama)*. New York: The Architectural League of New York.

Klopp, J. M., & Petretta, D. L. (2017). The Urban Sustainable Development Goal: Indicators, Complexity and the Politics of Measuring Cities. *Cities*, *63*, 92–97. https://doi.org/10.1016/j.cities.2016.12.019

Koch, F., & Krellenberg, K. (2018). How to Contextualize SDG 11? Looking at Indicators for Sustainable Urban Development in Germany. *ISPRS International Journal of Geo-Information*, *7*(12), 464.

Kumar, A. (2019). Dumping Yard-Free Society Through Sustainable Solid Waste Management: A Case Study of Varanasi City. In S. K. Ghosh (Ed.), *Waste Management and Resource Efficiency* (pp. 409–421). Springer. https://doi.org/10.1007/978-981-10-7290-1_35

Lempert, D. (2017). Testing the Global Community's Sustainable Development Goals (SDGs) Against Professional Standards and International Law. *Consilience*, *18*, 111–175.

Lucci, P. (2015). *"Localising" the Post-2015 Agenda: What Does It Mean in Practice?* London: Overseas Development Institute (ODI). https://www.uclg.org/sites/default/files/localization_targets_indicator_web.pdf

Magio, K. O., Lizama Aranda, L. L., González, L., & Alpuche, C. (2021). Analysis and Identification of Sustainable Public Policy for Management of Cultural and Natural Heritage in the Maya Region in line with the Sustainable Development Goals. *Heritage*, *4*(4), 4172–4183. https://doi.org/10.3390/heritage4040229 https://www.mdpi.com/2571-9408/4/4/229.

Morita, K., Okitasari, M., & Masuda, H. (2020). Analysis of National and Local Governance Systems to Achieve the Sustainable Development Goals: Case Studies of Japan and Indonesia. *Sustainability Science*, *15*(1), 179–202. https://doi.org/10.1007/s11625-019-00739-z

Moura, M. C. P. (2021). Historic Black Community Put at Risk by Truck Bridge. *Union of Concerned Scientists* (blog). https://blog.ucsusa.org/cecilia-moura/historic-black-community-put-at-risk-by-truck-bridge/

Nabiyeva, G. N., Wheeler, S. M., London, J. K., & Brazil, N. (2023). Implementation of Sustainable Development Goal 11 (Sustainable Cities and Communities): Initial Good Practices Data. *Sustainability*, *15*(20), Article 20. https://doi.org/10.3390/su152014810

Olawuyi, D. (2020). Sustainable Development and the Water-Energy-Food Nexus: Legal Challenges and Emerging Solutions. *Environmental Science & Policy*, *103*, 1–9. https://doi.org/10.1016/j.envsci.2019.10.009

Petti, L., Trillo, C., & Makore, B. N. (2020). Cultural Heritage and Sustainable Development Targets: A Possible Harmonisation? Insights from the European Perspective. *Sustainability*, *12*(3), 926. https://www.mdpi.com/2071-1050/12/3/926

Pradhan, P., Costa, L., Rybski, D., Lucht, W., & Kropp, J. P. (2017). A Systematic Study of Sustainable Development Goal (SDG) Interactions. *Earth's Future*, *5*(11), 1169–1179. https://doi.org/10.1002/2017EF000632

Rai, P. (2018). Problems of Sustainable Solid Waste Management in Urban Area: A Case Study of Varanasi City of India. *European Geographical Studies*, *4*. https://doi.org/10.13187/egs.2017.2.80

Sharpe, J. (2021). Africatown Concerns Remain Unresolved in Mobile Zoning Code, Activists Claim. *AL.com* (May 18, 2021). https://www.al.com/news/mobile/2021/05/africatown-concerns-remain-unresolved-in-mobile-zoning-code-activists-claim.html

Siegel, K. M., & Bastos Lima, M. G. (2020). When International Sustainability Frameworks Encounter Domestic Politics: The Sustainable Development Goals and Agri-Food Governance in South America. *World Development*, *135*, 105053. https://doi.org/10.1016/j.worlddev.2020.105053

Soergel, B., Kriegler, E., Weindl, I., Rauner, S., Dirnaichner, A., Ruhe, C., Hofmann, M., Bauer, N., Bertram, C., Bodirsky, B. L., Leimbach, M., Leininger, J., Levesque, A., Luderer, G., Pehl, M., Wingens, C., Baumstark, L., Beier, F., Dietrich, J. P., . . . Popp, A. (2021). A Sustainable Development Pathway for Climate Action within the UN 2030 Agenda. *Nature Climate Change*, *11*(8), 656–664. https://doi.org/10.1038/s41558-021-01098-3

Sterling, E. J., Pascua, P., Sigouin, A., Gazit, N., Mandle, L., Betley, E., Aini, J., Albert, S., Caillon, S., Caselle, J. E., Cheng, S. H., Claudet, J., Dacks, R., Darling, E. S., Filardi, C., Jupiter, S. D., Mawyer, A., Mejia, M., Morishige, K., . . . McCarter, J. (2020). Creating a Space for Place and Multidimensional Well-Being: Lessons Learned from Localizing the SDGs. *Sustainability Science*, *15*(4), 1129–1147. https://doi.org/10.1007/s11625-020-00822-w

Wang, X., Ren, H., Wang, P., Yang, R., Luo, L., & Cheng, F. (2018). A Preliminary Study on Target 11.4 for UN Sustainable Development Goals. *International Journal of Geoheritage and Parks*, *6*(2), 18–24. https://doi.org/10.17149/ijgp.j.issn.2577.4441.2018.02.002

23
SUSTAINABLE DEVELOPMENT AND COMPARATIVE CONTRACT LAW
Three Case Studies

Cristina Poncibò

23.1 Introduction: The EU Green Deal

The concepts, tools, and goals of SD (Dernbach and Cheever, 2015), including their institutional indicators issued by the UN (SDGs), have attracted significant attention in the area of public law and public international law. Scholars in these fields have widely explored and discussed the topic to date (Kanie and Biermann, 2017; Segger, Cordonnier, Saito and Weeramantry, 2017). The consequent effects of the controversial concept of SD on the legal world are inevitable and increasingly pervasive (Vogt and Weber, 2019).

Within the European Union, sustainability acts as a legally protected social value, being a principle by which member states will guide and align their policies. In particular cases, sustainability is assimilated to a legal standard which characterizes in qualitative terms certain European Union measures. The Treaty on European Union (TEU) provides in Article 3, paragraph 3(a) that the "European Union aims to establish an internal market that works for the sustainable development of Europe, based, among other things, on balanced economic growth and a high level of protection and the improvement of the quality of the environment." This rule, which is part of EU primary law, addresses only the economic and environmental dimensions of sustainability. Similarly, Article 11 of the Treaty on the Functioning of the European Union (TFEU) promotes sustainable development only from an environmental perspective. The European Union's social sustainability goals in international relations with third states are covered in Article 21, paragraph 2 of TEU. Lastly, sustainability sets a standard for economic and monetary policy, Article 119 of TFEU providing a sustainable balance of payments.

These goals are actually promoted through legislative provisions in the EU (e.g., EU regulations, EU directive) concerning specific obligations to enhance sustainability on states and private actors. In particular, the recent European Green Deal aims at making EU policy goals in this area effective, and the policy instruments well-defined, diverse, and at least equal to the number of policy goals.[1] The various action plans, strategies, regulations, and directives adopted in the implementation of the Green Deal meet this requirement. Conversely, the United Nations 2030 Agenda does not specify any effective measures to achieve the goals, or to what standard they should be implemented. The 169 targets set in 2015 are rather subdivisions of the 17 SDGs and do not provide for the member states available instruments for enforcement.

DOI: 10.4324/9781003285472-25

The European Green Deal not only declares sustainable development as a goal in itself but also sets a comprehensive roadmap of multi-annual strategies and top-to-bottom legislative acts to achieve it. Among these, the Corporate Sustainability Due Diligence Directive (CSDDD) stands as a cornerstone in embedding sustainability into corporate governance.[2] The CSDDD introduces binding obligations for companies to identify, prevent, mitigate, and account for adverse human rights and environmental impacts within their operations, supply chains, and value chains. By doing so, it shifts sustainability from a voluntary corporate responsibility to a legally enforceable duty. Additionally, the Directive supports the EU's overarching climate neutrality objectives by holding businesses accountable for their carbon footprints and encouraging long-term strategies that prioritize decarbonization and biodiversity protection (Ciacchi, 2024; Čulinović-Herc, 2023; Stella Richter and Passador, 2022).

Nevertheless, with regard to private law, by contrast, there has been less attention in the EU and, comparatively, although the SDGs concern not only governments and parliaments, the UN and other international institutions, but mainly business and the private sector (Akkermans and van Dijck, 2019). The Max Planck Institute for Comparative and International Law has organized a conference to launch this new research agenda, "The Private Side of Transforming the World—UN Sustainable Development Goals 2030 and the Role of Private International Law," on September 10–12, 2019 (Michaels, Ruiz Abou-Nigm and van Loon, 2021). Specifically, companies are gradually pushed by the EU toward the adoption of principles of both social and environmental responsibility within their strategies, structures, and management systems (Stella Richter and Passador, 2022; Sjåfjell, 2022).

For this chapter, I note that sustainability, and its indicators according to the UN (SDGs), interacts with private law in many aspects and, particularly, focus on contract law. As noted before, scholars in private law have started considering their interplay only recently (Akkermans and van Dijck, 2019). On the one hand, the indicators encourage freedom of contract when they call to "correct and prevent trade restrictions and distortions in world agricultural markets" (2.b) or "promote the development, transfer, dissemination and diffusion of environmentally sound technologies to developing countries on favourable terms [*omissis*] as mutually agreed" (17.7).

On the other hand, the same indicators may require to impose restrictions, for example, the "immediate and effective" eradication of forced labor, "modern slavery," and child trafficking (8.7, 16.2); "by 2030 [to] significantly reduce illicit financial and arms flows" (16.4); and to "substantially reduce corruption and bribery in all their forms" (16.5). There is clearly also a role for tort law, including its application to cross-border situations, for example, in order to fulfill goals regarding environmental protection and climate change (Mola and Poncibò, 2021; Poncibò, 2016; about property rights, see Ballardini, Kaisto, and Similä, 2021).

Within this context, the chapter questions how contract law offers support as a means of giving force to international standards related to sustainability by imposing obligations on companies in other countries. Moreover, because these public values are "embodied" in that quintessential private law instrument—the contract—they would more readily come within the purview of domestic courts or private arbitral bodies. In other words, it considers the contract governance of sustainability (Grundmann, Möslein and Riesenhuber, 2015; Brownsword, van Gestel and Micklitz, 2017; Zumbansen, 2007). This concept shows how legal departments of transnational corporations (TNCs) draw on the openness of contract as a framework for multiple forms of social cooperation to establish complex cascading webs of contracts. Ensuring control throughout the supply chain implies a new mode of usage for contract, beyond safeguarding and toward organizing cooperation. Yet the exact role of contract remains obscure for practical and conceptual reasons—practically because contracting is barely litigated upon in public, and conceptually because paradigms of contract, as they are taught in law schools around the world, largely stem from the 19th century, with few incremental developments since. In particular, from the perspective of privity, the idea that contracts are foundational building blocks in globally fragmented and centrally governed networks of production can hardly be grasped.

One may easily note that this interaction of public international law and national contract law constitutes a hybrid regulatory system for sustainability. It means a system in which different forms of regulation interact, such as public and private law, hard and soft law standards developed by international organizations and private regulation by companies.

Interestingly, only contracts between private parties can give a binding effect to standards that are otherwise not binding: In short, the chapter notes that public functions regarding the protection of our future are subject to a process of outsourcing to private actors (namely, transnational corporations and their global supply chains). The problem here is how feasible a move from soft to hard law (i.e., binding contractual obligations) of these goals without a direct regulation by states and/or international organizations. All this privatization in the international sphere raises many questions on whether private outsourcing will contribute to fundamental public law values, such as human rights regulations, regulations against corruption and waste, and environmental protection. Surely, to be successful, such a system needs the adoption of effective enforcement mechanisms to hold TNCs responsible for the way they act in addressing sustainability. To date, such mechanisms are still lacking.

In light of the preceding information, three case studies about the interplay of contract law and the goals of SD are considered in the following paragraphs. The case studies concern (1) the standardization and certification processes of SD indicators, (2) the agreement included in the Bangladesh Accord, and (3) the practice of incorporating sustainability contractual clauses into international supply chain agreements.

23.1.1 Standardization and Certification

First, contract law appears to play a central role in designing and governing the main private initiatives for SD, such as, just to provide an example, standardization, certification, and conduct codes (McCall-Smith and Rühmkorf, 2019). In fact, corporate sustainability is a new and evolving corporate management paradigm (Salminen, 2019). The term "paradigm" is used deliberately, since it is an alternative to the traditional growth and profit-maximization model. While corporate sustainability recognizes that corporate growth and profitability are important, it also requires the company to pursue societal goals, specifically those relating to SD: environmental protection, social justice and equity, and economic development (Sjåfjell, 2016).

Against this background, numerous sustainable supply chain governance initiatives have emerged across domestic jurisdictions and continents, including standards and certification initiatives, often filling the regulatory vacuum created by the lack of state regulation and effective global and national policies for sustainability (Eberlein, Abbott, Black, Meidinger, and Wood, 2004).

What the chapter stresses here is that these initiatives are primarily ground on international agreement networks among the interested parties, such as TNCs, suppliers, standardization networks, and auditors. In this sense, it is possible to argue that contract law is giving a substantial contribution in establishing, designing, and governing the said private sustainability mechanisms. However, the academic and scientific debate on the contribution of contract law to SD is still in its infancy.

By now, there is a substantial research body on SD and on sustainability standards and certification by third parties (i.e., auditing), above all in the social sciences, in business literature, as well as in development studies and agricultural economics (Brandi, 2016).

Firstly, the lead firms dominating importation into high-income economies employ a battery of standards to achieve multiple objectives. Directly and indirectly, standards and regulations determine not only the terms of market entry but also the extent to which different producers position themselves in GVCs in a manner that provides for socially and environmentally sustainable income growth. In general, they aim at reinforcing the objectives of the SDGs by promoting living standards, better forms of work, greater social inclusion, and the protection of the environment. In particular, it is possible to note that there are two families of regulations and standards affecting sustainability. Some of these standards affect the character of products (raw materials, intermediates, final goods, services), and others concern the character of the processes involved in the production of these products.

Secondly, in the social sciences, certification has been studied, for instance, as a non-state market-driven governance system, gaining rulemaking authority as private or transnational governance initiatives (Büthe and Mattli, 2017), and with a focus on various interactions between different transnational business governance schemes (Eberlein, Abbott, Black, Meidinger, and Wood, 2004). In the business literature, certification has been investigated in the context of environmental management systems as well as green or sustainable supply chain management or governance (Brandi, 2016), including in the context of global supply chains (De Marchi, di Maria, and Micelli, 2013). In addition, business literature and the literature on environmental management are increasingly addressing the potential link between economic and environmental benefits and have investigated, for example, how organizations define, interpret, and implement sustainability (Brandi, 2016).

Surely, TNCs and, more generally, private actors have gained substantial agency in enhancing sustainability—they have an increasing influence on outcomes, even on global issues, such as climate change (idem). Well, once private power becomes viewed as a delegation of sovereignty, it becomes quite difficult to distinguish between private and public law as ways of regulating and structuring markets (Peters, Köchlin, Förster, and Fenner Zinkernagel, 2009). The problem is that their agency exists, to quote an author, "beyond the state and its international organisations" and, I would add, without mechanisms to hold them liable in case of failure (Abbott, 2011). For our purposes, it is worth noting that this hybridization of public international law and contract law is a puzzle for legal scholars primarily because this process shows a variety of interactions crossing the public–private boundary, "entangling" public and private institutions, but these are largely ad hoc, not systematic.

23.1.2 Contracting for Sustainable Development

On May 15, 2013, a group of companies, mainly TNCs, agreed on legally enforceable commitments concerning sustainability by entering the Bangladesh Agreement ("the Accord"). The Bangladesh government ordered the organization to cease its operations by November 30, 2018, claiming that its national regulatory body was fully capable of taking over their work. The Accord disagreed, and a legal battle ensued. On May 19, 2019, the agreement was accepted by the Bangladesh Appellate Court, which has given permission for the Accord to continue during a transition period.

Basically, the Accord is a contract between apparel brands, retailers, and importers on the one hand, and international trade union federations on the other. It sits almost entirely apart from domestic law and regulation, or public international law. Funded by the private sector garment industry signatories, it provides for a system of oversight, credible inspections, mandatory remediation, and transparent reporting. Those contractual obligations are enforceable through international commercial arbitration, at the suit of the signatory unions (Salminen, 2018). Precisely, under the Accord, each signatory company committed, for example, to disclose its supplier factories in Bangladesh to recognized, independent fire safety experts and require these factories to submit to rigorous fire safety inspections led by these experts. The company also accepted a public disclosure of all the inspection reports of its supplier factories under the program, while requiring all its suppliers to implement all repairs and renovations necessary to make their factories safe, as determined through the inspection process. Also interesting, the company undertook to pay suppliers prices sufficient to make it possible for them to afford the necessary repairs and renovations and to operate in a safe manner and required suppliers to allow worker representatives into their factories to educate workers about workplace safety and worker rights.

Moreover, the Accord also provides a general remedy in case of breach of suppliers' obligations, by stating that each signatory company shall cease doing business with any supplier that fails to comply with any of these requirements.

One may argue that the Bangladesh Accord represents one of the successes of the freedom of contract in that it opens up vast possibilities for companies, especially TNCs, to come to act differently, even where regulators feel themselves constrained from acting. In fact, these commitments are not merely general statements of intent but binding, contractually enforceable obligations. Indeed, many of the Accord's signatory brands and retailers routinely include similar language in the legally binding contracts they make with suppliers, contractors, creditors, and consumers.

It is also important to note that the Accord establishes a private dispute resolution mechanism. Precisely, Article 5 of the Accord details the procedural aspects by providing that (1) disputes concerning implementation are first submitted to the seven-member oversight steering committee, which is comprised of three representatives chosen by the trade union signatories and three representatives chosen by the company signatories, with a representative of the ILO serving as a neutral chair. (2) Any decision of the steering committee may, at the request of either party, be appealed to a binding arbitration process, governed by the rules of the United Nations Commission on International Trade Law, which is the standard procedure for international arbitrations. As set out in the Accord, an arbitrator's award may be enforced in a law court of the domicile of the signatory party against whom enforcement is sought (e.g., an award against a German company may be enforced in a German court) according to the Convention on the Recognition and Enforcement of Foreign Arbitral Awards, also known as the New York Convention,[3] that was adopted by a United Nations diplomatic conference on June 10, 1958, and entered into force on June 7, 1959.[4] In this respect, it is worth noting that the commercial arbitration process provided for in the Accord is the same process relied upon by TNCs around the world to enforce the commitments made to them in commercial contracts with their ordinary business partners. In other words, under the Bangladesh Accord, the failure by a company to adhere to its obligations may result in an adverse, binding arbitral award, which may then be enforced by a court in the company's home country. Such an award could, for example, require the company to comply with its obligations to help pay for the installation of safety features, such as emergency exits, found necessary for one of its supplier factories to operate safely.

In this light, scholars noted that the Accord is a development of some significance. It is true that, for the first time, a number of TNCs have signed with global trade union federations what "looks like" a legally binding agreement, enforceable through the courts (Salminen, 2018). By entering the Accord, these companies commit to a range of measures aimed at transforming the working conditions at the premises of offshore suppliers who manufacture ready-made garments for them. On one hand, the Accord is, at the base, simply a contract. In this sense, the risks assumed by signatory companies are no different than those to which these same firms willingly expose themselves in the normal course of business in countries around the world, which routinely involves their entering into legally binding agreements with suppliers, contractors, creditors, and landowners, to name a few. On the other hand, in our view, the Accord produces few—if any—firm legal rights for or obligations on anyone (Salminen, 2018).

In summary, the Accord seems non-binding, aspirational, legally speaking, and weak. Thus, the chapter notes that contract theories—traditional, relational—struggle to make sense out of the legal nature of emergent complex contractual structures, like the Accord, that involve values and ethical considerations.

23.1.3 Incorporating Sustainability Goals Into Contractual Terms

The role of codes of conduct (as well as policies and guidelines issued by corporations) in advancing corporate social responsibility has attracted the attention of private law scholars in the last years (Rühmkorf, 2015; Vytopil, 2015; Beckers, 2015; Beckers, 2019). The implementation of supplier codes of conduct varies widely. Some companies only publish them on their website without implementing them through contractual mechanisms. Others implement a code of conduct through general terms and conditions. Another category of companies considers such codes of conduct to be a core element of their contracts with suppliers. Although the implementation of codes of conduct varies, the majority of the globally operating enterprises who have implemented a sustainability policy utilize a contractual mechanism to enhance compliance with their codes of conduct.

Within such a wider framework, the chapter focuses on the "contractualization of sustainability" process, namely, the implementation of the sustainability provisions of companies' codes of conduct and policies into contractual obligations.

In other words, sustainability contractual clauses (SCCs) would, in that way, represent a "juridical translation" of sustainable development objectives, leading to a contractual synthesis of disparate goals, such as fighting corruption, reducing poverty, protecting the environment for the present and future generations, and respecting diversity, just to mention a few.

Indeed, public law scholars were the first to discuss sustainability clauses in public procurement as well as in investment agreements (Cotula, 2010). Surprisingly, contract law scholars have largely ignored SCCs, with few exceptions (Mitkidis, 2015; Mitkidis, 2014). While this literature sometimes suggests that contract plays a role here, it never seriously considers what that means (see Cafaggi, 2013, who notes that "[t]ransnational commercial contracts have also become vehicles of implementation of public international regulation," and Verbruggen, 2014). The review of companies' practices in the literature helps in showing the different ways in which companies incorporate such standards into international commercial agreements.

Generally, legal counsels of TNCs usually design SCCs according to three main models, and their texts range from vague (non-binding) statements to more precise (binding) contractual obligations (Poncibò, 2022; Poncibò, 2016). Firstly, companies often include express clauses about sustainability goals into their agreements ("sustainability contractual clauses," SCCs). A model clause, which usually is very detailed and complex, requires thus:

> The Supplier hereby undertakes: [*omissis*] not to tolerate corruption in any way, shape or form in any jurisdiction, even if such activities are allowed, tolerated or non-prosecutable;—to assess and reduce the environmental impact of its own products and services throughout

their entire life cycle;—to use material resources responsibly, in order to achieve sustainable growth that respects the environment and the rights of future generations [*omissis*].

Pirelli Group, Sustainability Contractual Clause <www.pirelli.com/asset/index.php?idelement=59887> accessed 15.03.2022

Additionally, the contractual provision goes further by requiring the (weaker) party of the agreement (mainly a supplier) to address (by a miracle, one may think) a long list of obligations concerning social, economic, and environmental problems traditionally remitted to public law and agencies. They also include, just to mention a few, child labor, diversity and equal opportunities.

Secondly, companies may use international standards or their conduct codes, policies, and guidelines as a reference point (the reference model). The clause consists in referring to the companies' documents mentioned here. It sounds like this:

> The Supplier hereby declares to have read and understood the "Values and Ethical Code", the "Code of Conduct" and the [*omissis*] group policy "Social Responsibility for Occupational Health, Safety and Rights, and Environment", published on the web. The Code establishes [*omissis*] the principles that conduct the management of [*omissis*] group's business, as well as contractual relations and other relations with third parties.
>
> (Poncibò, 2016, *para. 1 and para. 2*)

More important, the clarity of the language in both the SCCs and the standards or codes referred to varies from more binding forms that require the contractual partners to meet these standards to more aspirational forms. In the latter case, particularly with respect to the reference model, the wording used is vague, and it can be said that it does not represent a binding provision but merely a statement. In this light, the issue of whether SCCs are binding and enforceable depends upon the way they are drafted in contracts between the parties.

Thus, the reference to international standards and codes of conduct often results in a mere statement because of the vagueness of the actual document. It would be different if the SCC at issue contained mandatory requirements and continuous improvement benchmarks or indicators. In the latter case, it is possible that the contractual provision is binding and enforceable.

23.2 Contract Law and Sustainable Development

The chapter underlines that the contracts and contractual provisions dealing with the goals of SD here examined have an evident public dimension because they deal with challenges such as environmental protection, equality, and distributive justice, just to mention a few (Poncibò, 2016, para. 2 and 3). Now, this shift of public interest within the private contract—which was intrinsically not meant to receive it—has not yet found a satisfying translation about the responsibility of the concerned actors. The debates around the idea of the many sustainability goals still need to find the proper answers. The question is how far it is appropriate and feasible to translate the new borders of the public interest—induced by the ever-growing intervention of private operators—into binding contractual obligations.

23.2.1 Promising Justice

The contract governance of SD concerns a number of environmental, economic, and societal goals of the present and future generations (Salas-Zapata and Ortiz-Muñoz, 2019). If the chapter refers to the UN indicators, the concept embraces, for example, those requirements with the goals of fighting poverty (SDG 1), of building effective, accountable institutions (SDGs 16.3, 16.5, 16.6, 16.7), and of ending all forms of discrimination against girls and women (particularly SDGs 5.2, 5a). Transparent and corruption-free institutional (SDG 16.5) proceedings, which have a sustainable effect, are also of highest priority.

In practice, the internet offers some examples of sustainability contractual clauses (SCCs) that TNCs often advertise before the public (Lomfeld and Wielsch, 2013; Zumbansen, 2013). Accordingly, a supplier undertakes, for example, to assess and reduce the environmental impact of its own products and services throughout their entire life cycle, and use material resources responsibly, in order to achieve sustainable growth that respects the environment and the rights of future generations (Pirelli, para. 2.5). In addition, the supplier is also required to fight corruption, ensure equal opportunities, and ensure that wages shall be sufficient to meet the basic needs of personnel (Pirelli, para. 2). These obligations do not complete the picture, because suppliers are generally subject to the principles, values, and commitments as expressed in the company's documents, such as policies, guidelines, and codes of conduct, to mention a few. In the words of an author, "these terms literally promise justice" (Lipson, 2019).

There is no need to point out that the traditional contract theory asserts the normative significance of the bilateral structure of a contract by limiting considerations of contract justice to the relationship between parties. Privity, as an expression of party autonomy, aims at capturing the significance of a contract's bilateral nature. Contractual rights and duties are limited to the parties who have consented to contract: Only they can pursue remedies for breach or seek to enforce the contract (see Rühmkorf, 2015, pp. 98–101).

On the contrary, in promising justice, SCCs and agreements concerning sustainability are often directed to the welfare of individuals who are not parties to the contract, such as employees and local communities. Thus, unlike traditionally third-party beneficiaries, the beneficiaries of these provisions are diffuse, undefined, and with respect to future generations, unknown to the parties of the contract. As a result, a non-contracting third party may often have an interest in enforcing a contractual provision, or an entire agreement, having the scope of advancing sustainability goals. For instance, the employees of third-world country suppliers who experience violations against their human rights and poor working conditions, or the general public affected by environmental pollution (Monteillet, 2017).

This demand for justice has not found a satisfactory solution to date: Only the party that a contractual clause expressly mentions and identifies may have a right to enforce clauses that are beneficial for them (for example, the employees and the local communities) (Rühmkorf, 2015, pp. 102–107, 106). In this respect, the chapter shows that the practical effects of SCCs, and other agreements concerning the sustainability goals, would be stronger if a contractual provision indicated these third parties as beneficiaries in an express way by granting them the right to enforce the agreement. Precisely, a clause should specify the name, membership of the beneficiary, or describe it.

In this respect, it should be pointed out that the doctrine of third-party beneficiaries is *per se* not adapted to protect the diffuse and unknown beneficiaries. The realistic option, especially in dealing with unidentified or diffuse parties, arises from tort law when it interplays with contract law (Poncibò, 2022). For example, the contract (or the contractual provision) represents a factual element that may give rise to a tortious liability.

Consequently, it is important to say that, in the case here examined, contract law interacts with many other fields of law, such as tort, trade and commercial law, competition and consumer law. It is necessary to have in mind the multidimensional nature of the problem. These contracts and clauses about the goals of SD are evidently producing side effects on third parties protected by tort law (Ulfbeck and Hansen, 2019).

23.2.2 The Transnational Context

The contract governance of SD clearly shows to have a transnational dimension (Poncibò, 2022). For example, it should be recalled that the Bangladesh Accord (paragraph 1.1) results from the agreement of a group of international players. In this way, Western TNCs with global supply chains increasingly incorporate various goals with suppliers (paragraph 1).

Clearly, the multi-jurisdictional environment of global trade strongly influences the design of these contractual provisions: For example, contract remedies mainly consist in private mechanisms (or arbitration in the case of the Bangladesh Accord) because of the lack of trust about enforcement before domestic jurisdictions (Cafaggi and Iamiceli, 2015).

Moreover, the contents of such contractual provisions are becoming standardized. In other words, the adoption of SCCs in the general terms and conditions of purchase is becoming a common practice for TNCs.

More important, companies can only enforce potential violations against their policies dealing with sustainability goals if they become aware that there is a breach of such policy along the supply chain in the first place. Therefore, it is important that transnational companies check their supply chains closely, namely, via the private mechanisms of coordination and subordination described in paragraphs 1.3 and 2.2. For this reason, the contractual provisions require the supplier to establish a mechanism of internal audit and reporting to the buyer.

Additionally, TNCs usually have a contract with its first-tier supplier alone, but not with the other suppliers further down the supply chain. However, it is clear that it is possible to overcome this problem by including a provision in the contract which demands that the first-tier supplier conclude the same or similar contractual obligations regarding corporate social responsibility with the other suppliers further down the supply chain. In this sense, sustainability clauses require the supplier "to ensure that any subcontractors and suppliers comply with the same conduct established above, and to regularly oversee observance of said obligations" (Pirelli's Sustainability Model Clause (Art. 2 lett. b)).

In this perspective, one may note that the contractual relationships governing the relationships between buyers and suppliers in global supply chains, while nominally contractual in the traditional sense, are establishing a sort of private order institutions. SCCs, for example, usually contain very detailed internal procedures for managing the relationship, monitoring, auditing, and dealing with failures. They also institutionalize auditing and certification by auditors.

Accordingly, some legal scholars have shown how to work around constraints imposed by privity of contract by exploring other contract doctrines, such as the choice in action doctrine that enables, to some extent, contractual enforcement of obligations across chains of suppliers as opposed to merely between two parties located within a chain (Cafaggi, 2011). In particular, some authors are researching for solutions for firms to enforce obligations against members of a supply chain outside of the restrictions of privity in order to increase commercial efficiency and facilitate the enforcement of higher labor and environmental standards (Sobel-Read, Anderson and Salminen, 2018).

Both authors seek to answer the question of how parties in the chain can ensure enforcement or pursue liability for breach of obligations that arise from contracts they are not party to. The point of view that underlies the question takes for granted the contractual right and obligations that contracts seek to impose.

In consideration of the aforementioned, the chapter concludes the section by stressing the transnational dimension of the contract governance of sustainability through the exercise of control over supply chains across jurisdictions. In such a context, it is evident that contract law assumes peculiars aims and features when it operates to manage chains of suppliers in different jurisdictions (Poncibò, 2016).

23.2.3 Fostering Social Change by Contract

It is worth noting that legal counsels are not drafting SCCs to create incentives for performance and breach primarily through courts or arbitration. It should also be underlined that the same Bangladesh Accord seems to be more a symbolic act than a contractual agreement of binding promises and subject to enforcement.

In particular, SCCs often refer to codes of conducts and policy documents mainly containing principles and guidelines not subject to contractual enforcement for their vagueness and lack of clarity. Moreover, SCCs expressly drafted into international supply agreements can result in a vague and imprecise statement, depending on their language and clarity. In summary, these contractual provisions—often—contain mere statements relevant for the parties and third parties, and for this reason, contract law faces difficulties in serving as a tool for the goals mentioned here.

This chapter argues that the fact that these contractual provisions are not enforceable (or under-enforceable) is not problematic. Some authors have already argued that promissory relationships of any significance often involve a mix of enforceable and unenforceable elements to achieve various goals (Bernstein, 2015).

For the purposes of this chapter, it emerges that contract law may serve sustainability by achieving social change within and outside the global supply chain. Indeed, SCCs are "sending" a message to suppliers and promoting a change in their behavior. Put differently, in analyzing the contract governance of sustainability, it is important to consider the expressive function of law. Indeed, legal scholars (and political ones) use it to make broad claims about the expressive dimension of statutes, judicial opinions, and other legal documents (McAdams, 2015; Sunstein, 1996). The nature of these claims varies. Some authors say that legal change will "send a message" of some sort for (morally) sanctioning and promoting behavioral change. Others claim that people comply with certain laws (the typical example: seat belt mandates and smoking restrictions) because of the expressive or symbolic effect of these rules (McAdams, 2015, p. 11 n.1, 2).

For example, someone noted how the law can work by changing the "social meaning" of behavior (the example is the law against dueling that created a new meaning to the decision to refuse a duel). Thus, there are so many normative claims about legal expression (and there is probably not one theory but a set of literature employing the same term).

Our point here follows: It is possible to understand these clauses (and the Bangladesh Accord) because of their symbolism. For example, the Bangladesh Accord has had a powerful symbolic impact, while it remains questionable whether the obligations contained in the Accord are enforceable before domestic jurisdictions. While McAdams (2015) has not included contract law into his analysis, the theory is very helpful in our case. Most surely, the literature in the field has rarely entered in the domain of contract law (Hadfield, 1998). The question is whether a statement in a contract (or the contract itself) is symbolic, expressive, or meaningful.

In reading contractual statements examined here, one may think of a case for scholars interested in semiotics. In the words of Eco, "[t]he sign is used to transmit information; to say or to indicate a thing that someone knows and wants others to know as well." The sign fits into this simplified canonical model of communication: source, sender, channel, message, receiver (Eco, 1978).

To be clear, our argument is that both the statements of the Bangladesh Accord and the SCCs—which are proliferating—in commercial contracting are signs, symbols that the parties, mainly TNCs as buyers, are sending to suppliers, traders, consumers.

A question is how signs are conveyed. It is therefore true that the expressive function generally implies an effective communication of the legal statement. Usually, contracts are subject to confidentiality and do not reach the public. Nevertheless, in our case, codes of conduct, the Bangladesh Accord, and SCCs are widely advertised in global supply chains, to traders, but also to the public. The large majority of companies publish these documents online to reach the public domain. The publication is, in reality, the cornerstone of an effective corporate sustainability.

This practice confirms that such statements are primarily directed at informing, signaling, and promoting social changes among suppliers and, more generally, in the community of traders. For this reason, TNCs, when adopting such provisions, are keen to reach a wider audience, also for reputational goals and risk-sharing strategies (Pirelli).

23.3 The Problem of the Enforcement

23.3.1 A Holistic Concept

If one desires to regulate the performance of suppliers with respect to SD Goals through contractual mechanisms, a question might be what sustainability is. The concept covers a broad spectrum of interlinked goals (i.e., holistic concept). In fact, sustainable development cannot be achieved through isolated initiatives but rather requires an integrated effort at various levels, comprising social, environmental, and financial aspects.

In recent years, the growing concerns for environmental and climate change, together with issues of poverty, increasing disparity between societies, and the tensions brought by social inequalities, have placed SD under the spotlight. National and international institutions, policymakers, and cross-country initiatives, as well as academics, have increased the attention given to social and environmental sustainability worldwide. Whereas everyone seems to agree on the importance of sustainable development, its very nature and meaning are discussed and analyzed.

As a result, the actual implementation of sustainability risks to be limited by the vagueness and ubiquity of its definition. For instance, due to the ubiquity of the notion of sustainability, different discourses (e.g., environmental, economic, social, cultural, ethical) have emerged over time, thereby associating this concept with social responsibility, environmental management, or business sustainability, which are often treated in separated ways. These different discourses have also revealed their appeal over corporations, whose role and responsibility for SD are becoming central because of the many shortcomings of public actors, such as states (Trubeck, Thomas and Santos, 2019a; Trubeck, Thomas and Santos, 2019b).

In this context, a sort of "sustainability rhetoric" is emerging in mission statements, internal codes, and external reporting systems. In some cases, this rhetoric was used in the attempt to reconstruct the eroded legitimacy of companies and did not necessarily involve the actual implementation of (or participation in) SD. Otherwise, such active implementation and participation would require organizations to alter their existing practices and to allow a concrete strategic move toward sustainability (Giovannoni and Fabietti, 2014).

Thus, the point here is whether the process of incorporation of SD Goals into a contract is really feasible. Clearly, the contractual arrangement should make clear what is expected from the supplier and which SD Goal the buyer is referring to.

23.3.2 Relational Contracting and Enforcement

This chapter also notes that the contract governance of SD is an example of relational contracting. To clarify, the three cases here examined rely on informal mechanisms of cooperation and trust (Macneil, 1985).

Firstly, the relational theory of contract has the merit of further extending the legal realist approach to provide a rich sociological account of the role of contract-making in the social order. It emphasizes how contracts are "embedded" in complex social relationships and, consequently, can vehicle not only the interest of the parties but also other (societal) values, such as solidarity and equality (Bernstein, 1996; Bernstein, 1992; Bernstein, 2001). On the contrary, contract theory—in its traditional conceptualization—is not capable to find an explanation of the increasing contractualization of public values.

Secondly, the contract governance of SD implies the establishment of a private order institution. It is also true that businesses have devised contractual structures to govern their relationships, especially when trading with their suppliers that, while nominally contractual in the traditional sense, are better understood as private order institutions. In this sense, they have structured their supplier relationships in ways that make the legal system largely irrelevant to their contracting relationships.

Thirdly, in our case, the formal legal infrastructure governing the contractual relationship appears to be of secondary importance compared to informal norms that are particularly relevant with respect to enforcement. To be clear, companies are not relying on mere promises to perform SD obligations or the shadow effects of potential court-imposed remedy to induce the level of performance. Rather, to increase the likelihood that their suppliers will perform as promised, they have developed a variety of methods that either recreate or approximate the core management techniques associated with intra-firm hierarchy (e.g., internal monitoring, training). To ensure that products meet their quality specifications, they go to great lengths to regulate the production processes used in suppliers' plants. In some instances, they also try to exert influence over sub-suppliers' plants. For example, they require suppliers to operate their plants in conformity with sustainability standards and may impose additional requirements that are designed to better tailor these standards to their individual company's particular needs.

23.3.3 Current Challenges in Global Value Chains

The enforcement of contractual provisions concerning SD Goals represents a challenge for contract law scholars. One might question the need to enhance contractual mechanisms as the current supplier codes of conduct contribute to improvement of the human and labor rights conditions. One might also consider it to be an illusion to assume that drafting more thorough contractual provisions will solve SD-related issues.

A big challenge is to assess the breach of a supplier with respect to SD Goals. Often, information on supplier performance is unavailable or insufficient. Therefore, buyers need access to this information and a contractual provision implementing a supplier obligation to provide that information is helpful. This provision might entail an obligation to provide information on the supply chains of the supplier as well as the most salient risks in these supply chains. Furthermore, it might also entail an obligation to implement a thorough process to engage with subcontractors and provisions on the cases in which subcontracting is permitted. In connection with this it might be worthwhile to impose mechanisms of self-evaluation on the supplier too. This creates an independent obligation for the (first-tier) supplier to conduct its own self-evaluation and to deal with its suppliers, notwithstanding the need for the buyer to monitor whether this is actually done. Moreover, implementing third-party audits, of which meaningful dialogue with local stakeholders is an important part, might also prove beneficial.

In particular, it emerges that the parties favor informal and private mechanisms to prevent and, when necessary, to react to a breach of contractual promises. Thus, when companies are committed to ensuring the sustainability of their businesses, they should be able to monitor the progress and evaluate the impacts so that they are able to ascertain in an objective way whether they have fulfilled their sustainability commitment or not. For example, it is possible to read thus:

> The Supplier acknowledges that the company has the right, at any time, to verify, either directly or through third parties, compliance by it with the obligations herein undertaken.
>
> (Pirelli Group, Sustainability Contractual Clause, para. 2.3)

Together, these oversight and audit should reduce the likelihood of a breach and thus set off a chain of reactions that either terminate or severely damage the parties' relationship. They also make it possible for these contract provisions to condition on information that in their absence would not be observable and would only be verifiable through the filing of a lawsuit and the conduct of civil discovery. Consequently, these provisions both expand the range of commitments that can be extra-legally enforced and significantly reduce the likelihood of a buyer filing a suit or terminating a supplier based only on his best guess of what civil discovery would reveal (Gilson, Sabel and Scott, 2010).

In addition, SCCs develop procedural tools to deal with their breach, including reporting, whistleblower strategies, corrective action plans, and blacklisting. For example:

> The Supplier may report to . . . any breach or suspected breach of the "Values and Ethical Code", the "Code of Conduct" and the . . . Group policy "Social Responsibility for Occupational Health, Safety and Rights, and Environment" or any applicable laws.
>
> (Pirelli Group, Sustainability Contractual Clause, para. 2.3)

In such a context, traditional remedies in both civil law and common law remain in the shadow. The model clause states:

> The Parties hereby agree that the company may terminate the Contract/s and/or the Order/s in the event that the Supplier should be held responsible for any breach of any of the provisions of paragraph . . . above.
>
> (Pirelli Group, Sustainability Contractual Clause, para. 2)

Thus, in case the party (often a supplier) is in breach of the SCC, the buyer can procure a remedy that, in our case, will most probably consists in demanding damages (Bair, 2009).

Making matters worse, the more likely form of damage for breach of SCCs will be to the reputation of the company. Thus, if the company learns the hard way that suppliers in its chain engaged in problematic labor practices, the harm would not be that breach necessarily impaired the value of the goods themselves but instead that breach damaged the corporation's reputation. In theory, courts in both common law and civil law may recognize claims for reputational damage as a consequence of a breach of contract, but they do not like to do so because, like all claims of consequential damage, courts worry that the harm may not have been in the contemplation of the parties and may be difficult to quantify (Lipson, 2019).

23.4 Conclusions

One of the biggest challenges of our time relates to the lack of political will. Many governments around the world are not fulfilling their duty to protect, as well as their duties to respect and fulfil the sustainability goals for both present and future generations (Sullivan, 2019).

In such a context, private law scholarship in the EU and globally is still at the beginning in researching how closely concepts, methodologies, and instruments of private law, including but not limited to contract law, are linked to sustainable development. In our case, private actors (TNCs, traders, suppliers) do not merely exacerbate the legitimacy crisis in public law (Trubeck, Thomas and Santos, 2019 a, b).

In discussing this topic, the chapter confirms that the distinction between public and private law is vanishing with respect to SD Goals. The achievement of SD Goals depends on a hybrid system of public international law and contract law, a system that evidently overcomes the traditional boundaries of disciplines (McCall-Smith and Rühmkorf, 2019). Thus, rather than a two-party model, it is necessary to focus on the context of the transaction, for example, the wider social body of social relations, ecosystems and sustainable development, and state or social regulatory structures. This paves the way for a socio-legal analysis of transactions that can draw from a wide and diverse array of approaches in the humanities and social sciences. Our case also reveals that a purely private realm exists only as a legal construction in contract law. Such an approach is also sensitive to the observation that our understanding as private lawyers remains too narrow, and it is necessary to take courage in reconsidering the public dimension of private contracts (Lomfeld and Wielsch, 2013).

On such basis, this chapter has offered an initial exploration of the contract governance of SD. Precisely, it argues that it consists of a network of contracts among private actors within global supply chains and across national jurisdictions. Such an argument becomes evident in considering the network of agreements behind the processes of creating standards, as well as monitoring and auditing their implementation. SCCs are becoming standard clauses of international supply agreements, with the hope that the contractualization of SD Goals will contribute in overcoming the mere rhetoric of sustainability.

Firstly, the investigation confirms that there are many cases where contract law interplays with SD. The cases examined in this chapter show that contracts can themselves serve as vehicles to promote public law values. Thus, although typically conceived as the quintessential private law form, contracts used in this way can be a tool to "publicize" the private relationship in part (e.g., the case of SCCs) or totally (e.g., the case of the Bangladesh Accord).

Secondly, the chapter discusses how such a relationship is essentially based on formal and informal contracting, a dynamic well explained in McNeil's work on the relational theory of contract law. Such theory is useful to understand the diversity of the ways in which contractual relationships occur, given the values, ideals, practices, and self-understandings of the participants in contractual relations and the wider relationships.

Thirdly, from our analysis, it is possible to conclude that contracts and contractual clauses dealing with the goals of SD could help create behavior standards, performance benchmarks, and a means for promoting a change within the global values chains. While such contractual provisions are not a panacea, they may also be useful because of their expressive function.

Finally, the focal point of this chapter consists in arguing that the contract law has a new social dimension consisting in promoting intergenerational justice as part of the SD Goals. In other words, this emerges by considering that the contractualization of SD sounds like promising justice across borders, and such process has a social impact, not necessarily supported by the enforcement of contractual obligations.

Notes

1 Communication from the Commission to the European Parliament, the European Council, the Council, the European Economic and Social Committee, and the Committee of the Regions, European Green Deal, Brussels, December 11, 2019, COM (2019) 640 final.
2 Directive (EU) 2022/2464 of the European Parliament and of the Council of 14 December 2022 amending Regulation (EU) No 537/2014, Directive 2004/109/EC, Directive 2006/43/EC, and Directive 2013/34/EU as regards corporate sustainability reporting, Brussels, December 14, 2022, OJ L 322, December 16, 2022, 15-80, at http://data.europa.eu/eli/dir/2022/2464/oj.
3 The Convention on the Recognition and Enforcement of Foreign Arbitral Awards, also known as the New York Convention, was Adopted by a United Nations Diplomatic Conference on (10 June 1958 and entered into force on 7 June 1959).
4 UNCITRAL, Convention on the Recognition and Enforcement of Foreign Arbitral Awards (New York, 1958), https://uncitral.un.org/sites/uncitral.un.org/files/media-documents/uncitral/en/new-york-convention-e.pdf.

References

Abbott K. W., 'Engaging the Public and the Private in Global Sustainability Governance' (30 November 2011) <https://ssrn.com/abstract=1966730>.

Akkermans B., and van Dijck G. (eds.), *Sustainability and Private Law*, Eleven International Publishing, Utrecht, 2019.

Bair J., 'Contextualising Compliance: Hybrid Governance in Global Value Chains' (2009) *New Political Economy*, 1.

Ballardini M., Kaisto J., and Similä J., 'Developing Novel Property Concepts in Private Law to Foster the Circular Economy' (2021) 279 *Journal of Cleaner Production*, 1.

Beckers A., *Enforcing Corporate Social Responsibility Codes on Global Self-Regulation and National Private Law*, Hart, Oxford, 2015.

Beckers A., 'Towards a Regulatory Private Law Approach for CSR Self-Regulation? The Effect of Private Law on Corporate CSR Strategies', Tilburg Private Law Working Paper Series No. 3/2019 <https://papers.ssrn.com/sol3/papers.cfm?abstract_id=3369596>.

Bernstein L., 'Beyond Relational Contracts: Social Capital and Network Governance in Procurement Contracts' (2015) 7 *Journal of Legal Analysis*, 561, 563.

Bernstein L., 'Merchant Law in a Merchant Court: Rethinking the Code's Search for Imminent Business Norms' (1996) 144 *University of Pennsylvania Law Review*, 1765.

Bernstein L., 'Opting Out of the Legal System: Extralegal Contractual Relations in the Diamond Industry' (1992) 21 *The Journal of Legal Studies*, 115.

Bernstein L., 'Private Commercial Law in the Cotton Industry: Creating Cooperation Through Rules, Norms, and Institutions' (2001) 99 *Michigan Law Review*, 1724.

Brandi C., 'Sustainability Standards and Sustainable Development—Synergies and Trade-Offs of Transnational Governance' (2016) *Sustainable Development*, 1–16.

Brownsword R., van Gestel A. J., and Micklitz H.-W., *Contract and Regulation*, Edward Elgar, Cheltenham, 2017.

Bukspan E., 'Extreme Makeover—Contract Law Edition: A New Home for Human Rights and Social Responsibility (Lessons from Israel)' (2012) 7 *Intercultural Human Rights Law Review*, 330–388.

Burand D., 'Contracting for Impact: Embedding Social and Environmental Impact Goals into Loan Agreements' (2017) 13 *New York University Journal of Law & Business*, 775–824.

Büthe T., and Mattli W., *The New Global Rulers: The Privatization of Regulation in the World Economy*, Princeton University Press, 2017.

Cafaggi F., 'The Regulatory Functions of Transnational Commercial Contracts: New Architectures' (2013) *Fordham International Law Journal*, 1557, 1564.

Cafaggi F., 'Transnational Governance by Contract-Private Regulation and Contractual Networks in Food Safety' (29 March 2011) <https://ssrn.com/abstract=1874749>.

Cafaggi F., and Iamiceli P., 'Contracting in Global Supply Chains and Cooperative Remedies' (2015) *Uniform Law Review*, 1–45.

Ciacchi S., 'The Newly-Adopted Corporate Sustainability Due Diligence Directive: An Overview of the Lawmaking Process and Analysis of the Final Text' (2024) 25 *ERA Forum*, 29–48.

Cotula L., *Investment Contracts and Sustainable Development. How to Make Contracts for Fairer and More Sustainable Natural Resource Investments*, IIED, 2010.

Čulinović-Herc E., 'Navigating the Corporate Sustainability Challenge-Proposal for a Directive on Corporate Sustainability Due Diligence in the EU Regulatory Arena' (17 April 2023). Young Universities for the Future of Europe (YUFE) Law Conference Proceedings No. 01/2022 < https://doi.org/10.2139>.

Cutler A. C., and Dietz T., *The Politics of Private Transnational Governance by Contract*, Routledge, Abingdon, 2017.

De Marchi V., di Maria I., and Micelli S., 'Environmental Strategies, Upgrading and Competitive Advantage in Global Value Chains' (2013) 22 *Business Strategy and the Environment*, 62–72.

Dernbach J. C., and Cheever F., 'Sustainable Development and Its Discontents' (2015) 4(2) *Transnational Environmental Law*, 247–287.

Eberlein B., Abbott K. W., Black J., Meidinger E., and Wood S., 'Transnational Business Governance Interactions: Conceptualization and Framework for Analysis' (2004) 8(1) *Regulation and Governance*, 1–21.

Eco U., *A Theory of Semiotics*, Indiana University Press, 1st ed., 1978.

Gilson R. J., Sabel C. F., and Scott R. E., 'Braiding: The Interaction of Formal and Informal Contracting in Theory, Practice and Doctrine' (11 January 2010). Stanford Law and Economics Olin Working Paper No. 389/2010 <https://ssrn.com/abstract=1535575>.

Giovannoni E., and Fabietti G., 'What Is Sustainability? A Review of the Concept and Its Applications', in C. Busco et al. (eds.), *Integrated Reporting*, Springer International Publishing, Switzerland, 2014, pp. 21–39.

Grundmann S., Möslein F., and Riesenhuber K., *Contract Governance Dimensions in Law and In-terdisciplinary Research*, Oxford University Press, Oxford, 2015.

Hadfield G. K., 'An Expressive Theory of Contract: From Feminist Dilemmas to the Conceptualization of Rational Choice in Contract Law' (1998) 146 *University of Pennsylvania Law Review*, 1235, 1267.

Kanie N., and Biermann F. (eds.), *Governing through Goals: Sustainable Development Goals as Governance Innovation*, MIT Press, Cambridge, MA, 2017.

Lin L. W., 'Legal Transplants Through Private Contracting: Codes of Vendor Conduct in Global Supply Chains as an Example' (2009) 57 *The American Journal of Comparative Law*, 711, 743.

Lipson J. C., 'Promising Justice: Contract (as) Social Responsibility' (3 March 2019). Temple University Legal Studies Research Paper No. 2019–21 <https://ssrn.com/abstract=3388312>.

Lomfeld B., and Wielsch D., 'Forward: The Public Dimension of Contract: Contractual Pluralism Beyond Privity' (2013a) 76(2) *Law and Contemporary Problems*, i–xii.

Lomfeld B., and Wielsch D., (eds.), 'The Public Dimension of Contract Law' (2013b) 2 *Law and Contemporary Problems*, I (Introduction to the special issue).

Macneil I. R., 'Reflections on Relational Contract' (1985) 141(4) *Zeitschrift Für Die Gesamte Staatswissenschaft/Journal of Institutional and Theoretical Economics*, 541–546.

McAdams R. H., *The Expressive Powers of Law. Theories and Limits*, Harvard University Press, Cambridge, MA, 2015.

McBarnet D., and Kurkchiyan M., 'Corporate Social Responsibility Through Contractual Control? Global Supply Chains and "Other-Regulation"', in D. McBarnet, A. Voiculescu, T. Campbell (eds.), *The New Corporate Accountability Corporate Social Responsibility and the Law*, Oxford University Press, 2009.

McCall-Smith K., and Rühmkorf A., 'From International Law to National Law: The Opportunities and Limits of Contractual CSR Supply Governance', in V. Ulfbeck, A. Andhov and K. Mitkidis (eds.), *Law and Responsible Supply Chain. Contract and Tort Interplay and Overlap*, Routledge, Abingdon, 2019.

Michaels R., Ruiz Abou-Nigm V., and van Loon H., *The Private Side of Transforming the World—UN Sustainable Development Goals 2030 and the Role of Private International Law*, Intersentia, 2021.

Mitkidis K. P., *Sustainability Clauses in International Business Contracts*, Eleven International Publishing, 2015.

Mitkidis K. P., 'Sustainability Clauses in International Supply Chain Contracts: Regulation, Enforceability and Effects of Ethical Requirements' (2014) 1 *Nordic Journal of Commercial Law*, 1–31.

Mola L., and Poncibò C., 'Trade and UN SDGs 2030: The Interplay between Public International Law and Contract Law', in A. Bahri, W. Zhou and D. Boklan (eds.), *Rethinking, Repackaging, and Rescuing World Trade Law in the Post-Pandemic Era*, Hart, 2021.

Monteillet V., *La Contractualisation du droit de l'environnement*, Dalloz, Paris, 2017, p. 11, n. 61.

Peters A., Köchlin L., Förster T., and Fenner Zinkernagel G., (eds.), *Non-State Actors as Standard Setters*, Cambridge University Press, Cambridge, 2009.

Pirelli Group, 'Sustainability Contractual Clause' <www.pirelli.com/asset/index.php?idelement=59887>.

Poncibò C., 'The Contract Governance Of Sustainable Development' (2022) 1 *Isaidat Law Review*, 35–57.
Poncibò C., 'The Contractualisation of Environmental Sustainability' (2016) 12 (4) *ERCL*, 335–355.
Rühmkorf A., *Corporate Social Responsibility, Private Law and Global Supply Chain*, Edward Elgar, Cheltenham, 2015.
Salas-Zapata W. A., and Ortiz-Muñoz S. M., 'Analysis of Meanings of the Concept of Sustainability' (2019) 1(27) *Sustainable Development*, 153–161.
Salminen J., 'The Accord on Fire and Building Safety in Bangladesh: A New Paradigm for Limiting Buyers' Liability in Global Supply Chains?' (2018) 66(2) *The American Journal of Comparative Law*, 411–451.
Salminen J., 'Sustainability and the Move from Corporate Governance to Governance Through Contract', in B. Sjåfjell and C. M. Bruner (eds.), *Cambridge Handbook of Corporate Law, Corporate Governance and Sustainability*, Cambridge University Press, Cambridge, 2019 and University of Oslo Faculty of Law Research Paper No. 2019-10 <https://ssrn.com/abstract=3383397>
Segger M. C., Cordonnier S., Saito Y., and Weeramantry C. G. (eds.), *Sustainable Development Principles in the Decisions of International Courts and Tribunals: 1992–2012*, Routledge, Abingdon, 2017.
Sinnig J., Zetzsche D. A., 'The EU's Corporate Sustainability Due Diligence Directive: From Disclosure to Prevention of Adverse Sustainability Impacts in Supply Chains' (14, 2024). <https://ssrn.com/abstract=4865488> or <http://dx.doi.org/10.2139/ssrn.4865488>.
Sjåfjell B., 'Regulating for Corporate Sustainability: Why the Public-Private Divide Misses the Point', University of Oslo Faculty of Law Research Paper No. 2016-03 <https://ssrn.com/abstract=2756410>
Sjåfjell B., 'Time to Get Real: A General Corporate Law Duty to Act Sustainably' (2022). Nordic & European Company Law Working Paper No. 2022-06.
Sobel-Read K., Anderson G., and Salminen J., 'Recalibrating Contract Law: Choses in Action, Global Value Chains and the Enforcement of Obligations Outside of Privity' (2018) 93 *Tulane Law Review*, 1–37.
Stella Richter M., and Passador M. L., 'Corporate Sustainability Due Diligence: Supernatural Superserious' (3 December 2022). Bocconi Legal Studies Research Paper No. 4293912 <https://doi.org/10.2139/ssrn.4293912>
Sullivan Z., 'The Real Reason the Amazon Is on Fire', *Time* (26 August 2019) <https://time.com/5661162/why-the-amazon-is-on-fire>.
Sunstein C., 'On the Expressive Function of Law' (1996) 144 *University of Pennsylvania Law Review*, 2021, 2046.
Trubeck D., Thomas C., and Santos A., 'World Trade and Investment Law in A Time of Crisis: Distribution, Development and Social Protection', Legal Studies Research Paper Series No.1470/2019
Trubeck D., Thomas C., and Santos A., *World Trade and Investment Law Reimagined: A Progressive Agenda for an Inclusive Globalization*, Anthem Press, 2019.
Ulfbeck V., and Hansen O., 'Interplay Between Contract and Tort in the Supply Chain', in V. Ulfbeck, A. Andhov and K. Mitkidis (eds.), *Law and Responsible Supply Chain. Contract and Tort Interplay and Overlap*, Routledge, Abingdon, 2019, pp. 133–145.
Verbruggen P., 'Regulatory Governance by Contract: The Rise of Regulatory Standards in Commercial Contracts' (2014) 35(3) *Recht der Werkelijkheid*, 79–100.
Vogt M., and Weber C., 'Current Challenges to the Concept of Sustainability' (2019) 2(4) *Global Sustainability*, 1–6.
Vytopil A. L., *Contractual Control in the Supply Chain on Corporate Social Responsibility, Codes of Conduct, Contracts and (Avoiding) Liability*, Eleven, The Netherlands, 2015.
Zumbansen P., 'The Law of Society: Governance Through Contract' (2007) 14(2) *Indiana Journal of Global Legal Studies*, 191–233.
Zumbansen P., 'Transnational Private Regulatory Governance: Ambiguities of Public Authority and Private Power' (2013) 76 *Law and Contemporary Problems*, 117–138.

24

PROCESSES AND CO-DESIGN METHODOLOGIES FOR THE TERRITORIALIZATION OF THE SDGS FOR THE COMMUNITIES IN TRANSITION

An Experimentation in the Metropolitan City of Reggio Calabria

Giuseppe Mangano

24.1 Introduction

The United Nations 2030 Agenda (2015) is the result of a long process characterized by key events in the international debate on sustainable development and based on the principle of integrating and implementing policies in the "three dimensions of sustainability" (Brundtland, 1987), to which a fourth one is added, which concerns the institutional capacity to pursue the principles of "no one left behind" in the plans and programs of member countries (ethical dimension). In fact, Agenda 2030 represents a global challenge which involves all components of civil society, businesses, and public bodies, adopting an integrated approach and concrete measures to trigger a major paradigm shift in development processes, capable of addressing the current environmental, economic, and social challenges. In order to trigger sustainable transformations and mitigate possible inconsistencies between Agenda 2030 and the sustainability policies of individual countries, the United Nations has encouraged national governments to build their own National Sustainable Development Strategies capable of initiating a process of "territorialisation" of the SDGs at the subnational and local levels. To reinforce the importance of this process, it is useful to recall the figure from the latest report of the Sustainable Development Solution Network (SDSN, 2021), according to which 105 of the 169 targets of the 17 SDGs (62%) will not be achieved without the involvement of all institutional levels (regions, provinces, metropolitan cities). With CIPE Deliberation 108/2017, Italy approved the National Strategy for Sustainable Development (NSSD) as a tool for coordinating the implementation of the 2030 Agenda, which incorporates the four principles of the 2030 Agenda (integration, universality, inclusion, transformation) to constitute the national reference framework for environmental and territorial planning, programming, and evaluation processes in the implementation of the SDGs at the local level. In its final version (MiTe, 2022), the NSSD is

DOI: 10.4324/9781003285472-26

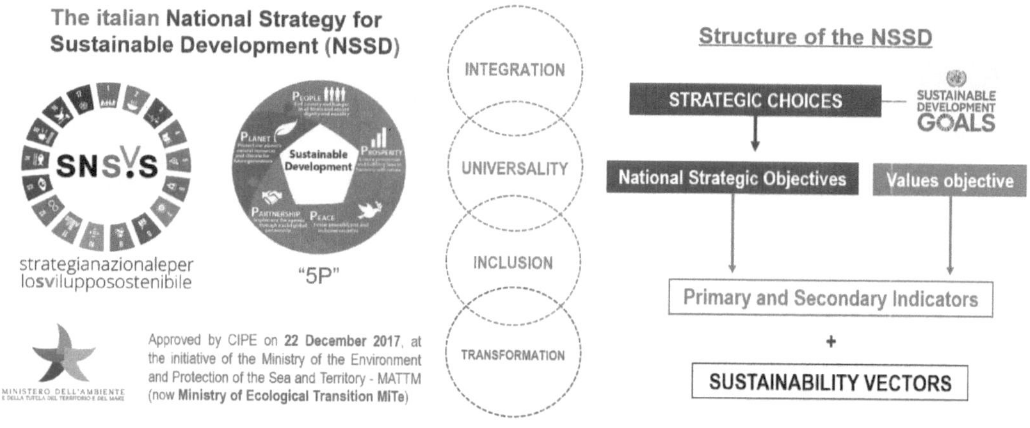

Figure 24.1 National Strategy for Sustainable Development (NSSD).
Source: Elaboration by G. Mangano.

articulated into Strategic Choices and National Strategic Objectives (NSSDs and NSOs) within six areas (people, planet, peace, prosperity, partnerships, and vectors of sustainability; Figure 24.1).

The NSSD has as its main aims the coherence of public policies on sustainable development, the territorialization of the SDGs and the local implementation of the 2030 Agenda, the multi-level involvement of stakeholders, and the information and education of civil society on sustainable development issues. The latter refers to the multi-actor, multi-sectoral, and multidisciplinary dimension as necessary conditions for the implementation of place-based sustainable development policies. In this sense, co-design and participation are the enabling factors (and precondition) for the implementation of the National Sustainable Development Strategy in territories. If, on the one hand, global initiatives "within the framework of Sustainable Development policies, change the paradigms of the transition towards the economic and social, as well as environmental, scenarios of the Fourth Industrial Revolution," on the other hand, it appears necessary to adopt an integrated approach to generate programs and projects for the "dissemination of urban and territorial issues, experimenting with good practices . . . , pursuing community involvement processes and committing public investments" (Nava, 2019). In this sense, the "new technical capabilities" and enabling technologies (global internet connectivity, peer-to-peer networks, 3D printing and digital manufacturing, IoT, big data and AI, abundant local renewable energy), as tools able to assist communities in responding to the "new social agendas" (no planet B, global and social equity, no one left behind, g-local collaborative communities, Agenda 2030, Next Generation EU, just transition mechanism), define new production and consumption models that, on the one hand, generate new business models and, on the other, make possible the transformation of knowledge into skills, through new design approaches (collaborative, circular, open, frugal, authentic, holistic, product–service thinking, designer as entrepreneur). This contribution illustrates the results of pilot actions and innovative experiences in the project for the implementation of the Metropolitan Agenda for Sustainable Development for the Metropolitan City of Reggio Calabria, "RC Metro Citizens in Transition" (MiTe, Sector 5 CM RC, ABITAlab).[1] In this project, the application of "quality education–oriented" co-design methodologies (SDG 4) supported by tools capable of making knowledge more accessible at several levels (of technologies and reference targets, from the academic world to citizens' organisations, etc.) implemented a pathway for increasing the knowledge and skills of the community involved. The project focus is on the issues of ecological and digital transition and enabled capacities for action with a view to sustainability and innovation (Mangano, 2021) and the integration of resources and skills of experts and citizens for the co-creation of knowledge and new project solutions (European Commission, 2006).

24.2 The Co-Design Methodology

The definition of *co-design* originated in Scandinavian countries in the 1960s as "co-operative design" in trade union organizations but evolved in the United States as "participatory design" and in the 1980s as "human-centered design" (Norman, 1990), indicating those approaches that actively involve multiple actors in design processes, from the idea definition phase to the generative phase, with a focus on the end user. Co-designing and co-programming activities are also defined in the ISO 13047:1999 standard ("human-centered design processes for interactive systems") and are part of the broader framework of "design thinking," an approach that invests several sectors, from the generation of digital services to the realization and design of components, from architecture to urban planning, from policymaking to social innovation. It is, therefore, a matter of evolving the concept of "participatory design" into "collaborative design" (co-design), according to Wates (Bocco, Cavaglià, 2008),

> referring to that act capable of involving people in decision-making processes that have an impact on their existence and on the surrounding environment and referring to innovative methodologies, in which it is not only the users who are involved in the fruition of a product or service, but the places themselves that are designed.

In this sense, co-design is a multi-level approach whose different phases in the individual levels must ensure a coherence of process and results achieved, in which the integration of resources and expert and non-expert competences (civic subjects) contributes to the co-creation of knowledge and new design solutions and in which, thanks to the ability to build competent and inclusive networks (community engagement), creative and multidisciplinary research approaches are achieved. It is also a "design-driven" approach, which involves the definition and application of a distributed model of action and "adopts participatory design tools within collaborative project forms and citizen engagement experiences" (De Salvo et al., 2021). It is important to emphasize that scientific research can accompany and assist policymakers and communities "in transition" with a "multidisciplinary" approach to issues, relying on "collective intelligence, since global catastrophes, such as climate change hazards, are of interest to entire communities and individuals simultaneously" (Nava, Melis, 2021), with the ultimate goal of generating "a net positive impact" on them, consistent with the aforementioned sustainability policies and social agendas (Sustainable Development Goals, Next Generation EU, etc.). In the "RC Metro Citizens in Transition" pilot project, it was a matter of experimenting with a co-design approach based on diffuse networks made up of users, citizens, and community organizations (Civic Ecosystem 2030) which co-create and collaborate in the design, realization, and delivery of new, innovative products and services referable to "knowledge city" tactics, where economies based on the exchange of material and immaterial values, the dissemination of knowledge, and the transfer of knowledge are triggered, which become the enabling elements for the construction of "resilient" and sustainable scenarios in real and everyday life contexts of "communities in transition" (Mangano, 2021). The construction of processes for knowledge and capacity building can be entrusted to co-design methodologies as a paradigmatic, generative, and operational approach that goes beyond the canons of participation and engagement, as an envisioning tool that aims at transparent forms of collaboration guided by designers, aimed at the realization of projects and scenarios that have the quality of life and collective well-being at their center (human-centered design). While participation and involvement are about capturing people's thoughts, needs, and requirements, co-design involves involvement in all design phases and in the delivery of products and services, with "multiple designers working together" (Sanders, 2008). The phases and operational tools of co-design processes constitute the most suitable methodological approach to the conception, implementation, prototyping of actions, and the achievement of the expected results of a project, according to a hybrid model of engagement of communities of young designers, citizens, territorial organizations, educational institutions, and public and private entities (Figure 24.2).

As C. Nava (2019) states, in this approach, knowledge becomes a common good (and objective) when it takes on "an important value in the transfer of its strategies and operations from small to large communities" (metaphor of the collective organism by E. Ostrom), and this certainly changes the relationship between the designers and the communities of reference themselves, so the project is no longer the product but a "complex system made up of tangible and intangible products, composed together of flows, physical and virtual interactions, which convey experiences, intimate and private, as well as social and relational" (Manzini, 2009).

In this sense, co-design, by opening up to collaboration of users in the creative stages of idea generation and application, enables people to be not just users but active and, above all, competent participants (Sander, Stopper, 2008). *Co-design* is a collaborative creative act aimed at the generation of open innovation ideas, processes, and activities: It is an "envisioning" tool that aims at transparent forms of collaboration led by "designers," aimed at the realization of projects and scenarios that have quality of life and collective well-being at their center (human-centered design). This working methodology takes the form of a pragmatic approach and a thinking and operational model for designing real, everyday contexts of people's lives. The phases and operational tools of co-design processes constitute the most suitable methodological approach for the conception, implementation, and prototyping of actions and the achievement of the expected results of

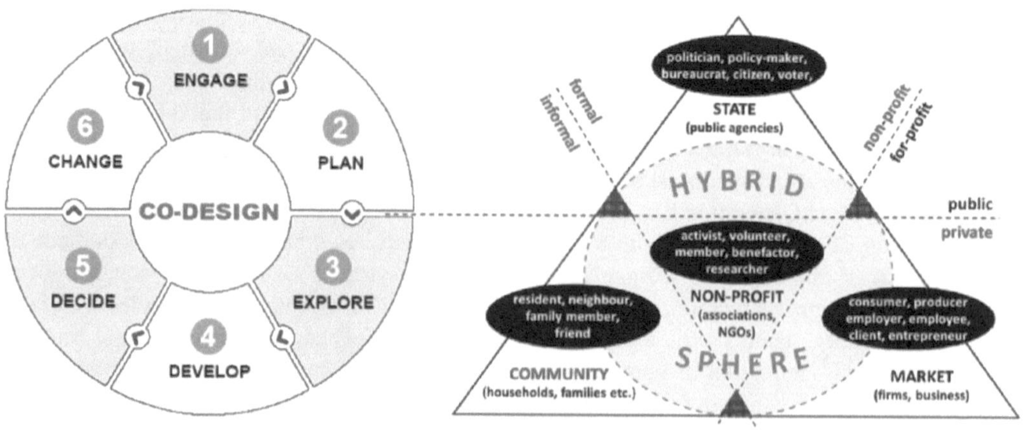

Figure 24.2 Framework explaining co-design methodology.

Source: Elaboration of G. Mangano on Avellino and Wittinger's adopted figure (2016).

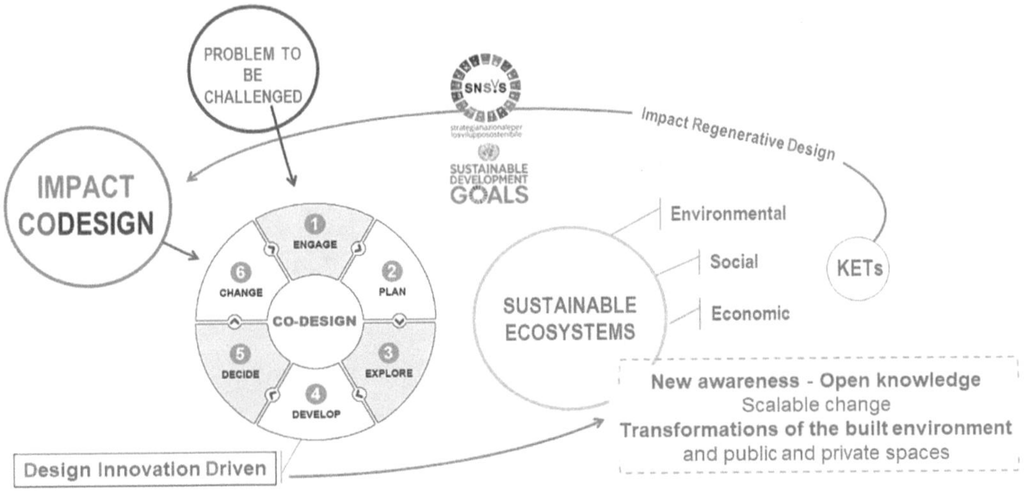

Figure 24.3 Graphic process of co-design methodology in sustainable project.
Source: Elaboration by G. Mangano.

a project with a high rate of engagement of communities of young designers, citizens, territorial organizations, educational institutions, and public and private entities (Mangano, 2021).

24.3 Co-Design in the Pilot Project RC Metro Citizens in Transition

"RC Metro Citizens in Transition,"[1] the pilot project for the Metropolitan Agenda for Sustainable Development (AMSvS) of the Metropolitan City of Reggio Calabria (Agreement ex MATTM DG-SVI and Sector 5, Professional Training, Metropolitan City of Reggio Calabria of 2019; Figure 24.4), has in its mission the will to transfer, through co-design, training, and knowledge paths, experiences and skills on the topics related to the Sustainable Development Goals of the UN 2030 Agenda and the areas ("5Ps"), the Strategic Choices and Objectives of the National Sustainable Development Strategy and the vectors of sustainability in their latest revision (MiTe, 2022).[2]

The project, which lasted 28 months (February 2019–May 2022), involved the ABITAlab laboratory of the Department of Architecture and Territory of the Mediterranean University of Reggio Calabria in the technical-scientific assistance to the National Table for Sustainable Development of the Ministry of Ecological Transition and in the planning and implementation activities of the pilot actions of co-design and knowledge transfer with the so-called "Civic Ecosystem 2030," the "in transition" community targeted by the same proposal, consisting of associations, civic organizations, individual citizens, high school and university students, and representatives and technical officials of public bodies.

The context in which the project took place is characterized by migration trends and indicators on professional employment with a negative balance. In fact, the Metropolitan City of Reggio Calabria, composed of 97 municipalities and with a total population of 526,586 inhabitants, has lost approximately 35,000 inhabitants in the last ten years. In Calabria, according to the latest ISTAT data, the unemployment rate registers four out of ten young people not in employment, with an increase in NEETs and a strong trend of school dropouts. A trend that leads about 35% of the population between 25 and 34 years of age to migrate to northern Italian regions or to Europe (Figure 24.5).

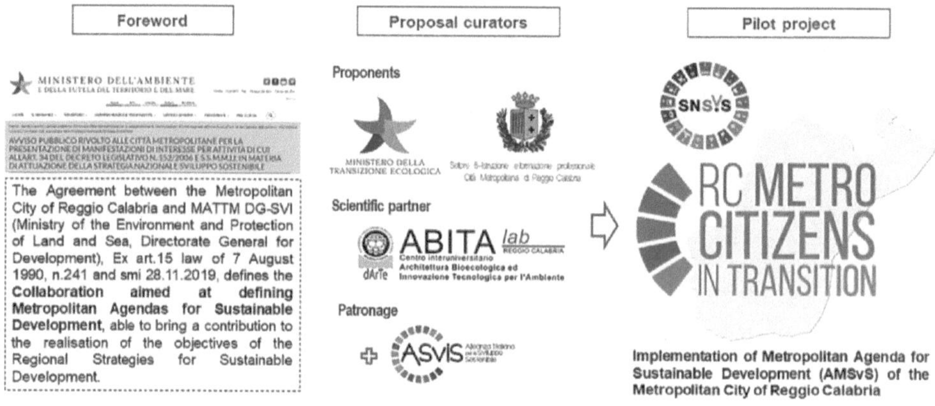

Figure 24.4 Graphic framework explaining the project structure and stakeholders.

Source: Elaboration by G. Mangano.

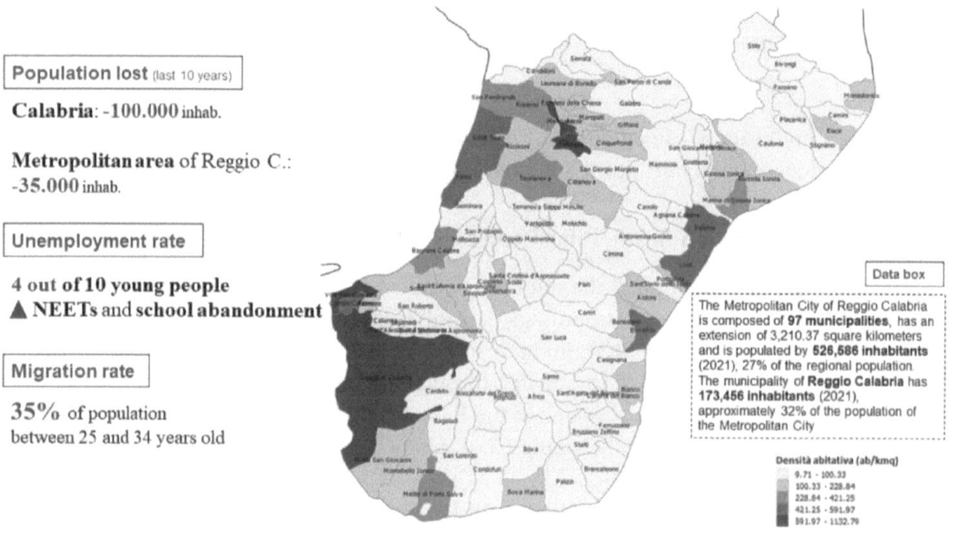

Figure 24.5 Map of inhabitants density in the province of Reggio Calabria as published in the Metropolitan Cities Dossier (DARA. Dipartimento per gli A[ari Regionali e le Autonomie, Presidenza del Consiglio dei Ministri and ISTAT, 2017).

Source: Data elaboration by G. Mangano.

Map source: Italy Metropolitan Cities Dossier, ANCE, 2014.

In this era of "global and local" transition, it is a question of wanting to generate a positive impact on the competitiveness rate of metropolitan territories and communities, acting with a view to overcoming the gaps in the minimum skills of young people, especially those who come from socio-economically disadvantaged families and live in the regions of Southern Italy, and to enhance the training of transversal and digital skills in the curricula through innovative and inclusive teaching (cf. data from the ISTAT 2021 SDGs report on goal 4). As a matter of fact, this approach is nourished by innovative actions and knowledge-based pilot experiences for education, information, and collaborative approaches with communities, which can trigger radical changes in people's behaviour and in their ability to act with a view to sustainability (Mangano, G., 2021). With reference to the construction of the contents of the Metropolitan Agenda, a multi-level and multidimensional approach was adopted (also with reference to the vectors of sustainability of the National Strategy for Sustainable Development), building experimental thematic projects that initiate actions on priorities of intervention on the 5Ps (people, planet, prosperity, peace, partnership).

The pilot actions of the project were implemented by acting on the three categories of intervention defined by the Ministry's Notice on Ecological Transition. For these reasons, through the project, three levels of co-design can be defined: (1) inter-institutional, multi-level, and multi-actor co-design; (2) co-design for actions and technologies for knowledge transfer on the issues of the National Sustainable Development Strategy and Agenda 2030; and (3) co-design for the construction of the Metropolitan Agenda for Sustainable Development to implement the visions and project ideas of the Metropolitan Strategic Plan (Mangano, 2022). In order to effectively achieve these objectives, the pilot project entrusts the Steering Committee for the Metropolitan Agenda for Sustainable Development of Reggio Calabria to Sector 5, "Professional Training," for the promotion and transfer of training and information on the issues, with the possibility of involving other institutions (e.g., district schools) and the metropolitan communities of local authorities throughout the territory. Through these actions, it was decided to act for the territorialization of the targets of SDG 4 (Quality Education), with an indirect impact also on goal 8 (Decent Work and Economic Growth) of Agenda 2030, on the choices and strategic objectives of the National Strategy for Sustainable Development, referring to the areas of people, prosperity, and peace and to the sustainability vectors 2 ("Education and Training") and 3 ("Participation for Sustainable Development") (Figure 24.6).

Figure 24.6 Figure explaining the multi-actor structure of the project.

Source: Elaboration by ABITAlab for the voluntary local review (VLR) of the Metropolitan City of Reggio Calabria (2022).

24.4 Co-Design and Knowledge Transfer Activities on Key Enabling Technologies and Sustainable Development: The Results

With the pilot actions 3 ("Making Metrocity"), 4 ("Knowledge Metrocity"), and 6 ("Co-Design Metrocity," the project started up with Ecosystem 2030 (*civic + educational*) co-design and co-management paths for the transfer of knowledge, training, and skills enhancement on enabling technologies (additive manufacturing and digital fabrication) and digital technologies (open platform and digital storytelling), as processes and tools aimed at the implementation of Agenda 2030 and the territorialization of the National Strategy for SDS. In particular, within the framework of the aforementioned actions are the following activities.

Cycle of Technical Seminars, "Knowledge Metrocity" (November 2020–June 2021), with the aim of transferring contents, methodologies, and experiences useful for the construction of the Metropolitan Agenda, with reference to the SNSvS areas "planet" and "prosperity," with invited experts and lecturers and technical targets of interested bodies, associations, university students, and individuals. The topics of the seminars covered issues associated with SDGs 8, 9, 11, 12, 13, and 15, with particular reference to climate change and the ecological and digital transition (Figure 24.7). Specifically, the issues addressed in these transfer paths concerned the geographical and environmental dimensions of the metropolitan territory in transition; acting for resilience: air and climate as a device for urban quality; the Neo-Anthropocene: designing augmented cities and communities; the circular metabolism of flows and resources: energy and waste, with the case of Peccioli (Belvedere S.p.A.) and civic actions; the 2030 Agenda; and the Strategic Plan of the Metropolitan City of Reggio Calabria. The alliance of sustainable visions for the territory and communities in transition.

Figure 24.7 Pictures from diverse RCMCiT project knowledge transfer activities.

Source: G. Mangano's elaboration.

In the school years 2020–2021 and 2021–2022, together with the UniRC Department of Architecture and Territory, two editions of Pathways for Transversal Skills and Orientation[2] (PCTO), "The Challenge of Agenda 2030," were carried out.[3] It targeted the students of high school institutes in the province of Reggio Calabria for the transfer of reference frameworks, experiences, and case studies. The themes included the challenge to climate change for resilient territories and communities, the conscientious use of material resources in the construction sector by activating recycling chains and circular economy processes, the design of integrated "green" systems for the sustainability of buildings, and smart devices and networks that consume and produce electricity from renewable sources only, thus contributing to the European objectives of "climate neutrality." Furthermore, experience and skills were transferred on the topics of integrated design and prototyping, with digital manufacturing applied to the architecture and construction sector. At the end of the knowledge transfer activities (lectures and seminars), the "Agenda 2030" Open School took place, workshops in which the students practiced experimental projects on the topics of digital fabrication (modelling and 3D printing) and circular design (projects using environmentally friendly materials and from recycled supply chains),[4] which were finally presented at an open public event (Figure 24.8).

At university level, in the teaching and training courses provided by ABITAlab at the "Sustainable Innovation Design" and Thesis Atelier,[3] in the 2021–2022 academic year, initiatives were held to directly transfer the topics of sustainable development and the ecological and digital

Figure 24.8 Pictures from knowledge transfer and capacity building, two of three activities with schools.

Source: Courtesy of G. Mangano, P. Bova ph.

transition, involving university students, local companies, lecturers, and experts at national level (Figure 24.9).

At the citizenship level, in May 2022, the workshop "Co-Design Metro City for the Metropolitan Agenda for Sustainable Development—Metropolitan City of Reggio Calabria" (Figure 24.10) aimed at the realization of a participation and co-design initiative on the strategic areas of the National Strategy for Sustainable Development, on the themes and knowledge transferred with

Figure 24.9 Pictures from knowledge transfer and capacity building activities with university students.
Source: Courtesy of G. Mangano, P. Bova ph.

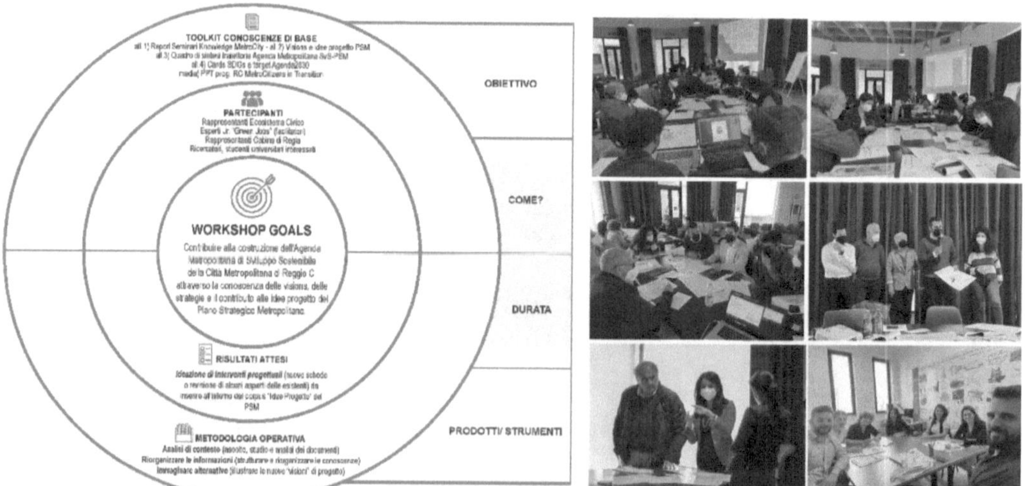

Figure 24.10 Co-design pictures and tool scheme.
Source: Courtesy of G. Mangano, P. Bova ph.

the pilot actions of the "RC Metro Citizens in Transition" project, and on the trajectories, visions, and project ideas contained in the Metropolitan Strategic Plan to be drafted, with the involvement of the civic ecosystem (associations, citizens, technicians of the authorities, etc.), the Authority's "green jobs" experts, the members of the Scientific Committee of the pilot project, and university students. The day's activity was carried out according to the methodology of co-design workshops, in groups divided into focus tables whose activities were structured and guided by one or more identified facilitators.[4]

The monitoring of actions and processes with co-design methodologies applied makes use of an impact assessment system based on the so-called "theory of change" or "process method" (Clark et al., 2004), which evaluates the change generated (outcome) in relation to the results (output) of the project actions (activities), for which material (financial investments) and immaterial (knowledge and skills) resources (input) have been employed. Activities and outputs represent the "performative" aspects of co-design processes, while outcomes, together with the changes that occurred independently of our actions (deadweight), represent the domain of the "impacts" generated. In addition, the monitoring of actions in co-design can also make use of the definition of logical-analytical tools, which aim to assess the project according to the criteria of "relevance," that is, how the objectives are adherent to real problems and needs of the territory, organizations, and/or reference communities; "internal coherence," that is, based on the analysis of the expected results with respect to the specific objectives defined; and "sustainability," that is, how the change generated can be considered lasting over time (Bussi, Russo, 2017). In this contribution, the co-design activities for knowledge transfer and skills enhancement in the pilot project "RC Metro Citizens in Transition" are evaluated in terms of effectiveness and efficiency in relation to the products,

Table 24.1 Impacts for the seminar cycle "Knowledge Metrocity"

No. of high schools (total)	*Time (total h)*	*No. of activities (total)*	*Participants (total)*	*No. of projects*
4 "green schools"	60 h lectures + 40 h workshop (first and second edition)	7 webinars + 1 Open School workshop (first edition) 11 seminars + 1 Open School workshop (second edition)	182 students	28 projects on the issues of SDGs 9, 11, 12, and 13; digital design; and circular materials

Table 24.2 Impacts for the transversal skills and orientation courses

No. of students (total)	*Time (total h)*	*No. of activities (total)*	*Researchers involved (total)*	*No. of projects*
102	155 (105 of workshop)	5 workshops 3 seminar cycles	6 + 1 startup	48 projects (40 about resilient urban design, LEED buildings, SDGs, and assessment; 8 of advanced design and 3D printing)

Table 24.3 Impacts for the seminars and workshops with university courses

No. of students (total)	*Time (total h)*	*No. of activities (total)*	*Researchers involved (total)*	*No. of projects*
140	210 (135 of workshop)	7 workshops 10 seminar cycles	6 + 1 startup	48 projects (40 about resilient urban design, LEED buildings, SDGs, and assessment; 8 of advanced design and 3D printing)

outputs, and community involvement capacity, which are useful factors for the achievement of the initial objectives and the demand expressed in the initial phase (European Commission, 2006). The following synoptic tables illustrate the "performative" levels of co-design activities illustrated in Section 4 and take into account indicators of an intangible type (e.g., time spent) and of a tangible type (human resources employed, community involved, projects and products realized).

24.5 Discussion

The co-design method was used among all technical and institutional levels of governance promotion, promoted among the different operators and stakeholders, as well as users of Ecosystem 2030, and not least for the level activated with new professional experiences aimed at green jobs, directly referred to the young experts involved in the Agenda 2030 process, within the metropolitan city and the university. The levels of assessment are pursued with the monitoring activities of effectiveness and efficiency of the pilot project (traceability of actions, activities, and results), with the technical-scientific activities on SNSvS for contribution to the work on the formulation of the SDG indicators and, finally, with the conditions accrued in the process of implementation, evaluation, and traceability for coherence policies, on territorial investments in the current experience (launched in April 2022), with the work conducted for the Metropolitan Strategic Plan. This multi-level strategy finalizes the tool of sustainability vectors (according to their latest structuring) and candidates them to become a strategic and open system of scenario assessment, in which it is possible to monitor and measure the level of implementation and impact of interventions of territorial investment projects for the competitive period 2021–2027. The multi-level co-design and transfer activities, within the framework of the National Sustainable Development Table, made it possible to contribute to the revision and final draft of the National Sustainable Development Strategy (MiTe, 2022). Through the "RC Metro Citizens in Transition" pilot actions and the results achieved, it was possible to "act" on the three sustainability vectors which represent the "enabling" conditions for the implementation of Agenda 2030 and the NSSD at the local level and the tool through which to initiate the monitoring and coherence of sustainable development policies, in order to guide public investments at regional and metropolitan level, also interfacing the goals of Agenda 2030, the National Strategic Choices and Objectives, with the missions and interventions of the National Recovery and Resilience Plan (NRP).

24.6 Conclusions

The project experimentation presented in this contribution allowed the research team to demonstrate how knowledge transfer–oriented actions with a high rate of involvement of the target community, based on "learning by doing" co-design methodologies (transfer + laboratory), turn out to be capable of producing a tangible positive impact, enabling all the actors involved to increase their competences, in the case of the "Knowledge Metrocity" and "Co-Design Metrocity" of the "RC Metro Citizens in Transition" pilot project, on the topics of ecological and digital transition, enabling technologies and sustainable development. These considerations, corroborated by the success in terms of the project's output, are also nourished by the ex post monitoring activity that the research team carried out through exercises and project experiments with the school and university students involved. In addition, the ability to produce positive impacts within the inter-institutional co-design activities (director cabin of the project with other departments of the Metropolitan City of Reggio Calabria) has allowed the activation of a further research and transfer project on the topics of the assessment of visions and project ideas of the Metropolitan Strategic Plan, in relation to the addresses of the Agenda 2030 and the implementation of the National Strategy for Sustainable Development (ABITAlab and Sector 10, "Planning and Special Laws," Metropolitan City of Reggio Calabria). In conclusion, a further opportunity to validate the results achieved is the selection of the project by the Ministry of Ecological Transition among the pilot experiences and good practices of territorialization of the SDGs to be presented at the High-Level Political Forum in New York (July 5–15, 2022), within the chapter dedicated to the implementation of the Agenda 2030 at territorial level in the voluntary national review of Italy.

Notes

1 "RC Metro Citizens in Transition. Project for the Metropolitan Agenda for Sustainable Development of the Metropolitan City of Reggio Calabria." Project financed within the framework of the public notice addressed to metropolitan cities for the submission of expressions of interest for activities under Article 34 of Legislative Decree No. 152/2006 and S.S.M.M.I.I. regarding the implementation of the National Sustainable Development Strategy: "Collaboration aimed at defining Metropolitan Agendas for Sustainable Development, capable of bringing a contribution to the realization of the objectives of the Regional Strategies for Sustainable Development" (green/digital strategic area and quality training). Partners: MiTe (Ministry of Ecological Transition) Sustainable Development Directorate, Metropolitan City of Reggio Calabria, with the head of Sector 5, Professional Training (responsible for the proposal), Dr. MariaLea Quattrone (technical responsible for the proposal), with scientific partner ABITAlab dArTe Prof.ssa C. Nava (scientific responsible) and Arch.RTdA G. Mangano (technical responsible). Website: www.rcmetrocitizensintransition.com

2 The "professional training" sector of the Metropolitan City of Reggio Calabria, in adherence to its institutional governance scope, and in the direction of policies for strengthening and inclusion at work, as outlined in the proposed agreement, has initiated the public selection of four junior "green jobs" experts, to be hired for collaborative activities in the project itself. Website: www.rcmetrocitizensintransition.com/knowledge-codesign-greenjobs.

3 PCTO (Pathways of Transversal Competence and Orientation), promoted by the ABITAlab center of the Department of Architecture and Territory (dArTe) of the Mediterranea University of Reggio Calabria, in agreement with schools in the Reggio Calabria metropolitan area. Website: www.rcmetrocitizensintransition.com/pcto-agenda2030.

4 Thesis workshop "Architecture of Transition on Climate Change Scenarios: Advanced design and enabling technologies for hybrid buildings." Pathway aimed at the development of the thesis for students of the fifth year, CdL Architecture LM-4 c.u., Department of Architecture and Territory, dArTe, University "Mediterranea" of Reggio Calabria, a.y. 2021/2022. Responsible: Prof. Consuelo Nava. Collaborators: Arch. RtdA G. Mangano, Arch. Ing. RtdA S. Cascone, Arch. PhD student A. Leuzzo, Arch. PhD student. D. Lucanto, Arch. eng. PhD student Z. Javanmard. Website: www.abitalab.unirc.it/didattica-formazione.

References

Bocco, A., Cavaglià, G. (2008). *Cultura Tecnologica dell'Architettura. Pensieri e parole, prima dei disegni*. Carocci Editore, Roma.

Brundtland, G.H. (1987). *Our common future: Report of the world commission on environment and development*. Geneva, UN-Dokument A/42/427. www.un-documents.net/ocf-ov.htm

Bussi, F., Russo, M. (2017). *I principali criteri per la valutazione della qualità del progetto*. Formez PA, Roma.

Clark, C., Rosenzweig, W., Long, D., Olsen, S. (2004). *Double bottom line project report: Assessing social impact in double bottom line ventures*. Center for Responsible Business, University of California, Berkeley.

De Salvo, G., Rossi, M., Bianchi, F. (2021). *Innovative approaches to sustainable development*. Milano: Springer.

European Commission (2006). *Impact assessment (IA) reports*. Bruxelles (BEL).

ISTAT (2021). *Rapporto SDGs. Informazioni statistiche su Agenda 2030 per l'Italia*. Roma.

Mangano, G. (2021). Innovazione e co-design per il trasferimento e il rafforzamento di conoscenze sui cambiamenti climatici con le comunità in transizione nella periferia sud di Reggio Calabria. In: Mangano G. (a cura di), *Knowledge vs climate change. Co-design e tecnologie abilitanti per la conoscenza ed azioni in risposta ai cambiamenti climatici*. Quaderno Collana SID n.3, Aracne Editore, Roma. ISBN: 978-88-255-4040-6, pp. 15–30.

Mangano, G., Leuzzo, A. (2022). *Co-design e tecnologie abilitanti.15 esperienze progettuali per la ricerca di frontiera e le comunità in transizione*. Aracne Editore, Roma. ISBN: 979-12-218-0197-2, pp. 83–92.

Manzini, E. (2009). Prefazione. In Rizzo F. (ed.), *Strategie di Co-design. Teorie, metodi e strumenti per progettare con gli utente*. Franco Angeli, Milano.

Ministero della Transizione Ecologica (2022). *Documento di Strategia Nazionale di Sviluppo Sostenibile*. ver. 06/2022. Roma.

Nava, C. (2019). *Ipersostenibilità e Tecnologie Abilitanti*. Teoria, Metodo, Progetto, Aracne Editore, Roma.

Nava, C., Melis, A. (2021). Un paradigma radicale bioecologico per le tecnologie progettanti con approccio transdisciplinare, *Technè—Journal of Technology for Architecture and Environment*, 21, Napoli.

Norman, D. A. (1990). *The design of everyday things*. New York: Basic Books.

Pianta, M., Lucchese, M. (2020). Rethinking the european green deal: An industrial policy for a just transition in Europe, *Review of Radical Political Economics*, 52(4).

Sachs, J., Kroll, C., Lafortune, G., Fuller, G., Woelm, F. (2021). *The decade of action for the sustainable development goals: Sustainable development report*. Cambridge University Press, Cambridge.

Sanders, E. B.-N. (2008). *An evolving map of design practice and design research*. Design Research Quarterly, 1(2), 1–17.

Sanders, E., Stappers, P.J. (2008). Co-creation and the new landscapes of design, *CoDesign*, 4(1), 5–18. https://doi.org/10.1080/15710880701875068

Schütz, F., Heidingsfelder, M.L., Schraudner, M. (2019). Co-shaping the future in quadruple helix innovation systems: Uncovering public preferences toward participatory research and innovation, *She Ji: The Journal of Design, Economics and Innovation*, 5(2).

United Nations (2015). *Transforming our world: The 2030 agenda for development sustainable*.

25

MEDIATING THE SDGS

The Role of (Mass) Media and Communication in Realizing the SDGs

Denise Voci and Matthias Karmasin

25.1 Introduction

The world has changed dramatically in the last two decades. The political, economic, and environmental problems current societies face will escalate rapidly if no action or radical shift is taken toward sustainability. Even though extreme poverty declined between 1990 and 2015 from nearly 36% to 10%, more than 1 billion people are still living in extreme poverty, while eight men alone have as much wealth as the bottom 3.5 million of the world's poorest people combined. More than 750 million people are starving, while about 2 billion people are overweight or obese. Inequalities between and among countries are steadily increasing, and human societies are causing more damage to the environment and the planet than ever before (SDG Zone, 2022). The Sustainable Development Goals (SDGs) were instituted in 2012 at the United Nations Conference on Sustainable Development in Rio de Janeiro to address such issues and give a plan to tackle these challenges. They address the global challenges we are currently facing, including climate change, environmental degradation, inequality, poverty, peace, and justice, and are meant to serve as a blueprint for a better and more sustainable future for all people on Earth (UNDP, 2022).

However, despite best intentions, the ambitious SDGs are repeatedly criticized because they are, in a way, inconsistent and difficult to quantify, and monitor, which makes their implementation (far) too slow. Indeed, while not denying that progress occurred and is evident in many areas addressed by the SDGs, the overall engagement and action toward the goals is not happening at the pace and scale needed to meet the deadline of 2030. This was one of the factors that led to the so-called "Decade of Action": A decade of even more ambitious measures started in 2020 to accelerate sustainable solutions and achieve the SDGs by 2030. A decade to transform (and save) our world (UN, 2022a).

Unfortunately, the multiple and interconnected global crises humanity is facing right now—the COVID-19 pandemic, the climate crisis, the impact of the Ukrainian and other wars—have understandably shifted society's priorities and urgencies into the recent past, resulting in further slowdowns and jeopardizing the achievement of the SDGs by 2030. However, these interconnected global crises have made the need for the achievement of the SDGs to change course rapidly even more relevant and urgent. Now, as never before, a collective commitment and engagement toward

DOI: 10.4324/9781003285472-27

the SDGs is needed to relaunch the realization of the 2030 Agenda during the Decade of Action (UN, 2022a, 2022b).

While the necessity of robust communication efforts for the successful relaunch of the "Decade of Action" to attain the Sustainable Development Goals (SDGs) by 2030 is evident, the SDGs and their associated targets have largely overlooked the role of the media and communication field. Despite the fundamental role of communication in fostering a global understanding of challenges and the imperative for sustainable development, as well as its significance as a prerequisite for driving substantial societal changes, the SDGs predominantly treat communication as a mere tool, a means to an end. Although the United Nations (UN) has recognized the importance of communication in certain initiatives, exemplified by events like the 2016 workshop "Communicating the Sustainable Development Goals—for Everyone" (UN, 2016) and publications such as the 2019 ESDN policy brief "Recommendations for Communicating Sustainable Development and the SDGs" (ESDN, 2019), the media and communication field continues to be viewed primarily as instrumental rather than an integral aspect of the sustainable development agenda.

This perspective is further reinforced by the SDGs' language and formulation, where the sole explicit and direct reference to the media and communication sector is found in *SDG 16*, under Target 10. This target emphasizes the need to "ensure public access to information," with a proposed indicator advocating for the establishment of a constitutional guarantee for the right to information (SDSN, 2023).

However, even within this reference, the full recognition of the essential role of the media and communication sector is not comprehensive. The subsequent wording of the target, which states, "[E]nsure public access to information and protect fundamental freedoms," employs the conjunction "and." This implies that while both objectives are necessary to achieve *goal 16* (Peace, Justice, and Strong Institutions), they are not interdependent. This oversight fails to acknowledge the fundamental role of media's publishing function for the creation of the public sphere, on which society relies to uphold democracies. Indeed, a well-informed public is crucial for functioning democracies, as societal decisions rely on accurate information (Kuklinski et al., 2000). In recent history, the issue of fake news and infodemics has highlighted how consistent misinformation can lead to suboptimal decisions and hinder collective understanding and engagement (Lewandowsky et al., 2017). Accordingly, Target 10 should have been crafted to underscore the inseparable link between guaranteeing public access to information and safeguarding fundamental freedoms, by stating, "Ensure public access to *reliable* information *to* protect fundamental freedoms," explicitly recognizing the foundational and practical contributions of media and communication in attaining these goals.

Such a revised target would consequently function both as an appeal for a robust, independent Fourth Estate and for promoting media literacy, as well as the responsible production and consumption of media products. Notably, these crucial elements are not addressed by the associated SDGs, namely, *SDG 16* (Peace, Justice, and Strong Institutions), *SDG 4* (Quality Education), and *SDG 12* (Responsible Consumption and Production). Within this chapter, we aim to address these overlooked aspects by emphasizing how the incorporation of media and communication can significantly contribute to the achievement of the SDGs.

25.2 SDG 12: Responsible Consumption and Production of Media Goods

To achieve the target of "ensur[ing] public access to *reliable* information *to* protect fundamental freedoms" for the realization of a peaceful and inclusive society for sustainable development, in line with *SDG 16*, responsible production and—in addition—distribution of information are required. However, this should go beyond the obvious need to communicate relevant research results, to use them for knowledge transfer, or to raise awareness by putting sustainability on the social and political agenda (Karmasin et al., 2021). Instead, the production and distribution of information to safeguard fundamental freedoms should be conceptualized as a comprehensive, responsible communication process. Such a process is essential to initiate social change toward a sustainable society, necessitating a more responsible lifestyle that, in turn, fosters the creation of a free and just society for all. However, although the principle of sustainability serves as a normative guide in the transition to sustainable development, it is not a rigid set of rules dictating specific actions. Rather, it should be viewed as guidance, as a rule indicating the direction for reflection and for how to communicate to differentiate the countless opportunities for action based on their "preferability" (Weder et al., 2019; Grunwald, 2016). Consequently, the interpretation of what *sustainability* and *sustainable development* mean and imply will vary based on individual experiences (Stables, 2013), making it perpetually subject to communicative negotiation processes (Weder et al., 2021; Fischer, 2019) of de- and reconstruction of cultural patterns (Weder et al., 2021).

At a societal level, this process primarily unfolds through mass media, playing a crucial role in fostering an understanding of the need for change, developing change initiatives, and implementing transformations on a whole-of-society level. The way in which this need for change is communicated holds significance in shaping societal perceptions and influencing the actions that are undertaken and supported. Consequently, how sustainable development is communicated is as important as the content of that communication itself (Voci & Karmasin, 2021).

Conceived in this way, communication is thus to be understood as both a prerequisite and a facilitator of sustainable development itself. It is indispensable for establishing a shared understanding of societal values regarding sustainability, which should guide the trajectory of sustainable development (Newig et al., 2013). Accordingly, engaging in a responsible production of information, understood as responsible communication *of*, *about*, and *for* sustainability (ebd.),[1] implies connecting the principles and values of sustainability to diverse aspects and fields of society by presenting, discussing, and reflecting on forward-looking arguments, possibilities for actions, and (opposing) viewpoints on societal development from a social, economic, environmental, and cultural perspective (Servaes, 2013).

In practical terms, this means that information about sustainability-related issues, including their causes, consequences, and viable solutions, should be conveyed in a more concrete, timely, and relatable manner. This is essential because the general public faces challenges in comprehending and processing highly complex scientific data and facts in their entirety (Garcia-Retamero & Cokely, 2013; Sterman, 2011). Additionally, individuals find it difficult to grasp and engage with concepts that are beyond "human scale" (Kaschak & Maner, 2009; Dahlstrom, 2014), such as the abstract idea of "anthropogenic climate change," which they cannot directly experience (Dahlstrom & Ritland, 2012). The connection to these broader concepts should be established in a subsequent step. This approach enables individuals to better understand how they can contribute to a more sustainable lifestyle through practical, small, and specific measures aimed at addressing or improving their local, "real" current issues.

Responsible sustainability communication, aiming to foster a peaceful and inclusive society, must move beyond one-way, purely explanatory, and top-down approaches. Instead, it should communicate and anchor sustainable development as a bottom-up process, as emphasizing a dialogue-oriented and reciprocal approach is crucial to engaging and involving diverse stakeholders. This can be achieved through the use of frames, metaphors, and storytelling techniques, to foster a collective framing, creating a "we-feeling." This shared sense of identity ensures a stronger willingness to support (also negative) measures, as they have not been ordered "top-down" but jointly shared in a bottom-up process (Mitchell et al., 2016; Moser, 2010). This can be reinforced by further contextualizing the communication message, integrating it within the recipients' personal experience and, importantly, their personal "worry budget," by employing emotional appeals (Wehrli et al., 2014). Consequently, this approach achieves a dual effect: It increases the perceived behavioral control, empowering recipients as the action needed—in this case, a more sustainable lifestyle—appears to be achievable (Villarino & Font, 2015). At the same time, the personal connection ensures the facilitation of behavioral change through the activation of protective actions (Mabon, 2020; Terpstra et al., 2009; Floyd et al., 2000; Neuwirth et al., 2000).

Moreover, responsible sustainability communication should adhere to principles of truthfulness, openness, transparency, and honesty. This necessitates addressing all concerns, uncertainties, and uncomfortable aspects associated with the shift toward sustainable development within the current capitalism-based world system. It goes beyond merely portraying the positive aspects—a narrative often focused on economic and technological progress, innovation, green jobs, and renewable energy (Weder et al., 2021). The "rainy side," involving personal sacrifices and relinquishing Western world standards, which not everyone may be willing to embrace at this point, should be communicated openly, honestly, and transparently. This process should involve collaborative efforts, a dialogue-oriented approach, and a bottom-up perspective that are essential for a shared vision of a free, just, and sustainable future (Figure 25.1).

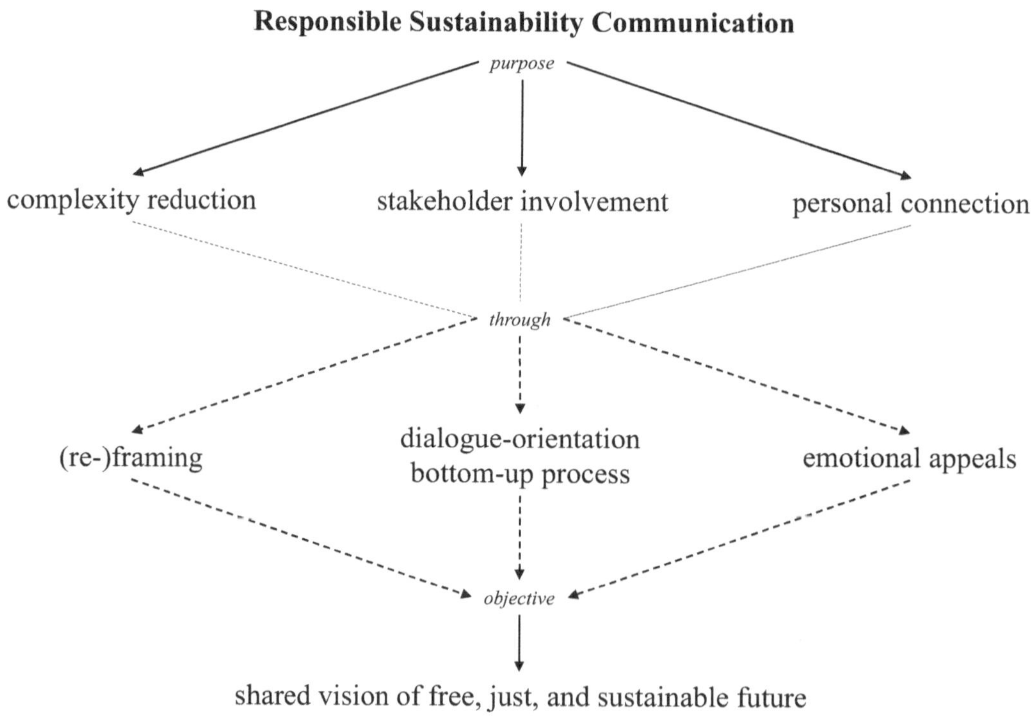

Figure 25.1 Process of responsible sustainability communication, depicting its purposes (complexity reduction, stakeholder involvement, personal connection), methods (dialogue-orientation, bottom-up process), and ultimate goal: a shared vision of a sustainable future

Source: Own illustration.

The purpose and objectives of responsible sustainability communication, aiming to "ensure public access to *reliable* information *to* protect fundamental freedoms" for the achievement of a peaceful and inclusive society for sustainable development, are currently significantly complicated by a deliberate misinformation campaign linked to (anthropogenic) climate change. This campaign undermines scientific findings, discredits researchers, and fosters doubt (McCright & Dunlap, 2017). Consequently, the media and communication sector must confront the issue of disinformation as part of its contribution to the responsible production of information.

As mentioned earlier, an informed public is indispensable for the proper functioning of a democracy, given that consistent and targeted misinformation can result in suboptimal decisions (Lewandowsky et al., 2017). Therefore, effectively addressing societal issues such as climate change and the transition to a sustainable society requires a focused effort to combat misinformation (West & Bergstrom, 2021). In terms of responsible information production and distribution, this involves rectifying scientifically inaccurate content, unveiling the generation and dissemination of misinformation, and addressing the institutional and political structures that facilitate its spread (Farrell et al., 2019). These measures are crucial to counteracting misinformation and ensuring that accurate, reliable, and responsible information prevails in public discourse. In concrete terms, this involves understanding and unveiling the communication techniques used by science deniers and educating the public on why denial arguments may initially seem convincing but are ultimately false and unfounded (Schmidt et al., 2013). Additionally, it entails revealing the individuals and organizations responsible for disseminating these messages, along with their financial interests and motivations (Voci & Karmasin, 2023).

Moreover, the responsible production and dissemination of information to support sustainable development necessitate ongoing efforts to combat the phenomenon known as "false balance." This issue is prevalent not only on social media, where controlling and rectifying user-generated content is often challenging, if not impossible, but also in professional settings, particularly in legacy media. In this context, journalists often present a topic as more balanced than the scientific evidence supports, a practice known as "bothsidesism." While well-intentioned in its attempt to avoid bias, this approach fosters the misleading notion that unfounded and questionable positions are respectable and valid (Grimes, 2019). This becomes particularly problematic when contrary views, based on personal opinions, are given equal weight to those grounded in scientifically incontrovertible facts (Cook et al., 2017). Such an approach has been demonstrated to significantly diminish audience certainty about scientific issues (Kortenkamp & Basten, 2015), creating the false impression that certain topics are scientifically controversial when, in fact, they are not. Consequently, addressing the issue of false balance is integral to the media and communication sector's commitment to contributing to SDG 12. This commitment involves actively participating in public debates, providing a robust counterpoint to the populist arguments of politicians and, simultaneously, offering guidance and reference points for the public to distinguish between opinions and scientifically based facts. Achieving this also entails substantial efforts in media education, training, and literacy, a topic that will be further discussed in relation to SDG 4.

Lastly, another dimension where the media and communication field can make a substantial contribution to SDG 12 regarding the responsible production of media goods for sustainable development lies in the practical domain—the day-to-day operations of media professionals and organizations. Media being dual goods encompassing both economic and cultural aspects renders media companies accountable on two fronts: as producers of both economic and cultural goods. In their role as economic entities, media companies are accountable for their economic actions and consequences, similar to any other business. In alignment with *SDG 12*, media companies must strive to produce their products in ways that promote sustainable management and the efficient utilization of natural resources. This could involve considerations such as monitoring the environmental footprint of materials used, reducing waste generation significantly, and adopting broader sustainable practices in organizational management, as outlined in the specific targets of goal 12 (UN, 2023b). However, the role of the media and communication field extends beyond mere efforts to reduce CO_2 emissions or integrate sustainable production models into the material creation of media goods. As socially relevant actors, the media bear an added responsibility in executing their watchdog role and portraying reality while articulating (or neglecting) ethical concerns and criticisms through their media outputs (Olkkonen, 2015; Sandoval, 2015). This means media companies, as producers of cultural goods, are subject to a further social responsibility—besides their physical, ecological footprint—that is, the "brain-print" they leave on their audience (Voci, 2022).

The notion of a "brain-print" left by media traces back to the well-known agenda-setting theory, which explains how media content shapes the issues the audience prioritizes (McCombs & Shaw, 1972). Moreover, extensive research underscores that the way the media report and convey certain subjects significantly influences how the public perceives these issues and reacts to them (Kitzinger, 1999; Ma, 2005; Holmes et al., 2009; Anderson, 2015; Janoušková et al., 2019). Notably, public concern about environmental matters, for instance, increases in tandem with heightened media coverage of these issues (Brown & Deegan, 1998). Hence, the media (companies), as prominent communicators to the public, carry a communicative responsibility (Weder & Karmasin, 2011), that is, a responsibility for *how* issues are communicated to the audience.

This dual responsibility of media (companies) has also been recognized recently by the United Nations itself, which has launched the SDG Media Compact in 2018. This initiative aims to encourage news and entertainment media companies to leverage their extensive outreach to the public to accelerate the transition toward sustainable development and achieve the SDGs through their media products and formats. What started in 2018 with 31 members of the media news and entertainment sector has grown to an alliance of over 200 members by 2022 who, together, (can) reach an audience of around 2 billion people worldwide. Members are expected to commit to the SDGs by using their creativity and resources to create content focused on the SDGs or some sustainability-related issues and/or to improve their sustainability practice in their everyday business and work practices (UN, 2022d).

Notably, when looking at how media (companies) interpret and implement their role in the spirit of the SDG Media Compact, it becomes apparent that the media sector—and particularly the film and entertainment sector—seems to be very committed to using its resources to highlight sustainability-related issues and advocate for the imperative shift toward sustainable development in its media products. Sustainability and its related issues are no longer a niche topic in fictional entertainment (Bilandzic & Kalch, 2021), as the "Cli-Fi" (climate change fiction) genre shows, which portrays post-(climate)apocalyptic future scenarios and extreme natural events on-screen. Prominent examples include films like *Soylent Green* (1973), *Waterworld* (1995), or *The Day After Tomorrow* (2004), to name a few, each exploring humanity's response to significant natural threats and catastrophes. Films like *Age of Stupid* (2009) and the recent Netflix hit *Don't Look Up* (2021) further underscore the portrayal and critique of humanity's perplexing "wait-and-see" stance in the face of crises. Moreover, the problematization of the relationship between humanity and the natural environment has extended to children's entertainment, as shown by various Disney/Pixar movies, such as *Happy Feet* (2006, 2011), the *Ice Age* saga (2002, 2006, 2009, 2012, 2016), *Moana* (2016), or the blockbuster *Frozen* (2013, 2019).

However, despite the growing numbers of fictional narratives addressing sustainability issues and their popularity, there is a notable lack of scientific research examining their effects on attitudes and behavior. This is surprising, given that fictional sustainability narratives can be seen as a promising way to raise awareness for sustainability issues and foster pro-sustainability attitudes and behaviors (Bilandzic & Kalch, 2021) primarily due to their emotive resonance. As previously discussed, the strategic use of emotional appeals can be a pivotal factor in sustainability communication, capable of guiding, and even influencing, motivation and intention for specific behaviors. Hence, media and communication scholars need to more effectively support the practical work undertaken by media companies by concentrating on the correlation between emotional appeals in media products and the audience's inclination toward adopting more sustainability-oriented behaviors. Understanding how to effectively communicate sustainability issues through journalistic or entertainment content to stimulate pro-sustainability behaviors could empower media organizations to align their techniques more precisely with the objectives of the SDG Media Compact. Therefore, it becomes crucial for media and communication professionals to be aware not only of their communicative responsibility to contribute to achieving the goals of both SDG 12 and the SDG Media Compact but also of the opportunities to translate this responsibility into practical action. Ideally, this awareness and application should not be limited to work activities and daily business as a "learning-by-doing" process but should be integrated into the education of media (and communication) professionals, particularly within the framework of *SDG 4*.

25.3 SDG 4: Media Education and Media Literacy

As previously mentioned, to "ensure public access to *reliable* information *to* protect fundamental freedoms" requires responsible production and consumption of information and media goods, aligning with *SDG 12*. However, in explaining the role of the media and communication field in advancing goal 12, it becomes evident that neither responsible production nor responsible consumption of media can be achieved without parallel emphasis on media education and media literacy.

Unfortunately, similar to other SDGs, SDG 4 does not explicitly mention the media and communication field. One might need to broaden the interpretation to include media education and media literacy, likely aligning with Target 7 of SDG 4: "[E]nsure that all learners acquire the knowledge and skills needed to promote sustainable development" (UN, 2023a). However, emphasizing media education holds significant importance for several reasons:

- To generate new theories, models, and methods to raise awareness of sustainability and to further explore the correlation between communication and behavioral patterns
- To bolster the communicative responsibility of media professionals, mitigating the issue of "false balance" and ensuring media products adhere to the principles of the SDG Media Compact.
- To align the dynamics of digitization with the SDGs. As media and communication science is one of the core disciplines in the realm of digitalization—a principal catalyst for contemporary societal transformations—it becomes imperative to align the dynamics of digitalization with the SDGs for steering social change toward sustainable development (Ordieres-Meré et al., 2020; TWI2050, 2019).

As a result, it appears essential that future media and communication professionals need to have a solid knowledge and understanding of sustainability concepts and challenges, while it is crucial to strengthen awareness of their pivotal role as communicators, media, and digitalization professionals in advancing the SDGs, enabling them to take responsibility in their professional pursuits.

Equally significant is the need for professionals in various sectors impacting sustainable development (e.g., engineering, architecture, finance, environment) to acquire at least foundational knowledge of media and communication science. This knowledge can be applied in their respective fields, underscoring the importance of interdisciplinary collaboration (Karmasin & Voci, 2021). To realize this, sustainability principles and related issues should assume a more prominent role in the educational curricula of media and communication sciences. Simultaneously, integrating media and communication science concepts into the educational programs of "sustainability sciences"[2] is crucial for fostering interdisciplinary knowledge exchange and collaboration.

Regarding the integration of sustainability knowledge, various studies confirm that despite the implementation of Education for Sustainable Development (ESD) initiatives in higher education institutions, graduates often perceive a lack of adequate knowledge of sustainability and related topics. Additionally, their degrees do not consistently influence their attitudes toward sustainability (Mintz & Tal, 2013; Yavetz et al., 2009; Kagawa, 2007). Similarly, this picture is quite confirmed when looking at the level of integration of sustainability aspects in, for example, media and communication curricula, as well as the level of integration of media and communication aspects in sustainability sciences study programs. Notably, the integration of sustainability aspects within media and communication curricula remains very low. Only 14% of analyzed curricula reference at least one sustainability-related aspect in their educational activities. This percentage further declines to 6% when searching for modules dedicated specifically to sustainability-related topics (Karmasin & Voci, 2021). While there is a slight increase in the integration of media and communication aspects in sustainability science curricula, the outcomes are still modest: 18% of analyzed curricula mention these aspects in their educational activities, yet only 10.6% integrate them at the module level (Voci & Karmasin, 2021) (Figure 25.2).

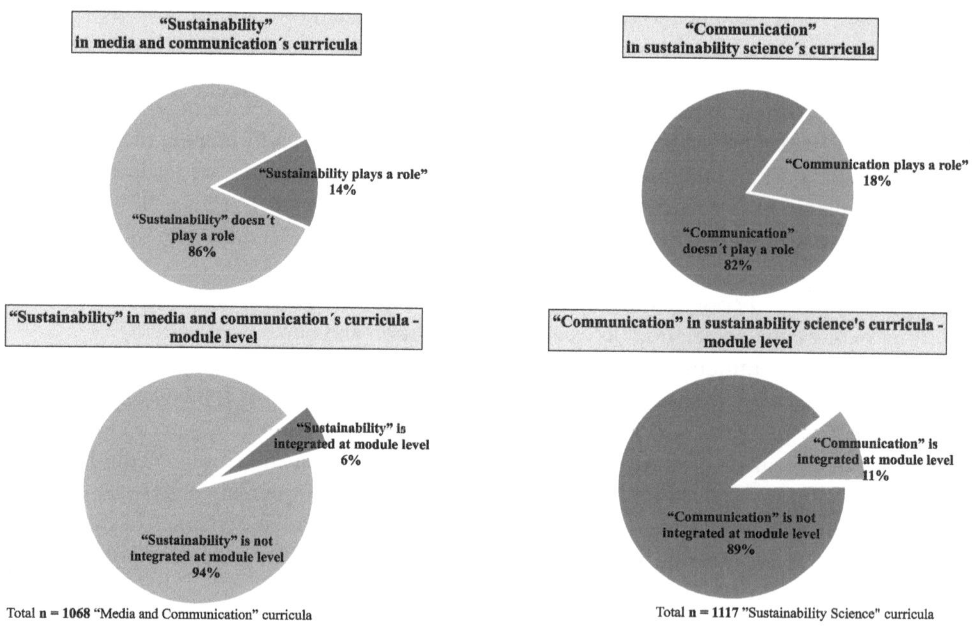

Figure 25.2 Four pie charts grouped into two categories: the role of "sustainability" in media and communication curricula on the left, and the role of "communication" in sustainability science curricula on the right. Each category includes data both overall and at the module level.

Source: Own illustration.

This indicates that students and graduates in media and communication sciences often lack fundamental knowledge about sustainability concepts and the imperative for sustainable development. Similarly, students and graduates of sustainability studies may have limited understanding of how to effectively share and communicate their research findings beyond mere information transfer, in the sense of a responsible sustainability communication. Bridging this knowledge gap necessitates a mutual integration of media and communication with sustainability aspects, promising substantial advantages in advancing the SDGs. Unfortunately, this reciprocal integration remains largely absent in practice. Yet the pressing need for a curricular reorientation and reorganization, in the sense of *SDG 4* (Quality Education), is essential to drive the necessary shift toward sustainable development.

In addition, it is necessary to include media literacy in the sense of goal *4* (Quality Education). This inclusion is crucial, not solely to address and mitigate the issue of false balance among media professionals, as mentioned earlier, but also to facilitate a more differentiated and responsible approach to reporting. Media literacy, encompassing practices like source evaluation, expert assessment, transparency, and contextualization, can play a pivotal role in fostering a balanced narrative rooted in factual evidence while upholding ethical standards. Furthermore, media literacy is an indispensable competency for the public. It involves the capacity to find, critically assess, and utilize information and media content from diverse sources (European Commission, 2007). It also involves the ability to critically reflect on the media's impact on information selection and opinion formation (De Abreu, 2020). Essentially, media literacy means the capability to navigate the media landscape, comprehend conveyed messages, and make informed decisions about encountered information. Therefore, media literacy plays a crucial role in responsible information consumption (SDG 12) by empowering individuals to engage critically with media content and fostering a more informed and engaged citizenry. By enhancing people's abilities to access reliable information, discern between truth and misinformation, and actively participate in social dialogues and initiatives, media literacy can significantly contribute to advancing the SDGs.

25.4 SDG 16: A Strong, Independent Fourth Estate

The preceding discussion emphasizes that to achieve the objective of guaranteeing "public access to *reliable* information *to* protect fundamental freedoms" for the realization of a peaceful and inclusive society for sustainable development (aligned with SDG 16), responsible production, dissemination, and consumption of information, along with media literacy, are imperative. Nevertheless, it is evident that the actual achievement of ensuring public access to reliable information and protecting fundamental freedoms relies significantly on the existence of strong and autonomous media institutions, embodying the concept of a Fourth Estate.

The concept of the Fourth Estate underscores the role of the media as an external overseer of governmental functions, serving as a counterbalance to the three traditional branches of power (legislative, executive, and judicial). This framework involves monitoring these powers from an external standpoint and representing the societal public interest. Widely regarded as a cornerstone of liberal democracies, the Fourth Estate holds the political system accountable, furnishes citizens with access to information for informed decision-making, and oversees the democratic process. Additionally, it serves a dual legitimizing role: confirming government actions as legitimate when functioning as a watchdog and, as a normative guide for journalists, legitimizing their role as society's overseer of governmental activities (Debatin, 2015). Hence, a strong and independent Fourth Estate plays a fundamental role in advancing sustainable development. It is not solely about delivering precise and responsible information—such as promoting health policies and holding authorities accountable for health care, aligning with *SDG 3* (Good Health and Well-Being), or advocating sustainable practices to combat climate change in accordance with *goal 13* (Climate Action), or challenging gender biases, spotlighting gender inequalities, and advocating for equitable representation across various domains, contributing significantly to *SDG 5* (Gender Equality).

Instead, the relevance of a strong Fourth Estate lies in fostering public participation. It empowers citizens to engage in decision-making processes, voice concerns, and actively contribute to shaping sustainable policies, aligning with Target 16.7's mandate to "[e]nsure responsive, inclusive, participatory and representative decision-making at all levels" (SDSN, 2023). An independent Fourth Estate holds governmental bodies, corporations, and institutions accountable for their actions, ensuring transparency in the execution of sustainable development initiatives. Moreover, it plays a pivotal role in unveiling corruption, human rights violations, and other injustices, thereby contributing to the establishment of accountable institutions as outlined in *SDG 16*.

25.5 Conclusion

The chapter aims to address the crucial absence of the media and communication field in the SDG implementation strategy, emphasizing the substantial advantages and potentials linked to its inclusion for the SDGs' achievement. The only mention of media and communication can currently be found in Target 16.10: "[E]nsure public access to information and protect fundamental freedoms." An unfortunate formulation, we argue, as it fails to recognize the fundamental role of responsible media and communications in the creation of the public sphere and—moreover—the maintenance of democracies. Thus, the aim to demonstrate that "public access to *reliable* information" is not just a prerequisite but also an enabler *to* "protect fundamental freedoms," by arguing that the media and communication's contribution to sustainable development extends beyond information transmission. Indeed, public access to *reliable* information requires responsible sustainability communication, including responsible information production and consumption. This, in turn, requires support from media education, media literacy, and the existence of a strong, independent Fourth Estate.

While the chapter effectively highlights the contribution of media and communication, there is always room for improvement. In the realm of responsible sustainability communication, fostering interdisciplinary collaboration, particularly with behavioral psychology, is essential. This collaboration aims to advance robust research exploring the interplay between communication, the use of specific appeals, and behavioral change. The objective is to gain a deeper understanding of the emotions that can trigger or promote sustainability-friendly behavior, ultimately contributing to sustainable development. In the field of education, a comprehensive restructuring of curricula is imperative. This includes a profound redesign of teaching and learning experiences, ensuring that students and graduates can comprehend, communicate, and implement the relevance and implications of sustainability in their spheres of influence. Acknowledging that this process involves overcoming institutional boundaries, it is anticipated to be a challenging yet necessary process (Tilbury, 2011).

Moreover, strengthening interdisciplinary media education is crucial to raising awareness among media professionals about their responsibility in producing and disseminating reliable information. This effort is especially important to counteract and mitigate the concerning issue of false balance—as mentioned earlier. This endeavor poses challenges not only in traditional media but also especially in the realm of social media, where the complexities are even more pronounced. In this digital landscape, false balance can be perpetuated by anyone, with the distinction that here it often stems from ideological or political motivations, serving as a disinformation strategy.

While fact-checking initiatives exist, they fall short of addressing the magnitude of the problem. To address this, a comprehensive approach is essential, encompassing stringent media policies that impose severe penalties on social media companies neglecting their duty to report, label, and remove disinformation and fake news. Such policies should hold these companies accountable for the content they produce and for third-party content they "only" publish or host. Additionally, a strong media literacy program is indispensable to equip individuals with the skills needed to discern and navigate the information landscape effectively.

Media literacy is also a crucial tool to combat the growing issue of politicization of the sustainability discourse. As the discussion around sustainability and its development transitions from scientific to political realms, the primary communicators shaping this narrative have shifted from scientists to politicians. Rather than leading the "SDG agenda," politicians often prioritize their political agendas, aiming to secure votes during elections. In this context, media literacy assumes a key role by ensuring that individual behavioral decisions within a media-literate society are founded on a balanced and thoughtful examination of information and scientific facts, rather than being influenced by political opinions and preferences.

Concluding, it is essential to underscore that responsible sustainability communication and media literacy are not realizable without the presence of a strong and independent Fourth Estate, as they are interlinked necessities. Placing these elements at the forefront of sustainable development strategies is crucial to fostering a media landscape that facilitates the dissemination of accurate information, encourages responsible citizen engagement, and holds stakeholders accountable. This, in turn, safeguards fundamental freedoms and significantly contributes to the achievement of the SDGs and the building of a more sustainable future.

Notes

1 Newig et al. (2013) distinguish three modes of sustainability communication: communication of sustainability (CoS), which involves one-way communication by experts to inform the public; communication about sustainability (CaS), a horizontal exchange of information and opinions to foster discourse and shared understanding; and communication for sustainability (CfS), which emphasizes the normative aspect, aiming to catalyze social change toward sustainable development. CoS is informative, CaS encourages discourse, and CfS seeks transformative action for sustainability.

2 *Sustainability science* is defined as a transdisciplinary academic field integrating research from natural, engineering, medical, and social sciences and humanities to contribute to solutions "for complex challenges that threaten the future of humanity and the integrity of the life support systems of the planet, such as climate change, biodiversity loss, pollution and land, and water degradation" (ESG, 2021).

References

Ajzen, I. (1991). The theory of planned behavior. *Organizational Behavior and Human Decision Processes*, *50*(2), 179–211.

Anderson, A. (2015). Reflections on environmental communication and the challenges of a new research agenda. *Environmental Communication*, *9*(3), 379–383.

Atkin, C. K. (2001). Theory and principles of media health campaigns. *Public Communication Campaigns*, *3*, 49–67.

Bennett, R. (1998). Shame, guilt & responses to non-profit & public sector ads. *International Journal of Advertising*, *17*(4), 483–499.

Bilandzic, H., & Kalch, A. (2021). Fictional narratives for environmental sustainability communication. In *The Sustainability Communication Reader* (pp. 123–142). Wiesbaden: Springer VS.

Bolsen, T., & Shapiro, M. A. (2018). The US news media, polarization on climate change, and pathways to effective communication. *Environmental Communication*, *12*(2), 149–163.

Boykoff, M. T., & Boykoff, J. M. (2004). Balance as bias: Global warming and the US prestige press. *Global Environmental Change*, *14*(2), 125–136.

Boykoff, M. T., & Roberts, J. T. (2007). Media coverage of climate change: Current trends, strengths, weaknesses. *Human Development Report*, *2008*(3), 1–53.

Brennan, L., & Binney, W. (2010). Fear, guilt, and shame appeals in social marketing. *Journal of Business Research*, *63*(2), 140–146.

Brown, N., & Deegan, C. (1998). The public disclosure of environmental performance information—a dual test of media agenda setting theory and legitimacy theory. *Accounting and Business Research*, *29*(1), 21–41.

Carvalho, A. (2009). Reporting the climate change crisis. In *The Routledge Companion to News and Journalism* (pp. 485–495). Routledge.

Castrechini, A., Pol, E., & Guàrdia-Olmos, J. (2014). Media representations of environmental issues: From scientific to political discourse. *European Review of Applied Psychology*, *64*(5), 213–220.

Cook, J., Lewandowsky, S., & Ecker, U. K. (2017). Neutralizing misinformation through inoculation: Exposing misleading argumentation techniques reduces their influence. *PLoS ONE*, *12*(5), e0175799.

Cortese, A. D. (2003). The critical role of higher education in creating a sustainable future. *Planning for Higher Education*, *31*(3), 15–22.

Cotte, J., Coulter, R. A., & Moore, M. (2005). Enhancing or disrupting guilt: The role of ad credibility and perceived manipulative intent. *Journal of Business Research*, *58*(3), 361–368.

Dahlstrom, M. F. (2014). Using narratives and storytelling to communicate science with non-expert audiences. *PNAS*, *111*(4), 13614–13620.

Dahlstrom, M. F., & Ritland, R. (2012). The problem of communicating beyond human scale. *Between Scientists and Citizens*, 121–130.

De Abreu, B. S. (2020). Global perspectives on media literacy. In B. S. De Abreu & W. G. Christ (Eds.), *Media Literacy in a Disruptive Media Environment*. New York: Routledge.

Debatin, B. S. (2015). Fourth estate. In G. Mazzoleni (Ed.), *The International Encyclopedia of Political Communication* (pp. 396–404). Wiley Blackwell.

Dispensa, J. M., & Brulle, R. J. (2003). Media's social construction of environmental issues: Focus on global warming–a comparative study. *International Journal of Sociology and Social Policy*, *23*, 74–104.

ESDN—European Sustainable Development Network. (2019). *Recommendations for Communicating Sustainable Development and the SDGs*. www.esdn.eu/publications/policy-briefs

European Commission. (2007). *A European Approach to Media Literacy in the Digital Environment*. https://eur-lex.europa.eu/LexUriServ/LexUriServ.do?uri=COM:2007:0833:FIN:EN:PDF

Fahy, D. (2017). Objectivity, false balance, and advocacy in news coverage of climate change. *Oxford Research Encyclopedia of Climate Science*. https://doi.org/10.1093/acrefore/9780190228620.013.345

Farrell, J., McConnell, K., & Brulle, R. (2019). Evidence-based strategies to combat scientific misinformation. *Nature Climate Change*, *9*(3), 191–195.

Fischer, D. (2019). Nachhaltigkeitskommunikation. In E. Zemanek & U. Kluvi (Eds.), *Nachhaltigkeit interdisziplinär: Konzepte, Diskurse, Praktiken* (pp. 51–66). Stuttgart: UTB.

Fischer, G. W., Morgan, M. G., Fischhoff, B., Nair, I., & Lave, L. B. (1991). What risks are people concerned about. *Risk Analysis*, *11*(2), 303–314.

Floyd, D. L., Prentice-Dunn, S., & Rogers, R. W. (2000). A meta-analysis of research on protection motivation theory. *Journal of Applied Social Psychology*, *30*(2), 407–429.

French, S., Argyris, N., Haywood, S. M., Hort, M. C., & Smith, J. Q. (2019). Communicating geographical risks in crisis management: The need for research. *Risk Analysis*, *39*(1), 9–16.

Garcia-Retamero, R., & Cokely, E. T. (2013). Communicating health risks with visual aids. *Current Directions in Psychological Science*, *22*(5), 392–399.

Gavin, N. T. (2007, September). Global warming and the British press: The emergence of an issue and its political implications. In *Elections Public Opinion and Parties Conference*. Bristol University.

Glik, D. C. (2007). Risk communication for public health emergencies. *Annual Review of Public Health*, *28*, 33.

Godemann, J., & Michelsen, G. (Eds.). (2011). *Sustainability Communication. Interdisciplinary Perspectives and Theoretical Foundations*. Springer.

Grimes, D. R. (2019). A dangerous balancing act: On matters of science, a well-meaning desire to present all views equally can be an Trojan horse for damaging falsehoods. *EMBO Reports*, *20*(8), e48706.

Grunwald, A. (2016). *Nachhaltigkeit verstehen. Arbeiten an der Bedeutung nachhaltiger Entwicklung*. München: Oekom Verlag.

Gupta, J. (2011). Communicating climate change to the wider public. *Climate Policy*, *11*(1), 834–835. https://doi.org/10.3763/cpol.2009.0686
Harrison, S., Macmillan, A., & Rudd, C. (2020). Framing climate change and health: New Zealand's online news media. *Health Promotion International*, *35*(6), 1320–1330.
Höijer, B. (2010). Emotional anchoring and objectification in the media reporting on climate change. *Public Understanding of Science*, *19*(6), 717–731.
Holmes, B. J., Henrich, N., Hancock, S., & Lestou, V. (2009). Communicating with the public during health crisis: Experts' experiences and opinions. *Journal of Risk Research*, *12*(6), 793–807. https://doi.org/10.1080/13669870802648486
Janoušková, S., Hák, T., Nečas, V., & Moldan, B. (2019). Sustainable development—a poorly communicated concept by mass media. Another challenge for SDGs? *Sustainability*, *11*(11), 3181.
Jarren, O., & Meier, W. A. (2002). Mediensysteme und Medienorganisationen als Rahmenbedingungen für den Journalismus. In *Journalismus—Medien—Öffentlichkeit* (pp. 99–163). VS Verlag für Sozialwissenschaften.
Kagawa, F. (2007). Dissonance in students' perceptions of sustainable development and sustainability: Implications for curriculum change. *International Journal of Sustainability in Higher Education*, *8*(3), 317–338.
Karmasin, M., & Voci, D. (2021). The role of sustainability in media and communication studies' curricula throughout Europe. *International Journal of Sustainability in Higher Education*, *22*(8), 42–68.
Karmasin, M., Voci, D., Weder, F., & Krainer, L. (2021). Future perspectives: Sustainability communication as scientific and societal challenge. In *The Sustainability Communication Reader* (pp. 585–591). Wiesbaden: Springer VS. https://doi.org/10.1007/978-3-658-31883-3_31
Kaschak, M. P., & Maner, J. K. (2009). Embodiment, evolution, and social cognition: An integrative framework. *European Journal of Social Psychology*, *39*(7), 1236–1244.
Kitzinger, J. (1999). Reaching risk and the media. *Health, Risk and Society*, *1*(1), 55–69.
Klaus, E. (2006). Von der Beschränktheit unserer Öffentlichkeitstheorien im europäischen Kontext. In *Europäische Öffentlichkeit und medialer Wandel* (pp. 93–106). VS Verlag für Sozialwissenschaften.
Kortenkamp, K. V., & Basten, B. (2015). Environmental science in the media: Effects of opposing viewpoints on risk and uncertainty perceptions. *Science Communication*, *37*(3), 287–313.
Kuklinski, J. H., Quirk, P. J., Jerit, J., Schwieder, D., & Rich, R. F. (2000). Misinformation and the currency of democratic citizenship. *The Journal of Politics*, *62*(3), 790–816.
Lewandowsky, S., Ecker, U. K., & Cook, J. (2017). Beyond misinformation: Understanding and coping with the "post-truth" era. *Journal of Applied Research in Memory and Cognition*, *6*(4), 353–369.
Lin, L., Rivera, C., Abrahamsson, M., & Tehler, H. (2017). Communicating risk in disaster risk management systems–experimental evidence of the perceived usefulness of risk descriptions. *Journal of Risk Research*, *20*(12), 1534–1553.
Luhmann, N. (1989). *Ecological Communication*. University of Chicago Press.
Ma, R. (2005). Media, crisis, and SARS: An introduction. *Asian Journal of Communication*, *15*(3), 241–246.
Mabon, L. (2020). Making climate information services accessible to communities: What can we learn from environmental risk communication research? *Urban Climate*, *31*, 1–14. https://doi.org/10.1016/j.uclim.2019.100537
McCombs, M. E., & Shaw, D. L. (1972). The agenda-setting function of mass media. *Public Opinion Quarterly*, *36*(2), 176–187.
McCright, A. M., & Dunlap, R. E. (2017). Combatting misinformation requires recognizing its types and the factors that facilitate its spread and resonance. *Journal of Applied Research in Memory and Cognition*, *6*(4), 389–396.
Mintz, K., & Tal, T. (2013). Sustainability in higher education courses: Multiple learning outcomes. *Studies in Educational Evaluation*, *41*, 113–123.
Mitchell, C. L., Burch, S. L., & Driscoll, P. A. (2016). (Mis) communicating climate change? Why online adaptation databases may fail to catalyze adaptation action. *Wiley Interdisciplinary Reviews: Climate Change*, *7*(4), 600–613.
Moser, S. C. (2010). Communicating climate change: History, challenges, process and future directions. *Wiley Interdisciplinary Reviews: Climate Change*, *1*(1), 31–53.
Neuwirth, K., Dunwoody, S., & Griffin, R. J. (2000). Protection motivation and risk communication. *Risk Analysis*, *20*(5), 721–734.
Newig, J., Schulz, D., Fischer, D., Hetze, K., Laws, N., Lüdecke, G., & Rieckmann, M. (2013). Communication regarding sustainability: Conceptual perspectives and exploration of societal subsystems. *Sustainability*, *5*(7), 2976–2990. https://doi.org/10.3390/su5072976

Olkkonen, L. (2015). Audience enabling as corporate responsibility for media organizations. *Journal of Media Ethics, 30*(4), 268–288.
Ordieres-Meré, J., Prieto Remon, T., & Rubio, J. (2020). Digitalization: An opportunity for contributing to sustainability from knowledge creation. *Sustainability, 12*(4), 1460.
Parton, K., & Morrison, M. (2011). *Communicating Climate Change: A Literature Review*. Paper presented at the 55th Annual AARES National Conference, Melbourne.
Pearce, W., Brown, B., Nerlich, B., & Koteyko, N. (2015). Communicating climate change: Conduits, content, and consensus. *Wiley Interdisciplinary Reviews: Climate Change, 6*(6), 613–626.
Pollard, S. J., Davies, G. J., Coley, F., & Lemon, M. (2008). Better environmental decision making—recent progress and future trends. *Science of the Total Environment, 400*(1–3), 20–31.
Pompper, D. (Ed.). (2018). *Climate and Sustainability Communication: Global Perspectives*. Routledge.
Renn, O. (2014). Four questions for risk communication: A response to Roger Kasperson. *Journal of Risk Research, 17*(10), 1277–1281.
Robertson, M. (2018). *Communicating Sustainability*. Routledge.
Rosenbaum, M. S., & Culshaw, M. G. (2003). Communicating the risks arising from geohazards. *Journal of the Royal Statistical Society: Series A (Statistics in Society), 166*(2), 261–270.
Sandoval, M. (2015). Corporate social (ir)responsibility in media and communication industries. In *Media and Left* (pp. 166–189). Brill.
Schäfer, M. S., & Bonfadelli, H. (2017). Umwelt-und Klimawandelkommunikation. In *Forschungsfeld Wissenschaftskommunikation* (pp. 315–338). Wiesbaden: Springer VS.
Scheufele, D. A. (2014). Science communication as political communication. *Proceedings of the National Academy of Sciences, 111*(supplement_4), 13585–13592.
Schmidt, A., Ivanova, A., & Schäfer, M. S. (2013). Media attention for climate change around the world: A comparative analysis of newspaper coverage in 27 countries. *Global Environmental Change, 23*(5), 1233–1248.
SDG Zone. (2022). *Why Do We Need the SDGs?* https://sdgzone.com/learn/why-do-we-need-sdgs/
SDSN—Sustainable Development Solutions Network. (2023). *Indicators and a Monitoring Framework*. https://indicators.report/goals/goal-16/
Servaes, J. (2013). Introduction: The kaleidoscope of text and context in communication. In *Sustainability, Participation & Culture in Communication: Theory and Praxis* (pp. 1–23). Intellect Ltd.
Shang, J., Basil, D. Z., & Wymer, W. (2010). Using social marketing to enhance hotel reuse programs. *Journal of Business Research, 63*(2), 166–172.
Smith, N. W., & Joffe, H. (2009). Climate change in the British press: The role of the visual. *Journal of Risk Research, 12*(5), 647–663.
Stables, A. (2013). The unsustainability imperative? Problems with "sustainability" and "sustainable development" as regulative ideals. *Environmental Education Research, 19*(2), 177–186. https://doi.org/10.1080/13504622.2012.729813
Staub, E. (2005). The roots of goodness: The fulfillment of basic human needs and the development of caring, helping and nonaggression, inclusive caring, moral courage, active bystandership, and altruism born of suffering. In G. Carlo & C. P. Edwards (Eds.), *Moral Motivation Through the Life Span* (pp. 33–72). University of Nebraska Press.
Stecula, D. A., & Merkley, E. (2019). Framing climate change: Economics, ideology, and uncertainty in American news media content from 1988 to 2014. *Frontiers in Communication, 4*, 6.
Sterman, J. D. (2011). Communicating climate change risks in a skeptical world. *Climatic Change, 108*(4), 811–826.
Stürmer, S., Snyder, M., & Omoto, A. M. (2005). Prosocial emotions and helping: The moderating role of group membership. *Journal of Personality and Social Psychology, 88*(3), 532.
Taute, H., & McQuitty, S. (2004). Feeling good! Doing good! An exploratory look at the impulsive purchase of the social good. *Journal of Marketing Theory and Practice, 12*(2), 16–27.
Terpstra, T., Lindell, M. K., & Gutteling, J. M. (2009). Does communicating (flood) risk affect (flood) risk perceptions? Results of a quasi-experimental study. *Risk Analysis: An International Journal, 29*(8), 1141–1155.
Tilbury, D. (2011). Higher education for sustainability: A global overview of commitment and progress. *Higher Education in the World, 4*(1), 18–28.
TWI2050—The World in 2050. (2019). The digital revolution and sustainable development: Opportunities and challenges. *Report Prepared by The World in 2050 Initiative, Laxenburg, International Institute for Applied Systems Analysis (IIASA)*, Austria.

UN. (2002). *Report of the World Summit on Sustainable Development*. https://digitallibrary.un.org/record/478154
UN. (2016). *Communicating the Sustainable Development Goals—For Everyone*. www.un.org/webcast/pdfs/160421pm-sdgs-com.pdf
UN. (2022a). *Decade of Action*. www.un.org/sustainabledevelopment/decade-of-action/
UN. (2022b). *Progress Toward the Sustainable Development Goals*. Report of the Secretary-General. https://unstats.un.org/sdgs/report/2022/
UN. (2022c). *The 17 Goals*. https://sdgs.un.org/#goal_section
UN. (2022d). *SDG Media Compact*. www.un.org/sustainabledevelopment/sdg-media-compact-about/
UN. (2022e). *SDG Media Zone*. www.un.org/sdgmediazone/
UN. (2023a). *Goals—4: Ensure Inclusive and Equitable Quality Education and Promote Lifelong Learning Opportunities for All*. https://sdgs.un.org/goals/goal4#targets_and_indicators
UN. (2023b). *Goals—12: Ensure Sustainable Consumption and Production Patterns*. https://sdgs.un.org/goals/goal12#targets_and_indicators
UN Documents. (2020). *Declaration of the United Nations Conference on the Human Environment*. http://un-documents.net/unchedec.htm
UNDP. (2022). *The SDGs in Action*. www.undp.org/sustainable-development-goals
Villarino, J., & Font, X. (2015). Sustainability marketing myopia: The lack of persuasiveness in sustainability communication. *Journal of Vacation Marketing*, *21*(4), 326–335.
Voci, D. (2022). Logos, ethos, pathos, sustainabilitos? About the role of media companies in reaching sustainable development. *Sustainability*, *14*(5), 2591.
Voci, D., Bruns, C. J., Lemke, S., & Weder, F. (2020). Framing the end: Analyzing media and meaning making during cape town's day zero. *Frontiers in Communication*, *84*.
Voci, D., & Karmasin, M. (2021). Sustainability and communication in higher education. In *7th International Conference on Higher Education Advances (HEAd'21)*. Universidad Politecnica de Valenica. http://doi.org/10.4995/HEAd21.2021.12831
Voci, D., & Karmasin, M. (2023). Sustainability communication: How to communicate an inconvenient truth in the era of scientific mistrust. *Journal of Communication Management*, *28*(1), 15–40. https://doi.org/10.1108/JCOM-05-2022-0060.
Voci, D., Karmasin, M., Nölleke-Przybylski, P., Altmeppen, K. D., Möller, J., & Rimscha, B. V. (2019). What is a media company today?: Rethinking theoretical and empirical definitions. *Studies in Communication and Media*, *8*(1), 29–52. https://doi.org/10.5771/2192-4007-2019-1-29.
Weder, F., & Karmasin, M. (2011). Corporate communicative responsibility: Kommunikation als Ziel und Mittel unternehmerischer Verantwortungswahrnehmung; Studienergebnisse aus Österreich. *Zeitschrift für Wirtschafts-und Unternehmensethik*, *12*(3), 410–428.
Weder, F., Hübner, R., & Voci, D. (2021). Ist nachhaltig "normal"?. In M. Jonas et al. (Eds.), *Reparieren, Selbermachen und Kreislaufwirtschaften* (pp. 147–169). Wiesbaden: Springer VS, https://doi.org/10.1007/978-3-658-31569-6_8
Weder, F., Karmasin, M., Krainer, L., & Voci, D. (2021). Sustainability communication as critical perspective in media and communication studies—An introduction. In *The Sustainability Communication Reader* (pp. 1–12). Wiesbaden: Springer VS.
Weder, F., Krainer, L., & Karmasin, M. (Eds.). (2021). *The Sustainability Communication Reader: A Reflective Compendium*. Springer Nature.
Weder, F., Lemke, S., & Tungarat, A. (2019). (Re) storying sustainability: The use of story cubes in narrative inquiries to understand individual perceptions of sustainability. *Sustainability*, *11*(19), 5264. https://doi.org/10.3390/su11195264
Wehrli, R., Priskin, J., Schaffner, D., Schwarz, J., & Stettler, J. (2014). Do sustainability experienced travellers prefer a more rational communication of the sustainability of a tourism product? *WIT Transactions on Ecology and the Environment*, *187*, 3–14.
West, J. D., & Bergstrom, C. T. (2021). Misinformation in and about science. *Proceedings of the National Academy of Sciences*, *118*(15), e1912444117.
Yavetz, B., Goldman, D., & Pe'er, S. (2009). Environmental literacy of pre-service teachers in Israel: A comparison between students at the onset and end of their studies. *Environmental Education Research*, *15*(4), 393–415.

26

CARBON NEUTRALITY AND THE ROLE OF CARBON DIOXIDE REMOVAL (CDR)

A Legal and Policy Analysis of China

Haomiao Du

26.1 Introduction

Global temperature rise, sea level rise, and more frequent extreme climate events require greater ambitions by countries in committing to the Paris Agreement. Scientific modelling demonstrates that the control of the global average temperature below 1.5°C requires net-zero emissions in the second half of the 21st century (Masson-Delmotte et al., 2018). The goal of net-zero emissions is in line with the Sustainable Development Goals (SDGs), mainly SDG 7 (Affordable and Clean Energy) and SDG 13 (Climate Action), because the net-zero target requires tremendous efforts globally on increasing energy efficiency and promoting renewable energies (SDG 7) and on reducing fossil fuel emissions and increasing carbon sinks (SDG 13).

The global net-zero target does not mean no emissions; rather, it requires that the remaining positive emissions be offset by removals from the atmosphere, or so-called negative emission technologies (NETs) (UNEP, 2021). Seeing the inadequate action on global climate crisis until now, the Intergovernmental Panel on Climate Change (IPCC) and the International Environmental Agency (IEA) have already concluded that it is impossible to reach net-zero emissions without NETs (UNEP, 2021). Carbon dioxide removal (CDR) or NETs offset fossil fuel emissions either by enhancing carbon storage in forests, oceans, and soils or by the technical means of direct air capture or bioenergy with carbon capture and storage (BECCS) (see Section 26.2.2) (McLaren et al., 2019). Although CDR and NETs are mostly used interchangeably in literature, this chapter prefers the term CDR than NETs as the overarching concept mainly because of the fact that nature-based carbon storage methods through, for example, afforestation are literally not "technologies."

Since 2019, the goal of "net-zero emissions" has proliferated globally, as an increasing number of countries have announced major commitments to significantly cut their GHG emissions, promising to reach net-zero in the coming decades (Energy and Climate, 2022). As a global pioneer in tackling climate change, the European Union (EU) announced the goal of climate neutrality by 2050 in its European Green Deal in 2019 (European Commission, 2019) and then legalized this policy goal in the new European Climate Law (European Parliament and Council, 2021). Developing countries are taking action as well. For instance, in September 2020, Chinese president Xi Jinping announced at the UN General Assembly that China will aim to achieve carbon neutrality before

DOI: 10.4324/9781003285472-28

2060 (UN Affairs, 2020). Since then, unprecedented actions on modifying or planning to modify climate regulations, policies, and industrial practices with an aim to integrate the 2060 carbon neutrality goal across hierarchical levels and economic sectors have been proliferating in China.

Among all available pathways that meet the target of net-zero emissions, CDR is already widely recognized as an indispensable strategy (Rogelj et al., 2021). However, the balance of emissions and removals achieved by using CDR is delicate: It is limited by the global removal capacity of CDR (UNEP, 2021) and affected by the proportion of emission reductions versus removals (Rogelj et al., 2021). Hence, it is worthwhile to investigate climate laws and policies governing CDR across jurisdictions for a better understanding of how to properly place its role in the portfolio of mitigation strategies.

The goal of net-zero emissions is particularly challenging for China, taking into account its role as the world's biggest emitter and the fact that carbon emissions are still yet to peak due to the continuous high demand for coal and natural gas consumption. The goal of net-zero emissions requires China to sharply cut emissions, particularly in the power sector, and offset the remaining CO_2 emissions with carbon capture, utilization, and storage (CCUS) and bioenergy with CCS (BECCS) technologies (He et al., 2022). In the discussions on various options to achieve carbon neutrality, Chinese scholars and practitioners have been focusing on the decarbonization of power generation, the transition to renewable energy, and electrifying end uses (Du and Zhang, 2022). Regarding the encouragement of technological innovations, most discussions have been limited to "low-carbon technologies," such as recycling wastes to generate energy and coupling burning coals with CCUS (MEE, 2021b, p. 30), while the relatively new concept of CDR remains underexplored in the Chinese context.

Given the preceding backdrop, it is not surprising that there is a research gap in Chinese laws and regulations on CDR techniques (Du, 2021). The legal research into the intersection of climate mitigation and technological innovations in China mainly surrounds the mechanisms for technology transfer under the United Nations Framework Convention on Climate Change (UNFCCC) (Zhou, 2019), the drafting of a new Chinese Climate Law that regulates mitigation technologies (Zhou, 2016), and policies on enforcing or encouraging technological upgrades for economy-wide decarbonization, in particular in the energy sector (Liu et al., 2022). It can be observed that the current policies are limited to incentivizing *incremental* innovations (upscaling the application of existing technologies) associated with climate mitigation, while policies for encouraging *disruptive* technologies (still at the early stage of development and no sufficient regulatory framework available) are underdeveloped (Xiangbai and Zahar, 2021).

This chapter aims to fill this research gap by exploring the possible integration and regulation of CDR for achieving carbon neutrality in China. This chapter will proceed as follows: Section 26.2 introduces the policy status quo of the 2060 carbon neutrality goal in China, the concept of CDR, and the current legal and policy frameworks for CDR in China. Section 26.3 examines the main regulatory barriers to integrate and regulate CDR. Section 26.4 explores the future development of CDR for contributing to delivering on carbon neutrality in China. Section 26.5 concludes.

26.2 The Carbon Neutrality Goal in China and the Role of CDR Techniques in Chinese Environmental Laws and Policies

26.2.1 The Goal of Achieving Carbon Neutrality by 2060

In the first nationally determined contribution (NDC) submitted to the secretariat to the UNFCCC in 2016, China pledged to peak CO_2 emissions "around 2030" and decrease carbon intensity

(CO_2 emissions per unit of GDP) by 60–65% below the 2005 level (NDRC, 2016). In October 2021, China submitted the Mid-Century Long-Term Low Greenhouse Gas Emission Development Strategy (LEDS) for the period until 2060 and updated its NDC for the implementation period until 2030 to the UNFCCC (MEE, 2021a, 2021b). The LEDS officially confirmed the goal of achieving carbon neutrality "before 2060," as announced by Chinese president Xi Jinping in September 2020. In the updated NDC, China aims to peak CO_2 emissions "before 2030" and decrease carbon intensity "over 65%."

In 2021, several domestic policy documents have been issued incompatible with China's international commitments. In March 2021, the National People's Congress approved the Outline of the 14th Five-Year Plan for National Economic and Social Development and Long-Range Objectives for 2035. The goals of peaking carbon emissions before 2030 and achieving carbon neutrality before 2060 have been integrated into the broader context of "ecological civilisation" (State Council, 2021, p. 94). In October 2021, the Action Plan for Carbon Dioxide Peaking before 2030 was issued to implement the strategies and development plans for peaking CO_2 emissions and achieving carbon neutrality (*Action Plan* in succeeding text) (NDRC, 2021).

It is expected that currently, the continuously decreased CO_2 intensity will enable the peaking of emissions before 2030 and even before 2025, but it is questionable how to implement the carbon neutrality goal after 2030 (Chen et al., 2022). Although the Action Plan introduced an "emissions cap," a quantified emissions cap has not yet been established. The main challenge lies in the difficulty of breaking down the goal into realistic targets for individual provinces (Carbon Brief, 2021). Following the launch of the 14th five-year plan, provinces or municipalities directly under the central government integrated their targets, tasks, and priorities regarding peaking carbon emissions before 2030 and carbon neutrality before 2060 in their own 14th five-year plans.

To date, the goal of carbon neutrality, and the related specific targets, has not been legalized yet by any binding instruments, and an independent climate law is not yet available in China. A new Law on Tackling Climate Change has been in the drafting process since 2010, but the progress is very slow due to the controversies over the legislative purposes and the relationship of this new climate law with other environmental and energy regulatory instruments. Given the absence of individual climate law, the current approach to regulating CO_2 emissions is to co-control air pollutants and GHGs under the recently revised Law on the Prevention and Control of Air Pollution (Jiang et al., 2013). Seeing the experiences of the countries and jurisdictions which already legalized their climate goals in law (e.g., the UK and the EU) (Energy and Climate, 2022), a new climate law that recognizes carbon peaking and neutrality will legalize the climate goals and facilitate the implementation and enforcement.

26.2.2 A Categorization of CDR Techniques

CDR, or carbon geoengineering, refers to a series of techniques that artificially increase carbon sinks on land or in water or remove existing CO_2 from the atmosphere by physical or chemical engineering methods (Shepherd, 2009). Main CDR methods include BECCS, direct air capture and storage (DACCS), afforestation and reforestation, soil carbon sequestration and biochar, land/ocean-based enhanced weathering, and ocean fertilization. Table 26.1 shows the definitions and categorization of CDR techniques. Based on the effects and effected environmental media, CDR methods can be categorized as (1) removing carbon sources by engineering methods (BECCS and DACCS), (2) increasing land carbon sinks (afforestation and reforestation, soil carbon sequestration and biochar, and soil-based enhanced weathering), and (3) increasing ocean carbon sinks (ocean alkalinization and ocean fertilization). Such a categorization is helpful for the investigation of relevant policies in later sections.

Table 26.1 Definitions and categorization of CDR techniques

CDR techniques	*Categorization*	*Definitions*
BECCS	Removing carbon sources by engineering methods	Carbon dioxide capture and storage (CCS) technology applied to a bioenergy facility
DACCS		Chemical process by which CO_2 is captured directly from the ambient air, with subsequent storage
Afforestation and reforestation	Increasing land carbon sinks	Planting of new forests on lands that historically have not contained forests Planting of forests on lands that have previously contained forests but that have been converted to some other use
Soil carbon sequestration and biochar		Land management changes which increase the soil organic carbon content, resulting in a net removal of CO_2 from the atmosphere Stable, carbon-rich material produced by heating biomass in an oxygen-limited environment, which reduce greenhouse gas emissions and sequester carbon emissions
Soil-based enhanced weathering; ocean-based enhanced weathering (ocean alkalization)	Increasing ocean carbon sinks	Enhancing the removal of CO2 from the atmosphere through dissolution of silicate and carbonate rocks by grinding these minerals to small particles and actively applying them to soils, coasts, or oceans
Ocean fertilization		Deliberate increase of nutrient supply to the euphotic layer of the sea in order to enhance biological production, through which additional CO_2 from the atmosphere is sequestered

Source: Compiled by the author.

Regarding the status quo of the deployment of CDR techniques in China, BECCS and DACCS are still at the stage of policy proposals and early demonstration. Most existing models of BECCS only incorporated bioelectricity with CCS (Weng et al., 2021), and coal bioenergy gasification systems with CCS have been identified as a cost-effective opportunity for China to decrease coal consumption (Lu et al., 2019). In comparison, research into the potential of DACCS in China's deep decarbonation pathway has not yet gained much attention. Afforestation and reforestation are not new methods for increasing carbon sinks in China, and the relevant research and practices regarding the effective management of forestry resources are becoming more important thanks to the potential of forests to significantly contribute to carbon neutrality (Cai et al., 2022). Research into soil carbon sequestration and biochar has been carried out for exploring soil fertility improvement for a long time (Han et al., 2022), while the investigation on the co-benefit of carbon sequestration in the context of pursuing carbon neutrality has recently begun (Song et al., 2022). The last two CDR techniques, that is, enhanced weathering and ocean fertilization, remain at the early stage of investigation, and only occasional mention of them can be found in the context of carbon neutrality (Wang et al., 2019).

26.2.3 An Overview of the Legal and Policy Frameworks for CDR in China

In general, all CDR techniques serve as options to mitigate climate change, as they either increase land or ocean carbon sinks or remove carbon sources by engineering methods. Regarding the legal and regulatory systems that apply to the research into and employment of CDR in China, the international legal agreements to which China is a contracting party and the national laws and regulations that enable the removal of carbon sources and the increase of carbon sinks on land or in the ocean will be examined in this section.

26.2.3.1 International Law: The Paris Agreement and the London Protocol

Regarding international engagement in climate change actions, China is a contracting party to the UNFCCC and the Paris Agreement. These two binding international legal instruments provide an enabling ground for China to develop CDR. The UNFCCC enables the development of CDR techniques because CDR can contribute to achieving the ultimate objective of "stabilization of greenhouse gas concentrations in the atmosphere at a level that would prevent dangerous anthropogenic interference with the climate system," as stated in Article 2. To achieve the long-term temperature goal set out in Article 2 of the Paris Agreement, Article 4.1 requires parties to "aim to reach global peaking of GHGs as soon as possible" and "to undertake rapid reductions thereafter by the best available science." In addition, the Paris Agreement adopted a bottom-up approach, which requires parties to pursue domestic mitigation measures to achieve the objectives of their NDCs (Article 4.2).

The legal implication of the Paris Agreement for mitigation is that China can select the most suitable mitigation measures, taking into account its continuous demand for fossil fuel consumption and rapid economic development. According to China's latest NDCs submitted in October 2021, the plans to decarbonize rapidly after 2030 heavily depend on substantial forestry sinks as well as the large-scale implementation of CCUS technologies that are currently not yet viable. Among the proposed new measures to implement the updated NDC goals, the ambition of increasing forestry sinks and the policies that synergize environmental protection, ecosystem conservation (both land and ocean ecosystems), and carbon sequestration (e.g., the promotion of nature-based solutions) would facilitate the implementation of soil-based and ocean-based CDR techniques as well as afforestation and reforestation. The new NDCs have no mention of the engineering CDR techniques, that is, DACCS and BECCS, but the demonstration and industrial application of CCUS in oil fields or the coal chemical industry will be rapidly pushed forward to support some regions, key industries, and key enterprises with favorable conditions to peak carbon emissions ahead of schedule (MEE, 2021b, p. 38).

Some multilateral environmental agreements on governing the marine environment apply to ocean-based CDR techniques, that is, ocean-based enhanced weathering and ocean fertilization, and also applicable to CO_2 ocean sequestration as a part of BE/DACCS. China is a contracting party to the United Nations Convention on Law of the Sea (UNCLOS) (UNGA, 1982), the 1972 London Convention on the Prevention of Marine Pollution by Dumping of Wastes and Other Matter (LC), and the 1996 London Protocol (LP) (IMO, 1972, 1996) and therefore shall comply with the relevant legal obligations. The UNCLOS enshrines general principles of marine environmental protection for emerging marine activities. Although the LC in general allows transboundary transport and storage of CO_2, it was superseded by the LP for parties to both. Therefore, the more restrictive approach under the LP applies to China. More specifically, one explicit regulation under the LP is the amended Annex I to the LP in 2006, which restricts CO_2 ocean sequestration to the injection of captured CO_2 streams in sub-seabed geological formations. However, this restriction is less of an issue for China, because the current demonstration projects on CCUS are onshore utilization and storage (Zhang, 2021). The other explicit regulation is the moratorium on ocean fertilization as required under the 2013 Amendment to the London Protocol on Marine Geoengineering (IMO, 2013) (*Amendment* in succeeding text), which limits lawful marine fertilization activities to "legitimate scientific research taking into account any specific placement assessment framework" (Annex 4, art. 1.3). The 2013 Amendment does not restrict ocean alkalinization in the same way but rather leaves the regulation open, until more scientific evidence of its impacts is available. Note that the moratorium has not yet entered into force, and thus, its legal effect remains non-binding. Regarding China's future implementation of offshore CO_2 storage or ocean-based CDR techniques, the regulatory gap of domestic law and its inconsistency with the LP will be discussed in Section 26.3.2.

26.2.3.2 Domestic Legal and Policy Frameworks for CDR Techniques

Currently, there is no overarching legal instrument to regulate CDR in China. The scattered legal and policy frameworks for CDR include three aspects: (1) explicit laws and policies on regulating different categories of CDR techniques, (2) general environmental law provisions that apply to CDR, and (3) laws and policies regarding the promotion of climate mitigation technologies and technological innovation more generally.

The *first* aspect are the laws and policies that are explicitly applicable to certain CDR techniques. Among the CDR techniques as listed in Table 26.1, explicit laws or policies are not available for regulating soil- and ocean-based enhanced weathering or ocean fertilization but are available for other CDR techniques in several ways. Regarding the first category of CDR techniques, that is, engineering methods, BECCS and DACCS are not visible in current climate policies, but CCUS equipped with coal plants has already been recognized as a significant supplement for peaking carbon emissions by 2030 and achieving carbon neutrality by 2060 (MEE, 2021b, p. 38). Although CCS is a key part of BE/DACCS, the distinction between the sources of the captured CO_2 emissions (i.e., from fossil fuel industries or bioenergy/the ambient atmosphere) reflects completely different mitigation strategies. The emphasis on CCUS while ignoring BE/DACCS in current climate policies exactly demonstrates the current prioritization of China's climate mitigation actions: Instead of rapidly phasing down fossil fuel consumptions and recapturing the existing emissions, China is utilizing CCUS to resolve the conflicts between the continuous high demand of coal energy and the high pressure of emissions reduction. Despite the enabling policies for rapid and large-scale implementation of CCUS in the 14th five-year period, the regulation of CCUS in China is still developing slowly. As put by Zhang (2021), the regulation of CCUS in China still lacks clarity on key aspects of CCUS, in particular regarding project approval and liability issues. Given the substantive difference in the rationale for selecting between CCUS and BE/DACCS as mitigation strategies (Haszeldine et al., 2018), more details about the regulation of CCUS will not be addressed in this chapter.

Regarding the second category, that is, increasing land carbon sinks, forestation plays a significant role in China's climate and environmental policies due to its versatility. Explicitly applicable policies supporting forestation can be found in multiple contexts of mitigation, adaptation, environmental protection, and ecosystem conservation (NDRC, 2021, para. 8; MEE, 2021b, p. 20; MEE, 2021c, p. 29). To enhance land carbon sinks in ecosystems, the Action Plan for peaking carbon emissions by 2030 specifies "the action for consolidating and enhancing carbon sink," which includes a target of expanding forest coverage to around 25% of China's total land area and increasing forest stock volume to 19 billion cubic meters by 2030 (NDRC, 2021, para. 8). In addition to forestation, the landscapes for promoting carbon sequestration are also expanded to grasslands, wetlands, and agriculture and rural areas, which arguably enables the inclusion of implementing soil carbon sequestration and biochar. Unlike forestation, quantitative targets for other land uses are not available.

Meanwhile, since 2019, forestation and sustainable land use management have been increasingly connected to the rapidly developed concept of nature-based solutions (NbS) in China (MF, MEE and MNR, 2020; MEE, 2021b, p. 37). NbSs have gained increasing attention because they can bring together actions on carbon sequestration, climate resilience, and ecosystem services restoration and therefore create co-benefits in many ways for both humans and nature. The proliferation of NbS not only further promotes the crucial role of forest ecosystem in China's terrestrial carbon sequestration (Huang et al., 2022) but also stimulates the implementation of many more nature-based land and ocean carbon sequestration methods (e.g., peatland, coastal areas, and recuperating corals).

The *second* aspect relates to environmental laws and policies that are generally applicable to CDR techniques, depending on the environmental media (land, ocean, or air) where a CDR technique is implemented and the potential harm or conflicts that a CDR technique might cause. General provisions under specific environmental legal and regulatory instruments are available regarding the prevention of soil, air, and water pollution (e.g., overdose on alkaline minerals in enhanced weathering on land or in water), the conservation of biodiversity, and the protection of the marine environment. Another thorny legal problem is probably the land use conflicts between, among others, forestation and other agricultural land uses. To deal with such conflicts, regulatory and policy instruments on territorial spatial planning legislation and legitimate spatial planning at different subnational levels and sectors are the most relevant ones. Given that land use conflicts are not unique in the context of tackling climate change but rather a broad and complicated issue associated with administrative rules, land ownership, property rights, and the planning system (Liu and Zhou, 2021), the legislation on territorial spatial planning will not be further addressed in this chapter.

The *third* aspect relates to the intersection between the governance of the research into and development of CDR and the policies for promoting carbon mitigation technologies and, more broadly, technological innovation. This aspect predominantly relates to BE/DACCS, soil- and ocean-based enhanced weathering, and ocean fertilization, which can be typically categorized as "disruptive technologies" (NDRC, 2021, para. 8I)). In contrast, afforestation, reforestation, soil carbon sequestration, and biochar are rather conventional methods for novel applications. In the current climate policies, "low-carbon," "zero-carbon," and "carbon-negative" technologies are commonly addressed in the context of "technological innovations" that support deep decarbonization for peaking carbon emission and achieving carbon neutrality (NDRC, 2021). Due to the high costs and a lack of financial incentives, disruptive CDR technologies aiming for negative emissions are still marginalized in China compared to "low-carbon technologies," with incremental technological improvement for decreasing coal consumption and "green technologies" that relate to various forms of zero-carbon renewable energy.

26.3 Main Regulatory Barriers to Integrate and Regulate CDR

26.3.1 Regarding the Explicit Regulation on CDR

As identified in Section 26.2.3.2, explicit laws and policies regulating CDR are mainly limited to afforestation and reforestation and CCUS as part of BE/DACCS. Regulatory gaps remain in governing other methods promoting land- and ocean-based carbon sinks, as well as the full supply chain of BE/DACCS. The main issues of the explicit regulation of CDR are identified in the following.

First, there is a lack of consistency in the definitions and categorization of CDR and its subcategories between different regulatory regimes and a lack of clarities in the roles of different CDR techniques in the climate mitigation portfolio for achieving carbon neutrality. From a legal perspective, clear definition and categorization will help delimitate whether the implementation of a technique is lawful, and the explicit recognition of the role of CDR in the mitigation portfolio will increase the predictability of relevant policies and stimulate stakeholders' involvement and investment. Currently, where CDR-related terms and their relationship with the goals of carbon peaking and neutrality are mentioned, the relevant legal and policy instruments demonstrate incorrect or inconsistent formulation of basic concepts. For instance, a similar term to CDR—NETs—is predominately used in recent policy documents related to carbon peaking and neutrality (NDRC, 2021). In general, it is not a problem to interchange the term NETs with CDR (Renforth and Wilcox, 2020), but the remaining confusion in Chinese policy documents is that CCUS is sometimes mixed with NETs, while CCUS coupled with oil fields or coal power plants can only reach zero emissions rather than negative emissions (Haszeldine et al., 2018).

Moreover, the terms "low-carbon technologies," "zero-carbon technologies," "negative-carbon technologies," and "green technologies" are capriciously picked in recent policy documents, while a clear delimitation of their connotations is not presented (MEE, 2021b). On the one hand, the booming use of those key terms by governments, non-governmental organizations, and civil societies demonstrates a campaign across sectors and provinces for promoting the deployment of new technologies for achieving climate goals. On the other hand, the unclear or inconsistent reference to the terminology reveals the unfamiliarity of policymakers with technologies and a considerable delay in the development of rules, regulations, and guidelines for governing those technologies compared to the development of their scientific aspects.

The reason for delaying the regulatory development of CDR is deeply rooted in China's regulatory approach and culture. Rather than exploring the regulatory issues in parallel with the technological issues, the more prevalent approach in China is to first develop the technology until realizing the impossibility of circumventing some regulatory issues (the development of CCUS is a typical example). If a technology is not yet promising for near-term implementation, it is unlikely to gain support for corresponding legal research into potential future implementation until a clear policy signal for prioritizing such a technology occurs. It can be observed that DACCS and BECCS are not included in the current policy documents (mainly referring to the Action Plan and the updated NDC in 2021), because they are not the prioritized mitigation options and are unlikely to be implemented before 2030. Similarly, current policies promoting methods for increasing carbon sinks mainly surround afforestation and reforestation, but policies for promoting other land- and ocean-based carbon sinks lag far behind. The rapid development of relevant legal research into, for example, ocean carbon sequestration will be unlikely until interest in near-term implementation emerges (probably not sooner than 2030).

26.3.2 Inconsistency With International Law in Regulating Ocean-Based CDR

As stated in Section 26.2.3.1, the LP is the international legal instrument with the most substantive relevance to ocean-based CDR techniques. As briefly mentioned in Section 2.3.2, specific domestic environmental legal instruments apply to CDR techniques insofar as the requirements for the protection and conservation of the marine environment under those instruments apply to a CDR activity that may harm the marine environment. The major legal inconsistency is that the LP adopted the precautionary approach, while China, as a contracting party to the LP, has not yet internalized this approach in its environmental law.

Article 3.1 of the LP stipulates that

> appropriate preventative measures are taken when there is reason to believe that wastes or other matter introduced into the marine environment are likely to cause harm even when there is no conclusive evidence to prove a causal relationship between inputs and their effects.

The 2013 Amendment to the LP also adopted the precautionary approach, requiring that no climate geoengineering activities be allowed in the absence of science-based, global, transparent, and effective control and regulatory mechanisms for geoengineering. Although the 2013 Amendment has not yet entered into force, it still has strong policy implications for China's implementation of marine CDR. The Chinese environmental law regime merely adopted the prevention principle that prevents the materialization of *known* or *anticipatable* risks but does not prohibit activities with scientific uncertainties. The prevention principle in the context of Chinese environmental law is typically formulated as the principle of "prevention in the first place and integrating prevention with control," according to Article 5 of the Law on Environmental Protection. This provision requires the prevention of the materialization of scientifically proven risks and the control of the severity of risks when the materialization is unavoidable. Furthermore, as the main domestic law mirroring the LP, the Law on Marine Environmental Protection (LMEP) does not include the precautionary approach either. Such a legal inconsistency especially exists in Article 12 and Chapter 7 of the LMEP on the prevention of the dumping of wastes. This inconsistency implies that domestic environmental law has lower criteria for regulating marine technologies than international law, which means, for instance, fewer requirements for risk and impact assessments for a marine CDR activity and a lower threshold of the approval of a marine CDR activity in China. In the future, if China plans to implement offshore CO_2 storage or ocean-based CDR techniques, it is necessary to integrate the precautionary approach into the relevant domestic environmental legal regime. Among others, the requirements related to risk assessments and project permits insurance need to take into account the specific assessment frameworks for marine geoengineering activities in Annex 5 of the 2013 Amendment.

26.3.3 Market Incentives for the Large-Scale Implementation of Different CDR Techniques

Even if the policy drivers that facilitate the development of CDR techniques exist, that is, the temperature goal under the Paris Agreement and the carbon neutrality goal in China, economic incentives would finally determine which CDR technique will be implemented at scale in the future. Both the policy support from governments (e.g., directly funding CDR projects or carbon tax) and the market mechanisms that allow the trading of carbon credits generated from CDR projects constitute such economic incentives for large-scale implementation of certain CDR techniques in the future. Regarding this, it is evident that the current climate policy promotes afforestation and reforestation for enhancing carbon sinks, and the policy support can arguably be extended to soil carbon sequestration and biochar, along with the proliferation of NbS in China. In comparison, the policy support including funding for other CDR methods is considerably weak and even a void. The debatable potential of producing bioenergy, the high cost of direct air capture, and the complexity of the overall supply chain of CCS collectively determine that BECCS and DACCS are unlikely to be deployed at a large scale without a deliberate effort by governments (UNCCTEC, 2021; Baus and Nehr, 2022; Beaumont, 2022).

Regarding the market mechanisms, China launched its national emissions trading scheme (ETS) in 2021 (MEE, 2021d). The current national ETS only covers the major emitting entities in the power sector, and it does not allow enterprises to offset their emissions with the reductions generated from carbon capture and sequestration. However, China's certified emission reductions (CCERs) have been playing a supplementary role to the national ETS to cover emission reductions outside the scope of the national ETS and enable enterprises to reduce the cost of complying with their obligations of mitigation (Li et al., 2019). *CCER* refers to the voluntary emission reductions recorded on the national trading registration system after the effects of renewable energy, forestry carbon sinks, methane utilization, and other projects in China are quantified and certified (MEE, 2021d). The national ETS allows 5% of verified emissions to be offset through the surrender of CCERs. Considering that the relevance of CCER for CDR is limited to the certified forestry carbon sink projects and the allowed offsets under the national ETS is merely 5%, the market incentive for developing CCER afforestation projects to increase carbon sinks seems low. However, the market incentive will increase in the near future when carbon sink methodologies are further developed to include diverse soil-based and water-based carbon removal methods (e.g., carbon farming and algae cultivation) and the market for voluntary emission offsets grows (see Section 26.4.1).

26.4 The Future of CDR in Contributing to Achieving Carbon Neutrality in China

26.4.1 CDR in the Policy Portfolio of Mitigation Strategies and Options

The role of CDR in the portfolio of mitigation strategies and options needs to be clearly defined in future climate policies if a new climate law is not yet mature. In general, it has been widely accepted that the climate benefits of CDR techniques are not equivalent to the impact of emissions (Kaupa, 2022); CDR will play a supplementary role in climate mitigation, which means that it should only be applied in situations where deep carbonization through emissions reduction is impossible or very difficult. Among the CDR techniques listed in Table 26.1, afforestation and reforestation will continuously play a vital role in increasing land carbon sinks in China, according to the Action Plan. The soil-based CDR techniques (soil carbon sequestration and biochar and soil-based enhanced weathering) are also likely to be increasingly implemented by following the policy encouragement of developing NbS, given the co-benefits of NbS for environmental protection, natural resource conservation, and climate resilience in addition to carbon sequestration.

BECCS and DACCS are the main engineering CDR methods for removing CO_2 from the atmosphere. For BECCS, the potential for large-scale implementation is determined by both the production of bioenergy and the supply chain of CCUS. Taking into account that the still high demand for coal for power generation is under the highest pressure of decarbonization, the technology of coupling biomass with coal for power generation can be an early opportunity for the development of the BECCS technology in China (Fan et al., 2021). In comparison, the large-scale implementation of DACCS seems less likely than BECCS in the coming decade due to its high cost and irrelevance with the decarbonization of the power sector.

To further incentivize the employment of BECCS and DACCS in China, it is essential to develop the regulations that recognize carbon offsets generated from carbon removals. Echoing the global trend of fulfilling corporate responsibilities for tackling climate change, more and more

Chinese enterprises that do not (yet) hold mandatory mitigation responsibilities will voluntarily take actions on offsetting their carbon emissions, which means that the market for voluntary carbon offsetting and certificate trading will probably grow in China. To date, the regulation on the voluntary GHG emission reduction transactions (NDRC, 2012) is still under revision, with an aim to promote the market potential of diverse carbon sinks and removals in a more inclusive and legitimate way. A valuable reference for China in this field would be the recent proposal adopted by the European Commission on the regulation on a Union certification framework for carbon removals (European Commission, 2022), which typically includes BECCS and DACCS.

Regarding the third category of CDR, as stated in Section 26.3.2 earlier, the legitimacy of applying ocean-based CDR techniques is impacted by international regulation and further complicated by inconsistent domestic law. Domestic rules, in particular, that regarding the protection of the marine environment and the dumping of waste into marine waters, need to be modified in consistency with the precautionary principle under the LP. In the future, likely, ocean fertilization will not receive much policy and market support in China due to the moratorium by the 2013 Amendment to the LP in addition to the already-existing skepticism by Chinese scientists about its potential ecological risks (Chinese Academy of Sciences, 2013). However, innovations in increasing coastal carbon sinks, or more generally, "blue carbon sinks" (such as the removal of CO_2 by algal and seaweed cultivation), are rapidly developing, although they are not typically recognized as CDR techniques. In this case, the development of ocean-based CDR techniques in China should not be isolated from the promotion of low-carbon fisheries. Even though the latter does not belong to marine CDR, based on the current definition and categorization, low-carbon fisheries can still contribute to carbon neutrality by increasing coastal carbon sinks, which essentially function the same as CDR. Moreover, since low-carbon fisheries create the co-benefits of increasing carbon sinks, marine environmental conservation, and economic gains, they could draw more market attention thanks to the non-climate co-benefits (Du, 2021).

Moreover, clear and consistent definitions and categorization of CDR techniques are required in future climate policies, as they will help delimitate the lawfulness of CDR activities and provide a stable and predictable policy environment for stimulating stakeholders' involvement and investment. More specifically, future climate policies and regulations need to explicitly define CDR or NETs and differentiate it/them from the similar terms "low-carbon technologies" or "green technologies." Taking into account that new CDR methods may still emerge, an exhaustive list of every possible CDR method may not be necessary. Instead, an open approach to merely defining the categories of CDR (as shown in Table 26.1) but leaving the lists of specific methods non-exhaustive for possibly including new and emerging technologies would provide more flexibility in regulating those CDR techniques with scientific uncertainties. A similar approach can be found as referred to the 2013 Amendment to the LP, which merely includes ocean fertilization under the list of marine geoengineering activities but leave the list open for new and emerging marine geo-engineering techniques (see Section 26.2.3.1).

26.4.2 The Employment of CDR and the Balancing Between Multiple SDGs

In China, the application of CDR is unavoidable for delivering carbon neutrality largely due to the difficulty of deep decarbonization. However, the application of CDR should not counteract the pace of decarbonization in the energy sector and jeopardize sustainable development (Mace et al., 2021), which means that CDR policies should be designed in conformity with multiple SDGs. How to pursue carbon neutrality, and how to apply CDR appropriately for carbon neutrality, essentially requires a balancing between multiple rights, benefits, and losses. An unbalancing between benefits and losses could occur when large-scale afforestation causes land use conflicts or fails to meet the expectation of CO_2 removals due to the inaccurate nature of forestry inventories. Large-scale land use changes for increasing carbon sinks would also negatively impact local peoples' livelihoods, especially the vulnerable and the poor ones (also related to SDG 12 on life on land). The balancing between the rights to economic development and the rights to a sustainable environment becomes even more difficult when the application of BECCS and DACCS is for the continuous consumption of coal (seeing the current proliferation of CCUS). A deeper discussion on how to balance the rights, benefits, and losses is beyond the scope of this chapter, but at least it is worth mentioning the significance of integrating a "just transition" into such a balancing (Carley and Konisky, 2020). In contrast to the EU's climate neutrality goal, which places "just transition" as an overarching principle (also in line with SDG 16 on justice) (seeing, for example, the new just transition mechanism and the new Regulation on Social Climate Fund proposed by the European Commission in 2021), such normative considerations are not explicitly integrated into the Chinese climate policies for carbon neutrality. Given that carbon neutrality is not and should not be the single objective of the society, a lack of normative considerations would cause a conflict when pursuing carbon neutrality counteracts other social, economic, and environmental objectives (Zhang et al., 2020). Moreover, when carbon peaking and neutrality targets are allocated to different subnational regions in China and they are required to follow a timetable for achieving the targets, a "carbon neutrality campaign" between provinces would possibly happen. Given the various socio-economic conditions at the subnational level, special attention should be given to the price of interprovincial equality behind achieving the goals (also related to SDG 12 on reducing inequalities). To a certain extent, it would be meaningful to apply BECCS and DACCS as a "time buffer" for deeper decarbonization and temporarily balancing the demand for economic development, but the deployment of those engineering methods should be managed proportionately by taking into account the capacity of emissions reduction in different provinces.

26.5 Conclusion

This chapter first concludes that, in the context of delivering China's targets of carbon peaking and neutrality, laws, regulations, and policies provide enablers and barriers for distinct CDR techniques. According to the current Chinese climate policies for carbon peaking and neutrality, a quantitative target of forest coverage by 2030 has already been made. It means that afforestation and reforestation will play a significant role in increasing land carbon sinks. Following the promotion of NbS, the landscapes for promoting carbon sequestration are also expanded to grasslands, wetlands, and agriculture and rural areas and, thus, arguably include soil carbon sequestration and biochar. Meanwhile, BECCS and DACCS are not likely to be implemented before 2030, but the employment of CCUS equipped with fossil fuel emission sources enjoys an increasing priority for achieving carbon peaking by 2030. In contrast, ocean alkalinization and ocean fertilization have received the least attention in China, but low-carbon fisheries are being promoted in the broad context of "blue carbon sinks." In addition, the supplementary role of carbon removals to carbon emissions reduction needs to be recognized by the Chinese climate policy. Also, the implementation of CDR techniques should be managed appropriately and proportionately to avoid the unbalancing between carbon neutrality and other social, economic, and environmental goals. Among others, it is worth escalating the principle of "just transition" to the overarching level of China's climate policies. The integration of "just transition" throughout the delivery of carbon neutrality in China would be essential for achieving other SDGs (e.g., SDGs 12, 15, and 16) hand in hand with SDG 13.

In the context of climate neutrality worldwide, the trend is the rapid growth of the market for voluntary carbon emission offsetting. One promising way for China to recognize carbon offsets generated from carbon removals would be setting up rules on the certification for carbon removals. Beyond voluntary carbon offsetting, carbon removal certificates have the potential to facilitate the development of carbon farming (e.g., soil-based CDR), sustainable building materials (permanent carbon storage in products), green financing, etc. and therefore boost sustainable development in multiple ways.

References

Baus, L., & Nehr, S. (2022). Potentials and limitations of direct air capturing in the built environment. *Building and Environment, 208*, 108629.

Beaumont, M. L. (2022). Making direct air capture affordable; Technology, market and regulatory approaches. *Frontiers in Climate, 4*, 756013.

Cai, W., He, N., Li, M., Xu, L., Wang, L., Zhu, J., . . . & Sun, O. J. (2022). Carbon sequestration of Chinese forests from 2010 to 2060: Spatiotemporal dynamics and its regulatory strategies. *Science Bulletin, 67*(8), 836–843.

Carbon Brief. (2021). *Q&A: What Does China's 14th 'Five Year Plan' Mean for Climate Change*? Available at: www.carbonbrief.org/qa-what-does-chinas-14th-five-year-plan-mean-for-climate-change/

Carley, S., & Konisky, D. M. (2020). The justice and equity implications of the clean energy transition. *Nature Energy, 5*(8), 569–577.

Chen, J., Xu, C., Gao, M., & Li, D. (2022). Carbon peak and its mitigation implications for China in the post-pandemic era. *Scientific Reports, 12*(1), 1–16.

Chinese Academy of Sciences. (2013). *Academician Jiao Nianzhi: "Fertilising" the Oceans Could Have Catastrophic Consequences*. Available at: www.cas.cn/ys/ysjy/201302/t20130220_3763860.shtml

Du, H. (2021). The international regulation of marine geoengineering—with policy implications for carbon neutrality in China. *Chinese Journal of Environmental Management, 2021*(6), 132–139 (in Chinese).

Du, H., & Zhang, H. (2022). Climate neutrality in the EU and China: An analysis of the stringency of targets and the adaptiveness of the relevant legal frameworks. *Review of European, Comparative & International Environmental Law*.

Energy & Climate Intelligence Unit. (2022). *Net Zero Scorecard*. Available at: https://zerotracker.net/

European Commission. (2019). *The European Green Deal (Communication)*. COM (2019) 640 final.

European Commission. (2022). *Proposal for a Regulation of the European Parliament and of the Council Establishing a Union Certification Framework for Carbon Removals*. COM (2022) 672 final.

European Parliament and Council. (2021). *Regulation 2021/1119 of 30 June 2021 Establishing the Framework for Achieving Climate Neutrality and Amending Regulations (EC) No 401/2009 and (EU) 2018/1999 ('European Climate Law')* [2021] OJ L243/1.

Fan, J., Li, J., Yan, S., Yu, C., Zhang, X., Xiao, P., . . . & Fang, M. (2021). Application potential analysis for bioenergy carbon capture and storage technology in China. *Thermal Power Generation, 50*(1).

Han, J., Zhang, A., Kang, Y., Han, J., Yang, B., Hussain, Q., . . . & Khan, M. A. (2022). Biochar promotes soil organic carbon sequestration and reduces net global warming potential in apple orchard: A two-year study in the Loess Plateau of China. *Science of the Total Environment, 803*, 150035.

Haszeldine, R. S., Flude, S., Johnson, G., & Scott, V. (2018). Negative emissions technologies and carbon capture and storage to achieve the Paris agreement commitments. *Philosophical Transactions of the Royal Society A: Mathematical, Physical and Engineering Sciences, 376*(2119), 20160447.

He, J., Li, Z., Zhang, X., Wang, H., Dong, W., Du, E., . . . & Zhang, D. (2022). Towards carbon neutrality: A study on China's long-term low-carbon transition pathways and strategies. *Environmental Science and Ecotechnology, 9*, 100134.

Huang, Y., Sun, W., Qin, Z., Zhang, W., Yu, Y., Li, T., . . . & Zhang, P. (2022). The role of China's terrestrial carbon sequestration 2010–2060 in offsetting energy-related CO_2 emissions. *National Science Review*.

International Maritime Organization (IMO). (1972). *Convention on the Prevention of Marine Pollution by Dumping of Wastes and Other Matter 1972*. Available at: www.imo.org/en/OurWork/Environment/Pages/London-Convention-Protocol.aspx.

IMO. (2006). *1996 Protocol to the London Convention on the Prevention of Marine Pollution by Dumping of Waste and Other Matter 1972*. Available at: www.imo.org/en/OurWork/Environment/Pages/London-Convention-Protocol.aspx

IMO. (2013). *Resolution LP.4(8) on the Amendment to the London Protocol to Regulate the Placement of Matter for Ocean Fertilization and Other Marine Geoengineering Activities*. Available at: www.imo.org/en/OurWork/Environment/Pages/geoengineering-Default.aspx

Jiang, P., Chen, Y., Geng, Y., Dong, W., Xue, B., Xu, B., & Li, W. (2013). Analysis of the co-benefits of climate change mitigation and air pollution reduction in China. *Journal of Cleaner Production*, *58*, 130–137.

Kaupa, C. (2022). Scrutinizing net zero: The legal problems of counting greenhouse gas emissions, removals and offsets together. *Review of European, Comparative & International Environmental Law*.

Li, L., Ye, F., Li, Y., & Chang, C. T. (2019). How will the Chinese certified emission reduction scheme save cost for the national carbon trading system? *Journal of Environmental Management*, *244*, 99–109.

Liu, Y., & Zhou, Y. (2021). Territory spatial planning and national governance system in China. *Land Use Policy*, *102*, 105288.

Liu, Z., Deng, Z., He, G., Wang, H., Zhang, X., Lin, J., . . . & Liang, X. (2022). Challenges and opportunities for carbon neutrality in China. *Nature Reviews Earth & Environment*, *3*(2), 141–155.

Lu, X., Cao, L., Wang, H., Peng, W., Xing, J., Wang, S., . . . & McElroy, M. B. (2019). Gasification of coal and biomass as a net carbon-negative power source for environment-friendly electricity generation in China. *Proceedings of the National Academy of Sciences*, *116*(17), 8206–8213.

Mace, M. J., Fyson, C. L., Schaeffer, M., & Hare, W. L. (2021). Large-scale carbon dioxide removal to meet the 1.5°C limit: Key governance gaps, challenges and priority responses. *Global Policy*, *12*(S1), 67–81. https://doi.org/10.1111/1758-5899.12921

Masson-Delmotte, V., Zhai, P., Pörtner, H. O., Roberts, D., Skea, J., Shukla, P. R., . . . & Waterfield, T. (2018). Global warming of 1.5 C. *An IPCC Special Report on the Impacts of Global Warming*, *1*(5).

McLaren, D. P., Tyfield, D. P., Willis, R., Szerszynski, B., & Markusson, N. O. (2019). Beyond "net-zero": A case for separate targets for emissions reduction and negative emissions. *Frontiers in Climate*, *1*, 4.

Ministry of Ecology and Environment (MEE). (2021a). *China's Mid-Century Long-Term Low Greenhouse Gas Emission Development Strategy*. Available at: https://unfccc.int/documents/307765

MEE. (2021b) *China's Achievements, New Goals and New Measures for Nationally Determined Contributions*. Available at: www4.unfccc.int/sites/ndcstaging/PublishedDocuments/China%20First/China%E2%80%99s%20Achievements,%20New%20Goals%20and%20New%20Measures%20for%20Nationally%20Determined%20Contributions.pdf

MEE. (2021c). *China's Policies and Actions for Addressing Climate Change (2020)*. Available at: www.ncsc.org.cn/yjcg/cbw/202111/P020211117418821628432.pdf

MEE. (2021d). *Measures for the Administration of Carbon Emissions Trading (for Trial Implementation)*. Available at: www.mee.gov.cn/xxgk2018/xxgk/xxgk02/202101/t20210105_816131.html

Ministry of Finance, Ministry of Ecology and Environment and Ministry of Natural Resources. (2020). *Guidelines on Ecosystem Restoration*. Available at: www.gov.cn/zhengce/2020-09/11/content_5542551.htm

National Development and Reform Commission (NDRC). (2012). *Notice of the National Development and Reform Commission on Issuing the Interim Measures for the Administration of Voluntary Greenhouse Gas Emission Reduction Transactions*. Available at: www.mee.gov.cn/ywgz/ydqhbh/wsqtkz/201904/P020190419527272751372.pdf

NDRC. (2016). *Enhanced Actions on Climate Change: China's Intended Nationally Determined Contributions*. Available at: www4.unfccc.int/sites/ndcstaging/PublishedDocuments/China%20First/China's%20First%20NDC%20Submission.pdf

NDRC. (2021). *Action Plan for Carbon Dioxide Peaking Before 2030*. Available at: https://en.ndrc.gov.cn/policies/202110/t20211027_1301020.html

Renforth, P., & Wilcox, J. (2020). The role of negative emission technologies in addressing our climate goals. *Frontiers in Climate*, *2*, 1.

Rogelj, J., Geden, O., Cowie, A., & Reisinger, A. (2021). Three ways to improve net-zero emissions targets. *Nature*, *591*(7850), 365–367.

Scott, V., & Geden, O. (2018). The challenge of carbon dioxide removal for EU policy-making. *Nature Energy*, *3*(5), 350–352.

Sharman, N. (2022). Inter-State climate technology transfer under the UNFCCC: A benefit-sharing approach. *Review of European, Comparative & International Environmental Law*.

Shepherd, J. G. (2009). *Geoengineering the Climate: Science, Governance and Uncertainty*. Royal Society.

Song, B., Almatrafi, E., Tan, X., Luo, S., Xiong, W., Zhou, C., . . . & Gong, J. (2022). Biochar-based agricultural soil management: An application-dependent strategy for contributing to carbon neutrality. *Renewable and Sustainable Energy Reviews, 164*, 112529.

State Council. (2021). *Outline of the People's Republic of China 14th Five-Year Plan for National Economic and Social Development and Long-Range Objectives for 2035*, English Translation. Available at: https://cset.georgetown.edu/wp-content/uploads/t0284_14th_Five_Year_Plan_EN.pdf (www.gov.cn/xinwen/2021-03/13/content_5592681.htm)

UN Affairs. (2020). *China Headed Towards Carbon Neutrality by 2060; President Xi Jinping Vows to Halt New Coal Plants Abroad*. Available at: https://news.un.org/en/story/2021/09/1100642

UN Environmental Programme (UNEP). (2021). *The Emissions Gap Report 2021: The Heat Is On*. Executive Summary. Available at: https://wedocs.unep.org/bitstream/handle/20.500.11822/36991/EGR21_ESEN.pdf

UN General Assembly (UNGA). (1982). *Convention on the Law of the Sea*, 10 December 1982. Available at: www.un.org/depts/los/convention_agreements/texts/unclos/unclos_e.pdf

United Nations Climate Change Technology Executive Committee (UNCCTEC). (2021). *Emerging Climate Technologies in the Energy Supply Sector*. Available at: https://unfccc.int/ttclear/tec/energysupplysector.html

Wang, X., Jiang, Z., Li, Y., Kong, F., & Xi, M. (2019). Inorganic carbon sequestration and its mechanism of coastal saline-alkali wetlands in Jiaozhou Bay, China. *Geoderma, 351*, 221–234.

Weng, Y., Cai, W., & Wang, C. (2021). Evaluating the use of BECCS and afforestation under China's carbon-neutral target for 2060. *Applied Energy, 299*, 117263.

Xiangbai, H., & Zahar, A. (2021). Innovation and the law in state reports on climate change action. In *The Transformation of Environmental Law and Governance* (pp. 129–146). Edward Elgar Publishing.

Zhang, C., Cai, W., Liu, Z., Wei, Y. M., Guan, D., Li, Z., . . . & Gong, P. (2020). Five tips for China to realize its co-targets of climate mitigation and sustainable development goals (SDGs). *Geography and Sustainability, 1*(3), 245–249.

Zhang, H. (2021). Regulations for carbon capture, utilization and storage: Comparative analysis of development in Europe, China and the Middle East. *Resources, Conservation and Recycling, 173*, 105722.

Zhou, C. (2016). The legal regulation of low-carbon technology—In the context of the relevant provisions under the 'Law on Tackling Climate Change' (Draft). *Law Review, 34*(5), 148–156 (in Chinese).

Zhou, C. (2019). Can intellectual property rights within climate technology transfer work for the UNFCCC and the Paris agreement? *International Environmental Agreements: Politics, Law and Economics, 19*(1), 107–122.

27

EDUCATION FOR SUSTAINABILITY IN THE UNIVERSITY–SOCIETY RELATIONSHIP

Possibilities and Challenges in the Contemporary Brazilian Context

Jaqueline Silva Melo, Armindo dos Santos de Sousa Teodósio, Carolina Costa Resende, and Virgínia Simão Abuhid

27.1 Introduction

Universities face major challenges in contemporary reality. Tackling fake news, denial of science, and the production of knowledge capable of leading societies to move forward in the transition to sustainability are some of these challenges.

The relationship between universities, civil society groups (social and environmental movements, NGOs, communities, and traditional groups), market players (large companies, small and medium-sized companies, and entrepreneurs), and government agencies is marked by the difficulty of recognizing valid knowledge, whether formal or informal, scientific or popular/communitarian, for the promotion of sustainability.

Faced with the complexity of the socio-environmental problems of contemporary reality, programs, projects, and interaction actions between the university and the society face renewed challenges related to the modernization of academic work as one of the university's social activities. Brazilian legislation establishes that the activities related to the interaction between university and society must seek the construction of collaborative projects which allow, through a dialogical interaction with society, an "exchange of knowledge, participation and contact with contemporary complex issues present in the social context." It is a collective construction that consolidates from dialogue with people, groups, and sectors outside the academic community seeking to contribute to social transformation (CNE, 2018, p. 2).

According to Oliveira (2013), Paulo Freire, the patron of education in Brazil, leads us to critically reflect on the term "university–society relations," which, in a first analysis, in its associative field, denotes that "the University-Society Relationship action involves, any whatever the sector in which it takes place, the need felt by those who make it to go to another part of the world, considered inferior, to, in their own way, normalize it." However, the author says that

DOI: 10.4324/9781003285472-29

"University-Society Relations is not this; that University-Society Relations is educational." For him, "[e]ducating and educating oneself, in the practice of freedom, is not to extend something from the seat of knowledge of the seat of ignorance to save, with this knowledge, those who inhabit it" (Oliveira, 2013, p. 16). For Freire, "University-Society Relations has been understood as transmission, transference, invasion and not as communication, as co-participation of subjects in the act of knowing" (Gadotti, 2017, p. 5). Thus, educational actions beyond the walls, promoted by programs, projects, and actions between the university and society, must constitute a training space in which more people learn the value of equality, freedom, autonomy, plurality, solidarity, and justice, principles that must be adopted by universities and reflected in their community and social projects and programs.

Allied to Paulo Freire's reading of the world and reality to face concrete problems, we can search for, according to Gadotti (2017, p. 12), "a more popular University-Society Relations." For this author, "University-Society Relations project in this perspective assumes a set of interrelated activities that take place in a given territory to solve certain problems through explicit strategies" (Gadotti, 2017, p. 12).

In addition to solving concrete problems, Edgar Morin brings up

> the importance of articulation between knowledge, considering that the disciplines, as they are structured, only serve to isolate objects from their environment and isolate parts of a whole. They eliminate existing clutter and contradictions, to give a false sense of order. Education, according to him, should break with that, showing the correlations between knowledge, the complexity of life and the problems that exist today.
>
> (Vieira, 2017, p. 10)

The university–society relations should also be analyzed from the perspective of the "ecology of knowledge," a concept introduced by Boaventura de Sousa Santos. For him,

> the university of the 21st century needs to recognize the existence of plural knowledge, which he called, highlighting the need for dialogue between the scientific and humanistic knowledge that the university produces with lay, popular, traditional, peasant knowledge, arising from of other non-Western cultures that circulate in society. They are knowledge developed by humanity and that need to find their points in common.
>
> (Morais, 2021, p. 71)

"The objective of the ecology of knowledge is to force scientific knowledge to confront other types of knowledge, in order to balance what was unbalanced in the first modernity: the relationship between the natural sciences and social practices" (Morais, 2021, p. 70).

Aware of the importance of university–society relations in the integral formation of students in the context of Brazilian higher education, the National Council of Education (CNE) of the federal government of Brazil, through Resolution No. 7 of 2018, established guidelines for the circularization of these interaction activities between universities and society in higher education. They become mandatory and must make up at least 10% of the total workload of undergraduate students, becoming part of the course curriculum. The aforementioned legal norm came into effect in 2023, and academic work began to be developed through the most diverse mechanisms.

According to Finnveden et al. (2020, p. 686), "research on the implementation of education for sustainable development (ESD) in higher education has been extensive and growing in the past 20 years." But we need more studies about how universities have been creating sustainable education and how academic organizations are structured to deal with sustainability in education. For Dobers et al. (2008), we need to "study and evaluate," or analyze, how, "on the level of organisations, . . . academic organizations and disciplines shape education for sustainability and research into sustainability issues" (p. 2,013).

The aim of this study is to analyze the experience of a higher education institution in Brazil which currently has more than 80,000 students distributed in seven municipalities in the state of Minas Gerais, Brazil. To achieve this objective, a case study of a qualitative nature was developed, focusing on the university–society relations modalities developed by this higher education institution (HEI) which point out the possibilities, risks, and pitfalls of modernizing the role of the university as one of the actors relevant in the promotion of sustainability in contemporary societies. Challenges linked to institutional inertia and the horizontal, democratic, and participatory articulation between academic and non-academic actors in the production of relevant knowledge for current socio-environmental challenges are discussed.

27.2 The University of the 21st Century

The second half of the 20th century witnessed the emergence of new information and communication technologies, bringing great advances in the scientific field and intensifying the flow of information and people. The so-called "information age" promoted greater integration between territories, thus creating a networked space.

This whole movement brought a context of breaking paradigms and social transformations that strongly impacted institutions, including higher education. For Santos and Filho (2008), this period triggered a series of challenges that culminated in three crises that affected the status quo of universities and forced them to rethink their role in society, as well as their way of acting.

The "crisis of hegemony" put the dominance of higher education attributed to the university to the test as the state began to allow other institutions to have other ways of building knowledge. The "legitimacy crisis," in turn, shook the university's consensual bias from the moment it faced the impasse of the hierarchy of specialized and elitist knowledge and the social and political demands for the democratization of education for other popular layers. And the last crisis, the "institutional" crisis, shook the autonomous roots of the university, while its objectives and values began to be strongly affected by the criteria of effectiveness of a business nature (Santos and Filho, 2008).

From then on, a concentration on the institutional crisis was perceived, with the intention of resolving the crises of hegemony and legitimacy in a secondary way. However, what happened was a growing intellectual mischaracterization of the university, the growing segmentation of the university system, and the commodification of education. As a consequence, according to Santos and Filho (2008, p. 40), scientific research originating from universities became distinct from the knowledge produced outside their walls, even though these actors hold the "university ethos." This distinction meant that the scientific problems to be solved were determined by the will of its investigators, fostering "predominantly disciplinary knowledge whose autonomy imposed a relatively decontextualized production process in relation to the urgencies of everyday life in societies." For Santos and Almeida Filho (2008), this homogeneous knowledge does not take into account the diversity, the multiplicity of actors and problems of society, which demands a new model of knowledge that the author calls "pluriversity." For the author, such knowledge is a "contextual knowledge insofar as the organizing principle of its production is the application that can be given to it," as it requires dialogue and confrontation between researcher and users of the science produced (Sousa, 2008, p. 40, 41).

It is in this context that the university–society relations consolidates itself as a channel of dialogue between the university and society, seeking to solve real problems with the potential to cause the much-needed social transformation. University–society relations is the core activity which, integrated with teaching and research, proposes student's protagonism, where the greatness of social justice and the pertinence of integral ecology are learned. According to the National Higher Education Assessment System (SINAES), it is to the extent that the construction and transfer of knowledge produced in universities take place, the assessment of impacts generated by scientific, technical, and cultural activities focused on local, regional, and national development. University–society relations favors spaces for the multiplication of university life, promoting the dissemination of academic knowledge, through permanent dialogue with society (Felippe, 2013, p. 19).

27.2.1 Interactions Between University and Society

University–society relations is one of the convergence spaces between technical-scientific knowledge, the understanding of reality, and the commitment to social transformation in the sense of autonomy, citizenship, and other values of the university.

The production of scientific knowledge, alongside teaching and research, can also be fostered by university–society relations, because, according to Figueiredo (2003, p. 1), "engaged research seeks to reflect on the praxis of action research, starting from the assumption that education cannot do without research," allowing us to share knowledge through "bridges built by meaningful relationships that can imply affective interactions, with real learning and production of partner knowledge" (Figueiredo, 2003, pp. 1–2).

For Santos and Filho (2008), the knowledge produced through this partnership has been developing a new and more intense relationship with science and technology and, therefore, requires greater participation in its production and in the evaluation of its impacts. For the author, "as science inserts itself more in society, society inserts itself more in science" (Santos and Filho, 2008, p. 43). However, the university suffered, and is suffering, as mentioned before, a crisis in its hegemony from the new technologies of communication and information, training, and citizenship. Therefore, it is necessary to find ways to adapt creatively to the changes that, according to Santos and Filho (2008), must start from the verification of the loss of hegemony and concentrate on the question of legitimacy, which goes through the definition of *university*. This definition is clear when there is, according to Santos and Filho (2008, p. 60), undergraduate and postgraduate training, research, and university–society relations—the rest is "higher education." It is through this path that the university will fight for its legitimacy, its most difficult fight, focusing on the domain of access, university–society relations, action research, ecology of knowledge, and university and public school (Santos and Filho, 2008, p. 61).

27.2.2 University–Society Relations and Sustainability

Faced with the growing and serious setbacks of civilization that prioritize individual interests to the detriment of the proper exercise of citizenship, "there is a need to repress, react and reformulate our way of being in the world" (Wisniewski, 2022, p. 32). The foundations on which economic growth is based in the globalized world have generated wealth and high productivity, but with a high social and environmental cost: Environmental degradation, conflicts, and exclusion are some of these bargaining chips. As Dobers et al. (2008) point out, "[t]here are strong arguments that the 'social paradigm' that influences universities nationally and internationally is not particularly concerned with education for sustainable development and post-normal science" (p. 212).

In many contemporary economic and social dynamics, natural heritage is seen as an object of property over which one has the right to exploit and destroy for the exclusive benefit of human beings, importing only monetary gains. Although there are expressions of concern about the impacts caused by environmental disorders, degradation and the deepening of environmental and climate problems continue to amplify (Martínez-Alier et al., 2010).

For Santos and Filho (2008), university–society relations activities should be "an alternative way to global capitalism, giving universities an active participation in building social cohesion, in deepening democracy, in the fight against social exclusion and environmental degradation and in the defense of cultural diversity" (Santos and Filho, 2008, p. 66), which go against society's urgent quest for sustainable development.

The complexity and diversity of global challenges are translated into different documents, agendas, and movements that, from different perspectives, reveal, warn, and signal demands and paths that are converging in many aspects. Among others, the Sustainable Development Goals and the World Conferences on Higher Education stand out. In the face of this complexity, higher education has a social responsibility to advance our multifaceted knowledge about various issues involving cultural, scientific, economic, and social dimensions and our ability to answer such questions. Higher education leads society to generate global knowledge to meet global challenges related to food security, climate change, conscious use of water, intercultural dialogue, renewable energy sources, and public health.

For Federico Mayor Zaragoza, former director general of UNESCO,

> from a global perspective, universities should be primarily responsible for mobilizing the world's citizens in search of a new way of approaching present trends. It would be a historic mistake not to act firmly and immediately to oppose the many ignorant and ideologically biased currents, sponsored by irresponsible leaders.
>
> (UNESCO, 2009, p. 75)

It is therefore necessary to change the system of values that serve as the background of the global economy in order to make it compatible with the demands of human dignity and environmental sustainability, which must be sought, in a relevant way, in the various university–society relations activities.

27.3 Methodology

To meet the objectives proposed in this research, a qualitative approach was established as a strategy, making it possible to explore the proposed problem, gathering a range of data and impressions that allowed analyzing the interaction of its variables.

As a method, action research was chosen, since the authors of the study are active participants in the strategic and operational processes of the dean of university–society relations of the investigated university. This choice was based on some assumptions inherent to the aforementioned method, facilitated by the experiences of the authors, acquired over the years in their individual and collective actions in teaching and in the administrative routines of the aforementioned pro-rector.

As advocated by Tripp (2005, p. 453), situational recognition and analysis provide a broader view of the practices and participants involved, which facilitates the proposition of improvements in order to plan an adequate change. In addition, the analyses, as well as the changes, are made in an iterative and incremental way, permeated by continuous reflections that go from the situational analysis, through the implementation phase of changes, as well as after the results were achieved (Tripp, 2005, pp. 453–454). Thus, the inclusion of researchers in the work routine of their object of study is an important factor for a deeper analysis of the studied phenomenon, as well as the possibilities of action and possible interventions after the research work.

The university–society relations activities of the investigated university are currently undergoing a process of transformations and adaptations due to legal requirements that have guided university–society relations arising from the National Council of Education (CNE), the regulatory body of university–society relations in Brazil. Such requirements have demanded the analysis of new possibilities for actions and questioned other practices already consolidated, which needed to be analyzed from information involving other actors and sectors of the university, which was possible through action research. For this, the authors of the present study participated in dialogues between the purpose pro-rectories (education, research, and university–society relations) and carried out the reading and in-depth interpretation of institutional documents, such as PDI (institutional development plan), PPI (institutional pedagogical project), PGE (strategic management plan), pedagogical projects of various undergraduate courses, opinions and technical reports from committees created for the analysis and elaboration of strategies for integrating university–society relations into the daily life of undergraduate courses, as well as internal guidance documents for the university–society relations curricula process. In addition, the authors participated in carrying out situational diagnoses and the consequent elaboration of proposals for the implementation of changes in the various actions that make up the university–society relations modalities of the investigated university, which allowed access and triangulation of the data to reach the results of this research: opinions and technical reports from commissions created for the analysis and elaboration of strategies to integrate university–society relations into the daily life of undergraduate courses, as well as internal guidance documents for the process of the making the curricula of the university–society relations.

According to Tripp (2005, p. 454), "action research works best with cooperation and collaboration because the effects of the practice of an isolated individual on an organization are never limited to that individual," which was evident in the trajectory of the researchers throughout the present work.

27.4 Data Analysis

27.4.1 University–Society Relations at the Investigated University

Changes in the contemporary world give higher education institutions (HEIs) an important role so that societies can face the contemporary challenges they face, especially those related to the promotion of sustainability. The national policy for university–society relations establishes that the activities of universities in their interaction with society seek "the construction of a corporate project, which allows, in an effective way, to implement an agenda of social inclusion, citizenship and humanist education, in the perspective of integral development of the human being."

It is important to emphasize the relevance of university–society relations as an agency for the production and application of knowledge in the solution of real problems, which has caused changes in university–society relations policies and regulations in Brazil, such as the one that occurred through Resolution No. 7 of 2018 of the National Council of Education (CNE), expanding academic work through university–society relations based on the most diverse mechanisms and modalities.

In 2012, 193 delegations and civil society representatives gathered to renew the global commitment to sustainable development, drawing up the Sustainable Development Goals (SDGs), composed of 70 countries and involving the most diverse stakeholders. The SDGs have as their prerogative the development of actions aimed at eradicating poverty, promoting prosperity and the well-being of all, protecting the environment, and tackling climate change by 2030.

Converging with these initiatives, Pope Francis also presented, in 2015, the encyclical letter "Laudato Si" on care for our common home. Known as the encyclical of the environment, the document presents a rich and well-founded diagnosis of the environmental ills that threaten life on the planet. Still based on the principle of integral ecology, translated by the idea that "everything is interconnected," Francis assumes that "there are not two separate crises: one environmental and the other social; but a single and complex socio-environmental crisis" (Francisco, 2015, n. 139). In another movement, the Pope called, for March 2020, a meeting with all the countries of the world to discuss a new economy, symbolically called "Economy of Francis" (Anec, 2021). The Economy of Francis brings in the core of its initiative the opening of revolutionary processes, because it focuses on "being more concerned with initiating processes than having spaces" (Francisco, 2013, n. 223). Thus, it becomes imperative to add, in a transdisciplinary perspective, the principles of integral ecology and the Economy of Francis in teaching, research, and university–society relations, through reflections and actions, collaborating to form agents of transformation in a critical way who are committed to directing the economy for the common good (Deyber, 2021).

All these initiatives are related to a perspective of life focused on the harmonious coexistence of man with nature, in a sustainable way, which makes no reference to the developmental paradigm based on the accumulation of wealth, economic growth, production, and unlimited consumption. Thus, based on the premises of the philosophy of "good living," and through the commitment to social transformation of the university–society relations, the present study will seek, in line with the objectives of sustainable development and supported by the premises of the "Economy of Francis and Clara," to show how the university–society relationship actions of the investigated university can contribute to the pursuit of dignity, professional development, and social inclusion, in a sustainable way, through the partnership between the university and the society.

The investigated university, faced with this reality, in line with its mission, and considering the national guidelines for HEIs, proposes to discuss a university–society relations policy articulated with teaching and research (Investigated University, 2006).

The management of university–society relations at the investigated university is carried out by the pro-rectory of university–society relations, one we call PROEX, who relies on the performance of an executive secretary; an academic, administrative, and communication advisory; a university–society relations observatory and project office; the university–society relations coordination of the campuses; and the units of the municipalities of Arcos, Betim, Contagem, Poços de Caldas, Serro, and Uberlândia, and of the metropolitan region of Belo Horizonte (Barreiro, Coração Eucarístico, Praça da Liberdade, and São Gabriel), in addition to the coordination of university–society relations of the institutes, the faculties, and the coordinators of the various undergraduate and graduate courses at the investigated university.

According to the university–society relations policy of the investigated university, "the university–society relations, as an end-activity integrated to Teaching and Research, is one of the places where the social function of the investigated university is exercised." By enabling the

> link between the academy and society, it works to promote citizenship, inclusion and social development. This is reflected in the citizen and humanist formation of students and teachers, in the perspective of integral development of the human being, the main mission of the University.
>
> (PPI from the investigated university, 2018)

University–society relationship practices compose, "allied to teaching and research, the principle of inseparability that governs universities, highlighted in Article 207 of the Federal Constitution of 1988. The articulation of these three areas enables the transforming relationship that links the University to society" (investigated university, 2018).

The university–society relations of the investigated university has sought to dialogue with the global environmental agenda through its projects and programs, in line with the objectives of sustainable development and with the propositions of Pope Francis in his last encyclical ("Laudato Si") in the search for a just and more sustainable society.

In accordance with its institutional development plan (PDI), the institutional pedagogical project (PPI) of the investigated university is constituted, foreseeing as part of its premises the responsibility for environmental preservation and the need for curricular practices conducive to the sustainable development of the country. The articulation of learning between theory and practice, through internship, research, and university–society relations, is also part of the guiding principles of the PPI at the investigated university. In 2011, understanding the complexity and dynamics of contextual changes due to technological and productive interference, as well as recent environmental regulation policies, the investigated university understood the need to change its PDI, proposing a specific chapter for the implementation of sustainability practices, with the proposal for an environmental responsibility program, explaining the theme as an item of its social commitment. Such a program presumes the involvement of the faculty, students, and employees as co-responsible for the sustainable future, considering that in order to initiate any proposal toward achieving the objectives, it is essential to plan and systematize actions in a political, strategic, and logically organized way, involving the entire organizational context.

Parallel to the new PDI, the strategic management plan (PGE) was prepared, which proposed to be the methodological document of the premises defined in the institutional development plan and in the institutional pedagogical project. Among the strategic themes idealized in the PGE, there is environmental sustainability. In view of this, the university–society relations at the investigated university presents in its actions what is advocated in the mission of the investigated university: "to implement an agenda of social inclusion, citizenship and humanist education, in the perspective of integral development of the human being" (investigated university, 2006).

For the investigated university, students' protagonism is the core of the dialogical relationship between academia and society, aiming at promoting citizenship, inclusion, sustainability, and social development, possible through its university–society relations guidelines, which are inseparability between teaching, research, and university–society relations; dialogic interaction between the university and other sectors of society; interdisciplinarity, multidisciplinarity, and transdisciplinarity; participation in the students' academic and professional training; contribution to the development of society, prioritizing actions aimed at the development of the social inclusion process, for the effectiveness of human rights, citizenship, and other values that guide the university's mission as a confessional and community institution, linked to the Catholic Church; and production and democratization of scientific and technological knowledge, linked to the areas of knowledge in which the university operates (investigated university, 2006).

The actions of the university–society relations of the investigated university began in the 1960s, with isolated initiatives of courses and departments aimed at assisting communities in vulnerable situations. These initiatives were consolidated through various university–society relations modalities, such as programs, projects, events, courses, and workshops that enabled dialogue between the university and other sectors of society (Resende et al., 2017, p. 91).

In recent years, the university has implemented important processes aimed at complying with MEC Resolution No. 7/2018, the norm that regulates university–society relations in Brazilian universities and, more specifically, establishes guidelines for the curricula of the core activity in undergraduate courses, process that has been structured based on an important articulation with the dean of undergraduate studies. Simultaneously, the rectory, through Ordinance No. 119/2022, established a commission to establish guidelines and conditions for the curricula of university–society relations. The work of the commission resulted in the elaboration of the guiding elements for the implementation of the university–society relations curriculum at the investigated university, with consequences in the alteration of the academic norms, which now contain references on the process of articulating the university–society relations to the curricular matrices of undergraduate courses (Resende et al., 2017).

To support the legal requirements and university–society relations activities, the PROEX's organizational structure is composed of (1) an executive secretary, responsible for advising the pro-rector, carrying out departmental and internal public management in its administrative processes; (2) academic assistants, responsible for the operational management of university–society relations modalities, through the planning and management of flows and processes and the articulation of the PROEX network; (3) the university–society relations observatory, responsible for data intelligence and monitoring of university–society relations activities, performing impact analysis, activity reports, monitoring, evaluation, quality management, project portfolio, implementation of technologies, and information systems; and (4) a coordinator of innovation and academic production.

From this context, we present in what follows an analysis of the university–society relations modalities developed at the investigated university from the watchful eye of an internal commission of the pro-rectory of university–society relations of the investigated university, formed by professors and technical administrative employees in the year 2021, having as a focus the analysis of the strengths and vulnerabilities of the activities carried out, the result of which will contribute to the reflection proposed by the present study.

Table 27.1 demonstrates the comparison of university–society relations indicators in the years 2020, 2021, and 2022, since the data for 2023 will still be investigated. Such information is collected by the monitoring and evaluation process, which constitutes an instrument to ensure the interaction between the planning and the execution of the university–society relations activities developed at the investigated university in its most diverse modalities. This process comprises the set of data collection, follow-up, and analysis procedures, with the purpose of getting to know each university–society relations activity better and checking whether the activities and results carried out by each one of them correspond to what was planned (investigated university portal, 2023).

The execution of the university–society relations activities of the investigated university relies on a wide infrastructure and a human capital composed of 22 employees of the technical administrative staff, 20 professors with workload, 76 coordinators of university–society relations projects of the undergraduate and postgraduate courses, 130 university–society relations coordinators of

Table 27.1 University–society relations performance indicators at the investigated university

Categories	*2020*	*2021*	*2022*
Projects	75	100	99
Students enrolled in university–society relations courses	13,430	13,006	16,334
Scholarship students	430	692	677
University–society relations courses	394	438	521
Teachers in university–society relations disciplines	483	469	698

Source: Dean of university–society relations/PROEX (2023).

undergraduate and postgraduate courses, and 18 university–society relations coordinators of institutes/colleges and of the units/campuses.

This entire set of people and processes is focused on supporting the various activities of the prorectory of university–society relations (PROEX), with emphasis on monitoring the coordination of pedagogical projects of courses, which includes guidance and follow-up to the coordination of undergraduate courses; the sector of follow-up, monitoring, and evaluation, which concentrates its work on reviewing the collection processes and processing data related to university–society relations; the coordination of courses and events which hosted the offer of university–society relations courses, working intensively to attract new proposals; the academic production coordination, which takes care of the publication of the university–society relations electronic magazine, *Conecte-se!* (*Qualis A4*), and the follow-up to the publication of other academic products, such as the university–society relations exhibition; the coordination of programs and projects which have fostered dialogue with *stricto sensu* graduate programs and monitors, with the support of the PROEX network, the execution of programs, and university–society relations projects.

27.4.2 Programs and Projects

The university–society relations projects of the investigated university are continuous actions of an educational, social, scientific, or technological nature, with a specific objective in the short and medium term. The programs, in turn, are an articulated set of pedagogical actions, of a theoretical or practical nature, systematically planned and organized, with a minimum workload and well-defined evaluation criteria (investigated university, 2023).

In 2021, 100 university–society relations projects were carried out at the investigated university, an average that has been maintained, as shown in Table 27.1. It should be noted that, over three consecutive years, the investigated university promoted university–society relations actions in Brumadinho, Minas Gerais, Brazil, and its surroundings, within the scope of the investigated program, "University and Brumadinho—Joining Forces," hoping to contribute to local communities in the recovery process after the rupture of dams operated by Vale S/A in 2019. The program had nine university–society relations projects and involved 28 teachers and 61 students from the university. Also noteworthy is the Paraopeba Project, through which the university carries out the coordination and methodological and final monitoring of independent technical advisories (ATIs), constituted by the judiciary as support institutions for the process of integral reparation of the damage caused by the rupture of dams of Vale S/A, and whose damage extended along the Paraopeba River and the Três Marias Lake. The initiative involved, throughout 2021, 40 professors and professionals from the technical-administrative staff, in addition to 46 university–society relations fellows from undergraduate and *stricto sensu* graduate courses. For the year 2022, within the scope of the call for university–society relations projects (74/2021), which has been published since 2005, 45 initiatives were approved to receive funding for carrying out actions in various areas of knowledge, in addition to others 54 projects approved without funding. In 2023, to date, more than 50 projects and one program are in progress.

Present in all units and *Campi* of the university, the university–society relations projects and programs have, in their guidelines, fundamental conditions for the fulfillment of the legal requirements pertinent to the national university–society relations policy. These guidelines go through the institutional link to the departments, institutes/faculties, unit/campus, deputy dean's office, dean of university–society relations, and articulation with teaching and research. As for the programs, these must be composed of at least two projects and provide for at least one more modality of university–society relations; have a coordinating professor and provide for the participation of university–society relations students, scholarship holders, and non-scholarship holders, undergraduate and graduate students, and direct and indirect beneficiaries; have a theoretical and academic foundation and the freedom to choose which SDG(s) the proposal will be linked to and to be directed to a thematic line defined by the institution; have an itinerant, regional, or national socio-spatial base; have a professor of the teaching career of the investigated university who has at least three hours of dedication to fulfill this attribution for the coordination of university–society relations actions; have cross-sector articulation (between the state (first), market (second), and civil society (third); carry out academic production with a minimum requirement of the publication of a scientific article and a presentation at a university–society relations event; have internal partnerships and external partnerships; have multilateral and/or inter-institutional funding with counterparts from internal and external partners; be registered in the project and agreement management system—GPC; allow the participation of an employee from the technical-administrative career (optional); have at least one intern or for the university–society relationship to have received a grant (to assist in the work of the program coordinator); perform partial monitoring (half-yearly) and final report (in the last year of the program); and be disclosed through public notice or extra public notice with its own promotion.

From the detailing of the guidelines and the clarity of the fundamentals presented in this modality, the importance and solidity of the projects and programs in the context of the university–society relations actions of the investigated university are evident. It is also important to highlight the alignment of actions in favor of sustainability, given the obligation to link projects and programs to the respective objectives of sustainable development of the UN. Consolidated projects and partnerships stand out, such as the sustainable university–society relations project, which for eight years has aimed to structure, implement, and monitor sustainable strategic actions at the investigated university, aiming at forming a cooperation network for sustainability at the university and surrounding communities, promoting a culture aligned with the principles of integral ecology, of the 2030 Agenda, and the concept of "sustainable university." In this perspective, and with innovative practices, it contemplates different guidelines that are articulated with almost all the Sustainable Development Goals and stimulate an integration center for environmental sustainability (CISAL) of the university.

Alignment with inclusive and affirmative public policies is guaranteed through various partnerships. One can cite as an example the strong relationship with the solidarity economy movement, which takes place through an annual seminar and solidarity economy fairs on campuses and units of the university, in which more than 30 exhibitors/small producers and artisans participate.

However, these modalities also present their challenges, especially regarding the need for improvements in their call notice and a better definition of the criteria for a project and other modalities to integrate the programs. In addition, it is necessary, based on internal studies, to better define flows for the presentation of proposals, greater integration of various university–society relations modalities to programs and projects within the information system for the project and program management (GPC), and finally the creation of criteria for evaluating processes and impacts arising from university–society relations programs and projects.

27.4.3 University–Society Relations Events

At the investigated university, public presentations and exhibitions are assumed as university–society relations events, with specific clientele, of the cultural, scientific, and technological knowledge or product developed, conserved, or recognized by society. Such events include congresses, seminars, meetings, conferences, debate cycles, exhibitions, shows, festivals, and sporting events, among others (investigated university portal, 2023).

Throughout 2021, the university accounted for 464 university–society relations events, with emphasis on the meeting of the "National Forum for University–Society Relations and Community Action," or FOREXT, held virtually on November 22 and 23, with the theme "The Sustainable Development Goals and University–Society Relations." The event was attended by prominent external professionals, such as a UNDP representative, professors, technicians, and students from educational institutions throughout Brazil.

In order to be carried out, university–society relations events at the investigated university must meet criteria pre-established by the dean of university–society relations, in view of the requirements of the university–society relations policy of the investigated university (2006) and the PROEX regulation (2022), which are to be operationally and economically viable; be open to the participation of other sectors of society, without excluding the internal community; address issues of relevance and social impact, if possible, open to the demands of external communities which enable debate and reflection that encourage the construction of a more just, fraternal, sustainable, democratic, inclusive, egalitarian, and safe society; be in line with at least a thematic area and university–society relations line and one of the Sustainable Development Goals (SDGs); be based on the principles of university–society relations and enable communication, preferably dialogical, between the university and other sectors of society; be articulated with teaching and/or research, at undergraduate and/or graduate levels, and contemplate interdisciplinarity; be open to student participation (undergraduate and graduate), contributing to their technical, professional, ethical, and humanistic improvement; meet the precepts of academic quality and encourage the publication of works by professors, technical-administrative staff, and students; be widely disseminated to other sectors of society, using the various means and forms of communication available; have university–society relations events planned to also meet the university–society relations curriculum be in accordance with the specific guidelines and regulations published for this purpose; have events be held in person, remotely, and in hybrid formats; ensure university–society relations events arising from programs, projects, courses, and provision of services, or carried out independently of the aforementioned modalities, are coordinated by at least one professor linked to the teaching career, or by substitute professors; make sure the events arising from the curricular university–society relations practices are referenced in the hours foreseen in the discipline and in the professor(s) who coordinate it; have their binding formalized in the SGE if the proposal foresees the active participation of students as university–society relations workers; and ensure all events provide for strategies to encourage participants to respond to the satisfaction and impact survey instrument(s) (social and on student training) prepared by the coordination of courses and events and monitoring and evaluation of PROEX, in dialogue with the evaluation committee (CPA). It will be up to the PROEX network to encourage and guide professors interested in preparing and submitting proposals and structuring them in accordance with current guidelines and regulations.

Bearing in mind the volume and capillarity of this university–society relations modality, some actions were proposed for the better management of the events, as well as the mitigation of the recurrent problems arising from these actions. In this sense, suggested is the creation of a specific commission to deal with the evaluation processes of university–society relations impact (for student training and for beneficiary audiences) with representatives of the PROEX advisory services, sectoral coordination, and CPA. In addition, also suggested is a greater depth and a better understanding of the flows and processes linked to the planning and feasibility of university–society relations events with external resources (partnerships, supporters, collection and payment of registrations) to think about the analysis flows of the financial viability of the proposals and their consequences from the point of view of informational systems.

27.4.4 Courses and Workshops

University–society relations courses and workshops are modalities provided for in the university–society relations policy of the investigated university (2006) and in the regulation of the dean of university–society relations (2022), defined as "an articulated set of pedagogical actions, of a theoretical or practical nature, planned and organized systematically, with a pre-established minimum duration and defined evaluation criteria" (investigated university, 2006).

In the pro-rectory of university–society relations, the modalities of courses and workshops are part of the same coordination that works with the purpose of standardizing, organizing, disseminating, and strengthening these fronts. It is up to the pro-rectory of university–society relations and the PROEX network to stimulate, encourage, assist, evaluate, and manage the offer and the realization of university–society relations courses and workshops at the university. Thus, any professor integrated into the teaching career can propose university–society relations courses and workshops in any of the areas of knowledge of the investigated university.

As of 2020, through a partnership with the virtual investigated university, PROEX started to propose university–society relations courses and distance workshops in synchronous, asynchronous, or hybrid modalities. These workshop courses are aimed at different audiences and aim to include and improve the quality of life, as well as to strengthen the university's dialogue with society, and, therefore, must necessarily be open to external audiences and may also include the academic community.

Courses can be categorized as (1) *diffusion/initiation*, with the purpose of publicizing a specific topic, which does not require previous schooling or work experience; (2) *update*, with the objective of continuing previously acquired knowledge or experience in one or more areas, which may or may not require previous schooling or work experience prerequisites; (3) *qualification/training*, which aims to meet a labor market demand and comprises training actions aimed at people or entities, public or private, including other HEIs, and which may or may not require previous schooling or professional experience prerequisites; (4) *professional/technological courses*, where proposals must consider the CNE resolution of January 5, 2021, which defines the general national curriculum guidelines for professional and technological education; and (5) *improvement*, which aims to deepen knowledge about a specific field. With regard to the workshops, the aim is to ensure continuous training, predominantly carried out according to practical or procedural know-how components and oriented toward the following objectives: (1) outlining or consolidating action procedures or producing intervention materials, concrete and identified, defined by the set of participants as the most appropriate response to the improvement of their educational interventions; (2) ensuring the functionality (utility) of the products obtained in the workshop, for the transformation of practices; (3) reflecting about the practices developed; and (4) building new procedural or technical means.

These modalities, unlike others, are present in a more constant way, with clearer procedures and in a transversal way, being commonly inserted in other university–society relationship practices, such as programs, projects, and events. However, the courses and workshops also present challenges, such as, for example, defining the workload for all categories; compliance with all or part of the guidelines contained in the PROEX regulation, namely, inseparability, dialogic interaction, multi/inter/transdisciplinarity, impact on student education, social impact, link with the areas of knowledge, and production and democratization of knowledge; establishing, in dialogue with the other commissions, whether it will be assumed that the reference will be the pedagogical project of the course, graduation, or post, or the area of knowledge; conducting transversal discussions on evaluation and monitoring mechanisms; and paying attention to general data protection law (LGPD) for support regarding enrollment and the acceptance of documents for the purpose of confirming enrollment.

Given these points of attention, the implementation of courses and workshops may take longer and compromise the offer of these modalities, especially in the short and medium term.

27.4.5 Provision of University–Society Relations Services

This modality of university–society relations concerns the carrying out of work offered by the university or contracted by third parties (community and/or companies), with articulation between teaching, research, and university–society relations. It must be characterized by intangibility, inseparability, variability, and perishability and may include advisory services, consultancies, interinstitutional and/or international cooperation, service to the public in spaces of culture, science and/or technology, technical diagnosis, institutional intervention, technical report, population assistance services (health care, sports and leisure, social assistance, legal assistance), and occasional service.

It should be noted that the provision of services at the university must always consider the pedagogical character of its action, eliminating the possibility of replacing the state in its functions and transforming itself into an agency for the sale of services (investigated university portal, 2023).

Among all the university–society relations modalities, the provision of services is the most incipient and with the greatest need for deepening on points inherent to teacher remuneration, fragility in the records that are carried out outside the university systems, the need for dialogue between PROEX and other sectors of the universities that also perform this type of service, slowness of internal processes to respond quickly to demands, lack of clarity in internal flows about this modality, and the need to format a form of cross-sectional evaluation that is integrated with other university–society relations modalities.

Bearing in mind the complexity of its execution and the bottlenecks that still exist, one can infer the reason this modality has not fully reached its full potential. Despite this, we can mention important initiatives, such as the "Brumadinho—Unindo Forças" program, the land regularization project for the quilombola population, the "Inclusive PUC," and the LER Project (which deals with the literacy of refugees), among others that are carried out with its own promotion or through a cooperation term, always aiming at the emancipation of excluded populations.

27.4.6 University–Society Relations Curriculum

The curricula of university–society relations practices at the Pontifical Catholic University of Minas Gerais (PUC MG) are part of a process that has been consolidating for some years; however, its effectiveness is more recent. Despite the recognition of university–society relations as an end activity in the various official documents of the investigated university, its effective integration with teaching and research has always been a challenge for the administrative dimension of the university (Resende et al., 2017).

According to Resende et al. (2017), one of the main strategies developed was the participation of PROEX in the analysis processes of the pedagogical projects of courses with the dean of undergraduate studies (PROGRAD), which ensured rethinking the uniqueness of each pedagogical project and bringing the university–society relations to permeate the training of students in a more organic way and not just as another mandatory item. In this sense, the partnership established between PROEX and PROGRAD was essential for the guidelines that guide university–society relations practices to be more appropriate in the development of activities in undergraduate courses at the investigated university.

This process was subsidized by the recent determination of the National Council of Education which, through Resolution No. 7 of December 18, 2018, "[e]stablishes the guidelines for university–society relations in Brazilian higher education and regulates the provisions of Goal 12.7 of Law No. 13.005/2014–2024 and makes other arrangements." In this sense, the actions that were now carried out by virtue of the university's internal regulations become mandatory by law:

> Art. 1 The Guidelines for University–Society Relations in Brazilian Higher Education are hereby instituted, which defines the principles, foundations and procedures that must be observed in the planning, policies, management and evaluation of higher education institutions of all education systems in the country. Art. 2nd The Guidelines for University–Society Relations in Brazilian Higher Education regulate the academic University–Society Relations activities of undergraduate courses, in the form of curricular components for the courses, considering them in their aspects that are linked to the training of students, as provided for in the Development Plans Institutional Projects (PDIs), and in Institutional Political Projects (PPIs) of educational entities, according to the egress profile, established in the Pedagogical Projects of the Courses (PPCs) and in other normative documents. Single paragraph. The Guidelines for University–Society Relations in Brazilian Higher Education can also be directed to postgraduate higher education courses, according to the Pedagogical Political Project (PPP) of the higher education institution.
>
> (CNE Resolution No. 7 of December 18, 2018)

In its Article 4, the referred Resolution determines that "[t]he university–society relations activities must compose, at least, 10% (ten percent) of the total student curricular workload of the undergraduate courses, which must be part of the curricular matrix of the courses" (Brasil, 2018). Therefore, the university is faced with a great challenge, since, despite already being part of the scope of its teaching–learning process, dealing with such provocation may seem to be an almost-ancestral issue (Resende et al., 2017). According to these authors:

> [I]t is known that the acknowledgment of university–society relations as a core training activity is set out in several national documents, as well as its necessary integration with teaching and research. However, the extent of rooting university–society relations in all university courses and guaranteeing the principle of inseparability at the same time is questioned.
>
> (Resende et al., 2017)

In this context, the investigated university has been undertaking, in recent years, many efforts to improve the implementation of curriculum based on legal requirements, but especially in its successful experience based on curricular university–society relations activities even before the force of law. According to Resende *et al.* (2027), "despite many advances, there are still many challenges for the identification, recognition, implementation, registration, validation and evaluation of this new university–society relations modality in a university of the size of the investigated university." In their study, these authors point out the need for "reflections on their theoretical and factual perspectives to root the university–society relations curriculum at the national level, in addition to research that evaluates its contribution to student and teacher education and to society" (Resende et al., 2017).

27.5 Final Remarks

Our chapter discusses education for sustainability in the university's relationship with society, highlighting the possibilities and challenges in contemporary Brazilian context.

To achieve the main objective of the study, we present an experience from a large university in Brazil. Considered one of the best private universities in Brazil, it is the only private institution in Minas Gerais to be part of the ranking of the best universities in the world in the *Times Higher Education* and elected six times by *Student Guide*, published by April Publisher, as the best among private ones in the country—the investigated university is recognized by the Vatican's Congregation for Catholic Education as the largest Catholic university in the world, housing more than 90,000 students in undergraduate and graduate courses. The institution maintains a permanent faculty of 1,500 professors and 2,900 employees (investigated university portal, 2023).

In its mission, the investigated university presents "the human and social development of the academic community based on ethical and solidary training, production and dissemination of knowledge, art and culture" (PDI, 2011). In this sense, the teaching, research, and university–society relations tripod is articulated in innovative projects aimed at transforming society. In this area,

> the commitment to social inclusion and the promotion of citizenship through education made the Investigated University one of the pioneers in the implementation of the federal government's scholarship program PROUNI. Between 2005 and the second half of 2022, the University granted more than 56,000 scholarships through the program.
>
> (investigated university portal, 2023)

Alongside this, it is evidenced in the results of the research that the university–society relationship actions of the investigated university have a prominent role in its most diverse documents, understanding that, as stated by Thiollent (2003), university–society relations is an important source of information for the academic world, difficult to obtain by other means. Through it, a wealth of contacts is established, with a wide variety of interlocutors.

The general results of this research allowed us to elucidate some issues of the analyzed phenomenon, for which we would like to list some considerations. In the first place, the institutional concern with full compliance with the legal assumptions inherent to university–society relations activities in higher education was evident. In addition, it should also be noted that university–society relations activities do not take place in isolation, guaranteeing the university–society relations guidelines advocated in the national university–society relations policy in Brazil, especially with regard to the inseparability of teaching, research, and university–society relations. As mentioned in the national university–society relations policy, "the limits and possibilities of the University directly affect the development of university–society relations" (PNE, 2012). This scenario presented in the PNE in 2012 is present today, which does not diminish the advances and achievements experienced in the university–society relations practices of the investigated university, despite the challenges presented here. Finally, the present study reinforced the institutional vocation of the researched university toward sustainable practices, subsidized by university–society relationship initiatives in light of the UN sustainable development objectives and the teachings of the Holy Father, Pope Francis, through his encyclical "*Laudato Si*," as well as the prerogatives of the "Economy of Francis and Clare" for a more just and fraternal economy.

We close our considerations here in the hope of a constant alignment of the university–society relations activities of the investigated university with "bem viver," whose precepts make up a worldview that, regardless of its etymology ("bem viver," "buen vivir/vivir bien," or "good living"), has a recent and growing debate that has been accentuated in recent years, in which it was intended to be used as a basic principle of the university–society relationist essence addressed in this research (Alcântara & Sampaio, 2017), instigating other researchers on this path toward a fairer, more inclusive, and supportive university/society.

References

Alcântara, LCS, & Sampaio, CAC (2017). *Bem Viver as a development paradigm: Utopia or a possible alternative*? Development and Environment, p. 40.

BRAZIL. Law 13.005 of June 25, 2014. *Approves the National Education Plan—PNE and Other Measures*. Official Gazette of the Union, Brasilia, 26 June. 2014. Available at: www.planalto.gov.br/CCIVIL_03/_Ato2011-2014/2014/Lei/L13005.htm. Accessed on 06/10/2023.

BRAZIL. National University-Society Relations Policy (2012). Available at: https://portal.pucminas.br/imagedb/documento/DOC_DSC_NOME_ARQUI20161028115851.pdf?_ga=2.53664744.1795712540.1687114794-1255515626.1659186298. Accessed on 06/15/2023.

BRAZIL. Superior, E. Resolution No. 7, of December 18, 2018. Available at: https://abmes.org.br/public/arquivos/legislacoes/Resol_7cne.pdf. Accessed on 06/20/2023.

Deyber, I (2021). Available at: www.pucminas.br/Pastoral/pensandobem/paginas/economia-defrancisco-uma-agenda-para-a-universidade.aspx. Accessed on 06/07/2023.

Dobers, P, Linderstrom, M, Mobjork, M (2008). Institutional entrepreneurship in an academic organization: Sustainability at Malardalen University. *International Journal of Innovation and Sustainable Development*, vol. 3, no. 3/4, pp. 202–216.

Felipe, WC (2013). *University-Society Relations in Higher Education Community Institutions: References for the Construction of a National University-Society Relations Policy in ICES*. Itajaí, SC: Editora UNIVALE.

Figueiredo, João BA (2003). *Dialogical Environmental Education and Social Representations of Water in the Northeastern Backlands Culture: A Contribution to Environmental Awareness in Irauçuba – CE (Brazil)*. Thesis (PhD in Biological Sciences/Ecology/Environmental Education) – UFSCar. São Carlos, SP, pp. 1–2.

Finnveden, G, et al. (2020). Evaluation of integration of sustainable development in higher education in Sweden. *International Journal of Sustainability in Higher Education*, vol. 21, no. 4, pp. 685–698.

Francisco, P (2013). Apostolic exhortation evangeli gaudium: Proclaiming the gospel in today's world. *Pomegranate*. Available at: www.vatican.va/content/francesco/pt/apost_exhortations/documents/pap a-—francesco_esortazione-ap_20131124_evangelii-gaudium.html. Accessed on 05/06/2023.

Francisco, P (2015). *Laudato Si': On Caring for Our Common Home*. Available at: http://w2.vatican.va/content/francesco/en/encyclicals/documents/papafrancesco_20150524_en cyclica-laudato-si.html. Accessed on 06/10/2023.

Freire, P (1921–1997). *University-Society Relations or Communication? [Electronic Resource]/Paulo Freire*, translation Rosiska Darcy de Oliveira, 1st ed. Rio de Janeiro: Paz e Terra, 2013.

Gadotti, M (2017). University-society relations: For what. *Instituto Paulo Freire*, vol. 15, pp. 1–18.

Goes, UTT (2021). *Training of the Educator-Co-investigator, Development of Co-learning and Pedagogical Practices: Convergences between Transdisciplinarity, Learning, Cognitive Mapping and Technologies*. Doctoral thesis. Fortress, p. 84.

Martínez-Alier, J, Pascual, U, Vivien, FD, & Zaccai, E (2010). Sustainable de-growth: Mapping the context, criticisms and future prospects of an emergent paradigm. *Ecological Economics*, vol. 69, no. 9, pp. 1741–1747.

Morais, MC (2008). *Ecology of Knowledge*. São Paulo.

Morais, MC (2021). *Ecology of Knowledge*. Hidelbrando dos Santos Soares, p. 70.

National Association of Catholic Schools (ANEC) (2021). https://anec.org.br/acao/economia-defrancisco-e-clara#informacoes. Accessed on 06/10/2023.

Oliveira, EM (2013). Social technology, university and society: Extension as a strategic space for intervention. *Technology and Society Journal*, vol. 9, no. 17.

PUC Minas (2006a). *PPI Institutional Pedagogical Project*. Belo Horizonte: Investigated University. Belo Horizonte/MG.

PUC Minas (2006b). *University-Society Relations Policy of the Investigated University*. Belo Horizonte/MG.

PUC Minas (2012). *Institutional Development Plan*. PDI. Belo Horizonte/MG.

Resende, MCF, de Albuquerque, LM, dos Reis Moreira, T, & de Oliveira Borges, BKN (2017). The curricularization of university-society relations practices at the investigated university. *Log in! Interdisciplinary Journal of University-Society Relations*, 1(2), 88–103.
Santos, BDS (2004). *Prudent Knowledge for a Decent Life: A Discourse on the Sciences*. Sao Paulo: Editora Cortez.
Santos, BDS, & Almeida Filho, ND (2008). *The University in the 21st Century: Towards a New University*. Coimbra: Almedina, pp. 107–257.
Thiollent, M (2003). *Action Research Methodology*. 12th ed.
Tripp, D (2005). Action research: A methodological introduction. *Education and Research*, 31(3), 443–466.
UNESCO (2009). *World Conference on Higher Education 2009*. The New Dynamics of Higher Education and Research for Social Change and Development. Announcement. Paris, France, July 5th to 8th.
Vieira, TM (2017). Philosophy and education: Contributions by Edgar Morin. *Completion of Course Work (Degree in Pedagogy)—Faculty of Education*, Federal University of Rio de Janeiro, Rio de Janeiro.
Wisniewski, E (2022). *The New Humanism: Civilizing Paradigms for the 21st Century From Pope Francis*. Available at: www.ihuunisinos.br/categorias/619997-o-novo-humanismo-paradigmas-civilizatorios-para-o-seculo-xxi-a-partir-dopapa-francisco Accessed on 06/17/2023.

28

ACHIEVING SUSTAINABLE MENSTRUAL HEALTH FOR ALL

Investigating Menstrual Cup Take-Up in India

Karan Babbar and Supriya Garikipati

28.1 Introduction

The global narrative around menstrual health has deep ramifications for the Sustainable Development Goals (SDGs), notably SDG 3, centered on ensuring health and well-being, and SDG 5, which underscores the essence of gender equality (Sommer et al., 2021). Period poverty, defined as a lack of access to period products, menstrual health and hygiene education, and sanitation infrastructure, particularly in low- and middle-income countries (LMICs), has emerged as a significant obstacle in the trajectory toward these SDGs (MacRae et al., 2019).

Over the past decade, India has made concerted efforts to improve menstrual health and hygiene. The Indian government, since 2011, has championed the Menstrual Hygiene Scheme (MHS) under the wider National Health Mission (NHM) (Khera, 2019). As state governments also participated, there has been an encouraging rise in the use of sanitary pads—growing from 12% in 2010 (*Plan India Annual Report*, 2012) to 78% by 2019–2021 (IIPS and Macro International, 2022) (see Figure 28.1).

However, an underlying challenge remains. These efforts have predominantly promoted disposable sanitary pads, often overlooking sustainable alternatives, like reusable pads and menstrual cups (Garikipati, 2020; Hennegan, 2019). Similar narratives continue to characterize the landscape

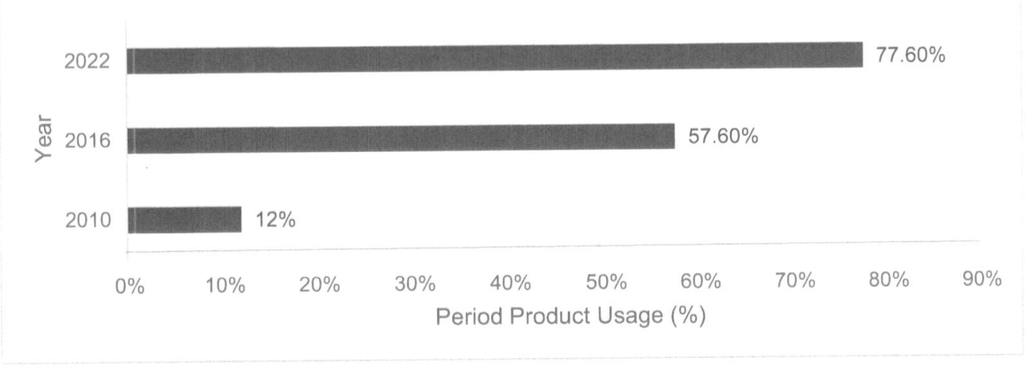

Figure 28.1 Period Product Usage in India from 2010 to 2022.

DOI: 10.4324/9781003285472-30

in the domain of menstrual health in India. Here, the Supreme Court of India directed the central government to establish a "uniform national policy" emphasizing free sanitary pads and separate toilets for girls in schools (Muralidharan et al., 2023). The repercussions are multifaceted. Disposables, while convenient, present both environmental and financial burdens. The recent global pandemic underscored the vulnerabilities of relying heavily on these products, particularly with the disruption of their supply chains (Babbar, Rustagi, et al., 2022; Garcon, 2020; Martin et al., 2022).

Despite over a decade of policies, period poverty remains troubling in India, especially among the urban poor (Babbar, Vandana, et al., 2023; Babbar and Dev, 2021, 2023). The environmental consequences are also stark. An annual generation of approximately 112,800 tons of menstrual waste underscores the ecological crisis that India grapples with (Bhor and Ponkshe, 2018).

Given this context, sustainable period products offer a promising solution. Among the available alternatives, reusable cloth pads and menstrual cups stand out. Their environmental friendliness, combined with their cost-effectiveness, makes them compelling choices. Menstrual cups, for instance, have a life cycle cost that is a fraction of the life cycle cost of disposables (Hait and Powers, 2019). Yet their adoption in India is hindered by cultural beliefs surrounding virginity and the act of vaginal insertion (Nappi et al., 2006). Consequently, despite their benefits, only 0.3% of Indian women aged 15–24 reportedly use menstrual cups (IIPS and Macro International, 2022).

Such data beckon a pivotal question: How can sustainable alternatives like menstrual cups, which have a mere fraction of the environmental and economic footprint of disposables, gain acceptance? This study seeks to engage with this complex landscape. Through an analytical lens, we aim to unravel the socio-demographic determinants influencing menstrual product choices in India. The overarching objective is to examine the determinants steering the choices between disposable and sustainable menstrual products.

In contributing to this discourse, our research intends to augment the scholarly conversation on menstrual health in LMICs (Almeida-Velasco and Sivakami, 2019; Anand et al., 2015). More broadly, by highlighting the sustainable dimensions of menstrual health, our insights resonate with the SDG agenda, emphasizing the urgency for environmentally resonant and resilient policies for menstrual health.

28.2 Menstrual Health and Hygiene Management (MHHM) in the Indian Context: Implications for SDGs 3 and 5

MHHM emerges as a pivotal component of women's and girls' health, shaping their capability to participate actively in societal roles with dignity and confidence. A cornerstone of efficacious MHHM is the informed selection and usage of menstrual absorbents, which encompass a variety of products ranging from cloth, sanitary napkins, and menstrual cups to tampons, among others (IIPS and Macro International, 2022). In regions grappling with the challenges of period poverty, a considerable fraction of the female population relies on used clothes or rags as a menstruation solution. Notably, in India, statistics indicate that over half of menstruating girls and women opt for cloth (IIPS and Macro International, 2022). Regrettably, the utilization of aged cloth, or the sheer absence of menstrual products, escalates the risk of health complications, including infections of the urinary and sexual tract (Almeida-Velasco and Sivakami, 2019; Anand et al., 2015).

Unhygienic practices, often stemming from insufficient knowledge of menstrual cloth maintenance and deeply rooted traditional taboos, further exacerbate these risks. For instance, sociocultural prohibitions often deter women from drying menstrual cloth under direct sunlight, compromising their hygiene (Baker et al., 2017).

Inextricably linked to MHHM is the access to sanitation and facilities for hand hygiene. The deficit in water, sanitation, and hygiene (WASH) infrastructure, particularly in educational institutions, bears profound implications for the retention rates of adolescent girls on a global scale (Adukia, 2017; Mahon and Fernandes, 2010). Disturbingly, comprehensive studies conducted by RICE between 2014 and 2018 unveiled that a staggering 44% of Indians practiced open defecation (RICE, 2019). The NSSO's 76th Round Report further accentuated this crisis, with findings that 29% lacked toilet access, while nearly 27% of women and girls in economically constrained nations did not have elementary handwashing amenities (UNICEF, 2018). Poor MHHM can inadvertently subject women to increased stress, societal ostracization, and psychological discomfort, impinging on their fundamental health and human rights (Babbar, Martin, et al., 2022; Sharma et al., 2020; Sommer, 2009).

A persistent lacuna in menstrual education, underscored by misinformation and harmful myths, continues to obstruct the optimal MHHM. Recent meta-analyses incorporating a broad spectrum of studies divulged that nearly half of Indian adolescent girls approach menarche ill-equipped and ill-informed (van Eijk et al., 2016). In many cases, families and communities perpetuate misleading beliefs, from barring women from kitchens to isolating them socially during menstruation (Babbar, 2021; Khanna et al., 2005).

Regrettably, even structured education does little to dispel myths or elevate menstrual knowledge (Garikipati, 2021; Hennegan and Montgomery, 2016). While commercial strategies have popularized disposable pads, sustainable options like menstrual cups remain obscured, attributed largely to restricted marketing budgets and financial constraints (Boyers et al., 2022; Garikipati and Phillips-Howard, 2022).

Interestingly, menstrual cups, despite the undercurrent of patriarchal taboos, have been evidenced to have superior acceptability amongst women acquainted with them. While the Global North has witnessed the proliferation of menstrual cup use, their acceptance in the Global South has been relatively muted, mainly due to cultural reservations regarding vaginal insertion (Nappi et al., 2006). Yet empirical studies, from diverse global settings—high-income regions to developing nations—reflect their favorable reception due to comfort, efficacy, and eco-friendliness (Beksinska et al., 2015; van Eijk et al., 2019).

This study endeavors to bridge existing knowledge gaps by delving into the socio-demographic nuances influencing the choice between sustainable and disposable menstrual products in India. By eschewing the influence of aggressive marketing, we seek to unearth the socio-economic determinants propelling women's awareness and adoption of sustainable menstrual solutions.

28.3 Data and Methodology

28.3.1 Data Source

This research draws from the most contemporary nationally representative, large-scale dataset available: the Demographic and Health Survey (DHS) of India, colloquially known as the National Family Health Survey (NFHS-5), carried out between 2019 and 2021. Orchestrated by the International Institute for Population Sciences, Mumbai, in alliance with the Ministry of Health and Family Welfare, the NFHS-5 acquires its insights through two primary instruments: (1) household questionnaires, elucidating data from all household residents, and (2) women-centric questionnaires, capturing vital socio-demographic details, as well as information pertaining to sexual and reproductive health, inclusive of menstrual cycles and related aspects. In a two-phase data collection, phase 1 (spanning June 2019 to January 2020) covered 17 states and five union territories, whereas phase 2 (from January 2020 to April 2021) incorporated 11 states and three union territories. Cumulatively, data from 636,699 households, encompassing 724,115 females and 101,839 males, was accrued (IIPS and Macro International, 2022). For the purposes of this research, the focus remains confined to the menstruation-related data of 241,112 women, aged between 15 and 24 years.

28.3.2 Dependent Variable Configuration

The dependent variables are framed using the NFHS-5's query, which probes the menstrual health and hygiene practices adopted by women to preclude visible staining. There were seven potential responses. Our research delves into discerning the socio-demographic catalysts steering the choice between sustainable products, such as menstrual cups, versus disposable alternatives. This considers females resorting to modern products, distinguishing between menstrual cup adopters and those preferring disposable options, that is, sanitary napkins, locally prepared sanitary napkins, and tampons. Consequently, the dependent variable becomes binary: 0 for modern disposables, and 1 for menstrual cups. Further investigation delves deeper into the driving factors behind sustainable product usage.

28.3.3 Independent Variables

A plethora of individual and household-level socio-demographic factors serves as the independent variables. On an individual scale, these include age at menarche, educational background, marital status, and frequency of engaging with various media (newspapers, radios, televisions). For a more holistic perspective, media engagement frequencies were amalgamated to forge a "mass media exposure" variable. At the household echelon, considerations comprise wealth index, toilet ownership, caste, religion, and residential locale (urban/rural). The selection of these variables has been guided by established literature delving into socio-economic and media influences on menstrual product decisions (Anand et al., 2015; Babbar et al., 2021; Babbar and Garikipati, 2023; Babbar and Sivakami, 2023; Goli et al., 2020; Roy et al., 2020).

28.3.4 Empirical Approach

We have opted for extensive descriptive statistics and pattern analysis, granting a clear portrayal of the various socio-demographic parameters influencing menstrual product preferences. These statistics provide a lucid overview of the patterns and tendencies within the data, aligning with the research's primary objective.

Firstly, we present the individual and household-level characteristics for the girls and women aged 15–24 from the NFHS-5 data. Subsequently, the next table focuses on our main variable of interest, that is, period product usage. Within this table, we have segregated period product usage into two categories: disposable and reusable products. We show the prevalence of disposable and sustainable period products by background characteristics. Lastly, we have taken a step further by dwelling on the prevalence of disposable vs. reusable period products, and to uncrate the influence of mass media factors into three components, that is, newspaper, radio, and television usage.

28.4 Results

Table 28.1 provides an overview of the study's key variables. In the first column, we observe that the majority of women surveyed in the NFHS-5 study are residents of rural areas (71%) and follow the Hindu faith (80%). A significant portion belongs to the category of other backward classes (45%). Approximately 87% of the women have completed at least secondary education. Interestingly, a substantial 64% of the participants are unmarried. Moreover, nearly 66% of the women use water and sanitation facilities that prioritize hygiene, and 58% have considerable exposure to mass media.

Shifting our focus to our variable of interest, that is, period products, we have parsed this category into two distinct types: disposable and sustainable period products. We present the prevalence

Table 28.1 Respondent's characteristics

Respondent characteristics	*Overall sample*
Variables	*Frequency (percentage)*
Individual level	
Age at menarche	
Less than 12 years	40,909 (16.96)
12 to 15 years	189,222 (78.46)
15 years and above	11,048 (4.58)
Education	
No education	15,833 (6.51)
Primary	15,028 (6.23)
Secondary	164,574 (68.24)
Higher	45,866 (19.02)
Marital status	
Married	86,702 (35.95)
Never married	153,228 (63.53)
Other	1,248 (0.52)
Mass media exposure	
No	47,661 (19.76)
Partial	54,486 (22.59)
Full	139,032 (57.65)
Wealth index	
Poor	101,142 (41.94)
Middle	50,656 (21.01)
Rich	89,375 (37.06)
Water and sanitation facility	
Unhygienic	83,009 (33.58)
Hygienic	158,171 (66.42)
Social category	
General	56,452 (19.68)
Scheduled caste	55,274 (24.04)
Scheduled tribe	23,191 (10.09)
Other backward classes	104,663 (45.52)
Don't know/unknown	1,532 (0.67)
Religion	
Hindu	193,622 (80.31)
Muslim	36,757 (15.24)
Others	10,733 (4.45)
Place of residence	
Urban	70,941 (29.42)
Rural	170,171 (70.58)

Source: NFHS-5 (2019–2021).

of both types based on participants' background characteristics, as shown in Table 28.2. The percentage of girls and women using disposable and sustainable period products decreases as the age of menarche increases. Similarly, usage of either product type rises with higher education levels. Women who are unmarried and have significant mass media exposure show the highest usage of either product.

Table 28.2 Prevalence of disposable and sustainable period products by background characteristics of women aged 15–24

Variables	*Disposable products (sanitary pads and tampons)*	*p-value*	*Sustainable products (menstrual cups)*	*p-value*
Individual variables				
Age at menarche				
Less than 12 years	32,096 (78.19)		156 (0.38)	
12 to 15 years	146,735 (77.29)	<0.01	547 (0.29)	<0.01
15 years and above	7,780 (70.18)		26 (0.24)	
Education				
No education	6,811 (43.21)		14 (0.09)	
Primary	7,997 (53.05)	<0.01	37 (0.25)	<0.01
Secondary	129,282 (78.29)		498 (0.30)	
Higher	42,518 (92.39)		180 (0.39)	
Marital status				
Married	62,284 (71.60)		203 (0.23)	
Never married	123,512 (80.34)	<0.01	524 (0.34)	<0.01
Other	815 (65.05)		2 (0.17)	
Mass media exposure				
No	25,796 (53.94)		70 (0.15)	
Partial	41,592 (76.08)	<0.01	187 (0.34)	<0.01
Full	119,223 (85.47)		472 (0.34)	
Household variables				
Wealth index				
Poor	63,119 (62.20)		236 (0.23)	
Middle	41,483 (81.61)	<0.01	125 (0.25)	<0.01
Rich	82,009 (91.45)		369 (0.41)	
Water and sanitation facility				
Unhygienic	54,551 (66.85)		249 (0.35)	
Hygienic	128,557 (82.31)	<0.01	562 (0.41)	<0.05
Social category				
General	38,442 (84.64)		139 (0.31)	
Scheduled caste	42,306 (76.26)	<0.01	144 (0.26)	>0.10
Scheduled tribe	15,168 (65.14)		74 (0.32)	
Other backward classes	81,065 (77.19)		321 (0.31)	
Don't know/unknown	964 (62.43)		6 (0.38)	

Table 28.2 (Continued)

Variables	*Disposable products (sanitary pads and tampons)*	*p-value*	*Sustainable products (menstrual cups)*	*p-value*
Religion				
Hindu	149,910 (77.14)		559 (0.29)	
Muslim	27,345 (74.14)	<0.01	127 (0.34)	>0.10
Others	9,356 (86.93)		44 (0.41)	
Place of residence				
Urban	63,334 (89.08)		344 (0.48)	
Rural	123,277 (72.14)	<0.01	386 (0.23)	<0.01
Period product usage	186,610 (77.12)		730 (0.3)	

Source: NFHS-5 (2019–2021).

Table 28.3 Prevalence of disposable and sustainable period products by mass media characteristics of girls and women aged 15–24

Mass media variables	*Disposable products (sanitary pads and tampons)*	*p-value*	*Sustainable products (menstrual cups)*	*p-value*
Reading newspaper				
Not at all	100,499 (69.50)		362 (0.21)	
Less than once a week	51,348 (86.58)	<0.01	242 (0.36)	<0.01
At least once a week	31,261 (90.37)		207 (0.57)	
Listening to radio				
Not at all	153,725 (76.09)		540 (0.23)	
Less than once a week	20,339 (82.06)	<0.01	160 (0.66)	<0.01
At least once a week	9,044 (85.82)		111 (0.80)	
Watching television				
Not at all®	35,708 (58.48)		146 (0.17)	
Less than once a week	42,294 (77.63)	<0.01	218 (0.37)	<0.01
At least once a week	105,106 (85.61)		447 (0.33)	

Source: NFHS-5 (2019–2021).

We discovered similar patterns in the prevalence of disposable and sustainable period products when considering household characteristics. The use of both disposable and sustainable products increases with greater wealth, access to clean water and sanitation, and urban residency. Among social categories, girls and women in the general category have the highest disposable product usage, followed by those in other backward classes, scheduled castes, and scheduled tribes. Similarly, women from other religions have the highest percentage of disposable product usage, followed by Hindus and Muslims. However, there is no significant difference in usage based on religion or social category for women using menstrual cups.

From the perspective of public policy, one important factor that emerges in these results as favoring the use of sustainable period products is the exposure to mass media. Thus, in Table 28.3

we dissect the prevalence of disposable and sustainable period products based on characteristics related to newspapers, radio, and television. We found that women who engage with newspapers, radio, or television at least once a week tend to have higher usage of both disposable and sustainable period products compared to those who do not engage with these media sources.

28.5 Discussion

In recent decades, a confluence of commercial interests and public policy has culminated in a marked ascent in the adoption of disposable sanitary pads within numerous LMICs. A salient exemplar is India, where the utilization of menstrual products observed an accelerated augmentation, ascending from a mere 12% in 2010 to a staggering 78% by 2021. This upsurge, however, has been overshadowed by apprehensions regarding both the environmental and fiscal longevity of advocating for disposable commodities. Aligning with the Sustainable Development Goals, particularly SDGs 3 (Good Health and Well-Being) and 5 (Gender Equality), there is an imperative to ensure that menstrual health management is both sustainable and promotes the well-being of all women. The recent pandemic starkly underscored the fragility in the supply chain of such commodities, due to an overarching dependence on international distribution channels (Babbar, Rustagi, et al., 2022; Babbar and Dev, 2023).

Consequently, the burgeoning sustainability discourse has spearheaded numerous innovations within the spectrum of eco-friendly menstrual products. Foremost among these is the menstrual cup, supplemented by innovations such as advanced reusable cloth pads and menstrual underwear, amongst others. These alternatives not only epitomize a diminutive environmental footprint in juxtaposition with disposable pads but also are characterized by economical long-term costs, aligning with the aims of SDGs 3 and 5. Comprehending the drivers behind the embracement of reusable menstrual commodities is quintessential for architecting policies that accentuate sustainable, cost-effective, and resilient menstrual hygiene for all.

Our empirical investigation delved into the socio-demographic determinants underpinning the usage of sustainable/reusable menstrual products in India. A multitude of the socio-demographic parameters scrutinized in our analysis exhibit significant associations with the adoption of modern menstrual commodities. This encompasses individual-centric parameters, such as the age of menarche, educational attainment, and exposure to mass media, and household-centric ones, including caste, religion, wealth index, water and sanitation facilities, and place of residence. Pertinently, for reusable menstrual commodities, a subset of these factors emerges as pivotal—notably, education, mass media engagement (primarily newspapers and radio), caste, and place of residence.

Our insights illuminate the cardinal role of educational attainment in shaping sustainable menstrual product choices. Such a correlation intimates that education potentially facilitates the adoption of menstrual cups in a dual manner: directly augmenting cognizance and understanding and indirectly dismantling deep-rooted taboos affiliated with vaginal insertion. Existing literature corroborates the assertion that education capacitates women in making enlightened menstrual choices (Goli et al., 2020; Khanna et al., 2005). Moreover, education acts as a catalyst in debunking entrenched myths surrounding menstruation, empowering women to maneuver autonomously, both physically and economically (Ram et al., 2020; Roy et al., 2020), thereby resonating with the tenets of SDG 5.

A salient determinant enhancing the propensity for sustainable menstrual product adoption is a woman's interaction with mass media, particularly the radio. Intriguingly, while our outcomes regarding radio engagement are emphatically robust, those linked to television exposure appear more attenuated. The radio, given its vast outreach, especially among semi-literate demographics, coupled with its capacity to disseminate information in vernacular languages, emerges as a potent policy instrument. Furthermore, the radio, with its inherent privacy, contrasts with mediums like television, which are often collectively used and placed within shared household spaces. Such nuances underline the radio's exceptional potential for transmitting salient public health messages, even those treading into culturally sensitive terrains (Ojha and Babbar, 2024; Ram et al., 2020).

A revelation from our data is the proclivity of women from socio-economically marginalized caste demographics (scheduled caste, scheduled tribe, and other backward classes) to exhibit higher menstrual cup adoption compared to their upper-caste counterparts. This observation warrants attention, especially juxtaposed against the broader trend wherein the former group exhibits a lower propensity toward modern menstrual products. Potential undercurrents for this pattern could be the prohibitive costs of disposables and the diminished prevalence of taboos surrounding vaginal insertion among these cohorts. Given that these groups encompass a significant proportion of India's populace, the implications of these findings for attenuating menstrual inequities and promoting gender equality, in line with SDG 5, are profound.

We also find a discernible urban–rural dichotomy, with urban women exhibiting augmented menstrual product adoption. Despite concerted governmental endeavors to ameliorate menstrual product uptake in rural geographies (Khera, 2019), the predominant focus on hard infrastructural provisions, such as subsidized or complimentary disposable pads, often eclipses the vital soft infrastructural facets. These encompass behavioral interventions targeting menstrual taboos and the dissemination of comprehensive menstrual product knowledge. A conceivable impediment to rural adoption could be the circumscribed mobility of women, compounded by extant societal taboos and the potential unavailability of products (Madhok, 2014). Our revelations resonate with antecedent studies highlighting the urban–rural chasm in menstrual product usage (Almeida-Velasco and Sivakami, 2019; Babbar et al., 2021; Singh et al., 2022).

From a policy vantage point, our findings underscore the imperative of recalibrating the narrative. Women endowed with awareness and information, emanating from progressive communities and residing in locales offering enhanced mobility and access, are poised to opt for menstrual solutions that are ecologically and fiscally prudent. Such choices, by virtue of their resilience, also offset supply chain perturbations witnessed during crises such as the COVID-19 lockdowns. Thus, there is a call for policy recalibration: transitioning from the endorsement of a monolithic product category to addressing informational lacunae among women, particularly those in rural precincts and from socio-economically marginalized castes. Amplifying product accessibility, especially for those with constrained mobility, is equally paramount.

Numerous NGOs have discerned the potential of enhancing awareness of sustainable menstrual alternatives as a linchpin strategy for menstrual health (Garikipati et al., 2019; Mahajan, 2019). Initiatives by entities such as the WSSCC and the Sikun Relief Foundation have pioneered innovative outreach methodologies, resulting in heightened menstrual awareness (WSSCC, 2018; Bhattacharya, 2018). Echoing these successful interventions, the Indian government unveiled a comprehensive menstrual education manual tailored for community health emissaries catering to rural cohorts (MHAI, 2022). Organizations like the AKDN, Sukhibhava, HasiruDala, and Tata Trusts have been even more ambitious, offering an array of products to the communities they serve, thereby catalyzing nuanced conversations around menstruation and waste management. Consequentially, their initiatives reinforce the credence that informed choices can arm women with the requisite knowledge and products, fostering a safe and sustainable menstrual experience. This approach, intrinsically aligned with SDGs 3 and 5, highlights the importance of holistic menstrual health, integrating sustainability and equality in the discourse.

28.6 Limitations

Every research effort has its limitations, and this study, despite its significance in the realm of menstrual health and its alignment with SDG objectives, is no exception.

Our primary limitation stems from the fact that the adoption rate of menstrual cups in India is notably low at present—merely 0.3% of our sample. Such a skewed sample can naturally lead to estimation challenges. While methodologies like propensity score matching might have been more apt, they would drastically shrink our sample, inhibiting our objective to discern the socio-economic contrasts between disposable and sustainable period product users. This limitation raises a bigger SDG-related question: How can we ensure universal access to sustainable period products (reflecting both SDGs 3 and 5) when the adoption rate is so low?

Using secondary data restricts our capacity to probe deeper into primary results. Thus, while we can hypothesize why some girls and women opt for sustainable menstrual products, the root mechanisms remain to be explored by further research in this area.

The fact that our study is confined to the 15–24 age bracket, representing only 12% of India's population, underscores the vast demographic that remains outside our purview. While understanding the menstrual practices of this group is crucial for SDG 5, the limited age range means that a large segment of menstruating individuals is excluded from our analysis, which could have SDG 3 implications.

The exclusion of trans and non-binary individuals who menstruate from our study indicates a broader societal and research blind spot (Babbar, Martin, et al., 2023). This glaring omission not only limits the comprehensiveness of our study but is also antithetical to the inclusive vision of the SDGs, especially SDG 5, which underscores the need for gender inclusivity in all aspects of life.

28.7 Concluding Comments

Nations around the globe have recognized menstrual health as an intrinsic right, with several offering free menstrual products as a cornerstone of their health and gender equality policies. Historically, there has been a significant dependence on disposable menstrual aids, mainly sanitary pads. This trend is slowly but assuredly shifting toward sustainability with the adoption of menstrual cups, in line with the sustainability ethos of the SDGs.

Our exploration, with India as its backdrop—a country accounting for 20% of the world's menstruating population—emerges as a pivotal contribution to the discourse on menstrual health within the framework of the SDGs.

Understanding the choices and challenges associated with menstrual health is not just a health concern (SDG 3) but intrinsically tied to gender equality (SDG 5). By ensuring accessible and sustainable menstrual products, we are not only improving the health and well-being of women but also empowering them by breaking down socio-economic barriers that arise due to menstrual poverty.

Our study's findings elucidate a noteworthy trend. Socio-economically marginalized caste communities, despite limited inclination for modern period products, show a heightened propensity toward menstrual cups. This underscores the potential to merge the aims of SDG 3 and SDG 5 through promoting sustainable menstrual practices which are health-conscious and empower women economically and socially.

Furthermore, the crucial role of knowledge dissemination becomes ever more relevant in the context of the SDGs. Radio, with its expansive reach, can be harnessed to advance both SDG 3 and SDG 5. By effectively communicating the benefits and debunking myths associated with sustainable menstrual products, we can foster better health practices and drive a gender-equal mindset, thereby challenging deep-rooted socio-cultural norms. Radio's historical success in health advocacy underscores its potency as a tool for change.

In summation, while addressing the menstrual health narrative of India, our study accentuates the broader global mission enshrined in the SDGs. By advocating for sustainable menstrual health practices, we are not only striving for better health outcomes but also fostering an environment where gender equality is more than just an aspirational goal—it is a lived reality.

References

Adukia, A. (2017). Sanitation and education. *American Economic Journal: Applied Economics*. https://doi.org/10.1257/app.20150083

Almeida-Velasco, A., and Sivakami, M. (2019). Menstrual hygiene management and reproductive tract infections: A comparison between rural and urban India. *Waterlines*, *38*(2), 94–112. https://doi.org/10.3362/1756-3488.18-00032

Anand, E., Singh, J., and Unisa, S. (2015). Menstrual hygiene practices and its association with reproductive tract infections and abnormal vaginal discharge among women in India. *Sexual and Reproductive Healthcare*. https://doi.org/10.1016/j.srhc.2015.06.001

Babbar, K. (2021). Taboos and myths as a mediator of the relationship between menstrual practices and menstrual health. *European Journal of Public Health*, *31*(Supplement_3). https://doi.org/10.1093/EURPUB/CKAB165.552

Babbar, K., and Dev, P. (2021). COVID-19 and period products usage among menstruating women in urban and rural India. *IIMA Working Papers*. https://ideas.repec.org/p/iim/iimawp/14666.html

Babbar, K., and Dev, P. (2023). Period products during the pandemic: The impact of lockdown on period products usage. *Applied Economics*, 1–17. https://doi.org/10.1080/00036846.2023.2257035

Babbar, K., and Garikipati, S. (2023). What socio-demographic factors support disposable vs. sustainable menstrual choices? Evidence from India's National Family Health Survey-5. *PLoS ONE*, *18*(8), e0290350. https://doi.org/10.1371/JOURNAL.PONE.0290350

Babbar, K., Martin, J., Ruiz, J., Parray, A. A., and Sommer, M. (2022). Menstrual health is a public health and human rights issue. *The Lancet Public Health*, *7*(1), e10–e11. https://doi.org/10.1016/S2468-2667(21)00212-7

Babbar, K., Martin, J., Varanasi, P., and Avendaño, I. (2023). Inclusion means everyone: Standing up for transgender and non-binary individuals who menstruate worldwide. *The Lancet Regional Health—Southeast Asia*, 100177. https://doi.org/10.1016/J.LANSEA.2023.100177

Babbar, K., Rustagi, N., and Dev, P. (2022). How COVID-19 lockdown has impacted the sanitary pads distribution among adolescent girls and women in India. *Journal of Social Issues*. https://doi.org/10.1111/JOSI.12533

Babbar, K., Saluja, D., and Sivakami, M. (2021). How socio-demographic and mass media factors affect sanitary item usage among women in rural and urban India. *Waterlines*, *40*(3), 160–178. https://doi.org/10.3362/1756-3488.21-00003

Babbar, K., and Sivakami, M. (2023). Promoting menstrual health: Towards sexual and reproductive health for all. *ORF*, *613*. www.orfonline.org/research/promoting-menstrual-health/

Babbar, K., Vandana, and Arora, A. (2023). Bleeding at the Margins: Understanding Period Poverty among SC and ST Women Using Decomposition Analysis. *Journal of Development Studies*. https://doi.org/10.1080/00220388.2023.2252139

Baker, K., Padhi, B., Torondel, B., Das, P., Dutta, A., and Sahoo, K. (2017). From menarche to menopause: A population-based assessment of water, sanitation, and hygiene risk factors for reproductive tract infection symptoms over life stages in rural girls and women in India. *PLoS ONE*, *12*(12).

Beksinska, M. E., Smit, J., Greener, R., Todd, C. S., Lee, M. L. T., Maphumulo, V., and Hoffmann, V. (2015). Acceptability and performance of the Menstrual Cup in South Africa: A randomized crossover trial comparing the menstrual cup to tampons or sanitary pads. *Journal of Women's Health*, *24*(2), 151–158. https://doi.org/10.1089/JWH.2014.5021

Bhattacharya, M. (2018). *Sikun Relief Foundation: An overview of informed choice of menstrual hygiene options* (unpublished).

Bhor, G., and Ponkshe, S. (2018). A decentralized and sustainable solution to the problems of dumping menstrual waste into landfills and related health hazards in India. *European Journal of Sustainable Development*, *7*(3), 334–334. https://doi.org/10.14207/EJSD.2018.V7N3P334

Boyers, M., Garikipati, S., Biggane, A., Douglas, E., Hawkes, N., Kiely, C., Giddings, C., Kelly, J., Exley, D., Phillips-Howard, P. A., and Mason, L. (2022). Period poverty: The perceptions and experiences of impoverished women living in an inner-city area of Northwest England. *PLoS ONE, 17*(7), e0269341. https://doi.org/10.1371/JOURNAL.PONE.0269341

Eijk, A. van, Sivakami, M., Thakkar, M., Bauman, A., Laserson, K., and Coates, S. (2016). Menstrual hygiene management among adolescent girls in India: A systematic review and meta-analysis. *BMJ Open, 6*(3).

Garcon, S. (2020, March). COVID-19 has made menstrual health more urgent than ever. *PSI.* www.psi.org/2020/03/covid-19-has-made-menstrual-health-more-urgent-than-ever/

Garikipati, S. (2020). Only 15% girls in India had access to sanitary pads during lockdown. *Scroll.* https://scroll.in/article/965984/only-15-girls-in-india-had-access-to-sanitary-pads-during-lockdown

Garikipati, S. (2021). Asymmetric information in menstrual health and implications for sustainability: Insights from India. *Sustainable Consumption and Production, I*, 391–412. https://doi.org/10.1007/978-3-030-56371-4_19

Garikipati, S., Docherty, R. J., and Phillips-Howard, P. A. (2019). *What's the bleeding problem? Policy and attitudes towards sustainable menstrual hygiene materials in India.* www.liverpool.ac.uk/management/people/economics

Garikipati, S., and Phillips-Howard, P. A. (2022). *What's the bleeding problem? Challenges and potential of informed choice for inclusive and sustainable menstrual outcomes.* Harvard Dataverse. https://doi.org/10.7910/DVN/PVGQD0

Goli, S., Sharif, N., Paul, S., and Salve, P. S. (2020). Geographical disparity and socio-demographic correlates of menstrual absorbent use in India: A cross-sectional study of girls aged 15–24 years. *Children and Youth Services Review, 117*, 105283. https://doi.org/10.1016/j.childyouth.2020.105283

Hait, A., and Powers, S. E. (2019). The value of reusable feminine hygiene products evaluated by comparative environmental life cycle assessment. *Resources, Conservation and Recycling, 150*, 104422. https://doi.org/10.1016/J.RESCONREC.2019.104422

Hennegan, J. (2019). Inserting informed choice into global menstrual product use and provision. *The Lancet Public Health, 4*(8), e361–e362. https://doi.org/10.1016/S2468-2667(19)30126-4

Hennegan, J., and Montgomery, P. (2016). Do menstrual hygiene management interventions improve education and psychosocial outcomes for women and girls in low and middle income countries? A systematic review. *PLoS ONE, 11*(2), e0146985. https://doi.org/10.1371/journal.pone.0146985

IIPS and Macro International. (2017). *National family health survey (NFHS—4), 2015–16 India report. In India.* Mumbai: IIPS. https://doi.org/kwm120[pii]10.1093/aje/kwm120

IIPS and Macro International. (2022). *National family health survey (NFHS—5), 2019–21 India report.* http://rchiips.org/nfhs/NFHS-5Reports/NFHS-5_INDIA_REPORT.pdf

Khanna, A., Goyal, R., and Bhawsar, R. (2005). Menstrual practices and reproductive problems: A study of adolescent girls in Rajasthan. *Journal of Health Management, 7*(1).

Khera, A. (2019). How can we make menstrual hygiene management more effective? *The Wire.* https://thewire.in/health/menstrual-hygiene-management-india

MacRae, E. R., Clasen, T., Dasmohapatra, M., and Caruso, B. A. (2019). "It's like a burden on the head": Redefining adequate menstrual hygiene management throughout women's varied life stages in Odisha, India. *PLoS ONE, 14*(8), e0220114. https://doi.org/10.1371/journal.pone.0220114

Madhok, D. (2014, August 21). The full extent of what urban India believes about menstruation is extraordinary. *The Quartz India.* https://timesofindia.indiatimes.com/city/bhopal/girls-dropping-out-of-mp-schools-double-in-3-years/articleshow/71354256.cms

Mahajan, T. (2019). Imperfect information in menstrual health and the role of informed choice. *Indian Journal of Gender Studies, 26*(1–2), 59–78. https://doi.org/10.1177/0971521518811169

Mahon, T., and Fernandes, M. (2010). Menstrual hygiene in South Asia: A neglected issue for WASH (water, sanitation and hygiene) programmes. *Gender and Development, 18*(1).

Martin, J., Babbar, K., and Maschette, U. (2022). Menstrual health for all requires wider high level commitment. *BMJ, 378*, o2222. https://doi.org/10.1136/BMJ.O2222

Menstrual Hygiene Alliance of India (MHAI) 2022. *Informed choice to promote sustainable menstrual health* (unpublished).

Muralidharan, A., Babbar, K., and Bhattacharjee, M. (2023). Does India need a 'uniform national policy' for menstrual health and hygiene? *The Quint.* https://www.thequint.com/fit/uniform-national-menstrual-health-and-hygiene-policy-india

Nappi, R. E., Liekens, G., and Brandenburg, U. (2006). Attitudes, perceptions and knowledge about the vagina: The International Vagina Dialogue Survey. *Contraception, 73*(5), 493–500. https://doi.org/10.1016/J.CONTRACEPTION.2005.12.007

Ojha, M., and Babbar, K. (2024). Power to choose? Examining the link between contraceptive use decision and domestic violence. *Economics & Human Biology*, *55*, 101416.

Plan India Annual Report. (2012). https://issuu.com/planindia/docs/eannual_report

Ram, U., Pradhan, M. R., Patel, S., and Ram, F. (2020). Factors associated with disposable menstrual absorbent use among young women in India. *International Perspectives on Sexual and Reproductive Health*, *46*, 223–234. https://doi.org/10.1363/46E0320

RICE. (2019). *Changes in open defecation in rural north India: 2014 – 2018*. https://riceinstitute.org/research/changes-in-open-defecation-in-rural-north-india-2014-2018-2/

Roy, A., Paul, P., Saha, J., Barman, B., Kapasia, N., and Chouhan, P. (2020). Prevalence and correlates of menstrual hygiene practices among young currently married women aged 15–24 years: An analysis from a nationally representative survey of India. *European Journal of Contraception and Reproductive Health Care*. https://doi.org/10.1080/13625187.2020.1810227

Sharma, S., Mehra, D., Brusselaers, N., and Mehra, S. (2020). Menstrual hygiene preparedness among schools in India: A systematic review and meta-analysis of system-and policy-level actions. *International Journal of Environmental Research and Public Health*, *17*(2). MDPI AG. https://doi.org/10.3390/ijerph17020647

Singh, A., Chakrabarty, M., Singh, S., Chandra, R., Chowdhury, S., and Singh, A. (2022). Menstrual hygiene practices among adolescent women in rural India: A cross-sectional study. *BMC Public Health*, *22*(1), 1–18. https://doi.org/10.1186/S12889-022-14622-7

Sommer, M. (2009). Ideologies of sexuality, menstruation and risk: Girls' experiences of puberty and schooling in northern Tanzania. *Culture, Health and Sexuality*. https://doi.org/10.1080/13691050902722372

Sommer, M., Torondel, B., Hennegan, J., Phillips-Howard, P. A., Mahon, T., Motivans, A., Zulaika, G., Gruer, C., Haver, J., and Caruso, B. A. (2021). How addressing menstrual health and hygiene may enable progress across the Sustainable Development Goals. *Global Health Action*, *14*(1), 1920315. https://doi.org/10.1080/16549716.2021.1920315

UNICEF. (2018). FAST FACTS: Nine things you didn't know about menstruation. *UNICEF*. www.unicef.org/press-releases/fast-facts-nine-things-you-didnt-know-about-menstruation

van Eijk, A. M., Zulaika, G., Lenchner, M., Mason, L., Sivakami, M., Nyothach, E., Unger, H., Laserson, K., and Phillips-Howard, P. A. (2019). Menstrual cup use, leakage, acceptability, safety, and availability: A systematic review and meta-analysis. *The Lancet Public Health*, *4*(8), e376–e393. https://doi.org/10.1016/S2468-2667(19)30111-2

Water Supply and Sanitation Collaborative Council (WSSCC) (2018). *Baseline assessment on MHM in 5 districts of Uttar Pradesh* (unpublished report).

29

THE ROLE OF PAYMENT FOR ECOSYSTEM SERVICES IN ACHIEVING THE SUSTAINABLE DEVELOPMENT GOALS

Lessons for Biodiversity in the Democratic Republic of the Congo

Clara My Lernborg and Mirey Atallah

29.1 Introduction

The tension between environment and economic development was first made explicit at the 1972 UN Conference on Environment and Development in Stockholm. Subsequent environmental conferences gave birth to the influential concept of sustainable development. In 1992, the signatories of the Convention of Biological Diversity recognized the necessity of finding new market-based mechanisms and including the private sector in the fight for biodiversity conservation. The reconciliation of environment and development is conceptualized in the sustainable development agenda with its 17 multistakeholder goals, but difficulties and trade-offs between nature and development remain.

Biodiversity enables the provision of multiple ecosystem services, varying from climate stabilization and improving food supply to providing medical ingredients. There is an increasing understanding that without the protection of biodiversity and the ecosystem services underpinning it, the SDGs, ranging from ending poverty and hunger to ensuring healthy lives and promoting well-being, cannot be met. Biodiversity losses have an impact on the security and value of investments in multiple sectors. The climate and biodiversity crises are intrinsically linked, thereby making reaching global climate and Sustainable Development Goals dependent on successful conservation, restoration, and management of biodiversity. It is therefore not surprising that biodiversity and its financing are attracting increasing attention from both public and private entities.

SDGs 14 (Life Below Water) and 15 (Life on Land) are 2 of the 17 transformative Sustainable Development Goals directly linked to biodiversity and ecosystem conservation. These two SDGs will therefore receive particular attention in this chapter, alongside SDG 1 (No Poverty), given the potential of payment for ecosystem services (PES) for poverty reduction in tandem with biodiversity goals. However, halfway to the looming 2030 deadline, these biodiversity-related SDGs remain in a precarious state. Biodiversity loss remains a grand societal challenge; WWF (2022)

DOI: 10.4324/9781003285472-31

reported a large decline in global populations of all major species, considering it a code red alert for humanity. Unfortunately, the rate of loss is likely accelerating.

The Democratic Republic of the Congo (DRC) is home to the second largest rainforest in the world and is one of 17 mega-diverse countries in terms of biodiversity. Thereby, the country is key to the fight against climate change and the realization of the SDGs overall, and the biodiversity-related goals in particular. However, it also has one of the poorest populations in the world, with 74.6% of Congolese people living on less than $2.15 a day (World Bank, 2023). This has implications for how to implement mechanisms and for addressing global biodiversity goals in the context of the DRC. One such mechanism is market-based programs for PES, aimed at regulating natural capital and ecosystem services.

Against this backdrop, this chapter sets out to examine the potential and drawbacks of the PES approach for contributing to the SDGs in the Global South, whilst simultaneously addressing the pressing issues in the DRC.

29.2 Understanding Biodiversity

29.2.1 Ecosystem Services

Natural capital is defined as "the world's stocks of natural assets, which includes geology, soil, air, water and all living things" (World Forum on Natural Capital, 2021). Natural capital offers a wide range of valuable services benefitting humankind, and ecosystem services play an important role here.

Ecosystem services (ES) are defined as natural processes by which "ecosystem, and the species that make them up, sustain and fulfil human life" (Daily, 1997) or the benefits obtained from people's use of ecosystem, according to the Millennium Ecosystem Assessment (MA). Biodiversity is instrumental to these ES' productivity and resilience. These uses can be refined in four categories: provisioning, regulating, supporting, and socio-cultural (see Table 29.1 for definitions and examples).

29.2.2 Updating the Biodiversity-Related SDGs

Biodiversity remains under threat. The identified conservation solutions have mainly included intergovernmental (i.e., global treaties) and governmental measures (regulation of quantity and price of natural capital through protected areas) and biodiversity finance (Flammer et al., 2023; Karolyi & Tobin-de la Puente, 2023).

Table 29.1 Types of ecosystem services and definition based on the Millennium Ecosystem Assessment

Type of ES	*Definition*
Provisioning	Water, food, wood, and other goods provided to populations by ecosystems
Regulating	Maintaining air and soil quality, controlling floods and diseases, or pollinating crops
Supporting	Providing living space for plants and animals, and preserving the diversity of plant and animal species
Socio-cultural	The intangible benefits that people derive from ecosystems, such as aesthetic inspiration, cultural identity, a sense of belonging, and spiritual experience linked to the natural environment

SDGs 14 and 15 are the two primary biodiversity-related goals in Agenda 2030. They are grounded in previous international agreements and soft law instruments (i.e., Aichi targets), prompting the SDGs' target date of 2020 rather than 2030. Target revisions were conditional on the Convention of Biological Diversity's (CBD) 15th Conference of the Parties (COP) in 2020, which was delayed because of COVID-19. It was finally held in 2022, and the new Kunming–Montréal Global Biodiversity Framework (GBF), with a global vision of "living in harmony with nature," was finally agreed upon. It formulates four goals by 2050, and 23 targets to be met by 2030 (UN, 2022). These partially overlap primarily with SDG 15 and provide new sector-specific guidance and indicators, particularly for those outdated sub-targets. When implementing Agenda 2030 biodiversity-related goals, it would therefore be useful for countries to include more qualitative indicators and align with the GBF. The DRC's SDG implementation report (OCDD, 2022) does not include details of any particular achievements or allocated governmental or donor funding; nevertheless, it does make reference to its other international climate engagements.

29.2.3 The Role of Ecosystem Services in the Congo Basin

The DRC is characterized by its abundance of natural resources, as well as by the overwhelming poverty of its population. It is home to 60% of the Congo Basin, the world's second largest rainforest, and a diverse fauna and flora with unique endemic species in the world. It is recognized as a "mega-diverse" country (CECM, 1998), that is, one of the richest countries on the planet in terms of biological diversity. This biodiversity is continuously threatened by increasing pressures on forest ecosystems.

Beyond being essential for biodiversity, forests are a vital aspect of the global climate change mitigation strategy and are key in closing the emissions gap (Busch et al., 2019). The DRC's forests constitute a majority of the Congo Basin, the world's second largest rainforest. Tropical forests are a cost-efficient means of preventing the negative consequences of climate change. Yet they remain under threat. In the DRC, the primary threats are slash-and-burn agriculture and high demand for wood fuel by local populations, which may be linked to the high national poverty level. This stands in sharp contrast to the primary drivers of tropical deforestation in the Amazon, which instead emanate from industrial agriculture.

The ES from tropical forests constitute a source of great economic and socio-cultural value (Fuss et al., 2021). These provide not only biodiversity preservation but also carbon sequestration, watershed, and erosion protection (Naidoo et al., 2019). We can divide the ES from tropical forests into two categories, both of which have global value:

- *Geographically defined ES*, that is, the conservation of a rich and unique biodiversity that does not exist elsewhere
- *Multiple-benefits ES*, that is, carbon sequestration, water regulation, soil retention, etc., possible to target through a single intervention

Both climate and biodiversity finance are crucial instruments for facilitating the maintenance and enhancement of such services. The use of payments for ecosystem services (PES) has increased rapidly in the past decade. The question is what instruments are useful for conserving DRC's unique mega-diverse rainforest and thereby contribute to the SDG Agenda and other climate goals. In order to examine the potential of PES for the DRC and identify the challenges standing in the way of its implementation, we will move to a literature review of its effectiveness and potential for poverty alleviation, livelihood, and environmental additionality.

29.3 Literature Review of PES

The current focus areas of PES research are PES design, effectiveness of PES outcomes, and market-based valuation of ES (Kolinjivadi et al., 2023). An important caveat comes from the limited geographical focus of such research. Most of the PES evidence comes from Latin America and China (see Table 29.3). Research on African programs is relatively scarce, with some examples from Mozambique (Jindal et al., 2012; Hegde & Bull, 2011), Uganda (Jayachandran et al., 2017), DRC (Froger et al., 2016), and Burkina Faso (Adjognon et al., 2019).

29.3.1 What Are Payment for Ecosystem Services?

29.3.1.1 Definitions

PES are voluntary transactions intended to encourage environmentally favorable activities through conditional cash transfers (Wunder, 2015). Its use has increased rapidly; Salzman et al. (2018) identified 550 PES programs worldwide. Its popularity reflects a move toward market-based environmental approaches in the Global South (Jones et al., 2020).

A PES is thus (Fripp, 2014)

- a *voluntary* transaction in which
- a *well-defined* environmental service (ES), or a form of land use likely to secure that service,
- is bought by at least one ES *buyer*
- from a minimum of one ES *provider*
- if and only if the provider continues to supply that service (*conditionality*).

A distinction can therein be made between the ES provider (participant receiving the benefits of the program in terms of direct gains or cost reductions), the beneficiary or user (consuming the ES), the payer (paying for improved land use/conservation without necessarily obtaining direct benefits from the provision of ES), the investor (allocating resources to ensure ES production), and the buyer (purchasing ES credits).

29.3.1.2 Typology Based on Buyer Status

A PES typology can be distinguished through the status of the participants and the payers of the ES. The latter may be voluntary (e.g., a donation to an environmental non-governmental organization) or coerced (e.g., when payment comes from the reallocation of a tax they have paid). Furthermore, payers may be direct or indirect users of the services rendered. The four main types of buyer-based PES are summarized in Table 29.2.

29.3.2 Key National PES Programs

Table 29.3, which follows, identifies the major PES programs (OECD, 2020) and the associated literature. As confirmed in PES meta-reviews, the Costa Rican, Mexican, and Chinese national

Table 29.2 Typology based on buyer status

	Voluntary payer	*Coerced payer*
Specific user	In the French water company Vittel's PES program, Vittel is the voluntary payer. It pays ES providers/farmers to adopt good practices to improve water quality in the watersheds from which Vittel draws its mineral water. The payer is a direct user of the service rendered.	The Mexican state's hydrological PES program involves payment by the government to forest owners to maintain forest cover in order to guarantee water consumption for citizens. Simultaneously, a compulsory contribution from the state's water distribution companies was set up. These must pay 3.5% of sales to contribute to the program. Payment is therefore not voluntary, but compulsory. The payer is a direct user of the service rendered.
Non-specific user	A bird's nest protection program in Cambodia was set up by the non-governmental organization Wildlife Conservation Society (WCS). The voluntary payers make donations to WCS, which pays local communities to protect bird's nests. The payers/donors are not necessarily direct users of the ES.	Programs in which funding comes from a heterogeneous group of taxpayers. The EU's environmental policy for agriculture is an example.

Source: Adapted from Leménager and Laurans (2015).

Table 29.3 Literature review of the major PES programs pertaining to poverty reduction and conservation

Country	*Program*	*Articles*
Brazil	Bolsa Verde	da Conceição et al. (2015);
China	Sloping Land Conversion Program	Liu et al. (2008), Liang et al. (2012), Yin et al. (2014), Trac et al. (2013), He and Sikor (2015), Qu et al. (2017), Lu and Yin (2020)
Costa Rica	Pagos por Servicios Ambientes (PSA)	Arriagada et al. (2011), Sanchez-Azofeifa et al. (2007), Locatelli et al. (2008), Robalino et al. (2014)
Ecuador	Socio Bosque	Raes et al. (2014), da Conceição et al. (2015)
Mexico	PSAH	Sims and Alix-Garcia (2017), Alix-Garcia et al. (2015), Garcia-Amado et al. (2011), Scullion et al. (2011)

Source: Selection adapted from OECD (2020).

PES programs are the most studied in relation to effectiveness, particularly regarding impacts on poverty reduction and conservation (Blundo-Canto et al., 2018).

29.3.3 Effectiveness of PES

The effectiveness of PES is difficult to measure, considering the wide range of measures involved in its assessment, ranging from economic to biophysical. Hitherto, research has shown mixed results in its examination of PES effectiveness—for example, carbon sequestration, biodiversity conservation, water quality.

Table 29.4 Articles identifying positive impacts of PES

Positive impacts	*Articles*
Poverty reduction	Sims and Alix-Garcia (2017)
Land tenure security	Lawlor et al. (2013)
Diversification	Pagiola et al. (2005), Wunder (2008)
Health	Wunder (2008)
Strengthened social networks	Tacconi et al. (2013)
Economic growth	Iftikhar et al. (2007), Tacconi et al. (2013)
Well-being	Jayachandran et al. (2017), Adjognon et al. (2019)

Several question PES effectiveness in achieving multiple objectives simultaneously (Muradian et al., 2013; Wunder, 2013). Studies have not been able to prove synergies between PES and livelihood (Pagiola et al., 2005), given the need for complex multi-actor and longitudinal evaluations involved (Hejnowicz et al., 2014).

29.3.4 Poverty Reduction and Livelihood

Poverty reduction is a key area for realizing sustainable development; SDG 1 is dedicated to it. However, it is also of utmost importance in achieving biodiversity conservation and ecosystem services goals. PES are expected to have moderate to no impact on poverty reduction. Sims and Alix-Garcia (2017) identify small but statistically significant impact in several cases. Studies show a more rigorous tie between poverty reduction and community forest management (Oldekop et al., 2019; Adjognon et al., 2019) and ecotourism (Naidoo et al., 2019).

Although PES are often conceived as primarily conservation-oriented, they are frequently combined with a secondary poverty reduction objective. PES typically produce livelihood changes (e.g., changes in incomes, employment, and ensuing welfare impacts). In the developing country context, a PES program may be primarily implemented in view of pro-poor impacts, with environmental impacts secondary (da Conceição et al., 2015).

Synergies between environmental conservation and poverty reduction should be sought. As seen in the preceding section, the impact on the latter is moderate to none. PES can, however, lead to negative livelihood impacts. Not only may social tensions arise from unequal benefit appropriation (Wunder, 2008), but land use restrictions may also affect cultural traditions (Yang et al., 2013). These impacts are seldom understood and rarely studied (Hejnowicz et al., 2014; Blundo-Canto et al., 2018). Furthermore, constraints in activities may also lead to market dependency (Pagiola et al., 2005).

29.3.5 Deforestation

PES were conceived to yield environmental *additionality* under cost-effective conditions. Additionality implies that the impacts would not have happened without the program (Engel et al., 2008). Several meta-analyses have assessed PES additionality of attained ecosystem goods and services of interest through project implementation (Börner et al., 2017; Grima et al., 2017). The results are mixed; two studies found that deforestation rates in Costa Rica were not significantly lower in areas receiving PES compared to those not receiving PES (Sierra & Russman, 2006).

A potential explanation was given by the success of previous conservation initiatives and ongoing changes in land use practices (Sanchez-Azofeifa et al., 2007).

In a review of meta-studies, Wunder et al. (2020) found that there was a limited number of rigorous systematic reviews of additionality; Pattanyak et al. (2010) could only identify six studies pertaining to forest cover, all from Costa Rica and Mexico. Samii et al. (2014) identified nine studies from four PES programs, all from the same Latin American countries. Only a modest effect on deforestation was found therein. Snilsveit et al. (2019) had a slightly larger sample with eight countries but remained pessimistic as to deforestation effects. As mentioned in the preceding section, Alix-Garcia et al. (2015) found low impact on deforestation but a high poverty reduction effect. The question remains whether this is always a necessary trade-off between potentially conflicting SDGs.

29.4 Lessons From the DRC

29.4.1 Past Experiences

29.4.1.1 PIREDD Plateaux

The PIREDD-Plateaux program was conducted between 2015 and 2019 and constitutes one of the first large-scale REDD+ investment projects in the DRC and is the precursor of the REDD+ program PIREDD Maï-Ndombe.

Activities to reduce pressure on the forests were carried out with local communities through the aegis of local development committees (CLD). This included reforestation activity and clearing savannahs for forest regeneration. These activities were eligible for ecosystem payments to be made through the CLD. Each hectare of savannah saved was valued at 5 USD per hectare and 150 USD per planted hectare; 20% of resulting funds were paid to land chiefs, while a majority of the funds were allocated to support community development projects, such as the construction or rehabilitation of schools, health centers, wells, and mills.

The five-year nature of the project may have made it difficult to incentivize continued efforts by local populations. Without additional financial support, research indicates that local communities may easily abandon such sustainable forest practices. Furthermore, the relatively low level of remuneration prompts questions on incentivization and remuneration. Large areas need to be committed to generate substantial benefits for local populations. Based on subsequent REDD+ experiences in the DRC, it is unclear whether prices per hectare correspond to the efforts required, the local populations' opportunity costs, or the sequestration capacity per hectare. Additional in-depth research on such programs is needed to draw practical lessons and guide policymakers to adapt PES programs to this important context. There is currently a dearth of PES research on the African context, particularly relating to the Congo Basin. This has implications for the optimal design of PES programs for decision-makers and, ultimately, the realization of the SDGs on the continent.

29.4.1.2 IBI Project

The Ibi-Batéké PES project (Ibi) was launched in 2008 and was the first reforestation project in the DRC to benefit from the UNFCCC Clean Development Mechanism. The plan was to plant 4,200 hectares of forest on the Batéké plateau to ensure satisfying the high demand for wood fuel in nearby capital Kinshasa.

It was implemented by Novacel, a local private company. It contracted sales of 500,000 tons of carbon credits to the World Bank's BioCarbon Fund and the Orbeo Carbon Fund, respectively; 12% of carbon revenues were to be invested in social activities for the benefit of local populations.

The focus was on reforestation and carbon sequestration. To improve local food security, the agro-forestry model included acacia and cassava. The business plan focused on sales of carbon credits, through manioc and wood fuel production stemming from sustainably managed forestry.

Some local populations were negatively impacted by Ibi, including the inhabitants of neighboring hamlets who, along with temporary plantation workers, had previously used the forest for wood fuel. Furthermore, women who relied on foraging non-timber forest products, for example, honey and mushrooms, were negatively impacted and were also excluded from the project benefits (Froger et al., 2016). Gendered and socio-cultural impacts were thereby not integrated in the PES design.

Ultimately, the results of the project were mixed, according to Froger et al. (2016). Furthermore, the production cost of wood fuel, or *makala*, has, in retrospect, been deemed too high compared to the market price in Kinshasa. Further speculations include the link between the market price of wood fuel with a poorly paid labor force. Thereby, the financial benefits were not able to cover all production costs in the Ibi project. An analysis on how to overcome these socio-cultural and economic challenges would be useful to draw lessons for the success of future ventures and thereby the achievement of the SDG Agenda.

29.5 PES Challenges

29.5.1 Financial Sustainability

One of the major challenges for the success of a PES program is to ensure sustainable, stable, sufficient, and long-term funding to achieve the desired impacts. This is especially a challenge in national PES programs, where funding is subject to project duration or policy cycles and where there are typically significant initial costs (e.g., feasibility study, identification of potential leaks, baseline studies, stakeholder consultations).

To ensure success, this type of program needs to adopt an incentive structure that promotes best practices and to integrate several types and sources of funding, notably to enable risk reduction. It is very important to design public financing mechanisms that encourage several sources of funding, particularly from the private sector (World Bank, 2012).

Notably, funding for Mexico's national PES program, established in 2003, was designed with annual funding of 20 million USD. Over time, this amount has increased to 30 million USD. Debate persists as to the sustainability of the program without government intervention. Moreover, the need to amend federal law to secure funding suggests that any law or agreement must be formulated in a way that prevents adaptation to devaluation over time.

29.5.2 Environmental Sustainability

While their objective is to ensure and strengthen ES, some PES programs are dependent on funding from sectors that have consequences quite contrary to this objective, such as funding from the exploitation of fossil fuels. For example, one of the main sources of funding for Costa Rica's National Forest Fund, which manages the national PES program for reducing emissions from deforestation, depends partially on government tax revenues from fossil fuels. Other sources of income include forestry sector taxes, World Bank grants and loans, and bilateral collaboration with Germany. In other words, reducing greenhouse gas emissions from deforestation is partially dependent on generating revenues from the fossil fuel sector, the latter being a major source of such emissions. This paradox can be seen as a way of offsetting emissions from the fossil fuel sector—or possibly as a kind of "greenwashing" of this sector. This may call into question the carbon-neutrality goals of countries employing a similar strategy to Costa Rica. It remains unclear how the latter intends to finance the PES program in the future. Nevertheless, it is counterintuitive to the SDG Agenda, and such trade-offs could potentially be avoided with a clearer SDG indicator and budgeting framework.

In the case of the DRC, potential sources of financing need to be carefully assessed, bearing in mind that (1) fossil energy resources are not—yet—exploited in the country, and their exploitation is strongly discouraged by international partners; (2) the prospects for the viability of such exploitation are dim in the face of a growing trend toward a decarbonized global economy; and (3) management of the reputational risk associated with financing virtuous circles with polluting activities is assured. Additional agreements with private companies extracting the country's rich stock of strategic minerals for the global green transition could be envisaged (i.e., lithium and cobalt), while ensuring that forest guardian communities receive payments commensurate with the value of the ES they provide and the lifestyles they enable to be perpetuated.

29.5.3 Social Sustainability

Existing evidence indicates a lack of gender sensitivity in PES programs. Women tend to experience a disproportionate loss of income in forest PES (Tuijnman et al., 2020). Furthermore, women's participation and benefits were found to be lower than those of men in Kenyan PES programs (Kariuki & Birner, 2016). Furthermore, consideration of power relations, including gender, between ES providers and users should thus be integrated in PES design in order to help avoid unequal benefit appropriation. Such consideration could have helped the Congolese Ibi project identify and overcome at least some of its biases and lack of inclusion (see further section later).

29.5.4 Supervision, Support, and Transaction Costs

While it is essential to define the objectives and expected results of a program at the outset, it is equally important to recognize that financial transactions alone will not necessarily bring about the expected change in behavior. Rarely does the incentivizing financial support fully compensate for ES providers' shifting cultivation and/or management techniques. In addition to financial support, a stable legal and institutional framework is a common denominator of all successful PES programs (Wunder et al., 2020). Furthermore, technical support to provide alternative solutions is a key element, at least at the outset of PES programs. Finally, monitoring parameters, proxy indicators, and the joint definition of monitoring–evaluation mechanisms need to be designed and thought through from the outset to bring clarity to provider communities. Clear communication, including in local languages, and a common definition of success parameters are also essential ingredients of a solid and viable long-term approach in line with the SDG Agenda and GBF.

29.6 Suggested Priorities for the DRC

29.6.1 Poverty Reduction

In order to avoid adverse effects, a majority of PES programs targeting deforestation link it with poverty reduction. Trade-offs, however, are sometimes inevitable. It is of utmost importance that providers of ES, such as the DRC, and the payers for such services (often public entities involved in bilateral or multilateral collaboration) consider the links between ES and economic development when setting up PES programs. PES may support economic development not only by providing income to people who previously provided ES without compensation but also by addressing the environmental factors that sustain poverty, such as degraded soils resulting in low agricultural productivity, and degraded water quality causing disease and health problems. Thereby, PES programs have the potential of promoting various SDGs.

In the context of forest-related PES programs, a successful design cannot avoid the livelihoods and poverty reduction objectives. The main assumption in setting up PES forest programs in a developing country context is that payments to the poorest populations will reduce deforestation. However, the poorest populations often do not have access to participation prerequisites (e.g., certificates, land ownership, land security, or the ability to absorb high opportunity costs). Indeed, the objective of poverty reduction is not necessarily always harmonized with that of reducing deforestation. The objectives of equality and poverty reduction can even be at odds with the deforestation objective itself. In the national Mexican PES program, the objective of giving equal payments to populations rather than focusing on areas at highest risk of deforestation ultimately undermined the deforestation reduction objective (Leménager & Laurans, 2015). This illustrates the continued tension between environmental and economic goals, intrinsic to the SDG Agenda.

Indeed, PES programs can exacerbate poverty in developing countries that rely heavily on the exploitation of natural resources. In this context, it is often the poorest populations who rely most heavily on the direct use of ES as a source of livelihood. Thus, a PES program that directs investments toward long-term environmental gains rather than short-term economic growth for local communities can result in high opportunity costs that impact negatively on the most vulnerable communities.

In the DRC context, the design of such a program would need to consider the multiple incomes emanating from the same geographic area, for example, the immediate income from charcoal burning, followed by more medium-term income from agriculture or timber sales. Currently, several such PES components are integrated and ongoing in DRC's integrated REDD+ programs. Furthermore, given the high level of natural resource dependence of poor communities, it is essential to ensure that such a system includes immediate payment mechanisms to avoid worsening communities' living conditions. Indeed, relying solely on payments deferred to verification of results could push communities deeper into poverty. However, it is important to note that disproportionately placing payments in the early years of a program may also be detrimental for financial sustainability of local communities.

29.6.2 Land Tenure

Land tenure and property rights are held up as a precondition of the effectiveness of PES programs (Wunder, 2008; Wunder et al., 2020). Their reform is expected to improve the livelihood and well-being aspects of communities as access to resources is secured and investments are made possible, thereby and, consequently, reduction of poverty and inequality (Lawry et al., 2017; Hajjar et al., 2021).

It is clear in the context of forests that property rights generally have positive impacts on poverty reduction (Miller et al., 2020). Overall, for the best results in terms of poverty reduction, tenure reform is best coupled with other policy results, namely, stable institutions, access to justice and the rule of law, as well as access to finance and basic infrastructure (Hajjar et al., 2021). However, most PES programs are implemented in the Global South (Salzman et al., 2018; Kolinjivadi et al., 2023), in which such institutional stability is often not available. This leads to specific issues and risks to mitigate.

In developing countries, while decisions on land use are taken at a decentralized level, the land rights of local communities often remain weak and ill-defined. However, in a resource-abundant context, companies may be interested in exploiting these resources commercially. Consequently, there are numerous examples of communities selling their forest use rights, in order to secure an income, to logging, agro-industrial, mining, and other companies. Local communities' level of remuneration remains low, while corporate use sometimes results in the degradation of the forest and the environment on which local communities depend.

In the DRC context, this remains a particular risk in terms of local community forest concessions. Risk mitigation for a PES program should aim to compensate forest communities if they manage the forest sustainably, with the establishment of management plans and application of clear standards, rather than selling logging rights. This would enable them to earn a higher income than they would have through contracts with logging companies and, at the same time, ensure the protection of their environment. This would thereby enable better synergies with other SDGs.

29.6.3 Inclusion of the Poorest Populations

Furthermore, the literature review found that poorer households are frequently underrepresented among PES providers (Pagiola et al., 2005). According to Miranda et al. (2003), these are often landowners who derive their income from urban activities; accordingly, participation is not at the expense of any income-generating activity. Zbinden and Lee (2005) confirm an overrepresentation of wealthier and more educated landowners in PES participation in Costa Rica. Additional design measures can be taken to target and include poor populations. For example, in order to improve access, Costa Rica replaced the need for a formal land title with a system for locally recognized tenure. This is a step that could be considered in the local DRC context.

29.7 Conclusion

Thanks to its associated ES, the conservation and restoration of DRC's forests and biodiversity are of utmost importance to reach both global climate goals and the SDG Agenda. For the past decades, PES have been hailed as part of the solution to forest and biodiversity conservation. An incentive-based, market-based conservation tool, it has considerable experience in Latin America and parts of Asia. This has enabled solid evaluations and suggestions for improvement. However, in the African, particularly sub-Saharan, context, this is not yet the case. This chapter has identified a selection of needs and challenges therein.

Agenda 2030 unfortunately does not provide a panacea for sustainable development. Similarly, the literature review herein identified that market-based PES approach does not provide easy solutions without trade-offs. The results on poverty reduction, livelihood, deforestation, and its environmental additionality are frequently mixed. However, the literature identifies a high level of potential for improved effectiveness, particularly in terms of PES design.

To date, a number of research gaps remain: (1) PES' social and cultural impacts, (2) assessment of environmental and economic additionality, and (3) evaluation of PES impacts related to trade-offs and synergies between multiple livelihood dimensions. The importance of socio-cultural impacts for PES design and efficiency is underlined by Van Hecken et al. (2015) and Kolinjivadi et al. (2023), who found that gender equality and power asymmetries remain understudied. Despite research calls for greater social contextualization of PES, most of it remains decontextualized from political, social, and ecological relations. A deepened understanding of such issues is necessary in order to successfully implement a national PES program in the DRC, a vast country with widely varying ethnic, social, and cultural practices.

Similarly to the overarching SDG Agenda, the choice between pro-poor and environmental objectives in PES mechanisms cannot be traded off in the DRC context. Its high level of climate vulnerability, its international climate engagements, as well as deep-rooted poverty do not allow for it. PES are not necessarily a cure-all for poverty reduction, or for carbon sequestration. However, a carefully considered PES design and approach can contribute to the achievement of the biodiversity-related SDGs, by more thoughtful design and socio-economic inclusion.

Nevertheless, the 2020 deadline for multiple sub-targets of SDG 15 is already past us. It would be beneficial, not least for developing countries' sometimes limited reporting resources, to link the SDG Agenda more clearly with international climate and biodiversity agendas. An integrated approach requires the recognition of such interlinkages. The GBF builds upon previous goals and already provides further sector-specific guidance and indicators. Given the difficulties in quantifying and monitoring the broad biodiversity-related SDGs, integrating the new GBF guidance would be helpful for the SDGs' achievement but also ease the burden of the countries of the Global South in terms of national reporting.

In conclusion, implementing a PES program in the DRC holds promise for contributing to multiple SDGs: SDG 1 for poverty reduction and SFG 15 by incentivizing conservation and sustainability land use. However, its long-term effectiveness remains uncertain, and substantial measurement challenges persist. Consideration of the national socio-economic context, equitable distribution of benefits, and a robust monitoring and evaluation framework are some of the necessary elements in allowing a PES program to fulfil its promise as a tool for achieving SDGs in the DRC and elsewhere.

References

Adjognon, G. S., van Soest, D., & Guthoff, J. C. (2019). *Reducing Hunger with Payments for Ecosystem Services (PES): Experimental Evidence from Burkina Faso*. World Bank Policy Research Working Paper (8974).

Alix-Garcia, J. M., Sims, K. R., & Yañez-Pagans, P. (2015). Only one tree from each seed? Environmental effectiveness and poverty alleviation in Mexico's payments for ecosystem services program. *American Economic Journal: Economic Policy*, *7*(4), 1–40.

Arriagada, R. A., Sills, E. O., & Pattanayak, S. K. (2011). Payments for environmental services and their impact on forest transition in Costa Rica. In *13th Annual Biodiversity and Economics for Conservation Conference*.

Biermann, F., Hickmann, T., Sénit, C. A., Beisheim, M., Bernstein, S., Chasek, P., . . . & Wicke, B. (2022). Scientific evidence on the political impact of the Sustainable Development Goals. *Nature Sustainability*, *5*(9), 795–800.

Blundo-Canto, G., Bax, V., Quintero, M., Cruz-Garcia, G. S., Groeneveld, R. A., & Perez-Marulanda, L. (2018). The different dimensions of livelihood impacts of payments for environmental services (PES) schemes: A systematic review. *Ecological Economics*, *149*, 160–183.

Börner, J., Baylis, K., Corbera, E., Ezzine-de-Blas, D., Honey-Rosés, J., Persson, U. M., & Wunder, S. (2017). The effectiveness of payments for environmental services. *World Development*, *96*, 359–374.

Busch, J., Engelmann, J., Cook-Patton, S. C., Griscom, B. W., Kroeger, T., Possingham, H., & Shyamsundar, P. (2019). Potential for low-cost carbon dioxide removal through tropical reforestation. *Nature Climate Change*, *9*(6), 463–466.

CPI. (2023). *Global Landscape of Climate Finance*. Available online at: www.climatepolicyinitiative.org/wp-content/uploads/2023/11/Global-Landscape-of-Climate-Finance-2023.pdf

da Conceição, H. R., Börner, J., & Wunder, S. (2015). Why were upscaled incentive programs for forest conservation adopted? Comparing policy choices in Brazil, Ecuador, and Peru. *Ecosystem Services*, *16*, 243–252.

Daily, G. C. (1997). *Nature's Services: Societal Dependence on Natural Ecosystems*. Covelo, CA: Island Press.

Engel, S. (2016). The devil in the detail: A practical guide on designing payments for environmental services. *International Review of Environmental and Resource Economics*, *9*(1–2), 131–177.

Engel, S., Pagiola, S., & Wunder, S. (2008). Designing payments for environmental services in theory and practice: An overview of the issues. *Ecological Economics*, *65*(4), 663–674.

Flammer, C., Giroux, T., & Heal, G. (2023). Biodiversity Finance (No. w31022). National Bureau of Economic Research.

Fripp. (2014). *Payments for Ecosystem Services (PES) A Practical Guide to Assessing the Feasibility of PES Projects*. Available online at: www.cifor.org/publications/pdf_files/Books/BFripp1401.pdf

Froger, G., Maizière, P. A., & Legrand, T. (2016). Paiements pour services environnementaux et vulnérabilité des populations locales dans les pays du Sud–Études de cas costaricien et congolais. *Développement durable et territoires. Économie, géographie, politique, droit, sociologie*, *7*(1).

Fuss, S., Golub, A., & Lubowski, R. (2021). The economic value of tropical forests in meeting global climate stabilization goals. *Global Sustainability*, *4*, e1.

García-Amado, L. R., Pérez, M. R., Escutia, F. R., García, S. B., & Mejía, E. C. (2011). Efficiency of payments for environmental services: Equity and additionality in a case study from a biosphere reserve in Chiapas, Mexico. *Ecological Economics*, *70*(12), 2361–2368.

Grima, N., Ringhofer, L., Singh, S. J., Smetschka, B., & Lauk, C. (2017). Mainstreaming biodiversity in development practice: Can the concept of PES deliver? *Progress in Development Studies*, *17*(4), 267–281.

Hajjar, R., Newton, P., Ihalainen, M., Agrawal, A., Alix-Garcia, J., Castle, S. E., . . . & Timko, J. A. (2021). Levers for alleviating poverty in forests. *Forest Policy and Economics*, *132*, 102589.

He, J., & Sikor, T. (2015). Notions of justice in payments for ecosystem services: Insights from China's Sloping Land Conversion Program in Yunnan Province. *Land Use Policy*, *43*, 207–216.

Hegde, R., & Bull, G. Q. (2011). Performance of an agro-forestry based payments-for-environmental-services project in Mozambique: A household level analysis. *Ecological Economics*, *71*, 122–130.

Hejnowicz, A. P., Raffaelli, D. G., Rudd, M. A., & White, P. C. (2014). Evaluating the outcomes of payments for ecosystem services programmes using a capital asset framework. *Ecosystem Services*, *9*, 83–97.

Iftikhar, U. A., Kallesoe, M., Duraiappah, A., Sriskanthan, G., Poats, S. V., & Swallow, B. (2007). Exploring the inter-linkages among and between Compensation and Rewards for Ecosystem Services (CRES) and human well-being. *CES Scoping Study Issue Paper*, *1*.

Jayachandran, S., De Laat, J., Lambin, E. F., Stanton, C. Y., Audy, R., & Thomas, N. E. (2017). Cash for carbon: A randomized trial of payments for ecosystem services to reduce deforestation. *Science*, *357*(6348), 267–273.

Jindal, R., Kerr, J. M., & Carter, S. (2012). Reducing poverty through carbon forestry? Impacts of the N'hambita community carbon project in Mozambique. *World Development*, *40*(10), 2123–2135.

Jones, K. W., Powlen, K., Roberts, R., & Shinbrot, X. (2020). Participation in payments for ecosystem services programs in the Global South: A systematic review. *Ecosystem Services*, *45*, 101159.

Kariuki, J., & Birner, R. (2016). Are market-based conservation schemes gender-blind? A qualitative study of three cases from Kenya. *Society & Natural Resources*, *29*(4), 432–447.

Karolyi, G. A., & Tobin-de la Puente, J. (2023). Biodiversity finance: A call for research into financing nature. *Financial Management*, *52*(2), 231–251.

Kolinjivadi, V., Van Hecken, G., & Merlet, P. (2023). Fifteen years of research on payments for ecosystem services (PES): Piercing the bubble of success as defined by a Northern-driven agenda. *Global Environmental Change*, *83*, 102758.

Lawlor, K., Myers Madeira, E., Blockhus, J., & Ganz, D. J. (2013). Community participation and benefits in REDD+: A review of initial outcomes and lessons. *Forests*, *4*(2), 296–318.

Lawry, S., Samii, C., Hall, R., Leopold, A., Hornby, D., & Mtero, F. (2017). The impact of land property rights interventions on investment and agricultural productivity in developing countries: A systematic review. *Journal of Development Effectiveness*, *9*(1), 61–81.

Legrand, L. (2013). *L'analyse institutionnaliste des Paiements pour Services Environnementaux (PSE): vers une nouvelle compréhension du cas costaricien*, Doctorate thesis in economics, Université Versailles Saint Quentin en Yvelines.

Leménager, T., & Laurans, Y. (2015). Les paiements pour services environnementaux: simplicité trompeuse de l'outil, hétérogénéité fondamentale des situations. In Mermet, L., & Leménager, T. (dir.), *Développement et biodiversité: comment négocier le tournant environnemental*? Paris: Agence Française de Développement.

Li, J., Feldman, M. W., Li, S., & Daily, G. C. (2011). Rural household income and inequality under the Sloping Land Conversion Program in western China. *Proceedings of the National Academy of Sciences*, *108*(19), 7721–7726.

Liang, Y., Li, S., Feldman, M. W., & Daily, G. C. (2012). Does household composition matter? The impact of the Grain for Green Program on rural livelihoods in China. *Ecological Economics*, *75*, 152–160.

Liu, J., Li, S., Ouyang, Z., Tam, C., & Chen, X. (2008). Ecological and socioeconomic effects of China's policies for ecosystem services. *Proceedings of the National Academy of Sciences*, *105*(28), 9477–9482.

Locatelli, B., Rojas, V., & Salinas, Z. (2008). Impacts of payments for environmental services on local development in northern Costa Rica: A fuzzy multi-criteria analysis. *Forest Policy and Economics*, *10*(5), 275–285.

Lu, G., & Yin, R. (2020). Evaluating the evaluated socioeconomic impacts of China's Sloping Land Conversion Program. *Ecological Economics*, *177*, 106785.

Miller, D. C., Ordoñez, P. J., Brown, S. E., Forrest, S., Nava, N. J., Hughes, K., & Baylis, K. (2020). The impacts of agroforestry on agricultural productivity, ecosystem services, and human well-being in low- and middle-income countries: An evidence and gap map. *Campbell Systematic Reviews*, *16*(1).

Miranda, M., Porras, I. T., & Moreno, M. L. (2003). *The Social Impacts of Payments for Environmental Services in Costa Rica: A Quantitative Field Survey and Analysis of the Virilla Watershed* (No. 1). IIED.

Muradian, R., Arsel, M., Pellegrini, L., Adaman, F., Aguilar, B., Agarwal, B., Corbera, B., Ezzine de Blas, D., Farley, J., Froger, G., Garcia-Frapolli, E., Gómez-Baggethun, E., Gowdy, J., Kosoy, N., Le Coq, J. F., Leroy, P., May, P., Méral, P., Mibielli, P., Norgaard, R., Ozkaynak, B., Pascual, U., Pengue, W., Perez, M., Pesche, D., Pirard, R., Ramos-Martin, J., Rival, L., Saenz, F., Van Hecken, G., Vatn, A., Vira, B., & Urama, K. (2013). Payments for ecosystem services and the fatal attraction of win-win solutions. *Conservation Letters*, *6*(4), 274–279.

Naidoo, R., Gerkey, D., Hole, D., Pfaff, A., Ellis, A. M., Golden, C. D., Herrera, D., Johnson, K., Mulligan, M., Ricketts, T. H., & Fisher, B. (2019). Evaluating the impacts of protected areas on human well-being across the developing world. *Science Advances*, *5*(4), eaav3006.

OCDD Report. (2022). *National Report on the Implementation of the SDGs*. Available online at: https://ocdd.cd/wp-content/uploads/2022/12/Rapport-National-sur-la-Mise-en-OEuvre-des-ODD-version-digitale.pdf

OECD. (2020). *Paying for Biodiversity: Enhancing the Cost-Effectiveness of Payments for Ecosystem Services*. Available online at: http://environment-ecology.com/biodiversity/625-enhancing-the-effectiveness-of-payments-for-ecosystem-services-pes.html

Oldekop, J. A., Sims, K. R., Karna, B. K., Whittingham, M. J., & Agrawal, A. (2019). Reductions in deforestation and poverty from decentralized forest management in Nepal. *Nature Sustainabil*, *2*(5), 421–429.

Pagiola, S., Arcenas, A., & Platais, G. (2005). Can payments for environmental services help reduce poverty? An exploration of the issues and the evidence to date from Latin America. *World Development*, *33*(2), 237–253.

Pattanayak, S. K., Wunder, S., & Ferraro, P. J. (2010). Show me the money: Do payments supply environmental services in developing countries? *Review of Environmental Economics and Policy*, *4*(2).

Qu, M., Liu, G., Lin, Y., Driedger, E., Peter, Z., Xu, X., & Cao, Y. (2017). Experts' perceptions of the sloping land conversion program in the Loess Plateau, China. *Land Use Policy*, *69*, 204–210.

Raes, L., Aguirre, N., D'Haese, M., & Van Huylenbroeck, G. (2014). Analysis of the cost-effectiveness for ecosystem service provision and rural income generation: A comparison of three different programs in Southern Ecuador. *Environment, Development and Sustainability*, *16*, 471–498.

Robalino, J., Sandoval, C., Villalobos, L., & Alpizar, F. (2014). Local effects of payments for environmental services on poverty. *Discussion Paper-Resources for the Future (RFF)*, 14–12.

Salzman, J., Bennett, G., Carroll, N., Goldstein, A., & Jenkins, M. (2018). The global status and trends of Payments for Ecosystem Services. *Nature Sustainability*, *1*(3), 136–144.

Samii, C., Lisiecki, M., Kulkarni, P., Paler, L., Chavis, L., Snilstveit, B., Vojtkova, M., & Gallagher, E. (2014). Effects of payment for environmental services (PES) on deforestation and poverty in low and middle income countries: A systematic review. *Campbell Systematic Reviews*, *10*(1), 1–95.

Sanchez-Azofeifa, G. A., Pfaff, A., Robalino, J. A., & Boomhower, J. P. (2007). Costa Rica's payment for environmental services program: Intention, implementation, and impact. *Conservation Biology*, *21*(5), 1165–1173.

Scullion, J., Thomas, C. W., Vogt, K. A., Perez-Maqueo, O., & Logsdon, M. G. (2011). Evaluating the environmental impact of payments for ecosystem services in Coatepec (Mexico) using remote sensing and on-site interviews. *Environmental Conservation*, *38*(4), 426–434.

Sierra, R., & Russman, E. (2006). On the efficiency of environmental service payments: A forest conservation assessment in the Osa Peninsula, Costa Rica. *Ecological Economics*, *59*(1), 131–141.

Sims, K. R., & Alix-Garcia, J. M. (2017). Parks versus PES: Evaluating direct and incentive-based land conservation in Mexico. *Journal of Environmental Economics and Management*, *86*, 8–29.

Snilsveit, B., Stevenson, J., Langer, L., Tannous, N., Ravat, Z., Nduku, P., Polanin, J., Shemilt, I., Eyers, J., & Ferraro, P. J. (2019). Incentives for climate mitigation in the land use sector—The effects of payment for environmental services on environmental and socioeconomic outcomes in low-and middle-income countries: A mixed-methods systematic review. *Campbell Systematic Reviews*, *15*(3), e1045.

Tacconi, L., Mahanty, S., & Suich, H. (2013). The livelihood impacts of payments for environmental services and implications for REDD+. *Society & Natural Resources*, *26*(6), 733–744.

Trac, C. J., Schmidt, A. H., Harrell, S., & Hinckley, T. M. (2013). Environmental reviews and case studies: Is the returning farmland to forest program a success? Three case studies from Sichuan. *Environmental Practice*, *15*(3), 350–366.

Tuijnman, W., Bayrak, M. M., Hung, P. X., & Tinh, B. D. (2020). Payments for environmental services, gendered livelihoods and forest management in Vietnam: A feminist political ecology perspective. *Journal of Political Ecology*, *27*(1), 317–334.

UN (2022). *Decision Adopted by the Conference of the Parties to the Convention on Biological Diversity. 15/4. Kunming-Montreal Global Biodiversity Framework*. Available online at: www.cbd.int/decisions/.

Van Hecken, G., Bastiaensen, J., & Windey, C. (2015). Towards a power-sensitive and socially-informed analysis of payments for ecosystem services (PES): Addressing the gaps in the current debate. *Ecological Economics*, *120*, 117–125.

van Noordwijk, M., Leimona, B., Jindal, R., Villamor, G. B., Vardhan, M., Namirembe, S., Catacutan, D., Kerr, J., Minang, P. A., & Tomich, T. P. (2012). Payments for environmental services: Evolution toward efficient and fair incentives for multifunctional landscapes. *Annual Review of Environment and Resources*, *37*, 389–420.

World Bank (2012). *Lessons learned for REDD+ from PES and Conservation Incentive Programs*. Available online at: https://documents1.worldbank.org/curated/en/279231468233935706/pdf/769340WP0Box0300March0201200PUBLIC0.pdf

World Bank (2023). *The World Bank in DRC*. Available online at: www.worldbank.org/en/country/drc/overview#:~:text=DRC%20is%20among%20the%20five,in%20SSA%20lives%20in%20DRC

Wunder, S. (2008). Payments for environmental services and the poor: Concepts and preliminary evidence. *Environment and Development Economics*, *13*(3), 279–297.

Wunder, S. (2013). When payments for environmental services will work for conservation. *Conservation Letters*, *6*(4), 230–237.

Wunder, S. (2015). Revisiting the concept of payments for environmental services. *Ecological Economics*, *117*, 234–243.

Wunder, S., Börner, J., Ezzine-de-Blas, D., Feder, S., & Pagiola, S. (2020). Payments for environmental services: Past performance and pending potentials. *Annual Review of Resource Economics*, *12*, 209–234.

WWF (2022). *Living Planet Report 2022*. Available online at: www.worldwildlife.org/pages/living-planet-report-2022.

Yang, W., Liu, W., Viña, A., Luo, J., He, G., Ouyang, Z., Zhang, H., & Liu, J. (2013). Performance and prospects of payments for ecosystem services programs: Evidence from China. *Journal of Environmental Management*, *127*, 86–95.

Yin, R., Liu, C., Zhao, M., Yao, S., & Liu, H. (2014). The implementation and impacts of China's largest payment for ecosystem services program as revealed by longitudinal household data. *Land Use Policy*, *40*, 45–55.

Zbinden, S., & Lee, D. R. (2005). Paying for environmental services: An analysis of participation in Costa Rica's PSA program. *World Development*, *33*(2), 255–272.

30
INTERLINKAGE BETWEEN SUSTAINABLE DEVELOPMENT AND HUMAN RIGHTS IN STATES REPORTS TO INTERNATIONAL MECHANISMS

Saionara Reis, Joana Soler Torramilans, Helene Møller Winterskov, and Stinne Skriver Jørgensen

30.1 Introduction

The 2030 Agenda for Sustainable Development and its Sustainable Development Goals (SDGs) are a groundbreaking and comprehensive plan that aims to build a more equal and peaceful society by 2030. It takes an interconnected approach to humanity's development challenges, addressing economic, social, and environmental issues in a holistic manner.

From a human rights perspective, the Agenda is groundbreaking because it translates and brings together internationally agreed-upon human rights laws and principles into one plan of action. This plan is easy and attractive to communicate, and it connects rights which are often dealt with in two separate categories as a consequence of geopolitical dynamics (Jensen, 2017; Roberts, 2015), namely, (1) economic, social, and cultural rights (ESCR), such as the rights to quality education, to food, and to work, and (2) civil and political rights (CPR), such as access to justice and participation in public affairs. In other words, this is an *agenda* which understands that integral development can only be achieved if the ambition to end poverty and inequalities is matched with the tools for the people to freely and safely claim their rights and have a voice in building their future.

The SDGs draw from internationally agreed-upon instruments, and most targets are directly connected to the core international human rights conventions, as well as environmental instruments and labor standards.[1] Therefore, identifying how they connect in practice is an important step toward more efficient and coherent use of government resources. It can help close gaps in the implementation of development and human rights plans, foster cross-fertilization of information and data between the two sectors, and lead to better reporting practices to national, regional, and global fora.

Despite the substantial interlinkage between the SDGs and human rights, little research has been conducted on how they intersect in practice and how different reporting mechanisms can support one another.

DOI: 10.4324/9781003285472-32

Through text mining analysis, this chapter approaches the interconnectedness between SDGs and human rights in two types of states reporting. The main focus of the investigation is on how states integrate human rights in SDG reporting through the Voluntary National Review (VNR). Complementarily, it looks into the national reports to the Universal Periodic Review (UPR) to highlight the interconnection between this human rights monitoring mechanism and the SDGs. The analysis is based on VNRs and UPRs from 151 countries between 2016 and 2023.

This study is the first of its kind to systematically analyze VNR and UPR reports to uncover trends, gaps, and connections between SDGs and human rights. The results show high convergence between issues related to human rights and SDGs in both VNRs and UPR reports, respectively. Nonetheless, it shows little coherence in how states address their human rights obligations in the VNRs and how they explicitly refer to the SDGs in UPR reports.

When investigating on the 2030 Agenda pledge to leave no one behind, it also reveals that selected rightsholder groups, such as children, women and girls, and youth, are mostly visible across the VNRs, while others—for example, Indigenous peoples and LBGTI+ groups—vary in the degree of absence across regions.

This work further demonstrates how the use of tailored artificial intelligence (AI) can aid the analysis of large amounts of text to uncover trends at global and regional levels and expose which rightsholder groups, sustainable development goals and targets, and human rights issues are most prominent and silent in state reporting. The analysis is not intended to be exhaustive but, rather, to shed light on examples of what can be extracted from the reports using AI to inspire future projects.

In the sequence, the chapter first describes the conceptual interlinkages between human rights and the SDGs. Afterward, it explains the methodology of the text mining algorithm and the data sources. Finally, it presents the findings and conclusions from the analysis.

30.2 Conceptual Interlinkages Between Human Rights and the Sustainable Development Goals

It is beyond discussion that there is a close connection between human rights and the SDGs (Kaltenborn et al., 2020).

When writing the preamble of the 2030 Agenda, states were explicit in the ambition for the SDGs to realize human rights for all without leaving anyone behind (United Nations, 2015). This means that achieving the Goals implies respecting internationally agreed-upon laws, principles, and standards to protect, promote, and fulfil the rights of all peoples.

30.2.1 Inherent Interconnectedness

According to the "Human Rights Guide to the SDGs"[2] developed by the Danish Institute for Human Rights (DIHR), more than 90% of the SDG targets align with the Universal Declaration of Human Rights (UDHR), the core international human rights treaties,[3] and other international instruments. Table 30.1 shows examples of these connections.

Scholars and practitioners have argued that the protection and promotion of human rights are essential for sustainable development and that achieving the SDGs contributes to the realization of human rights. Some studies discuss the overall connections (Forum-Asia, 2018; Feiring and König-Reis, 2020), while others investigate the relation between specific SDGs and human rights, such as in the field of poverty (Sano, 2020) and health (Dovlo, 2020; Kaltenborn, 2020).

By understanding and using interconnectedness, states can create a solid foundation for policies which contribute to states' commitments to both agendas (Migowe et al., 2024). For example, if

Table 30.1 Examples of how the SDGs connect to human rights instruments[4]

Sustainable Development Goals	*Examples of related human rights*
1. No Poverty	**Right to an adequate standard of living** [UDHR Art. 25; ICESCR Art. 11; CRC Art. 27]; **right to social security** [UDHR Art. 22; ICESCR Art. 9; CRPD Art. 28; CRC Art. 26]
2. Zero Hunger	**Right to food** [UDHR Art. 25; ICESCR Art. 11; CRC Art. 24(2)(c)]
3. Good Health and Well-Being	**Right to health** [UDHR Art. 25; ICESCR Art. 12; CEDAW Art. 12; CRC Art. 24]
4. Quality Education	**Right to education** [UDHR Art. 26; ICESCR Art. 13; CRC Arts. 28, 29; CRPD Art. 24]
5. Gender Equality	**Elimination of all forms of discrimination against women** [CEDAW Arts. 1–5; CRC Art. 2]
6. Clean Water and Sanitation	**Right to safe drinking water and sanitation** [ICESCR Art. 11]
7. Affordable and Clean Energy	**Right to an adequate standard of living** [UDHR Art. 25; ICESCR Art. 11]
8. Decent Work and Economic Growth	**Labor rights** [UDHR Art. 23; ICESCR Arts. 6, 7, 10; CRPD Art. 27]
9. Industry, Innovation, and Infrastructure	**Right to an adequate standard of living** [UDHR Art. 25; ICESCR Art. 11]
10. Reduce Inequality	**Right to equality and non-discrimination** [UDHR Art. 2; ICESCR Art. 2(2); ICCPR Arts. 2(1), 26; CERD Art. 2(2); CEDAW Art. 2]
11. Sustainable Cities and Communities	**Right to adequate housing** [UDHR Art. 25; ICESCR Art. 11]
12. Responsible Consumption and Production	**Right to food** and the **right to safe drinking water** [UDHR Art. 25(1); ICESCR Art. 11]
13. Climate Action	**Right to life** [UDHR Art. 3; ICESCR Art. 12]
14. Life Below Water	**Right of all peoples to freely dispose of their natural resources** [ICCPR, ICESCR Art. 1(2)]
15. Life on Land	**Right of all peoples to freely dispose of their natural resources** [ICCPR, ICESCR Art. 1(2)]
16. Peace, Justice, and Strong Institutions	**Right to life, liberty, and security of the person** [UDHR Art. 3; ICCPR Arts. 6(1), 9(1); ICPED Art. 1]; **right to access to justice and due process** [UDHR Arts. 8, 10; ICCPR Arts. 2(3), 14–15; CEDAW Art. 2(c)]; **right to participate in public affairs** [UDHR Art. 21; ICCPR Art. 25]
17. Partnerships for the Goals	**Right to development** [UDHR Art. 28; ICESCR Art. 2(1); CRC Art. 4; CRPD Art. 32(1)]

states treat SDG 3 (Health and Well-Being) and SDG 4 (Quality Education) as both a development goal and a universal human right, they can leverage the practices and standards established by longstanding human rights mechanism to draw stronger health and education policies.

30.2.2 The Reporting and Monitoring Mechanisms

Ever since the Millennium Development Goals (2000–2015), the distance between the human rights community and the development community has been discussed (Alston, 2005; Kaltenborn,

2020). This continues to affect the processes, reporting, and monitoring of the SDGs and human rights, which have been criticized for working in silos despite the clear interconnectedness (Migowe et al., 2024; Muñoz Carmona, 2023).

The international human rights monitoring mechanisms, established at global and regional levels, offer much guidance for the implementation of the SDGs and human rights. While the SDG progress is reviewed at the High-Level Political Forum (HLPF) through voluntary reporting on non-binding commitments, the international human rights treaties are binding and have their own accountability system, which includes the UPR. On the other hand, the SDG follow-up and review process was set up by the 2030 Agenda to be vibrant, triggering participatory approaches at subnational, national, regional, and global levels, including in the preparation of VNRs. Both processes have a lot to offer each other.

There are tools which facilitate the operationalization of the integrated approach. The "SDG–Human Rights Data Explorer"[5] helps identify country-specific guidance from the recommendations emitted by the human rights monitoring mechanisms with their connection to the SDG targets. This tool is also used to filter the recommendations which are relevant for specific rights-holder groups, such as Indigenous peoples, women and girls, and children, to be used as reference in development plans that affect them.

30.2.3 National Human Rights Institutions and the SDGs

National Human Rights Institutions (NHRIs) play a crucial role in understanding and implementing the synergies between the SDGs and the human rights obligations in practice (Rudolf, 2020; Feiring et al., 2017).

NHRIs are independent state institutions with a mandate to promote and protect human rights in a given country. In 1994, the United Nations General Assembly (UNGA) encouraged member states to establish or, "where they already exist, to strengthen national institutions for the promotion and protection of human rights and to incorporate those elements in national development plans" (A/RES/48/134). On the same occasion, UNGA adopted the Paris Principles, which set the standards for competences, independence, and operation of NHRIs.

In a significant move to integrate these institutions into the global sustainable development agenda, the "existence of independent [NHRIs] in compliance with the Paris Principles" became an SDG indicator (Indicator 16.a.1). In turn, in 2015, the Global Alliance of National Human Rights Institutions (GANHRI) adopted the Mérida Declaration,[6] "which outlines the functions and activities that NHRIs can undertake in order to promote a human-rights-compliant, transparent, participatory, and inclusive implementation of the SDGs" (Breuer and Leininger, 2021).

As of April 2023, there are 88 countries with a fully compliant NHRI accredited by GANHRI. Additionally, there are 32 countries with partially compliant institutions.[7] Many of these institutions have been engaging with national and global sustainable development processes.[8] They have inter alia been considered data sources for SDG indicators, offered advice and guidance to other state institutions, and brought them to account regarding their human rights obligations in the enactment of national development plans and programs (Breuer and Leininger, 2021).

30.3 Methodology and Data

This analysis was conducted with the support of AI. In this section, we describe the methodology behind the algorithm and the data sources used in the analysis.

30.3.1 Methodology: Text Mining of Large Amounts of Text

To process and draw insight from large amounts of qualitative data, such as the VNR and UPR reports, we use a text mining algorithm named Text Classify 8[9] (TC8), developed by the DIHR, in collaboration with Specialisterne.[10] The algorithm works on small extracts of text (text snippets), which typically correspond to a paragraph. Essentially, TC8 analyzes patterns in the text snippets to identify relevant thematic categories. A text snippet can be linked to none, one, or multiple categories. Specifically, TC8 can classify text into human rights themes, SDGs, and rightsholder groups.[11]

30.3.1.1 The Algorithm Categories Used in the Analysis

30.3.1.1.1 HUMAN RIGHTS CATEGORIES

The human rights categories of the algorithm were developed in collaboration with the Office of the High Commissioner for Human Rights (OCHCR).[12] OHCHR has classified 100 human rights themes, and 86 were used in this study. The human rights categories reflect themes related to (1) CPR, such as freedom of opinion, human traffic king, and arbitrary detention; (2) ESCR, such as the rights to social security, food, health, education, water and sanitation, work, and adequate housing; and (3) cross-cutting issues, such as equality and non-discrimination, or human rights and environment.[13] The human rights categories are used in the analysis of the VNRs.

30.3.1.1.2 RIGHTSHOLDER CATEGORIES

The algorithm is also trained to identify text related to selected rightsholder groups, specifically, women and girls; children; Indigenous peoples; persons with disabilities; migrants; refugees and asylum-seekers; internally displaced persons; human rights defenders; lesbian, gay, bisexual, transgender, and intersex (LGBTI+); older persons; and youth. The rightsholder categories are used in the analysis of the VNRs.

30.3.1.1.3 SUSTAINABLE DEVELOPMENT CATEGORIES

The sustainable development categories are based on the 17 SDGs and their 169 targets. The algorithm identifies sustainable development categories at target level.[14] The SDG categories are primarily used in the analysis of the UPR reports and in specific examples of the VNRs.

30.3.1.2 The Training Methodology Behind the Algorithm

The algorithm relies on *ground truth* (also known as training data). The ground truth consists of paragraphs of text that have been tagged by a human who is an expert within the field of sustainable development and human rights. Using the text tagged by the experts, TC8 identifies statistical patterns of words and standing expressions that have the most potential to be linked to a thematic category. For instance, the term *trafficking of women* is used to identify examples for SDG Target *5.2 (eliminate all forms of violence against all women and girls)*.

Using the *ground truth* in combination with an *expert dictionary* (list of words and expressions most relevant to a given theme created by an expert), the algorithm constructs the probability that any paragraph of text is annotated within a given category. If this probability is higher than the established threshold, the text is assigned to the category. Otherwise, the text is not tagged, as there is not enough certainty of the link created between the text and the category.[15]

30.3.1.3 Accuracy of the Algorithm

Using a machine learning algorithm like TC8 makes one able to classify a large amount of text in a very short time. It does, however, come with the possibility that some texts will be misclassified. *Misclassification* means that some snippets will be linked to a category even though they do not belong in that category (false positive) and, contrarily, that some snippets will not be linked to a category even though they should have been (false negative). Most misclassifications are related to ambiguous and/or rare categories.

To evaluate how well the algorithm is performing, a performance measure is calculated. This is an indication of the likelihood of false positives/false negatives within a category and aids in determining the need for category refinement and/or providing more training data.[16]

When working with supervised machine learning, one should also be aware that the tagging of the ground truth is done by humans, and therefore, it can introduce potential for disagreement, errors, and bias. Furthermore, TC8 works positively, meaning, that it identifies categories in text but does not confirm the absence of categories if none are detected. Instead, it indicates that TC8 could not find any categories in the text paragraph.[17]

30.3.2 The Data

This analysis looks at two sets of state reports, namely, the VNR and UPR national reports between 2016 and 2023.[18]

The **VNR** is the state report on the SDGs' progress. It serves as a basis for regular reviews by the HLPF and aims to facilitate sharing experiences, including successes, challenges, and lessons learned, in the implementation of the 2030 Agenda.[19] According to the "UN Secretary General Common Reporting Guidelines for the Production of VNRs,"[20] countries are encouraged to report on all SDGs, not just those under review in a given year. This approach ensures a comprehensive assessment of progress and challenges across the entire spectrum of Sustainable Development Goals. The VNR is voluntary by nature, and states do not receive institutionalized feedback on their report.

The **UPR** is a monitoring mechanism under the Human Rights Council (HRC) where states undergo a peer review regarding the progress and compliance vis-à-vis the international human rights obligations every 4.5 years. Based on the information submitted in the national report, as well as the input from the NHRI and civil society, states receive and are expected to follow up on recommendations from other UN member states to improve their human rights compliance.[21]

30.3.2.1 Country Representation in the Analysis

The analysis includes 151 UN member states. It looks at 265 VNRs and 210 UPR reports submitted in the period 2016–2023 (observe that the number of VNRs is higher due to a more frequent reporting practice by some states at the HLPF). The UPRs are, on average, 21.5 pages long, while the VNRs have, on average, 130 pages.

Not all countries which produced VNR and UPR reports are included in this analysis due to challenges related to PDF formatting and language (i.e., the algorithm can only analyze text written in English or Spanish, and some reports are exclusively available in Russian, French, or

Arabic). For consistency, countries that did not have both a VNR and a UPR report in accessible format and matching language have been excluded from this analysis.

It is important to highlight the resulting regional bias specifically for Africa, where many VNRs were only available in French. If we calculate the representativity of the states in this analysis, compared to the total number of UN member states in each region, there are only 55.5% of African states considered in this analysis, compared to 80.8% from Asia, 91.4% from the Americas, 90.7% from Europe, and 85.7% from Oceania (Figure 30.1).

30.3.2.2 Data Preparation—Text Snippets

To prepare the data for the analysis, we used a PDF extraction method which relies on PDF location markers. The text is extracted from the PDF in paragraph-sized "text snippets" based on predefined, logical paragraph criteria. In this chapter, we use both the number of reports and text snippets as analysis units.

30.3.2.3 Text Mining the VNR and UPR Reports

The algorithm is the main tool used for mining the data in this analysis. On some occasions, the tagging of the algorithm is complemented with the search of specific keywords. This is the case

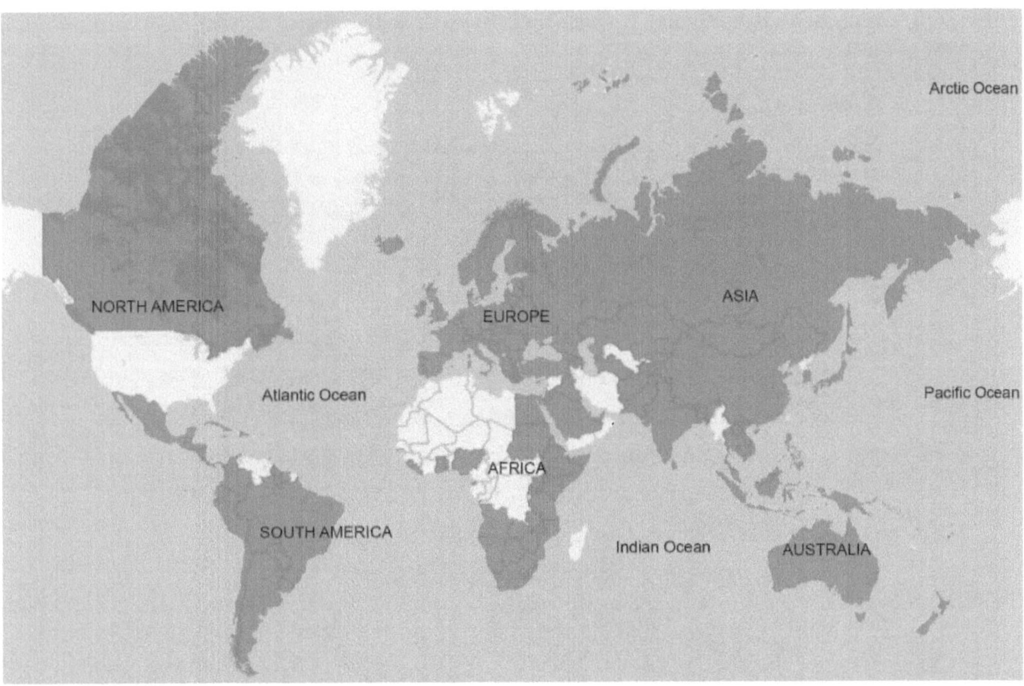

Figure 30.1 Countries included in this analysis.

Note: The dark color indicates the countries included in the analysis.

Source: DIHR 2024 team. © Tom Tom, © Microsoft Corporation, © OpenStreetMap.

when searching for specific human rights treaties and bodies, explicit mentions of NHRIs, as well as references to specific SDGs.

To examine the data, we assess the prevalence of categories and keywords per number of file and throughout the text of the reports, exploring differences among countries and regions over time. Our approach involves using descriptive statistics and exploratory analysis to identify patterns, relationships, and gaps.

30.4 Human Rights in the Voluntary National Reviews

The VNRs have emerged as a crucial tool for states to report and exchange insights on their progress towards realizing the SDGs. They offer valuable insights into implementation challenges, opportunities, and evolving perspectives on sustainable development within each country.

Considering the strong connection between the 2030 Agenda and human rights, VNRs should reflect on the latter (Muñoz Carmona, 2023). In this section, we investigate if and how this is taking place by assessing the prevalence of human rights themes, instruments, mechanisms, and institutions in the VNRs. We also explore which rightsholder groups are most prevalent or absent in the reports.

30.4.1 Prevalence of Human Rights Issues in the VNRs

Given the demonstrated synergies between the SDGs and human rights issues, it comes as no surprise that in the VNRs we find a prevalence of human rights themes throughout the reports. The ESCR are particularly prevalent, given their overlap with several SDGs. From Figure 30.2,

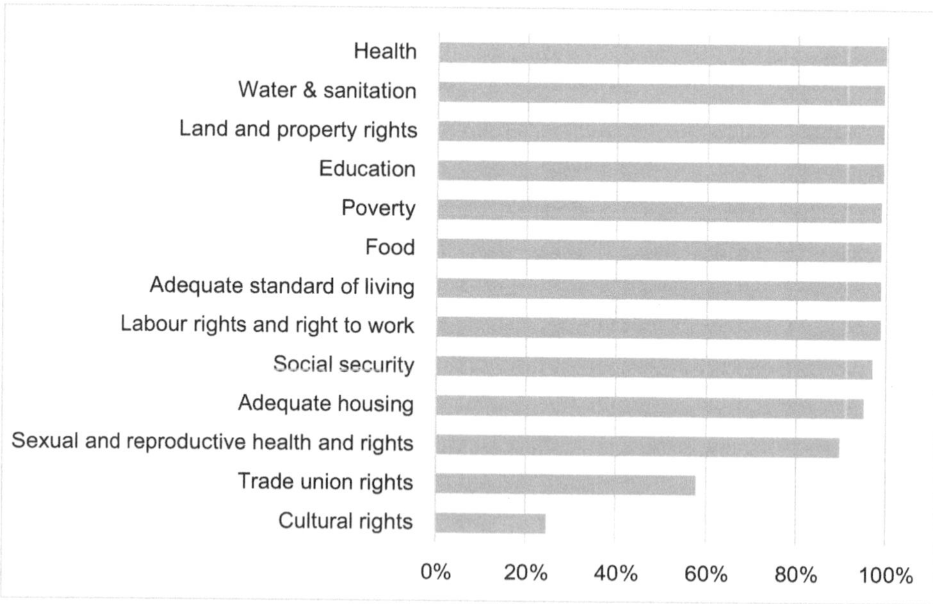

Figure 30.2 Percentage of VNRs linked to economic, social, and cultural rights themes.

Source: DIHR 2024 team calculation based on the application of the algorithm categories related to ESCR in 265 VNRs from 2016 to 2023. Unit of analysis: number of reports.

we see that issues related to the rights to health, water, food, work, and education, among others, are present in all or most of the 265 VNRs. Cultural rights are the exception, appearing in only one quarter of the reports, followed by trade union rights, which are reflected in just over half of them.

While the CPR themes comprise nearly double that of ESCR, the connection with the SDGs is, to some extent, concentrated around what is called SDG 16+, which comprises SDG 16 (on peaceful, just, and inclusive societies) and related targets.[22] Perhaps consequently, the prevalence of CPR in the VNRs is more diverse (Figure 30.3).

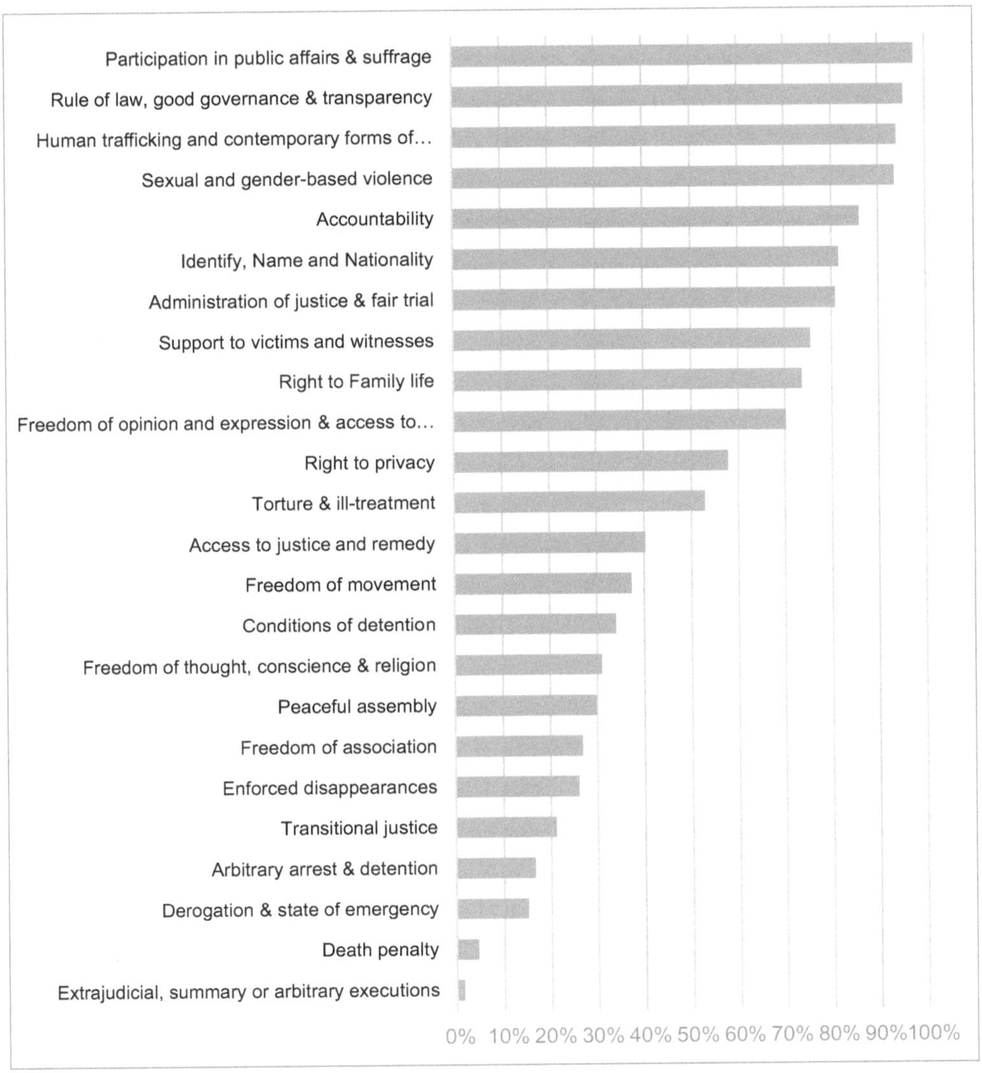

Figure 30.3 Percentage of VNRs linked to civil and political rights themes.

Source: DIHR 2024 team's calculation based on the application of the algorithm categories related to CPR in 265 VNRs from 2016 to 2023. Unit of analysis: number of reports.

Figure 30.4 Top five human rights themes most frequently mentioned throughout the VNRs, by region.

Note: The graph shows the top five human rights themes most mentioned in the VNRs in relation to the total number of text snippets (equivalent to paragraphs) in the reports.

Source: DIHR 2024 team's calculation based on the application of the algorithm categories related to human rights themes in 265 VNRs from 2016 to 2023. Unit of analysis: number of text snippets.

Issues related to participation, rule of law, human trafficking, and sexual and gender-based violence, among others, are reflected in most reports. On the other hand, issues related to the fundamental freedoms (of movement, association, thought, and opinion) are only reported to a certain extent by countries, despite clear overlap with SDG Target 16.10 (ensure public access to information and protect fundamental freedoms). A similar scenario appears for arbitrary detention and access to justice (related to SDG Target 16.3).

Additionally, we observe that the theme on the human rights principle of "equality and non-discrimination" is notably prevalent throughout 99% of the VNRs. This prominence can be attributed to its close alignment with SDGs 5, 10, and 16 and the crosscutting nature of these themes regarding the entire 2030 Agenda and the pledge to leave no one behind.

Some human rights themes with limited visibility in the VNRs include "extrajudicial, summary, or arbitrary executions" (four reports) and "arbitrary arrest and detention." The latter is found in less than 17% of VNRs, despite its explicit link to SDG issues such as rule of law, access to justice, and the protection of fundamental freedoms.

If we shift our focus to the prevalence of human rights themes within the reports in which they appear—meaning, how frequently they are discussed in relation to the number of paragraphs in the reports—we observe some regional differences (Figure 30.4). The right to health appears universally prominent across regions, on top of the prevalence list for all. In the sequence, reports from Europe emphasize themes related to "equality and non-discrimination" and "business and human rights," while "water and sanitation" comes high in the reports from Asia. Oceania's top concerns revolve around the environment, with categories like "environment," "water," "climate," and "land" taking precedence. In Africa, combatting "poverty" emerges high on the prevalence list, along with "water," "education," and "adequate standard of living." These trends offer valuable insights into the nuanced challenges and priorities shaping each region's human rights agenda.

30.4.2 Approaching Sustainable Development Issues as a Right

From the preceding section it is observable that human rights themes are strongly present in the VNRs, particularly those with an explicit connection to the Sustainable Development Goals and their targets. Nonetheless, this investigation merits the question of whether this prevalence is vague or whether states are intentionally approaching the SDGs as a matter of rights and in connection to their international human rights obligations. In this section we will also analyze how rightsholders and NHRIs are reflected in the reports.

30.4.2.1 Prevalence of Issues Specifically Referred to as "Rights"

When looking at examples in the VNR reports, we see that the human rights principles of "equality" and "non-discrimination" are most prominently being portrayed in connection to rights. In fact, among the text snippets identified by the algorithm as pertaining to this theme, 24% have the explicit word "right," often emphasizing "equal rights and freedoms."

The same applies to the themes related to "sexual and gender-based violence" and "violence against women," which are directly tied to SDG 5 (Gender Equality). Respectively, 24% and 21% of the text snippets within these categories contain reference to rights, particularly "women's rights."

Conversely, themes related to the "environment," which are very prevalent in the VNRs (99.2% of the reports), show minimum connection to rights. In fact, only 5.4% of the related texts contain the word "right." This is an area with ample opportunity for states to improve on, given that outside the reporting arena, environment-related issues, such as the impact of climate change on human rights, are raising the interest of courts, rightsholders, academia, and other stakeholders, with useful resources and lessons to help connect the dots.[23]

30.4.2.2 Connecting SDG Issues to International Human Rights Obligations

As showcased in the "Human Rights Guide to the SDGs,"[24] most Sustainable Development Goals and their targets are connected to international human rights treaties, labor standards, and environmental instruments. There is therefore an opportunity for states to approach them coherently in their implementation and reporting efforts. Intentionally connecting their human rights obligations with their commitments under the 2030 Agenda can foster efficiency in the programming, planning, and monitoring of both SDGs and human rights.

When analyzing the VNRs, we observe little coherence in how states connect SDG issues to their obligations under binding international instruments. Although most reports (60% of them) refer to at least one of the nine core international human rights treaties,[25] this is neither done systematically nor consistently with the average of states ratification of these instruments.[26]

SDG 5 (Gender Equality), for example, is among the most prominent goals in the VNRs (it appears in 88% of the reports). We observe nonetheless that only 36.6% of VNRs reference the Convention or the Committee on the Elimination of All Forms of Discrimination Against Women (CEDAW). This absence is particularly striking, considering that at the time of writing, 189 countries are party to the Convention.

Similarly, the Convention on the Rights of the Child (CRC) is perceived as of crosscutting relevance for all SDGs, in addition to aligning explicitly to targets under SDGs 5, 8, and 16. The CRC and the SDGs mutually reinforce each other, as the CRC establishes a comprehensive framework for children's rights, which are integral to all aspects of sustainable development.[27] However, despite this symbiotic relationship, the CRC is mentioned in only 27% of analyzed VNRs.

It does not go unnoticed that countries connecting their international obligations on children's rights in the SDG context bring strong elements for the realization of both SDGs and human rights. The VNR from Zimbabwe (2017), for instance, illustrates how the countries' obligations vis-à-vis the CRC and the African Charter on the Rights and Welfare of the Child have triggered the Constitutional Court in 2016 to pass "a judgement outlawing marriage of persons below the age of 18," therefore also contributing to SDG Target 5.3 (eliminate all harmful practices, such as child, early, and forced marriage and FGM).

In the Irish VNR (2023), the government recycled information received by and submitted to the human rights monitoring mechanism. They mention that "criticism was levelled [by the mechanism] against the Irish Government due to the lack of disaggregated data about human rights issues affecting children" and recognized that "[t]he absence of disaggregated data in this regard renders pupils living in Direct Provision almost invisible." The report reflects about the impact of this data gap for the implementation of the SDGs for this group.

These examples highlight an important reason for improved coherence between SDGs and human rights reporting. The accountability features of the human rights system require a certain level of detail from states in their reporting on progress vis-à-vis the Conventions. In addition, they provide feedback, guidance, and recommendation which support states in realizing and protecting rights. This mechanism is therefore a rich source of relevant information for states to use in both the implementation and reporting of related SDG issues.

30.4.2.3 Illustrating a Human Rights Perspective to SDG Implementation and Reporting

Generally, the ratification of binding international treaties triggers processes at national level to facilitate its implementation. This includes change in legislation, the adoption of national action plans, and the establishment of national mechanisms for reporting, implementation, and follow-up.

Perceiving the SDGs in light of human rights obligations can not only benefit more coherent and effective reporting but also opens access to resources, tools, data, and mechanisms which can help strengthen implementation. In the VNRs, some countries highlighted their efforts to promote an integrated approach.

Bosnia and Herzegovina (2023), for example, reported on its efforts to "enhance cooperation among relevant institutions responsible for monitoring human rights, statistical agencies and institutes, civil society organizations," among others.

Similarly, Mozambique (2020), highlighted the importance of having a human rights perspective and abiding to the international human rights framework to contribute to building peaceful societies regarding, among others, "the rights of women, children, the elderly and people with disabilities" and to "address elements related to peace and violence—such as torture, human trafficking and political and civil rights."

In the example of Costa Rica (2017) we observe the state leveraging the human rights–based approach when developing policies to implement SDG 3 (on health). It specifically mentions the "installation of specialized units with a human rights perspective to provide dignified care to LGBTIQ persons."[28] Similarly, it emphasizes the efforts to "improve the care and treatment of people with mental health disorders in the community from a human rights perspective."[29]

30.4.3 Rightsholder Groups in the Voluntary National Review

The 2030 Agenda articulates the unequivocal commitment of member states to leave no one behind. This commitment implies addressing discrimination and inequalities and providing for policy solution which addresses the specific needs of all persons. To operationalize this commitment at country level, it is essential to identify who is being left behind and which are the root causes and systemic barriers preventing progress for these groups.

The VNRs are not the only monitoring process countries use to track progress on the sustainable development, but it certainly is one which encourages a participatory approach and generates national and international visibility to issues prioritized in the report.

Visibility for rightsholder groups in the VNRs can be interpreted as a sign that their needs and claims are on the radar of governments. Invisibility, on the other hand, indicates a gap in how groups are addressed at the national level. The silence about some groups in the VNRs could also be related to a literal interpretation of the 2030 Agenda by some countries, which then report only on groups which are specifically mentioned in the SDG targets and indicators. This represents, naturally, a big gap in the pledge to leave no one behind and speaks to the importance of acknowledging beyond the usual suspects in the global sustainable development agenda.

According to Figure 30.5, "youth" (98.5%), "children" (96%), and "women and girls" (94%) are mentioned in nearly all VNRs. "Migrants," "persons with disabilities," and "older persons" come in the sequence with relatively high prevalence. "Indigenous peoples" and "refugees and asylum-seekers" appear in about half of the reports. On the lower side appears "human rights defenders" (28%), "internally displaced persons" (42%), and "LGBTI+" persons (47%).

When zooming into an example of regional prevalence, we observe that "Indigenous peoples" is the 7th (of 11) most mentioned group in the Americas, appearing in 40% of the reports in the region. It is noticeable that among the reports mentioning this group, the frequency is quite high: Indigenous peoples are the 4th most mentioned groups within the reports in which it appears.

One can also explore which SDGs are mostly reported on in relation to different rightsholder groups. For instance, Figure 30.6 shows the top 10 SDG targets which appear in connection to Indigenous groups in the VNRs from the Americas.

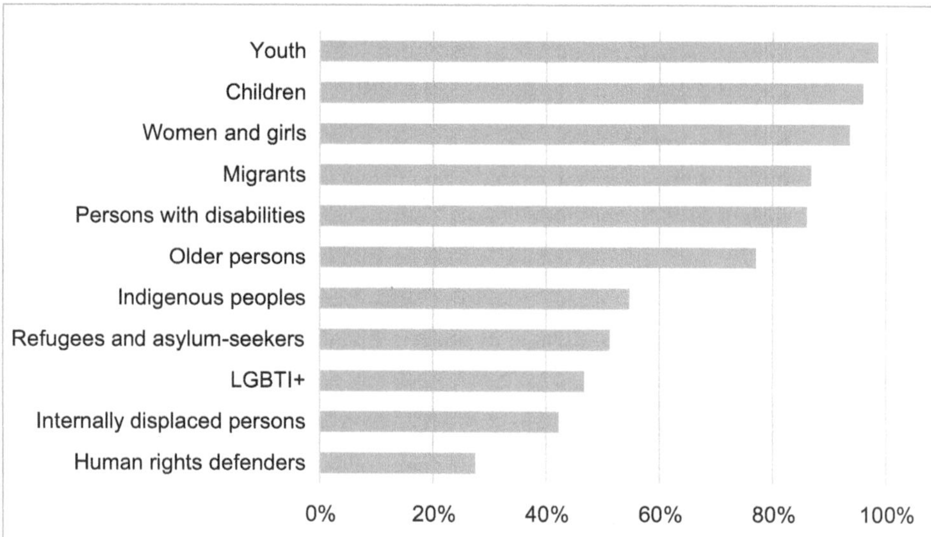

Figure 30.5 Percentage of VNRs linked to the rightsholder groups.

Source: DIHR 2024 team's calculation based on the application of the algorithm categories related to rightsholder groups in 265 VNRs from 2016 to 2023. Unit of analysis: number of reports.

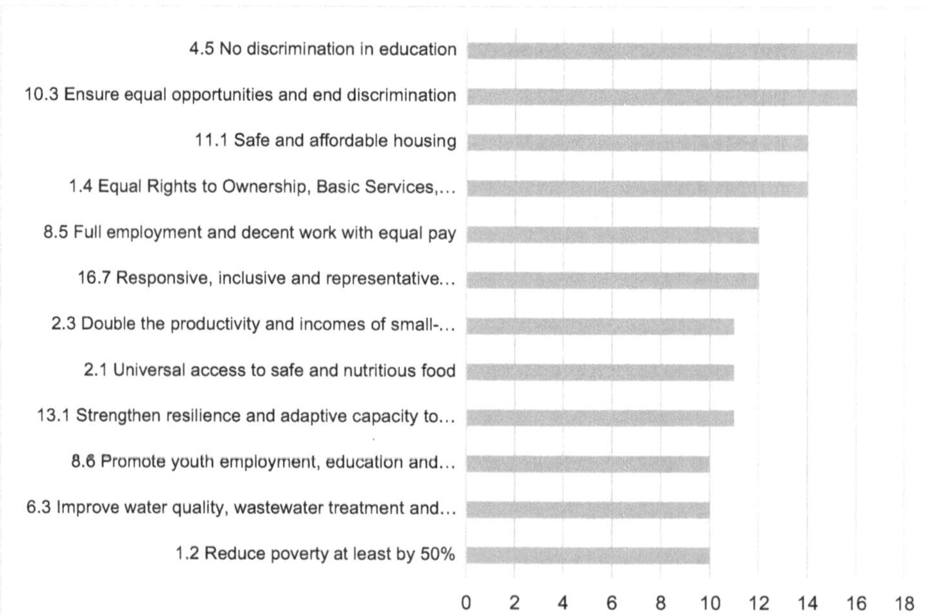

Figure 30.6 Top 10 SDG targets in connection to Indigenous peoples in the VNRs from the Americas region.

Note: The graph shows the number of VNRs that contain at least one text snippet tagged simultaneously to the algorithm categories on indigenous peoples and on sustainable development (SDG target).

Source: DIHR 2024 team's calculation based on the application of the algorithm in 57 VNRs from the Americas from 2016 to 2023.

When exploring examples in the prevalence of groups per number of VNRs by region, "LGBTI+" persons appear among the three least mentioned groups in Africa, Asia, and Oceania, contrary to the Americas, where it appears in the top five. "Human rights defenders" appears consistently among the three most invisible groups in all regions, despite being the focus of one important SDG indicator for human rights (SDG Indicator 16.10.1).

30.4.4 Leveraging the Mandate of National Human Rights Institutions for the Sustainable Development Goals

UNGA and HRC resolutions[30] have encouraged member states and the UN system to leverage the mandate of NHRIs in support of SDG implementation. Given their independent nature, mandate, and expertise in the human rights field, there is a unique role for them to play in support of the "leave no one behind" pledge.

Considering the specific SDG indicator on the existence of NHRIs (SDG Indicator 16.a.1) and the fact that 120 countries have a fully or partially compliant NHRI, there is a relatively low level of reporting on these institutions in the VNRs (Figure 30.7). Additionally, there are varied practices among states regarding how they report on their NHRI in connection with the SDG processes.

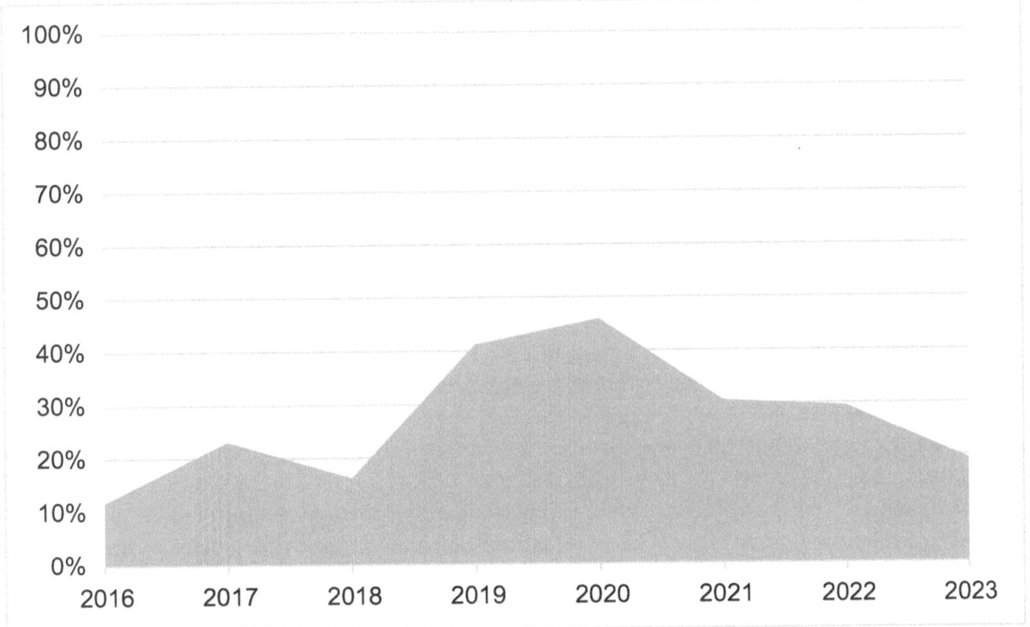

Figure 30.7 Percentage of VNRs with explicit mention of their own NHRI over time.

Note: This figure should be interpreted with attention to the fact that not all countries have an NHRI. As of April 2023, there are 120 countries with a fully or partially compliant NHRI, and they might not be equally represented throughout the years. Specifically, 19 countries with an NHRI are not included in this analysis due to the limitations explained in section "3.2.1 Country representation in the analysis."

Source: DIHR 2024 team's calculation based on a word search of the GANHRI's list of NHRIs[31] in 265 VNRs from 2016 to 2023. Unit of analysis: number of reports.

While countries like Rwanda (2019 and 2023) and Malaysia (2021) simply mention the existence of their NHRI, we found good practices of substantive reporting on the contribution of these institutions to advance the SDGs in reports from other countries. Ireland (2023), for instance, included in its VNR the integral parallel report with recommendations and assessments on the SDGs from the Irish Human Rights and Equality Commission.

Ghana (2019 and 2022) highlighted the role of the Commission on Human Rights and Administrative Justice in the development of legislation and policies on gender equality (SDG 5) and in addressing human rights violations through receiving complaints from citizens and conducting investigations.

Finally, Hungary (2018) acknowledged "the engagement of the Office of the Commissioner for Fundamental Rights . . . in the process of preparing the VNR." It also highlighted the importance of realizing the recommendations from human rights institutions for the national implementation of the SDGs, particularly in areas "related to law-making and to the conduct of public authorities."

30.5 Sustainable Development in UPR Reporting

There is ample potential for using human rights monitoring to assess progress and gaps in the SDG implementation: from identifying key issues and vulnerable populations to improving efficiency, coherence, and accountability. To foster this, the UN created a repository on good practices on how the universal human rights monitoring process (the UPR) can be used to maximize impact for sustainable development.[32]

Complementing this effort, in this section we use AI to assist in identifying content that can be useful for the SDGs from one element of the UPR process, namely, the national reports. Firstly, we analyze the presence of SDG-related issues in the reports and, secondly, we search for explicit mentions of the SDGs.

30.5.1 SDG-Related Issues in National Reports to the Universal Periodic Review

The striking convergence between SDGs and human rights is exposed when running the algorithm on UPR reports: Several SDGs are tagged in 100% of them (Figure 30.8). In addition, 70% of all text snippets (equivalent to the report paragraphs) address SDG-related issues, signifying that these are prominently discussed throughout the text.

Not surprisingly, SDG 16 (Peace, Justice, and Strong Institutions) appears as the most mentioned throughout the reports (in 29% of text snippets). SDG Targets 16.3 (rule of law and access to justice) and 16.7 (responsive, inclusive, and participatory decision-making) are the most prevalent. SDG 5 (Gender Equality) and SDG 4 (Quality Education) come in the sequence, appearing in 16.5% and 13% of the text snippets, respectively.

When assessing trends over time, we observe that issues related to SDG 13 (Climate Action) are increasingly being picked up in UPR reports, with a steady rise from 52% in 2016 to 71% of the reports in 2023. This is consistent with the notable surge in awareness and attention toward issues related to human rights and the environment over the past decade, culminating in the UN formally recognizing the *right to live in a clean, healthy, and sustainable environment for everyone, everywhere*, in 2021 (A/HRC/RES/48/13).[33]

At the lowest side of prevalence appears SDG 14 (Life Below Water), SDG 7 (Affordable and Clean Energy), SDG 15 (Life on Land), and SDG 9 (Industry, Innovation, and Infrastructure).

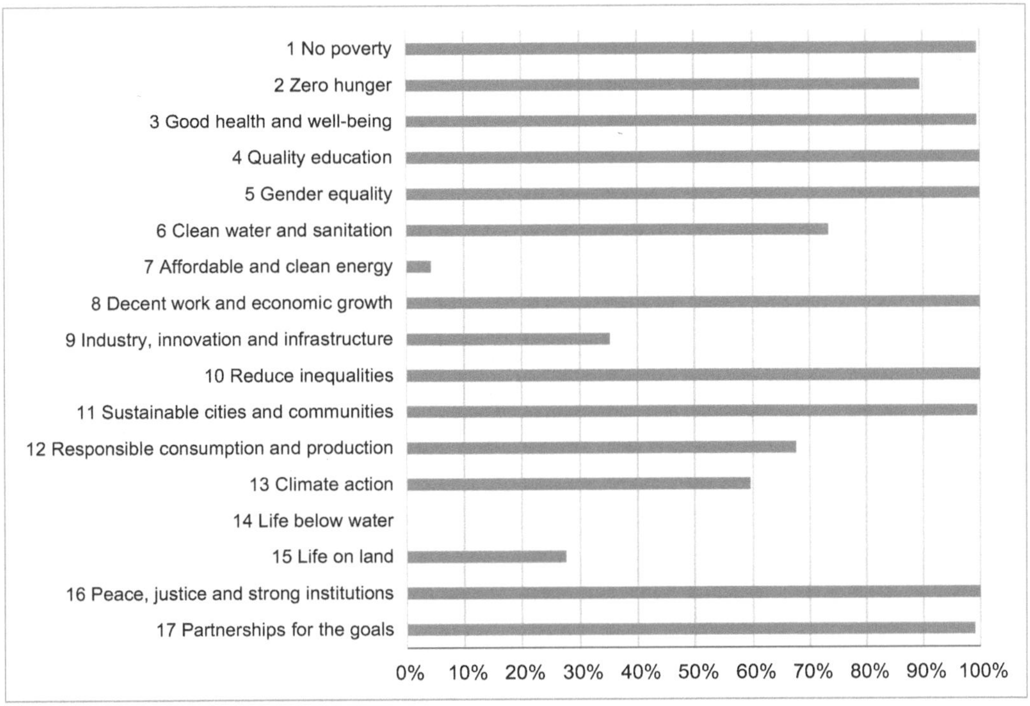

Figure 30.8 Percentage of UPR reports linked to the SDGs.

Source: DIHR 2024 team's calculation based on the application of the "sustainable development" categories of the algorithm in 210 UPR national reports from 2016 to 2023. Unit of analysis: number of reports.

The small attention dedicated in the UPRs to these issues could indicate an oversight of this body regarding the human rights impact to specific rightsholder groups. This opens an opportunity for the human rights system to learn from sustainable development knowledge and practices in these areas. For example, extensive studies from DIHR on SDG 14 demonstrate the devastating consequences of the big fishery and aquaculture industry to the rights and livelihoods of small fishery communities.[34]

30.5.2 SDGs Mentioned Explicitly in the UPR Reports

From early days, the human rights system undertook the task of understanding how the 2030 Agenda intersected with human rights and which role the system and the human rights actors could play in leveraging it as an opportunity to advance rights. For instance, by 2024, the Human Rights Council had organized six Intersessional Meetings on Human Rights and the 2030 Agenda[35] to, among other things, provide a multistakeholder space for dialogue and exchange of good practices, achievements, challenges, and lessons learned.

The increased awareness of human rights actors to the SDGs is observed in the UPR reports, where we capture a trend of explicit mentions of SDGs and in a higher number of reports over time. The trend, nonetheless, appears to be losing tracking in most recent years (Figure 30.9).

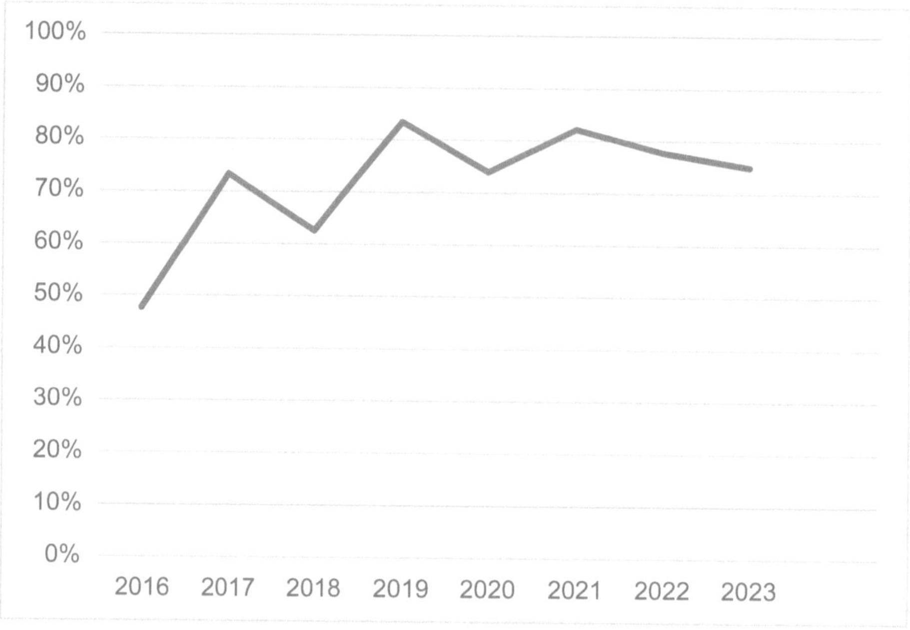

Figure 30.9 Percentage of UPRs with explicit mention of the SDGs over time.

Source: DIHR 2024 team's calculation based on word search of the SDGs in 210 UPR national reports from 2016 to 2023. Unit of analysis: number of reports.

When analyzing the content of the reports, we observe that the SDG–human rights connections are often related to how measures triggered by the human rights system contribute directly to the implementation of specific SDGs and targets. Lesotho (2019), for instance, reported on how the "training of teachers that are conversant in teaching methods for the different types of disabilities" is partially helping it meet SDG 4.5 "by ensuring that persons with disabilities have equal access to education premises."

We also observe states using data from the SDG monitoring processes to substantiate their UPR report. Botswana (2023), for example, cites the UN Water SDG 6 snapshot to report that "78% of monitored water bodies in Botswana had good ambient water quality in 2020."

The explicit mention of the SDGs in UPR reports adds an important layer of accountability to the 2030 Agenda. Differently from the HLPF, where states do not receive formal feedback on the status of their SDG implementation, the UPR brings states to account based on their own report, as well as those presented by civil society organizations and the NHRI.

According to the UN good practices report on how the UPR process supports sustainable development, "on average each state receives over two hundred [UPR] recommendations covering all human rights including specific issues relating to women and children, health, education, labour, the environment, climate, migration, displacement" (UNDP, 2022).[36] The recommendations received by states in the UPR process not only bring attention to shortcomings in human rights and related SDG implementation but also shows the way forward.

30.6 Conclusion

The interlinkages between the SDG commitments and human rights obligations are substantial, but there is a perceived lack of intentionality in how states address them in implementation and reporting. With the support of AI, this analysis attempted to shed light on examples of practical interconnectedness in states' reports to prominent monitoring systems in both fields, namely, the VNR and UPR reports.

The chapter investigated the prevalence of human rights in the VNRs, both intentionally and as a mere consequence of thematic overlap, to better understand how these connections are made in practice. Moreover, it examined the prevalence of the SDGs in the UPR national reports to highlight how this mechanism can support and provide information to SDG reporting, implementation, and prioritization.

The study showed that human rights issues, in particular, but not limited to, those related to ESCR, are highly reflected in the VNRs. Nonetheless, further examination revealed little coherence in how states connect them to their obligations under binding international instruments. Even so, there were interesting examples of countries highlighting the role of their NHRIs and explicitly reporting on the human rights–based approach used in SDG-related processes and programs. When examining the prevalence of rightsholder groups, children, women and girls, and youth appear significantly prominent in the VNRs, while human rights defenders are mostly invisible in the reports across regions. Other groups, including LGBTI+ and Indigenous peoples, vary from region to region, reflecting the different priorities around the world.

As for the UPR reports, the analysis reflected an enormous prevalence of SDG-related issues and exposed a vast potential to utilize this mechanism for enhancing SDG accountability and for building on the synergies to demonstrate how human rights further the advancement of the SDGs.

This work showcased how tailored AI opens opportunities for understanding regional and global trends and gaps that can help inform policies and decision-making processes. AI can help extract meaningful patterns and point to insights beyond the human capacity, to help digest the ever-increasing amount of information on SDGs and human rights out there. In this study, the use of AI complemented the human intervention, which remained indispensable for a deeper analysis and extraction of detailed information.

The analysis was not exhaustive, but it illustrated the potential of delving deeper into the findings across different thematic categories of the algorithm. The study aims to inspire further investigations, specifically to gain a more detailed understanding of how specific rightsholder groups are addressed in connection with SDGs and human rights issues in documents beyond the VNRs and UPR reports. This type of analysis can shed light on the imbalance in how certain groups and human rights issues are prioritized within states' agendas. Ultimately, it adds a layer of accountability to states' commitments regarding human rights obligations and their efforts to leave no one behind.

Notes

1 The instruments linked to the SDGs by the Danish Institute for Human Rights are available at: https://sdg.humanrights.dk/list-goals-targets.

2 Explore the links between the SDGs and human rights on the Human Rights Guide to the SDGs, available at: https://sdg.humanrights.dk.

3 The list of the nine core international human rights treaties are available at: www.ohchr.org/en/core-international-human-rights-instruments-and-their-monitoring-bodies.
4 Adapted from OHCHR, available at: https://www.ohchr.org/sites/default/files/Documents/Issues/Development/SR/AddisAbaba/SDG_HR_Table.pdf.
5 Explore the recommendations from the human rights monitoring mechanisms applicable to countries and connected to the SDGs at the SDG–Human Rights Data Explorer, available at: https://sdgdata.humanrights.dk/.
6 The Mérida Declaration, available at https://cfnhri.org/wp-content/uploads/2020/01/Merida-Declaration-FINAL.pdf.
7 Chart on accreditation of NHRIs as of April 2023, available at: https://ganhri.org/wp-content/uploads/2023/05/StatusAccreditationChartNHRIs.pdf.
8 See the GANHRI website for information and resources related to NHRIs engagement with SDG processes: https://ganhri.org/ganhri-hrc54-statement-sdgs/.
9 TC8 is the AI used in the SDG–Human Rights Data Explorer (reference 5), which classifies about 220,000 human rights recommendations to the different SDG targets. OHCHR has also adopted it to aid expert thematic tagging of recommendations in their Universal Human Rights Index (https://uhri.ohchr.org/en/).
10 Specialisterne is a Danish social enterprise harnessing the talents of people with a diagnosis on the autistic spectrum to secure meaningful employment opportunities for them.
11 For this analysis, if a text snippet within a report is linked to a certain category, the entire report is classified as related to this category.
12 OHCHR human rights categories, available at: https://uhri.ohchr.org/assets/search-human-rights-recommendations.
13 See all categories at "Performance calculations of the Algorithm," available at: https://sdgdata.humanrights.dk/sites/sdgdata.humanrights.dk/files/media/document/Performance%20calculation%20notes.XLSX.
14 For a list of SDGs and targets, visit "The 17 Goals," available at: https://sdgs.un.org/goals.
15 For more information on the technicalities of the algorithm, please refer to "Textclassify algorithm and its use," available at: https://sdgdata.humanrights.dk/files/media/document/Technical%20note%20-%20TextClassify%20algorithm%20and%20its%20use.pdf.
16 For detailed performance assessments of each category using the training data, see "Performance calculations of the Algorithm," available at: https://sdgdata.humanrights.dk/sites/sdgdata.humanrights.dk/files/media/document/Performance%20calculation%20notes.XLSX.
17 For more information on the technical aspects of the algorithm, please visit Methodology | SDG–Human Rights Data Explorer, at https://sdgdata.humanrights.dk/en/methodology, or contact the authors.
18 Access the VNRs at https://hlpf.un.org/countries, and the UPR reports at www.ohchr.org/en/hr-bodies/upr/documentation.
19 More information on VNR process: https://sustainabledevelopment.un.org/vnrs/.
20 Voluntary common reporting guidelines for voluntary national reviews, available at: https://sustainabledevelopment.un.org/content/documents/27171SG_Guidelines_2021.final.pdf.
21 More information on the UPR process: www.ohchr.org/en/hr-bodies/upr/upr-home.
22 Pathfinders for Peaceful, Just and Inclusive Societies, *The Roadmap for Peaceful, Just and Inclusive Societies—A Call to Action to Change our World* (New York: Center on International Cooperation, 2019). Available at www.sdg16.plus/resources/the-roadmap-for-peaceful-just-and-inclusive-societies/.
23 The Submission of the OHCHR to the 21st Conference of the Parties to the United Nations Framework Convention on Climate Change offers guidance to states in connecting human rights and climate change. Available at: www.ohchr.org/sites/default/files/Documents/Issues/ClimateChange/COP21.pdf.
24 Human Rights Guide to the SDGs, available at: https://sdg.humanrights.dk.
25 The nine core international human rights treaties are available at: www.ohchr.org/en/core-international-human-rights-instruments-and-their-monitoring-bodies.
26 Number of states which ratified the core human rights treaties, by their acronyms: CCPR: 173; CESCR: 171; CERD: 182; CMW: 58, CED: 69; CAT: 173; CEDAW: 189; CRC: 196 CRPD: 186.
27 Women's Major Group, "Sustainable Development Goals and UN Human Rights Mechanisms from a Feminist Perspective," 2021. Available at https://sdgs.un.org/sites/default/files/2021-07/SDG%20from%20a%20Feminist%20Perspective%20-%20CRC.pdf.
28 Translated from the original: "se han instalado unidades especializadas con enfoque de derechos humanos, que atienden dignamente a personas LGBTIQ."

29 Translated from the original: "mejorar la atención y tratamiento de personas con trastornos mentales en el ámbito comunitario bajo un enfoque de derechos humanos."
30 For example: HRC Resolution A/HRC/RES/51/31 and UNGA Resolution A/RES/78/204.
31 Chart on accreditation of NHRIs as of December 2023, available at: https://www.ohchr.org/en/documents/tools-and-resources/current-chart-accreditation-nhris.
32 UNDP, "UN Good Practices: How the Universal Periodic Review Process Supports Sustainable Development" (New York, 2022). Available at: www.undp.org/publications/un-good-practices-how-universal-periodic-review-process-supports-sustainable-development.
33 Resolution adopted by the Human Rights Council on October 8, 2021 (A/HRC/RES/48/13).
34 See DIHR page dedicated to the work on SDG 14: www.humanrights.dk/promoting-human-rights-fisheries-aquaculture.
35 Information about the latest and previous Intersessional Meetings on human rights and the 2030 Agenda at the Human Rights Council is available at: www.ohchr.org/en/hr-bodies/hrc/sixth-intersessional-meeting-human-rights-and-the-2030-agenda.
36 UNDP, "UN Good Practices" (2022). Available at: https://www.undp.org/publications/un-good-practices-how-universal-periodic-review-process-supports-sustainable-development.

References

Alston, P. (2005). Ships passing in the night. The current state of the human rights and development debate seen through the lens of the millennium development goals. *Human Rights Quarterly*, 27(3).

Breuer, A., and Leininger, J. (2021). Horizontal accountability for SDG implementation: A comparative cross-national analysis of emerging national accountability regimes. *Sustainability*, 13(13), 7002. https://doi.org/10.3390/su13137002

Dovlo, D. (2020). People and their health systems: The right to universal health coverage and the SDGs in Africa. *Sustainable Development Goals and Human Rights*. Interdisciplinary Studies in Human Rights Series, pp. 53–71. Available at: https://link.springer.com/chapter/10.1007/978-3-030-30469-0_4

Feiring, B., and König-Reis, S. (2020). *Leveraging Human Rights Mechanisms to Improve SDG Follow-up and Review. Making the HLPF More Inclusive*, No. 2. Available at: www.daghammarskjold.se/wp-content/uploads/2020/03/hlpf_2_birgitte_saionara.pdf

Feiring, B., Thornberry, F., and Hassler, A. (2017). *Human Rights and Data—Tools and Resources for Sustainable Development*. Available at Microsoft Word—DIHR Human Rights and Data Oct 2017 Final.

Forum-Asia (2018). *Sustainable Development Goals—A Human Rights Based Approach*. Working Paper, No. 4. Bangkok: Asian Forum for Human Rights and Development.

Jensen, S. L. B. (2017). *The Making of International Human Rights: The 1960s, Decolonization, and the Reconstruction of Global Values*. Cambridge University Press.

Kaltenborn, M. (2020). *The Human Rights Framework for Establishing Social Protection Floors and Achieving Universal Health Coverage. Sustainable Development Goals and Human Rights*. Interdisciplinary Studies in Human Rights Series. Springer Nature, pp. 29–51. Available at: https://link.springer.com/chapter/10.1007/978-3-030-30469-0_

Kaltenborn, M., Krajewski, M., and Kuhn, H. (2020). *Sustainable Development Goals and Human Rights* (p. 239). Available at: https://library.oapen.org/bitstream/handle/20.500.12657/22942/1007219.pdf?sequence=1&isAllowed=y

Migowe, B., Shelton, C., and Rattray, S. (2024). *Human Rights and SDG Systems Integration: Beyond Linkages, Data and Efficiency to Leave No One Behind. Development Future Series No. 66*. United Nations Development Programme Global Policy Network Brief.

Muñoz Carmona, L. (2023). *Synergies or Silos? Exploring Human Rights Considerations in Sustainability Reporting Practices in the Nordics. Netherlands Quarterly of Human Rights*. Available at: https://doi.org/10.1177/09240519231223692

Roberts, C. N. J. (2015). *The Contentious History of the International Bill of Human Rights*. Cambridge University Press, New York, p. 213.

Rudolf, B. (2020). *Freedom from Violence, Full Access to Resources, Equal Participation, and Empowerment: The Relevance of CEDAW for the Implementation of the SDGs. Sustainable Development Goals and Human Rights*. Interdisciplinary Studies in Human Rights Seriespp, pp. 73–94. Available at: https://link.springer.com/chapter/10.1007/978-3-030-30469-0_5

Sano, H.-O. (2020). *How Can a Human Rights-Based Approach Contribute to Poverty Reduction? The Relevance of Human Rights to Sustainable Development Goal One*. *Sustainable Development Goals and Human Rights*. Interdisciplinary Studies in Human Rights Series, pp. 11–27. Available at: How Can a Human Rights-Based Approach Contribute to Poverty Reduction? The Relevance of Human Rights to Sustainable Development Goal One | SpringerLink.

Shah, S., and Sivakumaran, S. (2021). The use of international human rights law in the Universal Periodic Review. *Human Rights Law Review*, 21(2), 264–301.

United Nations. (2015). *Transforming Our World: The 2030 Agenda for Sustainable Development*. Available at: https://sustainabledevelopment.un.org/content/documents/21252030%20Agenda%20for%20Sustainable%20Development%20web.pdf

INDEX

Note: Numbers in **bold** indicate a table. Numbers in *italics* indicate a figure on the corresponding page.